TABLES FOR
NORMAL TOLERANCE LIMITS,
SAMPLING PLANS,
AND SCREENING

STATISTICS: Textbooks and Monographs

A SERIES EDITED BY

D. B. OWEN, Coordinating Editor
Department of Statistics
Southern Methodist University
Dallas, Texas

OTHER VOLUMES IN PREPARATION

TABLES FOR
NORMAL TOLERANCE LIMITS, SAMPLING PLANS, AND SCREENING

ROBERT E. ODEH

Department of Mathematics
University of Victoria
Victoria, B.C., Canada

D. B. OWEN

Department of Statistics
Southern Methodist University
Dallas, Texas

MARCEL DEKKER, Inc.   New York and Basel

Library of Congress Cataloging in Publication Data

Odeh, Robert E.
  Tables for normal tolerance limits, sampling plans and screening

  (Statistics, textbooks and monographs ; v. 32)
  Bibliography: p.
  Includes index.
  1. Quality control--Tables, calculations, etc.
2. Sampling (Statistics)--Tables.
I. Owen, Donald B., joint author. II. Title.
TS156.026   001.4'22'0212   79-27905
ISBN 0-8247-6944-9

MARCEL DEKKER, INC.

270 Madison Avenue, New York, New York 10016

Current printing (last digit)

10 9 8 7 6 5 4 3 2 1

PRINTED IN THE UNITED STATES OF AMERICA

To Barbara, Karen, David, and Lynne
R.E.O.

To Ellen, Mary Ellen, David, Matthew, and Mark
D.B.O.

PREFACE

This book was designed to assist in the use of acceptance sampling plans and tolerance limits for widely differing requirements. Tables, based on the univariate normal distribution, are given for a wide range of sample sizes, confidence levels, and desired reliabilities.

A second feature of the book is a collection of tables for screening based on the bivariate normal distribution. That is, screening procedures and tables are provided to allow decisions to be made on acceptability of items, where specifications on the items are on one variate and measurements are made on a second correlated variate. These techniques are useful, for example, in applications of nondestructive testing where it is not economically feasible to measure the variate of interest.

The material in this book was designed for use by industrial statisticians, quality control engineers, nondestructive testing engineers, and reliability engineers. In addition, it is hoped that the book will be of interest to teachers and research workers in the fields of statistics, operations research, civil engineering, and industrial engineering.

It was our purpose to present here a set of tables that were extremely comprehensive. In order to accomplish this goal, some previously computed tables have been expanded. In addition, to make this volume self-contained, several new tables have been computed especially for this volume. Details are given in the section "Sources of the Tables" at the back of this book.

The tables have been very carefully computed and we believe all of them are accurate to the number of places printed. They have all been freshly computed using the latest available algorithms and reproduced directly from computer output.

All computations were done at the University of Victoria on an IBM 370/158 using double-precision arithmetic. Camera-ready copy for the tables was produced on a typewriter terminal. We wish to thank Mr. K.B. Wilson, scientific programmer, who assisted with the computation, formatting, and checking of the tables, and with the preparation of camera-ready tables.

We also wish to express our appreciation to Mrs. Georgina Smith for her meticulous typing of the camera-ready copy of the text.

Research for this project was partially supported by grants from the University of Victoria, and from Natural Sciences and Engineering Research Council Canada (to R.E. Odeh under Grant No. A-5203); and from the United States Office of Naval Research (to D.B. Owen under Contract No. N00014-76-C-0613). We are grateful to these agencies for their assistance.

Robert E. Odeh
D.B. Owen

CONTENTS

PART C
Mathematical Derivations

TABLES FOR
NORMAL TOLERANCE LIMITS,
SAMPLING PLANS,
AND SCREENING

INTRODUCTION AND NOTATION

This book is divided into three parts. Part A, which includes Tables 1 through 7, covers one-sided and two-sided sampling plans and tolerance limits associated with the univariate normal distribution. Part B, including Tables 8, 9, and 10, deals with screening procedures (with both known and unknown parameters) based on the bivariate normal distribution.

Since Parts A and B supply tabular material for the procedures without justification, we have tried in Part C to present enough of the mathematical background so that the reader can understand how the tables were computed, and what methods of formulation were employed. In addition, within each section of Part C references are given for the reader who desires to pursue the methods further.

Descriptions of all the tables and examples of their use have been placed together in a separate section immediately preceding the tables. This arrangement has given us the opportunity to add comments that relate one set of tables to another.

The following notation is used for the table headings: Except for Table 5 and Table 6, which are each one page long, each table is divided into subtables. For example, Table 1 is comprised of 13 subtables, labelled TABLE 1.1-TABLE 1.13, corresponding to the 13 values of the parameter γ. In addition, some subtables contain more than one page. For example, TABLE 1.1 contains four pages labelled TABLE 1.1.1-TABLE 1.1.4. If a subtable consists of only one page we omit the third figure. Thus, TABLE 2.1 contains only one page.

DESCRIPTIONS OF THE TABLES

PART A
UNIVARIATE NORMAL PROCEDURES

In *variables acceptance sampling* we assume that we have either a *lower specification limit*, L, or an *upper specification limit*, U, or we have both L and U. We call items outside the specification limits *defective*. That is, if we have a lower specification limit, L, and the measurement on the item, X, falls below L then we call the item defective. The measurement, X, is assumed to be normally distributed with mean, μ, and standard deviation, σ. Usually μ and σ are unknown and have to be estimated from a sample. The estimators are based on a sample, $x_1, x_2, x_3, \cdots, x_n$. The *sample mean* is $\bar{x} = \frac{1}{n} \sum\limits_{i=1}^{n} x_i$ which estimates μ and has a normal distribution with mean μ and variance σ^2/n. The *sample standard deviation* is

$s_x = \sqrt{\frac{1}{n-1} \sum\limits_{i=1}^{n} (x_i - \bar{x})^2}$ which estimates σ. The quantity $(n-1)s_x^2/\sigma^2$ has

a chi-squared distribution with $n-1$ degrees of freedom. In case μ is

known and σ is unknown we estimate σ by $s_x' = \sqrt{\frac{1}{n} \sum\limits_{i=1}^{n} (x_i - \mu)^2}$ and

$n\, s_x'^2/\sigma^2$ has a chi-squared distribution with n degrees of freedom.

When both μ and σ are known the proportion defective can be read directly from a table of the normal probability distribution. For example, if we have an upper specification limit U only, then $(U-\mu)/\sigma$ may be computed and if this should be 1.64485, for instance, then the *fraction defective* is 0.05. Similarly, the quantity $(L-\mu)/\sigma$ may be computed and the percentage defective below L obtained. With two-sided limits, i.e., both L and U specified, the fractions defective in both tails may be added together so that the fraction non-defective is in the center of the distribution. We indicate this in what follows by the words *control center*. Alternatively, each tail of the distribution may be controlled separately and we indicate this in what follows by *control both tails*. The quantities $(U-\mu)/\sigma$ or $(L-\mu)/\sigma$ or $(X-\mu)/\sigma$ are called *standardized deviates*. We will use the notation K_p for these where p is the proportion of the normal distribution below the deviate. For example, if $(U-\mu)/\sigma = 1.64485 = K_p$ then $p = 0.95$. Since the percentage defective in the upper tail is defined to be the percentage above U, the percentage defective is $100(1-p)\% = 5\%$. If $(L-\mu)/\sigma = -1.64485 = K_p$, then $p = 0.05$, and since the percentage defective in the lower tail is defined to be the percentage below L, the percentage defective is $100p\% = 5\%$. Mathematically, the definition of K_p is $p = \int_{-\infty}^{K_p} (2\pi)^{-\frac{1}{2}} e^{-t^2/2} \, dt$.

When μ and σ are unknown, then the underlying distribution becomes the noncentral t-distribution. The form this usually takes is $\Pr\{T_f \leq k\sqrt{n} \mid \delta = K_p\sqrt{n}\} = \gamma$ where k is some constant, δ is the noncentrality parameter, γ is the confidence we require, and f is the degrees of freedom of the estimator for σ used. Tables 1, 2 and 6 give the solution to this equation for k when all of the other quantities are given. Table 7 gives the solution for p when all of the other parameters are given. Tables 3, 4 and 5 contain solutions to the problem (not based on noncentral t) when there are both upper and lower specification limits, or where both upper and lower limits are to be computed.

TABLE 1. *Factors for one-sided sampling plans and tolerance limits*

Variables acceptance sampling arises when we are given a specification limit (L for a lower limit or U for an upper limit), and based on the outcome of a random sample, $x_1, x_2, \cdots, x_n$, from a large lot we are asked to decide if the lot should be accepted or rejected. The measurements are assumed to be normally distributed with unknown mean, μ, and unknown variance, σ^2. The lot is accepted if $\bar{x} + ks \leq U$ for an upper specification limit or if $\bar{x} - ks \geq L$ for a lower specification limit, where k is the quantity tabulated in Table 1. The choice of γ depends upon whether a consumer protection plan is wanted or one which protects the producer. The consumer would generally choose $\gamma = 0.10$. In this table P is the proportion of acceptable items required.

For example, from Table 1.9.1 corresponding to $\gamma = 0.1$, for $N = 40$, $P = 0.975$ the value of k is 1.655. Suppose, that based on a sample of size $n = 40$ we use the plan: accept the lot if $\bar{x} + 1.655s \leq U$; reject the lot if $\bar{x} + 1.655s > U$. Then, if the lot contains 100P% = 97.5% of items below U, the probability is $1.0 - \gamma = 0.9$ that the lot will be accepted.

Tolerance limits are values of the measured variable which are computed from the sample such that we can say with probability γ that at least 100P% of the population lies below the value $\bar{x} + ks$ for an upper limit or above $\bar{x} - ks$ for a lower limit.

For example, we can be 90% sure that at least 99% of the normal population is below $\bar{x} + 3.212s$, where $\bar{x}$ and s are computed from a sample of size 15. The number 3.212 is read from Table 1.5.1 with $\gamma = 0.90$, $P = 0.99$ and $N = 15$.

TABLE 2. *Sample size requirements for one-sided sampling plans*

If a plan is required which protects both the consumer and the producer then the sample size must be chosen to do this properly. The *producer's risk* is usually labeled α and is usually 0.05 (but is not always taken to be 0.05). This occurs at a proportion defective $P = p_1$. The *consumer's risk* is labeled β and is usually taken to be $\beta = 0.10$. This occurs at a proportion defective p_2. Hence, the probability of acceptance of the lot is $1 - \alpha = 0.95$ when the proportion defective is p_1, and is $\beta = 0.10$ when the proportion defective is p_2. Approximate formulae for k and n are:

$$k = \frac{K_\alpha K_2 + K_\beta K_1}{K_\alpha + K_\beta}$$

$$n = \left(1 + \frac{k^2}{2}\right)\left(\frac{K_\alpha + K_\beta}{K_1 - K_2}\right)^2$$

where the capital K's are normal deviates as defined above with K_1, K_2 corresponding to $1 - p_1$, $1 - p_2$, respectively. For further information the reader is referred to Duncan (1974), p. 269, or Eisenhart, Hastay and Wallis (1947), p. 17.

However, Table 2 gives the values of n and k based on an exact calculation of the noncentral t-probabilities involved. The method of selecting the value of k is illustrated by the following example: For the value of n given in the table if nominal $\alpha = 0.025$, nominal $\beta = 0.05$, $p_1 = 0.005$ and $p_2 = 0.075$, $k = 1.958$ gives true $\alpha = 0.022$ and true $\beta = 0.05$ whereas $k = 1.972$ gives true $\alpha = 0.025$ and true

β = 0.0459. The value of k as given in Table 2.4 is (1.958 + 1.972)/2 = 1.965. The value of n = 31 is the smallest sample size which gives true $\alpha \leq$ nominal α and true $\beta \leq$ nominal β.

TABLE 3. *Two-sided tolerance limit factors for a normal distribution (Control center)*

The tables give factors, k, such that we can be 100γ% sure that at least 100P% of the population is between $\bar{x}$ - ks and $\bar{x}$ + ks.

For example, if γ = 0.99, P = 0.75 and n = 40, we can be 99% sure that at least 75% of the population is contained between $\bar{x}$ - 1.576s and $\bar{x}$ + 1.576s. (See Table 3.2.1).

TABLE 4. *Two-sided tolerance limit factors for a normal distribution (Control both tails)*

The tables give factors, k_1 and k_2, such that we can be 100γ% sure that no more than 100p_1% of the population is below $\bar{x}$ - k_1s and no more than 100p_2% of the population is above $\bar{x}$ + k_2s.

For example, if γ = 0.9, p_1 = 0.05, p_2 = 0.01, and n = 15 we can be 90% sure that there is no more than 5% of the population below $\bar{x}$ - 2.526s and no more than 1% of the population above $\bar{x}$ + 3.428s. The values of k_1 and k_2 were obtained by entering Table 4.5.1 with P = p_1 = 0.05 to obtain 2.526 and with P = p_2 = 0.01 to obtain 3.428. Note that this is a stronger statement than saying 94% of the population is between two limits.

TABLE 5. *Two-sided sampling plan factors for a normal distribution (Control center)*

The sampling plan considered here calls for acceptance of a lot if $\bar{x}$ - ks $\geq$ L and $\bar{x}$ + ks $\leq$ U, where k is the factor read in the table. If either inequality is violated the lot is rejected. We can then be 100γ% = 90% sure that the lot has at most 100P% defective. Only the table for γ = 0.90 is given and hence this is for a consumer's risk of β = 0.10.

A plot of the probability of acceptance versus the proportion defective is called an *operating characteristic curve* (OC curve for short). In this case the OC curve becomes a narrow band rather than a single curve, depending on how the percentage defective is split between the two tails of the distribution. For most sample sizes, the split which gives one of the boundaries of the OC band is for all of the percentage defective to be in one tail, while the other boundary corresponds to an equal split between the tails. For some sample sizes, however, different splits than these two give the boundaries of the OC band. The value of k given in this table correspond to the lower boundary of the OC band. Hence, it is possible to say that we are <u>at least</u> 90% sure that no more than 100P% defective occurs in accepted lots.

If sample sizes need to be computed for given α and β the formulae given for Table 2 may be used as an approximation, with α replaced by $\alpha/2$. Alternatively, Table 2 may be used with entry $(\alpha/2, \beta, p_1, p_2)$, i.e., using the same parameters as with a one-sided plan, except using only half the α level wanted. For example, if $\alpha = 0.05$, $\beta = 0.10$, $p_1 = 0.02$, $p_2 = 0.07$, we enter Table 2.5 to obtain n = 79, k = 1.708. Then we take a sample of size 79 and accept the lot if $\bar{x} - 1.708s \geq L$ and $\bar{x} + 1.708s \leq U$. Otherwise we reject the lot. We can then be approximately 95% sure of accepting if the true proportion defective is 0.02 and approximately 90% sure of rejecting if the true proportion defective is 0.07. We say approximately 90% assurance because we are using a one-sided plan to approximate what is required for a two-sided plan. It is, however, a very good approximation.

TABLE 6. *Two-sided sampling plan factors (Control both tails)*

This sampling plan differs from that given in Table 5 only in that each tail of the distribution is controlled separately. We accept a lot if $\bar{x} - k_L s \geq L$ and $\bar{x} + k_U s \leq U$ where k_L is read from Table 6 with $P = p_L$ and k_U is read from Table 6 with $P = p_U$. Then we can be 90% sure that there is <u>at most</u> $100p_L$% defective in the lower tail of the distribution and <u>at most</u> $100p_U$% defective in the upper tail of the distribution for accepted lots.

Note that in reality this table is the same as Table 1 with $\gamma = 0.90$ and the P of this table equal to $1 - P$ of Table 1. This fact enables one to use Table 1 to expand on the entries here.

When α, β, p_1, p_L, and p_U are given then approximate formulae for k and n are

$$k = \frac{K_{\alpha/2} K_L + K_\beta K_1}{K_{\alpha/2} + K_\beta}$$

$$n = \left(1 + \frac{k^2}{2}\right) \left(\frac{K_{\alpha/2} + K_\beta}{K_1 - K_L}\right)^2$$

which are computed as in the description for Table 2 with α replaced by $\alpha/2$ and K_2 replaced with K_L or K_U (normal deviates corresponding to $1 - p_L$ or $1 - p_U$). In this case p_1 represents the sum of the fraction defective in the two tails which should be acceptable $100(1-\alpha)\%$ of the time while p_L and p_U represent the fraction defective in each tail which should be accepted only $100\beta\%$ of the time. Two calculations are made, one with p_L and one with p_U and the largest n chosen. Table 6 corresponds to $\beta = 0.10$ only, but Table 2 can be used for additional values of γ.

For example, if $\alpha = 0.05$, $\beta = 0.10$, $p_1 = 0.02$ and $p_L = 0.05$, $p_U = 0.06$ we enter Table 2.5 with $\alpha = 0.025$, $p_1 = 0.02$, $p_2 = 0.05$ and obtain $n = 168$, $k_L = 1.808$. If we enter with $\alpha = 0.025$, $p_1 = 0.02$, $p_2 = 0.06$ we obtain $n = 109$, $k_U = 1.755$. In order to have the guarantee required for the lower tail we will need $n = 168$. Now go to Table 1.5.3 to obtain the k for the upper tail. We have to interpolate using the values of K_p (instead of P) as a linear interpolate:

P	K_p	k
0.90	1.28155	1.429
0.94	1.55477	
0.95	1.64485	1.813

The result, $k = 1.718$ for $P = 0.94$, is correct to all three decimal places.

Hence, we take a sample of size 168 and accept the lot if $\bar{x} - 1.808s \geq L$ and $\bar{x} + 1.718s \leq U$; and reject otherwise. We can then be 90% sure that there is no more than 5% defective in the lower tail of the distribution and no more than 6% defective in the upper tail of the distribution while at the same time we are 95% sure of accepting the lot if its total percentage defective does not exceed 2%.

TABLE 7. *Confidence limits on the proportion in one tail of the normal distribution*

This table gives lower confidence limits on $\gamma = \Pr\{Y \geq L\}$ where Y is a performance variable and L is the lower specification limit. The procedure is to take a preliminary sample of size n and compute $K = (\bar{y}-L)/s$. For example, if $K = 1.8$, $\eta = 0.90$, and $n = 27$, we can be 90% sure that at least 91.657% of the Y's are greater than or equal to L. (See Table 7.3.4).

An upper limit on Y can be obtained by taking one minus the table entry for n and $-K$. Hence, for $K = 1.8$ and $n = 27$ we can be 80% sure that $0.91657 \leq \gamma \leq 0.98566$. Note that we obtained this result from Table 7.3.4 and that the confidence is $2(0.9) -1 = 0.8$.

If confidence limits on $1 - \gamma = \Pr\{Y \leq L\}$ are needed, the easiest approach is to find them on γ as above, and then subtract them from one.

PART B
SCREENING PROCEDURES BASED ON THE BIVARIATE NORMAL DISTRIBUTION

If a lower specification limit, L, is given, all product with a performance characteristic above L are considered acceptable while product below L are considered defective. Suppose we wish to raise the proportion of acceptable product from γ to δ where $\delta > \gamma$, but we cannot measure the performance variable directly. This may be due to the fact that the performance variable is the lifetime or that measuring the performance may in some way degrade the item. Let the performance variable be Y and suppose a correlated variable, X is available which may be measured without degrading the product. Also assume X and Y have a joint bivariate normal distribution with means μ_x, μ_y and standard deviations σ_x, σ_y respectively, and correlation ρ.

TABLE 8. *Screening proportion for normal conditioned on normal distribution*

Table 8 gives the proportion, β, of product to be accepted based on measuring X in order to raise the proportion of Y's greater than L from γ to δ for given values of ρ, γ and δ, i.e., our rule is to select all products for which $X \geq \mu_x - K_\beta \sigma_x$ where K_β is defined, as before, as

$$\beta = \int_{-\infty}^{K_\beta} (2\pi)^{-\frac{1}{2}} e^{-t^2/2} \, dt.$$ If this is done for the entire population then

the proportion of Y's in the selected population will be δ.

Thus the screening procedure separates the product into two groups, one containing 100β% of the original population (called the selected group) and one containing $100(1-\beta)$% of the original population (called the rejected group). In the selected group, the proportion δ have Y values above L, and the proportion $1 - \delta$ have Y values below L. In the rejected group, the proportion $(\gamma-\delta\beta)/(1-\beta)$ have Y values above L, while the proportion $(1-\gamma-\beta + \delta\beta)/(1-\beta)$ have Y values below L.

The procedure is also applicable for the situations where X and Y are negatively correlated, or if there is an upper specification limit, U, on Y. In particular these situations can be handled as follows:

(1) positive correlation and upper specification limit U on Y:
    a. accept all units for which $X \leq \mu_x + K_\beta \sigma_x$
    b. reject all other submitted units.

(2) Negative correlation and lower specification limit, L on Y:
    a. enter Table 8 with the absolute value of the correlation,
    b. accept all units for which $X \leq \mu_x + K_\beta \sigma_x$
    c. reject all other submitted units.

(3) Negative correlation and upper specification limit U on Y:
    a. enter Table 8 with the absolute value of the correlation,
    b. accept all units for which $X \geq \mu_x - K_\beta \sigma_x$
    c. reject all other submitted units.

In case this procedure is applied when only a small sample is to be selected, say n items, the analysis of the results in the sample will come from the binomial distribution with parameters δ and n. For example, suppose 10 items are selected in the case where the present stockpile contains 85% acceptable units, with a lower specification limit L, but a particular customer requires 95% acceptable units. A variable correlated to the performance with $\rho = 0.75$ is available, but we cannot measure performance directly because it degrades the item. Hence we enter Table 8.1 with $\delta = 0.95$, $\gamma = 0.85$ and $\rho = 0.75$ and obtain $\beta = 0.745$. Since $K_{0.745} = 0.659$, the selected group is then made up of those items for which $X \geq \mu_x - 0.659\sigma_x$ and if this were done for a sample of size 10, the probability of Z or less defectives in the group of 10 is given by

$$\Pr\{Z \text{ or less defectives}\} = \sum_{i=0}^{Z} \binom{10}{i} \delta^i (1-\delta)^{10-i}$$

and for the example under consideration this is

Z	0	1	2	3
$\Pr\{Z \text{ or less defectives}\}$	0.599	0.914	0.988	0.999

On the other hand, without screening these probabilities would have been

Z	0	1	2	3
Pr{Z or less defectives}	0.197	0.544	0.820	0.950

In other words, without screening there is only about a 20% chance of no defectives in the same of 10 items, while with screening there is almost a 60% chance of no defectives.

TABLE 9. *Screening factors for normal conditioned on t-distribution*

All of the parameters were assumed known in the description for Table 8. However, this is seldom the case. For Table 9 we assume that a preliminary sample of size n has been taken to estimate μ_x and σ_x. The rule then becomes to select all (future) product for which $X \geq \bar{x} - t_\beta s_x[(n+1)/n]^{\frac{1}{2}}$ with known γ, δ and ρ where t_β is read from Table 9 with $f = n - 1$. In case μ_x is known the procedure is to select items for which $X \geq \mu_x - t_\beta s'_x$ where $\bar{x}, s_x$ and s'_x are as defined on page 3.

In a preceding section we gave tables to take care of the cases where γ is unknown (Table 7) and in the following section we will give tables for the case where ρ is unknown (Table 10). Then we will summarize the screening procedure when all parameters are unknown.

TABLE 10. *Confidence limits on the correlation coefficient*

The sample correlation coefficient is computed from

$$r = \sum_{i=1}^{n} (x_i - \bar{x})(y_i - \bar{y})/[(n-1)s_x s_y]$$

If $n = 5$, and $r = 0.6$, the 95% lower confidence limit on ρ is -0.3376. (From Table 10.3). If $n = 5$, and $r = 0.6$ then the 95% confidence limits on ρ are $(-0.5069, 0.9409)$. If $n = 5$, and $r = 0.6$ then the 95% upper confidence limit on ρ is 0.9122.

This illustrates what is meant by the symbols ↓+ and ↑-. The first means to read down the column and leave the sign alone. The second means to read up (use the listings for R on the right) and change the sign of the table entry. Note that for one-sided limits we have confidence $\eta = 1 - \alpha$, and for two-sided limits we have confidence $\eta = 1 - 2\alpha$.

SCREENING PROCEDURES WHEN ALL OF THE PARAMETERS ARE UNKNOWN

The screening procedure will be illustrated for the case where a lower specification L is given. For an upper specification limit see the instructions for Table 8.

(1) The sample correlation is computed from a preliminary sample of size n. A lower $100\eta\%$ confidence limit on ρ is obtained from Table 10 and if this is positive we proceed to step 2. Call this lower limit ρ^*. If this lower confidence limit is negative, an upper $100\eta\%$ limit is also computed. If this is positive the process stops since the two variables X and Y could then be independent. If the $100\eta\%$ upper confidence limit on ρ is also negative, then we proceed to step 2 using this limit and modify the following steps according to the instructions for Table 8 with a negative correlation.

(2) A $100\eta\%$ lower confidence limit on $\gamma = \Pr\{Y \geq L\}$ is obtained as in the instructions for Table 7. Call this lower limit γ^*.

(3) Enter Table 9 with $(f = n - 1, \gamma^*, \rho^*[n/(n+1)]^{\frac{1}{2}}, \delta)$ in place of $(f, \gamma, \rho, \delta)$ where δ is the proportion of the selected population of performance variates which are to be above L. If $\delta < \gamma^*$ stop the procedure. The entire population meets the criterion. Obtain t_β from Table 9.

(4) All product are accepted (selected) for which $X \geq \bar{x} - t_\beta[(n+1)/n]^{\frac{1}{2}}s_x$

(5) We can then be at least $100(2\eta-1)\%$ sure that at least $100\delta\%$ of the Y's are above L in the selected population.

For example, if a preliminary sample of size 17 is taken and r = 0.94, then choosing $\eta = 0.95$ we obtain a 95% lower confidence limit on ρ to be $\rho^* = 0.8558$ by linear interpolation in r in Table 10.15. If $K = (\bar{y}-L)/s_y = 2.0$, then from Table 7.4.4, a 95% lower confidence limit on γ is $\gamma^* = 0.89977$.

We enter Table 9.2.3 with (16, 0.90, 0.8317, 0.95) for $(f, \gamma, \rho, \delta)$ and obtain $t_\beta = 1.384$ by linear interpolation in ρ. Our criterion is then to select all items for which $X \geq \bar{x} - 1.424s_x$. Then in the selected group we can be at least 90% sure that at least 95% of the items have values of the performance variable, Y, which are greater than the lower specification limit L.

If this screening is performed on a finite group of, say M items then the items in that group follow a binomial distribution with parameters M and δ. The situation is the same as given in the description of Table 8, except that we say we are at least $100(2\eta-1)\%$ sure that the probability of Z or less defectives is <u>at least</u> that given by the binomial distribution. Hence, if $M = 10$ for the example above with $\delta = 0.95$ and $\eta = 0.95$, then we are at least 90% sure that the probability of zero defectives in this group is 0.599.

PART A

TABLES FOR UNIVARIATE NORMAL PROCEDURES

TABLE 1

Factors for one-sided sampling plans and tolerance limits (for a normal distribution) for values of P = 0.75, 0.90, 0.95, 0.975, 0.99, 0.999, 0.9999 and N = 2(1)100(2)180(5)300(10)400(25)650(50)1000, 1500, 2000, 3000, 5000, 10000, ∞.

TABLE 1.1.1. FACTORS FOR ONE-SIDED SAMPLING PLANS AND TOLERANCE LIMITS.

GAMMA = 0.995

N \P →	0.750	0.900	0.950	0.975	0.990	0.999	0.9999
2	117.887	206.065	262.861	312.857	371.243	493.126	593.465
3	12.381	19.833	24.609	28.848	33.846	44.396	53.152
4	5.997	9.360	11.512	13.426	15.688	20.478	24.465
5	4.172	6.448	7.902	9.195	10.725	13.969	16.671
6	3.339	5.143	6.293	7.316	8.526	11.093	13.233
7	2.864	4.409	5.392	6.266	7.300	9.494	11.322
8	2.556	3.938	4.817	5.597	6.520	8.478	10.110
9	2.338	3.610	4.416	5.133	5.979	7.775	9.272
10	2.176	3.367	4.121	4.790	5.581	7.258	8.656
11	2.050	3.180	3.894	4.527	5.276	6.862	8.184
12	1.949	3.030	3.712	4.318	5.033	6.547	7.809
13	1.865	2.907	3.564	4.147	4.834	6.291	7.504
14	1.795	2.805	3.441	4.004	4.669	6.078	7.251
15	1.735	2.718	3.336	3.883	4.530	5.898	7.037
16	1.683	2.643	3.246	3.780	4.409	5.743	6.853
17	1.638	2.577	3.167	3.689	4.305	5.608	6.693
18	1.598	2.520	3.098	3.610	4.213	5.490	6.552
19	1.562	2.468	3.037	3.539	4.132	5.385	6.428
20	1.530	2.423	2.982	3.476	4.059	5.292	6.317
21	1.500	2.381	2.932	3.419	3.993	5.207	6.217
22	1.474	2.344	2.887	3.368	3.934	5.131	6.127
23	1.449	2.309	2.846	3.321	3.880	5.062	6.045
24	1.427	2.278	2.809	3.278	3.831	4.998	5.969
25	1.407	2.249	2.774	3.239	3.785	4.940	5.900
26	1.387	2.222	2.743	3.202	3.743	4.886	5.837
27	1.370	2.197	2.713	3.169	3.705	4.837	5.778
28	1.353	2.174	2.686	3.137	3.668	4.790	5.723
29	1.338	2.153	2.660	3.108	3.635	4.747	5.672
30	1.323	2.132	2.636	3.081	3.603	4.707	5.624
31	1.309	2.114	2.614	3.055	3.574	4.669	5.580
32	1.297	2.096	2.593	3.031	3.546	4.634	5.538
33	1.284	2.079	2.573	3.008	3.520	4.601	5.498
34	1.273	2.063	2.554	2.987	3.495	4.569	5.461
35	1.262	2.048	2.536	2.966	3.472	4.540	5.426
36	1.252	2.034	2.519	2.947	3.450	4.511	5.393
37	1.242	2.020	2.503	2.929	3.429	4.485	5.361
38	1.232	2.007	2.488	2.912	3.409	4.459	5.331
39	1.223	1.995	2.474	2.895	3.390	4.435	5.302
40	1.215	1.983	2.460	2.879	3.372	4.412	5.275
41	1.207	1.972	2.447	2.864	3.355	4.390	5.249
42	1.199	1.962	2.434	2.850	3.339	4.369	5.224
43	1.191	1.951	2.422	2.836	3.323	4.349	5.201
44	1.184	1.941	2.410	2.823	3.308	4.330	5.178
45	1.177	1.932	2.399	2.810	3.293	4.311	5.156
46	1.170	1.923	2.389	2.798	3.279	4.294	5.135
47	1.164	1.914	2.378	2.787	3.266	4.277	5.115
48	1.158	1.906	2.368	2.775	3.253	4.260	5.096
49	1.152	1.898	2.359	2.764	3.241	4.244	5.077
50	1.146	1.890	2.350	2.754	3.229	4.229	5.059

TABLE 1.1.2. FACTORS FOR ONE-SIDED SAMPLING PLANS AND TOLERANCE LIMITS.

GAMMA = 0.995 (CONTINUED)

N \P →	0.750	0.900	0.950	0.975	0.990	0.999	0.9999
51	1.141	1.882	2.341	2.744	3.217	4.215	5.042
52	1.135	1.875	2.332	2.734	3.206	4.200	5.025
53	1.130	1.868	2.324	2.725	3.196	4.187	5.009
54	1.125	1.861	2.316	2.716	3.185	4.174	4.993
55	1.120	1.855	2.308	2.707	3.175	4.161	4.978
56	1.115	1.848	2.301	2.699	3.165	4.148	4.964
57	1.111	1.842	2.294	2.690	3.156	4.137	4.950
58	1.106	1.836	2.287	2.682	3.147	4.125	4.936
59	1.102	1.830	2.280	2.675	3.138	4.114	4.923
60	1.098	1.825	2.273	2.667	3.129	4.103	4.910
61	1.094	1.819	2.267	2.660	3.121	4.092	4.897
62	1.090	1.814	2.260	2.653	3.113	4.082	4.885
63	1.086	1.809	2.254	2.646	3.105	4.072	4.873
64	1.082	1.804	2.249	2.639	3.097	4.062	4.862
65	1.078	1.799	2.243	2.633	3.090	4.053	4.851
66	1.075	1.794	2.237	2.626	3.083	4.043	4.840
67	1.071	1.789	2.232	2.620	3.076	4.035	4.829
68	1.068	1.785	2.226	2.614	3.069	4.026	4.819
69	1.065	1.781	2.221	2.608	3.062	4.017	4.809
70	1.061	1.776	2.216	2.603	3.056	4.009	4.799
71	1.058	1.772	2.211	2.597	3.049	4.001	4.790
72	1.055	1.768	2.206	2.592	3.043	3.993	4.780
73	1.052	1.764	2.202	2.586	3.037	3.985	4.771
74	1.049	1.760	2.197	2.581	3.031	3.978	4.762
75	1.046	1.756	2.193	2.576	3.025	3.970	4.754
76	1.043	1.752	2.188	2.571	3.020	3.963	4.745
77	1.041	1.749	2.184	2.566	3.014	3.956	4.737
78	1.038	1.745	2.180	2.561	3.009	3.949	4.729
79	1.035	1.742	2.176	2.557	3.003	3.943	4.721
80	1.033	1.738	2.172	2.552	2.998	3.936	4.713
81	1.030	1.735	2.168	2.548	2.993	3.930	4.706
82	1.028	1.732	2.164	2.543	2.988	3.923	4.698
83	1.025	1.728	2.160	2.539	2.983	3.917	4.691
84	1.023	1.725	2.157	2.535	2.979	3.911	4.684
85	1.021	1.722	2.153	2.531	2.974	3.905	4.677
86	1.018	1.719	2.149	2.527	2.969	3.900	4.670
87	1.016	1.716	2.146	2.523	2.965	3.894	4.663
88	1.014	1.713	2.143	2.519	2.961	3.888	4.657
89	1.012	1.710	2.139	2.515	2.956	3.883	4.650
90	1.010	1.708	2.136	2.512	2.952	3.878	4.644
91	1.007	1.705	2.133	2.508	2.948	3.872	4.638
92	1.005	1.702	2.130	2.505	2.944	3.867	4.632
93	1.003	1.699	2.127	2.501	2.940	3.862	4.626
94	1.001	1.697	2.123	2.498	2.936	3.857	4.620
95	0.999	1.694	2.121	2.494	2.932	3.852	4.614
96	0.998	1.692	2.118	2.491	2.928	3.847	4.609
97	0.996	1.689	2.115	2.488	2.925	3.843	4.603
98	0.994	1.687	2.112	2.484	2.921	3.838	4.598
99	0.992	1.684	2.109	2.481	2.917	3.834	4.592

TABLE 1.1.3. FACTORS FOR ONE-SIDED SAMPLING PLANS AND TOLERANCE LIMITS.

GAMMA = 0.995 (CONTINUED)

N \P → ↓	0.750	0.900	0.950	0.975	0.990	0.999	0.9999
100	0.990	1.682	2.106	2.478	2.914	3.829	4.587
102	0.987	1.678	2.101	2.472	2.907	3.820	4.577
104	0.983	1.673	2.096	2.466	2.900	3.812	4.567
106	0.980	1.669	2.091	2.461	2.894	3.804	4.557
108	0.977	1.665	2.086	2.455	2.888	3.796	4.548
110	0.974	1.661	2.081	2.450	2.882	3.788	4.539
112	0.971	1.657	2.077	2.445	2.876	3.781	4.530
114	0.968	1.653	2.072	2.440	2.870	3.774	4.522
116	0.965	1.649	2.068	2.435	2.864	3.767	4.514
118	0.962	1.646	2.064	2.430	2.859	3.760	4.506
120	0.960	1.642	2.060	2.426	2.854	3.753	4.498
122	0.957	1.639	2.056	2.421	2.849	3.747	4.490
124	0.955	1.635	2.052	2.417	2.844	3.741	4.483
126	0.952	1.632	2.048	2.413	2.839	3.735	4.476
128	0.950	1.629	2.045	2.408	2.834	3.729	4.469
130	0.947	1.626	2.041	2.404	2.830	3.723	4.462
132	0.945	1.623	2.038	2.401	2.825	3.717	4.455
134	0.943	1.620	2.034	2.397	2.821	3.712	4.449
136	0.941	1.617	2.031	2.393	2.817	3.707	4.443
138	0.939	1.615	2.028	2.389	2.813	3.701	4.437
140	0.937	1.612	2.025	2.386	2.809	3.696	4.431
142	0.934	1.609	2.022	2.382	2.805	3.691	4.425
144	0.932	1.607	2.019	2.379	2.801	3.686	4.419
146	0.931	1.604	2.016	2.376	2.797	3.682	4.413
148	0.929	1.602	2.013	2.373	2.793	3.677	4.408
150	0.927	1.599	2.010	2.369	2.790	3.672	4.403
152	0.925	1.597	2.007	2.366	2.786	3.668	4.397
154	0.923	1.595	2.005	2.363	2.783	3.664	4.392
156	0.921	1.592	2.002	2.360	2.780	3.659	4.387
158	0.920	1.590	1.999	2.357	2.776	3.655	4.382
160	0.918	1.588	1.997	2.355	2.773	3.651	4.378
162	0.916	1.586	1.994	2.352	2.770	3.647	4.373
164	0.915	1.584	1.992	2.349	2.767	3.643	4.368
166	0.913	1.582	1.990	2.346	2.764	3.640	4.364
168	0.912	1.580	1.987	2.344	2.761	3.636	4.359
170	0.910	1.578	1.985	2.341	2.758	3.632	4.355
172	0.909	1.576	1.983	2.339	2.755	3.628	4.351
174	0.907	1.574	1.981	2.336	2.752	3.625	4.347
176	0.906	1.572	1.978	2.334	2.749	3.621	4.343
178	0.904	1.570	1.976	2.331	2.747	3.618	4.339
180	0.903	1.568	1.974	2.329	2.744	3.615	4.335
185	0.900	1.564	1.969	2.323	2.738	3.607	4.325
190	0.896	1.560	1.964	2.318	2.732	3.599	4.316
195	0.893	1.556	1.960	2.313	2.726	3.591	4.307
200	0.890	1.552	1.955	2.308	2.720	3.584	4.299
205	0.887	1.549	1.951	2.303	2.715	3.577	4.291
210	0.885	1.545	1.947	2.299	2.709	3.571	4.283
215	0.882	1.542	1.943	2.294	2.704	3.565	4.276
220	0.880	1.538	1.939	2.290	2.700	3.558	4.268

TABLE 1.1.4. FACTORS FOR ONE-SIDED SAMPLING PLANS AND TOLERANCE LIMITS.

GAMMA = 0.995 (CONTINUED)

N \P →	0.750	0.900	0.950	0.975	0.990	0.999	0.9999
225	0.877	1.535	1.936	2.286	2.695	3.553	4.262
230	0.875	1.532	1.932	2.282	2.690	3.547	4.255
235	0.872	1.529	1.929	2.278	2.686	3.541	4.248
240	0.870	1.526	1.926	2.274	2.682	3.536	4.242
245	0.868	1.524	1.922	2.271	2.678	3.531	4.236
250	0.866	1.521	1.919	2.267	2.674	3.526	4.230
255	0.864	1.518	1.916	2.264	2.670	3.521	4.225
260	0.862	1.516	1.913	2.261	2.666	3.517	4.219
265	0.860	1.513	1.911	2.258	2.663	3.512	4.214
270	0.858	1.511	1.908	2.254	2.659	3.508	4.209
275	0.857	1.509	1.905	2.252	2.656	3.504	4.204
280	0.855	1.507	1.903	2.249	2.653	3.500	4.199
285	0.853	1.504	1.900	2.246	2.650	3.496	4.194
290	0.851	1.502	1.898	2.243	2.646	3.492	4.190
295	0.850	1.500	1.895	2.240	2.643	3.488	4.185
300	0.848	1.498	1.893	2.238	2.641	3.484	4.181
310	0.845	1.495	1.889	2.233	2.635	3.477	4.173
320	0.842	1.491	1.885	2.228	2.630	3.470	4.165
330	0.840	1.487	1.881	2.224	2.625	3.464	4.157
340	0.837	1.484	1.877	2.220	2.620	3.458	4.150
350	0.835	1.481	1.873	2.215	2.615	3.452	4.144
360	0.832	1.478	1.870	2.212	2.611	3.447	4.137
370	0.830	1.475	1.866	2.208	2.607	3.441	4.131
380	0.828	1.472	1.863	2.204	2.602	3.436	4.125
390	0.826	1.470	1.860	2.201	2.599	3.431	4.119
400	0.824	1.467	1.857	2.198	2.595	3.427	4.114
425	0.819	1.461	1.851	2.190	2.586	3.416	4.101
450	0.815	1.456	1.844	2.183	2.578	3.406	4.089
475	0.811	1.451	1.839	2.177	2.571	3.397	4.079
500	0.807	1.446	1.833	2.171	2.564	3.388	4.069
525	0.804	1.442	1.829	2.165	2.558	3.381	4.060
550	0.801	1.438	1.824	2.160	2.553	3.374	4.051
575	0.798	1.435	1.820	2.156	2.547	3.367	4.043
600	0.795	1.431	1.816	2.151	2.542	3.361	4.036
625	0.793	1.428	1.812	2.147	2.538	3.355	4.029
650	0.790	1.425	1.809	2.143	2.533	3.349	4.023
700	0.786	1.420	1.803	2.136	2.525	3.339	4.011
750	0.782	1.415	1.797	2.130	2.518	3.330	4.000
800	0.779	1.410	1.792	2.124	2.512	3.322	3.991
850	0.775	1.406	1.787	2.119	2.506	3.315	3.982
900	0.772	1.403	1.783	2.114	2.501	3.308	3.974
950	0.770	1.399	1.779	2.110	2.496	3.302	3.967
1000	0.767	1.396	1.776	2.106	2.491	3.296	3.961
1500	0.750	1.374	1.751	2.078	2.460	3.257	3.914
2000	0.740	1.362	1.736	2.062	2.441	3.234	3.887
3000	0.727	1.347	1.719	2.043	2.419	3.207	3.855
5000	0.715	1.332	1.702	2.024	2.398	3.180	3.824
10000	0.703	1.317	1.685	2.005	2.377	3.153	3.793
∞	0.674	1.282	1.645	1.960	2.326	3.090	3.719

TABLE 1.2.1. *FACTORS FOR ONE-SIDED SAMPLING PLANS AND TOLERANCE LIMITS.*

GAMMA = 0.990

$N \backslash P \rightarrow$	0.750	0.900	0.950	0.975	0.990	0.999	0.9999
2	58.940	103.029	131.426	156.424	185.617	246.557	296.726
3	8.728	13.995	17.370	20.365	23.896	31.348	37.533
4	4.715	7.380	9.083	10.598	12.387	16.176	19.327
5	3.454	5.362	6.578	7.660	8.939	11.649	13.906
6	2.848	4.411	5.406	6.290	7.335	9.550	11.396
7	2.491	3.859	4.728	5.500	6.412	8.346	9.957
8	2.253	3.497	4.285	4.985	5.812	7.564	9.024
9	2.083	3.240	3.972	4.622	5.389	7.014	8.368
10	1.954	3.048	3.738	4.351	5.074	6.605	7.881
11	1.853	2.898	3.556	4.140	4.829	6.288	7.503
12	1.771	2.777	3.410	3.971	4.633	6.035	7.201
13	1.703	2.677	3.290	3.832	4.472	5.827	6.954
14	1.645	2.593	3.189	3.716	4.337	5.652	6.747
15	1.595	2.521	3.102	3.616	4.222	5.504	6.571
16	1.552	2.459	3.028	3.531	4.123	5.377	6.419
17	1.514	2.405	2.963	3.456	4.037	5.265	6.287
18	1.481	2.357	2.905	3.390	3.960	5.167	6.170
19	1.450	2.314	2.854	3.331	3.892	5.079	6.066
20	1.423	2.276	2.808	3.278	3.832	5.001	5.974
21	1.399	2.241	2.766	3.230	3.777	4.931	5.890
22	1.376	2.209	2.729	3.187	3.727	4.867	5.814
23	1.355	2.180	2.694	3.148	3.681	4.808	5.745
24	1.336	2.154	2.662	3.111	3.640	4.755	5.681
25	1.319	2.129	2.633	3.078	3.601	4.706	5.623
26	1.303	2.106	2.606	3.047	3.566	4.660	5.569
27	1.287	2.085	2.581	3.018	3.533	4.618	5.519
28	1.273	2.065	2.558	2.992	3.502	4.579	5.473
29	1.260	2.047	2.536	2.967	3.473	4.542	5.429
30	1.247	2.030	2.515	2.943	3.447	4.508	5.389
31	1.236	2.014	2.496	2.922	3.421	4.476	5.351
32	1.225	1.998	2.478	2.901	3.398	4.445	5.315
33	1.214	1.984	2.461	2.881	3.375	4.417	5.281
34	1.204	1.970	2.445	2.863	3.354	4.390	5.249
35	1.195	1.957	2.430	2.846	3.334	4.364	5.219
36	1.186	1.945	2.415	2.829	3.315	4.340	5.191
37	1.177	1.934	2.402	2.814	3.297	4.317	5.164
38	1.169	1.922	2.389	2.799	3.280	4.296	5.138
39	1.161	1.912	2.376	2.784	3.264	4.275	5.113
40	1.154	1.902	2.364	2.771	3.249	4.255	5.090
41	1.147	1.892	2.353	2.758	3.234	4.236	5.068
42	1.140	1.883	2.342	2.746	3.220	4.218	5.046
43	1.133	1.874	2.331	2.734	3.206	4.201	5.026
44	1.127	1.865	2.321	2.722	3.193	4.184	5.006
45	1.121	1.857	2.312	2.711	3.180	4.168	4.987
46	1.115	1.849	2.303	2.701	3.168	4.153	4.969
47	1.110	1.842	2.294	2.691	3.157	4.138	4.952
48	1.104	1.835	2.285	2.681	3.146	4.124	4.935
49	1.099	1.828	2.277	2.672	3.135	4.110	4.919
50	1.094	1.821	2.269	2.663	3.125	4.097	4.903

TABLE 1.2.2. FACTORS FOR ONE-SIDED SAMPLING PLANS AND TOLERANCE LIMITS.

GAMMA = 0.990 (CONTINUED)

N \P →	0.750	0.900	0.950	0.975	0.990	0.999	0.9999
51	1.089	1.814	2.261	2.654	3.115	4.084	4.888
52	1.084	1.808	2.254	2.645	3.105	4.072	4.874
53	1.080	1.802	2.247	2.637	3.096	4.060	4.860
54	1.075	1.796	2.240	2.629	3.087	4.049	4.846
55	1.071	1.790	2.233	2.622	3.078	4.038	4.833
56	1.067	1.785	2.226	2.614	3.070	4.027	4.821
57	1.063	1.779	2.220	2.607	3.061	4.017	4.808
58	1.059	1.774	2.214	2.600	3.053	4.007	4.797
59	1.055	1.769	2.208	2.594	3.046	3.997	4.785
60	1.052	1.764	2.202	2.587	3.038	3.987	4.774
61	1.048	1.759	2.197	2.581	3.031	3.978	4.763
62	1.045	1.755	2.191	2.575	3.024	3.969	4.752
63	1.041	1.750	2.186	2.569	3.017	3.960	4.742
64	1.038	1.746	2.181	2.563	3.010	3.952	4.732
65	1.035	1.741	2.176	2.557	3.004	3.944	4.722
66	1.032	1.737	2.171	2.552	2.998	3.936	4.713
67	1.028	1.733	2.166	2.546	2.991	3.928	4.704
68	1.025	1.729	2.162	2.541	2.985	3.920	4.695
69	1.023	1.725	2.157	2.536	2.980	3.913	4.686
70	1.020	1.722	2.153	2.531	2.974	3.906	4.677
71	1.017	1.718	2.148	2.526	2.968	3.899	4.669
72	1.014	1.714	2.144	2.521	2.963	3.892	4.661
73	1.012	1.711	2.140	2.516	2.958	3.885	4.653
74	1.009	1.707	2.136	2.512	2.952	3.878	4.645
75	1.006	1.704	2.132	2.508	2.947	3.872	4.638
76	1.004	1.701	2.128	2.503	2.942	3.866	4.630
77	1.002	1.698	2.125	2.499	2.938	3.860	4.623
78	0.999	1.694	2.121	2.495	2.933	3.854	4.616
79	0.997	1.691	2.117	2.491	2.928	3.848	4.609
80	0.995	1.688	2.114	2.487	2.924	3.842	4.602
81	0.992	1.685	2.110	2.483	2.919	3.836	4.596
82	0.990	1.682	2.107	2.479	2.915	3.831	4.589
83	0.988	1.680	2.104	2.475	2.911	3.826	4.583
84	0.986	1.677	2.100	2.472	2.907	3.820	4.577
85	0.984	1.674	2.097	2.468	2.902	3.815	4.571
86	0.982	1.672	2.094	2.465	2.898	3.810	4.565
87	0.980	1.669	2.091	2.461	2.895	3.805	4.559
88	0.978	1.666	2.088	2.458	2.891	3.800	4.553
89	0.976	1.664	2.085	2.454	2.887	3.795	4.547
90	0.974	1.661	2.082	2.451	2.883	3.791	4.542
91	0.972	1.659	2.080	2.448	2.880	3.786	4.537
92	0.970	1.657	2.077	2.445	2.876	3.782	4.531
93	0.969	1.654	2.074	2.442	2.872	3.777	4.526
94	0.967	1.652	2.071	2.439	2.869	3.773	4.521
95	0.965	1.650	2.069	2.436	2.866	3.769	4.516
96	0.963	1.647	2.066	2.433	2.862	3.764	4.511
97	0.962	1.645	2.064	2.430	2.859	3.760	4.506
98	0.960	1.643	2.061	2.427	2.856	3.756	4.501
99	0.959	1.641	2.059	2.424	2.853	3.752	4.497

TABLE 1.2.3. FACTORS FOR ONE-SIDED SAMPLING PLANS AND TOLERANCE LIMITS.

GAMMA = 0.990 (CONTINUED)

N \P →	0.750	0.900	0.950	0.975	0.990	0.999	0.9999
100	0.957	1.639	2.056	2.422	2.850	3.748	4.492
102	0.954	1.635	2.052	2.416	2.844	3.741	4.483
104	0.951	1.631	2.047	2.411	2.838	3.733	4.474
106	0.948	1.627	2.043	2.406	2.832	3.726	4.466
108	0.945	1.624	2.038	2.402	2.827	3.719	4.458
110	0.942	1.620	2.034	2.397	2.821	3.712	4.450
112	0.940	1.617	2.030	2.392	2.816	3.706	4.442
114	0.937	1.613	2.026	2.388	2.811	3.699	4.435
116	0.935	1.610	2.023	2.384	2.806	3.693	4.427
118	0.932	1.607	2.019	2.380	2.802	3.687	4.420
120	0.930	1.604	2.015	2.375	2.797	3.682	4.413
122	0.928	1.601	2.012	2.372	2.792	3.676	4.407
124	0.925	1.598	2.008	2.368	2.788	3.670	4.400
126	0.923	1.595	2.005	2.364	2.784	3.665	4.394
128	0.921	1.592	2.002	2.360	2.780	3.660	4.388
130	0.919	1.589	1.999	2.357	2.776	3.655	4.382
132	0.917	1.587	1.996	2.353	2.772	3.650	4.376
134	0.915	1.584	1.993	2.350	2.768	3.645	4.370
136	0.913	1.582	1.990	2.347	2.764	3.640	4.365
138	0.911	1.579	1.987	2.344	2.761	3.636	4.359
140	0.909	1.577	1.984	2.340	2.757	3.631	4.354
142	0.908	1.575	1.982	2.337	2.754	3.627	4.349
144	0.906	1.572	1.979	2.334	2.750	3.623	4.344
146	0.904	1.570	1.976	2.331	2.747	3.618	4.339
148	0.902	1.568	1.974	2.329	2.744	3.614	4.334
150	0.901	1.566	1.971	2.326	2.740	3.610	4.329
152	0.899	1.564	1.969	2.323	2.737	3.606	4.325
154	0.898	1.562	1.966	2.320	2.734	3.602	4.320
156	0.896	1.560	1.964	2.318	2.731	3.599	4.316
158	0.894	1.558	1.962	2.315	2.728	3.595	4.312
160	0.893	1.556	1.960	2.313	2.726	3.591	4.307
162	0.891	1.554	1.957	2.310	2.723	3.588	4.303
164	0.890	1.552	1.955	2.308	2.720	3.584	4.299
166	0.889	1.550	1.953	2.306	2.717	3.581	4.295
168	0.887	1.548	1.951	2.303	2.715	3.578	4.291
170	0.886	1.547	1.949	2.301	2.712	3.575	4.287
172	0.885	1.545	1.947	2.299	2.710	3.571	4.284
174	0.883	1.543	1.945	2.297	2.707	3.568	4.280
176	0.882	1.542	1.943	2.294	2.705	3.565	4.276
178	0.881	1.540	1.941	2.292	2.702	3.562	4.273
180	0.880	1.538	1.940	2.290	2.700	3.559	4.269
185	0.877	1.535	1.935	2.285	2.694	3.552	4.261
190	0.874	1.531	1.931	2.280	2.689	3.545	4.253
195	0.871	1.527	1.927	2.276	2.684	3.539	4.245
200	0.868	1.524	1.923	2.271	2.679	3.532	4.238
205	0.866	1.521	1.919	2.267	2.674	3.526	4.230
210	0.863	1.518	1.916	2.263	2.669	3.520	4.224
215	0.861	1.515	1.912	2.259	2.665	3.515	4.217
220	0.859	1.512	1.909	2.255	2.660	3.509	4.211

TABLE 1.2.4. FACTORS FOR ONE-SIDED SAMPLING PLANS AND TOLERANCE LIMITS.

GAMMA = 0.990 (CONTINUED)

N \P →	0.750	0.900	0.950	0.975	0.990	0.999	0.9999
225	0.856	1.509	1.905	2.252	2.656	3.504	4.205
230	0.854	1.506	1.902	2.248	2.652	3.499	4.199
235	0.852	1.504	1.899	2.245	2.648	3.494	4.193
240	0.850	1.501	1.896	2.242	2.645	3.490	4.187
245	0.848	1.499	1.894	2.238	2.641	3.485	4.182
250	0.847	1.496	1.891	2.235	2.638	3.481	4.177
255	0.845	1.494	1.888	2.232	2.634	3.476	4.172
260	0.843	1.492	1.886	2.229	2.631	3.472	4.167
265	0.841	1.490	1.883	2.227	2.628	3.468	4.162
270	0.840	1.487	1.881	2.224	2.625	3.464	4.158
275	0.838	1.485	1.878	2.221	2.622	3.461	4.153
280	0.837	1.483	1.876	2.219	2.619	3.457	4.149
285	0.835	1.482	1.874	2.216	2.616	3.453	4.145
290	0.834	1.480	1.872	2.214	2.613	3.450	4.141
295	0.832	1.478	1.870	2.211	2.611	3.447	4.137
300	0.831	1.476	1.868	2.209	2.608	3.443	4.133
310	0.828	1.473	1.864	2.205	2.603	3.437	4.126
320	0.826	1.469	1.860	2.201	2.598	3.431	4.119
330	0.823	1.466	1.856	2.197	2.594	3.425	4.112
340	0.821	1.463	1.853	2.193	2.590	3.420	4.106
350	0.819	1.461	1.850	2.189	2.585	3.415	4.100
360	0.816	1.458	1.847	2.186	2.581	3.410	4.094
370	0.814	1.455	1.844	2.182	2.578	3.405	4.088
380	0.812	1.453	1.841	2.179	2.574	3.401	4.083
390	0.811	1.451	1.838	2.176	2.571	3.396	4.078
400	0.809	1.448	1.836	2.173	2.567	3.392	4.073
425	0.805	1.443	1.830	2.166	2.560	3.382	4.062
450	0.801	1.438	1.824	2.160	2.553	3.374	4.051
475	0.797	1.434	1.819	2.155	2.546	3.365	4.042
500	0.794	1.430	1.814	2.149	2.540	3.358	4.033
525	0.791	1.426	1.810	2.144	2.535	3.351	4.025
550	0.788	1.422	1.806	2.140	2.530	3.345	4.017
575	0.786	1.419	1.802	2.136	2.525	3.339	4.010
600	0.783	1.416	1.799	2.132	2.520	3.333	4.004
625	0.781	1.413	1.795	2.128	2.516	3.328	3.998
650	0.779	1.411	1.792	2.125	2.512	3.323	3.992
700	0.775	1.406	1.787	2.118	2.505	3.314	3.981
750	0.771	1.401	1.782	2.113	2.499	3.306	3.972
800	0.768	1.397	1.777	2.108	2.493	3.299	3.963
850	0.765	1.394	1.773	2.103	2.488	3.292	3.956
900	0.763	1.390	1.769	2.099	2.483	3.286	3.949
950	0.760	1.387	1.766	2.095	2.479	3.281	3.942
1000	0.758	1.385	1.762	2.091	2.475	3.276	3.936
1500	0.742	1.365	1.740	2.066	2.446	3.240	3.895
2000	0.733	1.354	1.727	2.052	2.430	3.219	3.870
3000	0.722	1.340	1.712	2.034	2.410	3.195	3.842
5000	0.711	1.327	1.696	2.017	2.391	3.171	3.814
10000	0.700	1.313	1.681	2.000	2.372	3.147	3.785
∞	0.674	1.282	1.645	1.960	2.326	3.090	3.719

TABLE 1.3.1. *FACTORS FOR ONE-SIDED SAMPLING PLANS AND TOLERANCE LIMITS.*

GAMMA = 0.975

N \P →	0.750	0.900	0.950	0.975	0.990	0.999	0.9999
2	23.565	41.201	52.559	62.558	74.234	98.607	118.672
3	5.469	8.797	10.927	12.816	15.043	19.741	23.639
4	3.398	5.354	6.602	7.710	9.018	11.785	14.087
5	2.660	4.166	5.124	5.975	6.980	9.107	10.877
6	2.279	3.568	4.385	5.111	5.967	7.781	9.291
7	2.044	3.206	3.940	4.592	5.361	6.989	8.344
8	1.883	2.960	3.640	4.243	4.954	6.460	7.712
9	1.764	2.783	3.424	3.992	4.662	6.079	7.259
10	1.673	2.647	3.259	3.801	4.440	5.791	6.916
11	1.600	2.540	3.129	3.650	4.265	5.565	6.646
12	1.540	2.452	3.023	3.528	4.124	5.382	6.428
13	1.490	2.379	2.936	3.427	4.006	5.230	6.248
14	1.447	2.317	2.861	3.342	3.907	5.103	6.096
15	1.410	2.264	2.797	3.268	3.822	4.993	5.966
16	1.378	2.218	2.742	3.204	3.749	4.898	5.853
17	1.349	2.177	2.693	3.148	3.684	4.815	5.754
18	1.324	2.141	2.650	3.099	3.627	4.741	5.667
19	1.301	2.108	2.611	3.054	3.575	4.675	5.588
20	1.280	2.079	2.576	3.014	3.529	4.616	5.518
21	1.261	2.053	2.544	2.978	3.487	4.562	5.454
22	1.243	2.028	2.515	2.945	3.449	4.513	5.397
23	1.227	2.006	2.489	2.914	3.414	4.469	5.344
24	1.213	1.985	2.465	2.887	3.382	4.428	5.295
25	1.199	1.966	2.442	2.861	3.353	4.390	5.250
26	1.186	1.949	2.421	2.837	3.325	4.355	5.209
27	1.174	1.932	2.402	2.815	3.300	4.322	5.170
28	1.163	1.917	2.384	2.794	3.276	4.292	5.134
29	1.153	1.903	2.367	2.775	3.254	4.263	5.101
30	1.143	1.889	2.351	2.757	3.233	4.237	5.069
31	1.134	1.877	2.336	2.740	3.213	4.212	5.039
32	1.125	1.865	2.322	2.723	3.195	4.188	5.011
33	1.116	1.853	2.308	2.708	3.178	4.166	4.985
34	1.109	1.843	2.296	2.694	3.161	4.145	4.960
35	1.101	1.833	2.284	2.680	3.145	4.125	4.937
36	1.094	1.823	2.272	2.667	3.131	4.106	4.914
37	1.087	1.814	2.262	2.655	3.116	4.088	4.893
38	1.081	1.805	2.251	2.643	3.103	4.071	4.873
39	1.075	1.797	2.241	2.632	3.090	4.054	4.853
40	1.069	1.789	2.232	2.621	3.078	4.039	4.835
41	1.063	1.781	2.223	2.611	3.066	4.024	4.817
42	1.058	1.774	2.214	2.601	3.055	4.010	4.801
43	1.052	1.767	2.206	2.592	3.044	3.996	4.784
44	1.047	1.760	2.198	2.583	3.034	3.983	4.769
45	1.042	1.753	2.190	2.574	3.024	3.970	4.754
46	1.038	1.747	2.183	2.566	3.014	3.958	4.740
47	1.033	1.741	2.176	2.558	3.005	3.946	4.726
48	1.029	1.735	2.169	2.550	2.996	3.935	4.713
49	1.025	1.730	2.163	2.543	2.988	3.924	4.700
50	1.021	1.724	2.156	2.535	2.980	3.914	4.688

TABLE 1.3.2. FACTORS FOR ONE-SIDED SAMPLING PLANS AND TOLERANCE LIMITS.

GAMMA = 0.975 (CONTINUED)

N \P →	0.750	0.900	0.950	0.975	0.990	0.999	0.9999
51	1.017	1.719	2.150	2.529	2.972	3.904	4.676
52	1.013	1.714	2.144	2.522	2.964	3.894	4.664
53	1.009	1.709	2.139	2.515	2.957	3.885	4.653
54	1.006	1.704	2.133	2.509	2.950	3.875	4.642
55	1.002	1.700	2.128	2.503	2.943	3.867	4.632
56	0.999	1.695	2.123	2.497	2.936	3.858	4.622
57	0.996	1.691	2.117	2.491	2.929	3.850	4.612
58	0.993	1.687	2.113	2.486	2.923	3.842	4.602
59	0.989	1.683	2.108	2.480	2.917	3.834	4.593
60	0.986	1.679	2.103	2.475	2.911	3.826	4.584
61	0.984	1.675	2.099	2.470	2.905	3.819	4.576
62	0.981	1.671	2.094	2.465	2.899	3.812	4.567
63	0.978	1.668	2.090	2.460	2.894	3.805	4.559
64	0.975	1.664	2.086	2.456	2.889	3.798	4.551
65	0.973	1.661	2.082	2.451	2.883	3.791	4.543
66	0.970	1.657	2.078	2.447	2.878	3.785	4.536
67	0.968	1.654	2.074	2.442	2.873	3.779	4.528
68	0.965	1.651	2.070	2.438	2.869	3.773	4.521
69	0.963	1.648	2.067	2.434	2.864	3.767	4.514
70	0.961	1.645	2.063	2.430	2.859	3.761	4.507
71	0.958	1.642	2.060	2.426	2.855	3.755	4.500
72	0.956	1.639	2.056	2.422	2.850	3.750	4.494
73	0.954	1.636	2.053	2.418	2.846	3.744	4.487
74	0.952	1.633	2.050	2.415	2.842	3.739	4.481
75	0.950	1.630	2.047	2.411	2.838	3.734	4.475
76	0.948	1.628	2.044	2.408	2.834	3.729	4.469
77	0.946	1.625	2.041	2.404	2.830	3.724	4.463
78	0.944	1.623	2.038	2.401	2.826	3.719	4.458
79	0.942	1.620	2.035	2.398	2.822	3.714	4.452
80	0.940	1.618	2.032	2.394	2.819	3.710	4.447
81	0.938	1.615	2.029	2.391	2.815	3.705	4.441
82	0.936	1.613	2.026	2.388	2.812	3.701	4.436
83	0.935	1.611	2.024	2.385	2.808	3.696	4.431
84	0.933	1.608	2.021	2.382	2.805	3.692	4.426
85	0.931	1.606	2.019	2.379	2.802	3.688	4.421
86	0.930	1.604	2.016	2.377	2.798	3.684	4.416
87	0.928	1.602	2.014	2.374	2.795	3.680	4.412
88	0.926	1.600	2.011	2.371	2.792	3.676	4.407
89	0.925	1.598	2.009	2.368	2.789	3.672	4.402
90	0.923	1.596	2.006	2.366	2.786	3.668	4.398
91	0.922	1.594	2.004	2.363	2.783	3.664	4.394
92	0.920	1.592	2.002	2.361	2.780	3.661	4.389
93	0.919	1.590	2.000	2.358	2.777	3.657	4.385
94	0.918	1.588	1.998	2.356	2.775	3.654	4.381
95	0.916	1.586	1.995	2.353	2.772	3.650	4.377
96	0.915	1.585	1.993	2.351	2.769	3.647	4.373
97	0.913	1.583	1.991	2.349	2.767	3.644	4.369
98	0.912	1.581	1.989	2.346	2.764	3.640	4.365
99	0.911	1.579	1.987	2.344	2.761	3.637	4.361

TABLE 1.3.3. FACTORS FOR ONE-SIDED SAMPLING PLANS AND TOLERANCE LIMITS.

GAMMA = 0.975 (CONTINUED)

N \P →	0.750	0.900	0.950	0.975	0.990	0.999	0.9999
100	0.909	1.578	1.985	2.342	2.759	3.634	4.357
102	0.907	1.574	1.981	2.338	2.754	3.628	4.350
104	0.904	1.571	1.978	2.333	2.749	3.622	4.343
106	0.902	1.568	1.974	2.329	2.745	3.616	4.336
108	0.900	1.565	1.971	2.325	2.740	3.610	4.330
110	0.898	1.562	1.967	2.322	2.736	3.605	4.323
112	0.895	1.559	1.964	2.318	2.732	3.599	4.317
114	0.893	1.557	1.961	2.314	2.728	3.594	4.311
116	0.891	1.554	1.958	2.311	2.724	3.589	4.305
118	0.889	1.551	1.955	2.308	2.720	3.584	4.299
120	0.887	1.549	1.952	2.304	2.716	3.580	4.294
122	0.885	1.546	1.949	2.301	2.712	3.575	4.288
124	0.884	1.544	1.946	2.298	2.709	3.571	4.283
126	0.882	1.542	1.944	2.295	2.705	3.566	4.278
128	0.880	1.539	1.941	2.292	2.702	3.562	4.273
130	0.878	1.537	1.938	2.289	2.699	3.558	4.268
132	0.877	1.535	1.936	2.286	2.696	3.554	4.263
134	0.875	1.533	1.934	2.283	2.692	3.550	4.259
136	0.873	1.531	1.931	2.281	2.689	3.546	4.254
138	0.872	1.529	1.929	2.278	2.686	3.542	4.250
140	0.870	1.527	1.927	2.276	2.684	3.539	4.245
142	0.869	1.525	1.924	2.273	2.681	3.535	4.241
144	0.867	1.523	1.922	2.271	2.678	3.532	4.237
146	0.866	1.521	1.920	2.268	2.675	3.528	4.233
148	0.865	1.520	1.918	2.266	2.673	3.525	4.229
150	0.863	1.518	1.916	2.264	2.670	3.522	4.225
152	0.862	1.516	1.914	2.261	2.667	3.518	4.222
154	0.861	1.514	1.912	2.259	2.665	3.515	4.218
156	0.859	1.513	1.910	2.257	2.663	3.512	4.214
158	0.858	1.511	1.908	2.255	2.660	3.509	4.211
160	0.857	1.510	1.906	2.253	2.658	3.506	4.207
162	0.856	1.508	1.905	2.251	2.656	3.503	4.204
164	0.854	1.507	1.903	2.249	2.653	3.501	4.201
166	0.853	1.505	1.901	2.247	2.651	3.498	4.197
168	0.852	1.504	1.900	2.245	2.649	3.495	4.194
170	0.851	1.502	1.898	2.243	2.647	3.492	4.191
172	0.850	1.501	1.896	2.241	2.645	3.490	4.188
174	0.849	1.499	1.895	2.240	2.643	3.487	4.185
176	0.848	1.498	1.893	2.238	2.641	3.485	4.182
178	0.847	1.497	1.892	2.236	2.639	3.482	4.179
180	0.846	1.495	1.890	2.235	2.637	3.480	4.176
185	0.843	1.492	1.886	2.230	2.632	3.474	4.169
190	0.841	1.489	1.883	2.226	2.628	3.468	4.163
195	0.839	1.486	1.880	2.223	2.623	3.463	4.156
200	0.836	1.484	1.876	2.219	2.619	3.458	4.150
205	0.834	1.481	1.873	2.216	2.615	3.453	4.144
210	0.832	1.478	1.870	2.212	2.612	3.448	4.139
215	0.830	1.476	1.867	2.209	2.608	3.443	4.133
220	0.828	1.473	1.865	2.206	2.604	3.439	4.128

TABLE 1.3.4. FACTORS FOR ONE-SIDED SAMPLING PLANS AND TOLERANCE LIMITS.

GAMMA = 0.975 (CONTINUED)

N \P →	0.750	0.900	0.950	0.975	0.990	0.999	0.9999
225	0.827	1.471	1.862	2.203	2.601	3.435	4.123
230	0.825	1.469	1.859	2.200	2.598	3.430	4.118
235	0.823	1.467	1.857	2.197	2.595	3.426	4.113
240	0.822	1.465	1.854	2.195	2.591	3.423	4.109
245	0.820	1.463	1.852	2.192	2.589	3.419	4.105
250	0.818	1.461	1.850	2.189	2.586	3.415	4.100
255	0.817	1.459	1.848	2.187	2.583	3.412	4.096
260	0.815	1.457	1.846	2.185	2.580	3.408	4.092
265	0.814	1.455	1.844	2.182	2.578	3.405	4.088
270	0.813	1.453	1.842	2.180	2.575	3.402	4.085
275	0.811	1.452	1.840	2.178	2.573	3.399	4.081
280	0.810	1.450	1.838	2.176	2.570	3.396	4.077
285	0.809	1.449	1.836	2.174	2.568	3.393	4.074
290	0.808	1.447	1.834	2.172	2.566	3.390	4.071
295	0.806	1.445	1.832	2.170	2.563	3.387	4.067
300	0.805	1.444	1.831	2.168	2.561	3.385	4.064
310	0.803	1.441	1.827	2.164	2.557	3.379	4.058
320	0.801	1.439	1.824	2.161	2.553	3.374	4.052
330	0.799	1.436	1.821	2.157	2.549	3.370	4.047
340	0.797	1.434	1.819	2.154	2.546	3.365	4.042
350	0.795	1.431	1.816	2.151	2.543	3.361	4.037
360	0.793	1.429	1.813	2.148	2.539	3.357	4.032
370	0.792	1.427	1.811	2.146	2.536	3.353	4.027
380	0.790	1.425	1.809	2.143	2.533	3.349	4.023
390	0.789	1.423	1.806	2.141	2.530	3.346	4.019
400	0.787	1.421	1.804	2.138	2.528	3.342	4.015
425	0.784	1.417	1.799	2.133	2.521	3.334	4.005
450	0.780	1.413	1.795	2.127	2.515	3.327	3.997
475	0.777	1.409	1.790	2.123	2.510	3.320	3.989
500	0.775	1.406	1.787	2.118	2.505	3.314	3.981
525	0.772	1.402	1.783	2.114	2.501	3.308	3.975
550	0.770	1.400	1.780	2.110	2.496	3.303	3.968
575	0.768	1.397	1.776	2.107	2.492	3.298	3.963
600	0.766	1.394	1.774	2.104	2.489	3.293	3.957
625	0.764	1.392	1.771	2.101	2.485	3.289	3.952
650	0.762	1.390	1.768	2.098	2.482	3.285	3.947
700	0.759	1.386	1.764	2.093	2.476	3.278	3.939
750	0.756	1.382	1.759	2.088	2.471	3.271	3.931
800	0.753	1.379	1.756	2.084	2.466	3.265	3.924
850	0.751	1.376	1.752	2.080	2.462	3.259	3.917
900	0.749	1.373	1.749	2.076	2.458	3.254	3.911
950	0.747	1.370	1.746	2.073	2.454	3.250	3.906
1000	0.745	1.368	1.743	2.070	2.451	3.246	3.901
1500	0.732	1.352	1.725	2.049	2.427	3.216	3.866
2000	0.724	1.342	1.714	2.037	2.413	3.199	3.846
3000	0.715	1.331	1.701	2.022	2.397	3.178	3.822
5000	0.705	1.319	1.688	2.008	2.381	3.158	3.798
10000	0.696	1.308	1.675	1.994	2.365	3.138	3.775
∞	0.674	1.282	1.645	1.960	2.326	3.090	3.719

TABLE 1.4.1. *FACTORS FOR ONE-SIDED SAMPLING PLANS AND TOLERANCE LIMITS.*

GAMMA = 0.950

N \P →	0.750	0.900	0.950	0.975	0.990	0.999	0.9999
2	11.763	20.581	26.260	31.257	37.094	49.276	59.304
3	3.806	6.155	7.656	8.986	10.553	13.857	16.598
4	2.618	4.162	5.144	6.015	7.042	9.214	11.019
5	2.150	3.407	4.203	4.909	5.741	7.502	8.966
6	1.895	3.006	3.708	4.329	5.062	6.612	7.901
7	1.732	2.755	3.399	3.970	4.642	6.063	7.244
8	1.618	2.582	3.187	3.723	4.354	5.688	6.796
9	1.532	2.454	3.031	3.542	4.143	5.413	6.469
10	1.465	2.355	2.911	3.402	3.981	5.203	6.219
11	1.411	2.275	2.815	3.292	3.852	5.036	6.020
12	1.366	2.210	2.736	3.201	3.747	4.900	5.858
13	1.328	2.155	2.671	3.125	3.659	4.787	5.723
14	1.296	2.109	2.614	3.060	3.585	4.690	5.609
15	1.268	2.068	2.566	3.005	3.520	4.607	5.510
16	1.243	2.033	2.524	2.956	3.464	4.535	5.424
17	1.220	2.002	2.486	2.913	3.414	4.471	5.348
18	1.201	1.974	2.453	2.875	3.370	4.415	5.281
19	1.183	1.949	2.423	2.841	3.331	4.364	5.221
20	1.166	1.926	2.396	2.810	3.295	4.318	5.167
21	1.152	1.905	2.371	2.781	3.263	4.277	5.118
22	1.138	1.886	2.349	2.756	3.233	4.239	5.073
23	1.125	1.869	2.328	2.732	3.206	4.204	5.031
24	1.114	1.853	2.309	2.710	3.181	4.172	4.994
25	1.103	1.838	2.292	2.690	3.158	4.142	4.959
26	1.093	1.824	2.275	2.672	3.136	4.115	4.926
27	1.083	1.811	2.260	2.654	3.116	4.089	4.896
28	1.075	1.799	2.246	2.638	3.098	4.066	4.868
29	1.066	1.788	2.232	2.623	3.080	4.043	4.841
30	1.058	1.777	2.220	2.608	3.064	4.022	4.816
31	1.051	1.767	2.208	2.595	3.048	4.002	4.793
32	1.044	1.758	2.197	2.582	3.034	3.984	4.771
33	1.037	1.749	2.186	2.570	3.020	3.966	4.750
34	1.031	1.740	2.176	2.559	3.007	3.950	4.730
35	1.025	1.732	2.167	2.548	2.995	3.934	4.712
36	1.019	1.725	2.158	2.538	2.983	3.919	4.694
37	1.014	1.717	2.149	2.528	2.972	3.904	4.677
38	1.009	1.710	2.141	2.518	2.961	3.891	4.661
39	1.004	1.704	2.133	2.510	2.951	3.878	4.646
40	0.999	1.697	2.125	2.501	2.941	3.865	4.631
41	0.994	1.691	2.118	2.493	2.932	3.854	4.617
42	0.990	1.685	2.111	2.485	2.923	3.842	4.603
43	0.986	1.680	2.105	2.478	2.914	3.831	4.591
44	0.982	1.674	2.098	2.470	2.906	3.821	4.578
45	0.978	1.669	2.092	2.463	2.898	3.811	4.566
46	0.974	1.664	2.086	2.457	2.890	3.801	4.555
47	0.971	1.659	2.081	2.450	2.883	3.792	4.544
48	0.967	1.654	2.075	2.444	2.876	3.783	4.533
49	0.964	1.650	2.070	2.438	2.869	3.774	4.523
50	0.960	1.646	2.065	2.432	2.862	3.766	4.513

TABLE 1.4.2. FACTORS FOR ONE-SIDED SAMPLING PLANS AND TOLERANCE LIMITS.

GAMMA = 0.950 (CONTINUED)

N \P →	0.750	0.900	0.950	0.975	0.990	0.999	0.9999
51	0.957	1.641	2.060	2.427	2.856	3.758	4.504
52	0.954	1.637	2.055	2.421	2.850	3.750	4.494
53	0.951	1.633	2.051	2.416	2.844	3.742	4.485
54	0.948	1.630	2.046	2.411	2.838	3.735	4.477
55	0.945	1.626	2.042	2.406	2.833	3.728	4.468
56	0.943	1.622	2.038	2.401	2.827	3.721	4.460
57	0.940	1.619	2.034	2.397	2.822	3.714	4.452
58	0.938	1.615	2.030	2.392	2.817	3.708	4.445
59	0.935	1.612	2.026	2.388	2.812	3.701	4.437
60	0.933	1.609	2.022	2.384	2.807	3.695	4.430
61	0.930	1.606	2.019	2.380	2.802	3.689	4.423
62	0.928	1.603	2.015	2.376	2.798	3.684	4.416
63	0.926	1.600	2.012	2.372	2.793	3.678	4.410
64	0.924	1.597	2.008	2.368	2.789	3.673	4.403
65	0.921	1.594	2.005	2.364	2.785	3.667	4.397
66	0.919	1.591	2.002	2.361	2.781	3.662	4.391
67	0.917	1.589	1.999	2.357	2.777	3.657	4.385
68	0.915	1.586	1.996	2.354	2.773	3.652	4.379
69	0.913	1.584	1.993	2.351	2.769	3.647	4.373
70	0.911	1.581	1.990	2.347	2.765	3.643	4.368
71	0.910	1.579	1.987	2.344	2.762	3.638	4.362
72	0.908	1.576	1.984	2.341	2.758	3.633	4.357
73	0.906	1.574	1.982	2.338	2.755	3.629	4.352
74	0.904	1.572	1.979	2.335	2.751	3.625	4.347
75	0.903	1.570	1.976	2.332	2.748	3.621	4.342
76	0.901	1.568	1.974	2.329	2.745	3.617	4.337
77	0.899	1.565	1.971	2.327	2.742	3.613	4.333
78	0.898	1.563	1.969	2.324	2.739	3.609	4.328
79	0.896	1.561	1.967	2.321	2.736	3.605	4.323
80	0.895	1.559	1.964	2.319	2.733	3.601	4.319
81	0.893	1.557	1.962	2.316	2.730	3.597	4.315
82	0.892	1.556	1.960	2.314	2.727	3.594	4.310
83	0.890	1.554	1.958	2.311	2.724	3.590	4.306
84	0.889	1.552	1.956	2.309	2.721	3.587	4.302
85	0.888	1.550	1.954	2.306	2.719	3.583	4.298
86	0.886	1.548	1.952	2.304	2.716	3.580	4.294
87	0.885	1.547	1.950	2.302	2.714	3.577	4.291
88	0.884	1.545	1.948	2.300	2.711	3.574	4.287
89	0.882	1.543	1.946	2.297	2.709	3.571	4.283
90	0.881	1.542	1.944	2.295	2.706	3.567	4.279
91	0.880	1.540	1.942	2.293	2.704	3.564	4.276
92	0.879	1.538	1.940	2.291	2.701	3.561	4.272
93	0.877	1.537	1.938	2.289	2.699	3.559	4.269
94	0.876	1.535	1.937	2.287	2.697	3.556	4.266
95	0.875	1.534	1.935	2.285	2.695	3.553	4.262
96	0.874	1.532	1.933	2.283	2.692	3.550	4.259
97	0.873	1.531	1.931	2.281	2.690	3.547	4.256
98	0.872	1.530	1.930	2.279	2.688	3.545	4.253
99	0.871	1.528	1.928	2.278	2.686	3.542	4.250

TABLE 1.4.3. FACTORS FOR ONE-SIDED SAMPLING PLANS AND TOLERANCE LIMITS.

GAMMA = 0.950 (CONTINUED)

N \P →	0.750	0.900	0.950	0.975	0.990	0.999	0.9999
100	0.870	1.527	1.927	2.276	2.684	3.539	4.247
102	0.868	1.524	1.923	2.272	2.680	3.534	4.241
104	0.866	1.521	1.920	2.269	2.676	3.530	4.235
106	0.864	1.519	1.917	2.266	2.672	3.525	4.229
108	0.862	1.517	1.915	2.262	2.669	3.520	4.224
110	0.860	1.514	1.912	2.259	2.665	3.516	4.219
112	0.858	1.512	1.909	2.256	2.662	3.511	4.214
114	0.856	1.510	1.907	2.253	2.658	3.507	4.209
116	0.855	1.507	1.904	2.251	2.655	3.503	4.204
118	0.853	1.505	1.902	2.248	2.652	3.499	4.199
120	0.851	1.503	1.899	2.245	2.649	3.495	4.195
122	0.850	1.501	1.897	2.242	2.646	3.492	4.190
124	0.848	1.499	1.895	2.240	2.643	3.488	4.186
126	0.847	1.497	1.893	2.237	2.640	3.484	4.182
128	0.845	1.496	1.890	2.235	2.638	3.481	4.178
130	0.844	1.494	1.888	2.233	2.635	3.478	4.174
132	0.843	1.492	1.886	2.230	2.632	3.474	4.170
134	0.841	1.490	1.884	2.228	2.630	3.471	4.166
136	0.840	1.489	1.882	2.226	2.627	3.468	4.162
138	0.839	1.487	1.880	2.224	2.625	3.465	4.159
140	0.837	1.485	1.879	2.222	2.622	3.462	4.155
142	0.836	1.484	1.877	2.220	2.620	3.459	4.152
144	0.835	1.482	1.875	2.218	2.618	3.456	4.148
146	0.834	1.481	1.873	2.216	2.616	3.453	4.145
148	0.833	1.479	1.872	2.214	2.613	3.451	4.142
150	0.832	1.478	1.870	2.212	2.611	3.448	4.139
152	0.830	1.476	1.868	2.210	2.609	3.445	4.136
154	0.829	1.475	1.867	2.208	2.607	3.443	4.133
156	0.828	1.474	1.865	2.207	2.605	3.440	4.130
158	0.827	1.472	1.864	2.205	2.603	3.438	4.127
160	0.826	1.471	1.862	2.203	2.601	3.435	4.124
162	0.825	1.470	1.861	2.201	2.600	3.433	4.121
164	0.824	1.469	1.859	2.200	2.598	3.431	4.118
166	0.823	1.467	1.858	2.198	2.596	3.428	4.116
168	0.822	1.466	1.856	2.197	2.594	3.426	4.113
170	0.822	1.465	1.855	2.195	2.592	3.424	4.111
172	0.821	1.464	1.854	2.194	2.591	3.422	4.108
174	0.820	1.463	1.852	2.192	2.589	3.420	4.106
176	0.819	1.462	1.851	2.191	2.587	3.418	4.103
178	0.818	1.460	1.850	2.189	2.586	3.416	4.101
180	0.817	1.459	1.849	2.188	2.584	3.414	4.098
185	0.815	1.457	1.846	2.185	2.580	3.409	4.093
190	0.813	1.454	1.843	2.181	2.577	3.404	4.087
195	0.811	1.452	1.840	2.178	2.573	3.400	4.082
200	0.809	1.450	1.837	2.175	2.570	3.395	4.077
205	0.808	1.447	1.835	2.172	2.566	3.391	4.072
210	0.806	1.445	1.832	2.170	2.563	3.387	4.068
215	0.804	1.443	1.830	2.167	2.560	3.384	4.063
220	0.803	1.441	1.828	2.164	2.557	3.380	4.059

TABLE 1.4.4. FACTORS FOR ONE-SIDED SAMPLING PLANS AND TOLERANCE LIMITS.

GAMMA = 0.950 (CONTINUED)

$N \backslash P \rightarrow$	0.750	0.900	0.950	0.975	0.990	0.999	0.9999
225	0.801	1.439	1.825	2.162	2.555	3.376	4.055
230	0.800	1.437	1.823	2.160	2.552	3.373	4.051
235	0.798	1.436	1.821	2.157	2.549	3.370	4.047
240	0.797	1.434	1.819	2.155	2.547	3.367	4.043
245	0.796	1.432	1.817	2.153	2.544	3.363	4.040
250	0.795	1.431	1.815	2.151	2.542	3.361	4.036
255	0.793	1.429	1.814	2.149	2.540	3.358	4.033
260	0.792	1.428	1.812	2.147	2.537	3.355	4.029
265	0.791	1.426	1.810	2.145	2.535	3.352	4.026
270	0.790	1.425	1.809	2.143	2.533	3.349	4.023
275	0.789	1.423	1.807	2.141	2.531	3.347	4.020
280	0.788	1.422	1.805	2.140	2.529	3.344	4.017
285	0.787	1.421	1.804	2.138	2.527	3.342	4.014
290	0.786	1.419	1.802	2.136	2.525	3.340	4.012
295	0.785	1.418	1.801	2.135	2.524	3.337	4.009
300	0.784	1.417	1.800	2.133	2.522	3.335	4.006
310	0.782	1.415	1.797	2.130	2.518	3.331	4.001
320	0.780	1.412	1.794	2.127	2.515	3.327	3.996
330	0.778	1.410	1.792	2.124	2.512	3.323	3.992
340	0.777	1.408	1.790	2.122	2.509	3.319	3.988
350	0.775	1.406	1.787	2.119	2.506	3.316	3.983
360	0.774	1.404	1.785	2.117	2.504	3.312	3.980
370	0.772	1.403	1.783	2.115	2.501	3.309	3.976
380	0.771	1.401	1.781	2.113	2.499	3.306	3.972
390	0.770	1.399	1.780	2.111	2.496	3.303	3.969
400	0.769	1.398	1.778	2.109	2.494	3.300	3.965
425	0.766	1.394	1.774	2.104	2.489	3.294	3.957
450	0.763	1.391	1.770	2.100	2.484	3.288	3.950
475	0.761	1.388	1.766	2.096	2.480	3.282	3.944
500	0.758	1.385	1.763	2.092	2.475	3.277	3.938
525	0.756	1.382	1.760	2.089	2.472	3.272	3.932
550	0.754	1.380	1.757	2.086	2.468	3.268	3.927
575	0.752	1.378	1.755	2.083	2.465	3.264	3.922
600	0.751	1.376	1.752	2.080	2.462	3.260	3.918
625	0.749	1.374	1.750	2.077	2.459	3.256	3.913
650	0.748	1.372	1.748	2.075	2.456	3.253	3.910
700	0.745	1.368	1.744	2.071	2.451	3.247	3.902
750	0.743	1.365	1.741	2.067	2.447	3.241	3.896
800	0.740	1.363	1.737	2.063	2.443	3.236	3.890
850	0.738	1.360	1.734	2.060	2.439	3.232	3.885
900	0.736	1.358	1.732	2.057	2.436	3.227	3.880
950	0.735	1.356	1.729	2.054	2.433	3.224	3.875
1000	0.733	1.354	1.727	2.052	2.430	3.220	3.871
1500	0.722	1.340	1.712	2.035	2.411	3.195	3.842
2000	0.716	1.332	1.703	2.024	2.399	3.181	3.825
3000	0.708	1.323	1.692	2.012	2.385	3.164	3.805
5000	0.700	1.313	1.681	2.000	2.372	3.147	3.786
10000	0.693	1.304	1.670	1.988	2.358	3.130	3.766
∞	0.674	1.282	1.645	1.960	2.326	3.090	3.719

TABLE 1.5.1. *FACTORS FOR ONE-SIDED SAMPLING PLANS AND TOLERANCE LIMITS.*

GAMMA = 0.900

$N \backslash P \rightarrow$	0.750	0.900	0.950	0.975	0.990	0.999	0.9999
2	5.842	10.253	13.090	15.586	18.500	24.582	29.587
3	2.603	4.258	5.311	6.244	7.340	9.651	11.566
4	1.972	3.188	3.957	4.637	5.438	7.129	8.533
5	1.698	2.742	3.400	3.981	4.666	6.111	7.311
6	1.540	2.494	3.092	3.620	4.243	5.556	6.646
7	1.435	2.333	2.894	3.389	3.972	5.202	6.223
8	1.360	2.219	2.754	3.227	3.783	4.955	5.927
9	1.302	2.133	2.650	3.106	3.641	4.771	5.708
10	1.257	2.066	2.568	3.011	3.532	4.629	5.538
11	1.219	2.011	2.503	2.935	3.443	4.514	5.402
12	1.188	1.966	2.448	2.872	3.371	4.420	5.290
13	1.162	1.928	2.402	2.820	3.309	4.341	5.196
14	1.139	1.895	2.363	2.774	3.257	4.273	5.116
15	1.119	1.867	2.329	2.735	3.212	4.215	5.047
16	1.101	1.842	2.299	2.701	3.172	4.164	4.986
17	1.085	1.819	2.272	2.670	3.137	4.119	4.932
18	1.071	1.800	2.249	2.643	3.105	4.078	4.884
19	1.058	1.782	2.227	2.618	3.077	4.042	4.841
20	1.046	1.765	2.208	2.596	3.052	4.009	4.802
21	1.035	1.750	2.190	2.576	3.028	3.979	4.767
22	1.025	1.737	2.174	2.557	3.007	3.952	4.734
23	1.016	1.724	2.159	2.540	2.987	3.927	4.704
24	1.007	1.712	2.145	2.525	2.969	3.903	4.677
25	1.000	1.702	2.132	2.510	2.952	3.882	4.652
26	0.992	1.691	2.120	2.496	2.937	3.862	4.628
27	0.985	1.682	2.109	2.484	2.922	3.843	4.606
28	0.979	1.673	2.099	2.472	2.909	3.826	4.585
29	0.973	1.665	2.089	2.461	2.896	3.810	4.566
30	0.967	1.657	2.080	2.450	2.884	3.794	4.548
31	0.961	1.650	2.071	2.440	2.872	3.780	4.531
32	0.956	1.643	2.063	2.431	2.862	3.766	4.514
33	0.951	1.636	2.055	2.422	2.852	3.753	4.499
34	0.947	1.630	2.048	2.414	2.842	3.741	4.485
35	0.942	1.624	2.041	2.406	2.833	3.729	4.471
36	0.938	1.618	2.034	2.398	2.824	3.718	4.458
37	0.934	1.613	2.028	2.391	2.816	3.708	4.445
38	0.930	1.608	2.022	2.384	2.808	3.698	4.434
39	0.926	1.603	2.016	2.377	2.800	3.688	4.422
40	0.923	1.598	2.010	2.371	2.793	3.679	4.411
41	0.919	1.593	2.005	2.365	2.786	3.670	4.401
42	0.916	1.589	2.000	2.359	2.780	3.662	4.391
43	0.913	1.585	1.995	2.354	2.773	3.654	4.381
44	0.910	1.581	1.990	2.348	2.767	3.646	4.372
45	0.907	1.577	1.986	2.343	2.761	3.638	4.363
46	0.904	1.573	1.981	2.338	2.756	3.631	4.355
47	0.901	1.570	1.977	2.333	2.750	3.624	4.347
48	0.899	1.566	1.973	2.329	2.745	3.617	4.339
49	0.896	1.563	1.969	2.324	2.740	3.611	4.331
50	0.894	1.559	1.965	2.320	2.735	3.605	4.324

TABLE 1.5.2. FACTORS FOR ONE-SIDED SAMPLING PLANS AND TOLERANCE LIMITS.

GAMMA = 0.900 (CONTINUED)

N \P →	0.750	0.900	0.950	0.975	0.990	0.999	0.9999
51	0.891	1.556	1.962	2.316	2.730	3.599	4.317
52	0.889	1.553	1.958	2.312	2.726	3.593	4.310
53	0.887	1.550	1.955	2.308	2.721	3.587	4.303
54	0.884	1.547	1.951	2.304	2.717	3.582	4.297
55	0.882	1.545	1.948	2.301	2.713	3.577	4.291
56	0.880	1.542	1.945	2.297	2.709	3.571	4.285
57	0.878	1.539	1.942	2.294	2.705	3.566	4.279
58	0.876	1.537	1.939	2.290	2.701	3.562	4.273
59	0.875	1.534	1.936	2.287	2.697	3.557	4.267
60	0.873	1.532	1.933	2.284	2.694	3.552	4.262
61	0.871	1.530	1.931	2.281	2.690	3.548	4.257
62	0.869	1.527	1.928	2.278	2.687	3.544	4.252
63	0.868	1.525	1.925	2.275	2.683	3.539	4.247
64	0.866	1.523	1.923	2.272	2.680	3.535	4.242
65	0.864	1.521	1.920	2.269	2.677	3.531	4.237
66	0.863	1.519	1.918	2.267	2.674	3.527	4.233
67	0.861	1.517	1.916	2.264	2.671	3.524	4.228
68	0.860	1.515	1.913	2.261	2.668	3.520	4.224
69	0.858	1.513	1.911	2.259	2.665	3.516	4.219
70	0.857	1.511	1.909	2.256	2.662	3.513	4.215
71	0.855	1.509	1.907	2.254	2.660	3.509	4.211
72	0.854	1.508	1.905	2.252	2.657	3.506	4.207
73	0.853	1.506	1.903	2.249	2.654	3.503	4.203
74	0.851	1.504	1.901	2.247	2.652	3.499	4.200
75	0.850	1.503	1.899	2.245	2.649	3.496	4.196
76	0.849	1.501	1.897	2.243	2.647	3.493	4.192
77	0.848	1.499	1.895	2.241	2.644	3.490	4.189
78	0.846	1.498	1.893	2.239	2.642	3.487	4.185
79	0.845	1.496	1.892	2.237	2.640	3.484	4.182
80	0.844	1.495	1.890	2.235	2.638	3.482	4.179
81	0.843	1.493	1.888	2.233	2.635	3.479	4.175
82	0.842	1.492	1.887	2.231	2.633	3.476	4.172
83	0.841	1.490	1.885	2.229	2.631	3.473	4.169
84	0.840	1.489	1.883	2.227	2.629	3.471	4.166
85	0.839	1.488	1.882	2.226	2.627	3.468	4.163
86	0.838	1.486	1.880	2.224	2.625	3.466	4.160
87	0.837	1.485	1.879	2.222	2.623	3.463	4.157
88	0.836	1.484	1.877	2.220	2.621	3.461	4.154
89	0.835	1.483	1.876	2.219	2.619	3.459	4.151
90	0.834	1.481	1.874	2.217	2.618	3.456	4.149
91	0.833	1.480	1.873	2.216	2.616	3.454	4.146
92	0.832	1.479	1.871	2.214	2.614	3.452	4.143
93	0.831	1.478	1.870	2.213	2.612	3.449	4.141
94	0.830	1.477	1.869	2.211	2.611	3.447	4.138
95	0.829	1.475	1.867	2.210	2.609	3.445	4.136
96	0.828	1.474	1.866	2.208	2.607	3.443	4.133
97	0.827	1.473	1.865	2.207	2.606	3.441	4.131
98	0.827	1.472	1.864	2.205	2.604	3.439	4.129
99	0.826	1.471	1.862	2.204	2.602	3.437	4.126

TABLE 1.5.3. FACTORS FOR ONE-SIDED SAMPLING PLANS AND TOLERANCE LIMITS.

GAMMA = 0.900 (CONTINUED)

N \P →	0.750	0.900	0.950	0.975	0.990	0.999	0.9999
100	0.825	1.470	1.861	2.203	2.601	3.435	4.124
102	0.823	1.468	1.859	2.200	2.598	3.431	4.119
104	0.822	1.466	1.857	2.197	2.595	3.428	4.115
106	0.820	1.464	1.854	2.195	2.592	3.424	4.111
108	0.819	1.462	1.852	2.192	2.589	3.420	4.107
110	0.818	1.460	1.850	2.190	2.587	3.417	4.103
112	0.816	1.459	1.848	2.188	2.584	3.414	4.099
114	0.815	1.457	1.846	2.186	2.582	3.411	4.095
116	0.814	1.455	1.844	2.183	2.579	3.408	4.092
118	0.812	1.454	1.842	2.181	2.577	3.405	4.088
120	0.811	1.452	1.841	2.179	2.574	3.402	4.085
122	0.810	1.451	1.839	2.177	2.572	3.399	4.081
124	0.809	1.449	1.837	2.175	2.570	3.396	4.078
126	0.808	1.448	1.835	2.173	2.568	3.393	4.075
128	0.806	1.446	1.834	2.172	2.566	3.391	4.072
130	0.805	1.445	1.832	2.170	2.564	3.388	4.069
132	0.804	1.444	1.831	2.168	2.562	3.386	4.066
134	0.803	1.442	1.829	2.166	2.560	3.383	4.063
136	0.802	1.441	1.828	2.165	2.558	3.381	4.060
138	0.801	1.440	1.826	2.163	2.556	3.378	4.057
140	0.800	1.439	1.825	2.161	2.554	3.376	4.055
142	0.799	1.437	1.823	2.160	2.552	3.374	4.052
144	0.799	1.436	1.822	2.158	2.551	3.372	4.049
146	0.798	1.435	1.821	2.157	2.549	3.370	4.047
148	0.797	1.434	1.819	2.155	2.547	3.368	4.044
150	0.796	1.433	1.818	2.154	2.546	3.366	4.042
152	0.795	1.432	1.817	2.153	2.544	3.364	4.040
154	0.794	1.431	1.816	2.151	2.543	3.362	4.037
156	0.793	1.430	1.815	2.150	2.541	3.360	4.035
158	0.793	1.429	1.813	2.149	2.540	3.358	4.033
160	0.792	1.428	1.812	2.147	2.538	3.356	4.031
162	0.791	1.427	1.811	2.146	2.537	3.354	4.029
164	0.790	1.426	1.810	2.145	2.535	3.352	4.027
166	0.790	1.425	1.809	2.144	2.534	3.351	4.025
168	0.789	1.424	1.808	2.142	2.533	3.349	4.023
170	0.788	1.423	1.807	2.141	2.531	3.347	4.021
172	0.787	1.422	1.806	2.140	2.530	3.346	4.019
174	0.787	1.421	1.805	2.139	2.529	3.344	4.017
176	0.786	1.420	1.804	2.138	2.528	3.343	4.015
178	0.785	1.420	1.803	2.137	2.526	3.341	4.013
180	0.785	1.419	1.802	2.136	2.525	3.339	4.011
185	0.783	1.417	1.800	2.133	2.522	3.336	4.007
190	0.782	1.415	1.797	2.131	2.519	3.332	4.003
195	0.780	1.413	1.795	2.128	2.517	3.329	3.999
200	0.779	1.411	1.793	2.126	2.514	3.326	3.995
205	0.778	1.410	1.791	2.124	2.512	3.322	3.991
210	0.776	1.408	1.789	2.122	2.509	3.319	3.988
215	0.775	1.406	1.788	2.120	2.507	3.317	3.985
220	0.774	1.405	1.786	2.118	2.505	3.314	3.981

TABLE 1.5.4. FACTORS FOR ONE-SIDED SAMPLING PLANS AND TOLERANCE LIMITS.

GAMMA = 0.900 (CONTINUED)

N \P →	0.750	0.900	0.950	0.975	0.990	0.999	0.9999
225	0.773	1.403	1.784	2.116	2.503	3.311	3.978
230	0.772	1.402	1.783	2.114	2.501	3.308	3.975
235	0.771	1.401	1.781	2.112	2.499	3.306	3.972
240	0.769	1.399	1.780	2.111	2.497	3.304	3.969
245	0.768	1.398	1.778	2.109	2.495	3.301	3.966
250	0.767	1.397	1.777	2.107	2.493	3.299	3.964
255	0.767	1.396	1.775	2.106	2.491	3.297	3.961
260	0.766	1.394	1.774	2.104	2.489	3.295	3.959
265	0.765	1.393	1.773	2.103	2.488	3.293	3.956
270	0.764	1.392	1.771	2.101	2.486	3.290	3.954
275	0.763	1.391	1.770	2.100	2.485	3.289	3.952
280	0.762	1.390	1.769	2.099	2.483	3.287	3.949
285	0.761	1.389	1.768	2.097	2.482	3.285	3.947
290	0.761	1.388	1.767	2.096	2.480	3.283	3.945
295	0.760	1.387	1.766	2.095	2.479	3.281	3.943
300	0.759	1.386	1.765	2.094	2.477	3.280	3.941
310	0.758	1.384	1.762	2.091	2.475	3.276	3.937
320	0.756	1.383	1.761	2.089	2.472	3.273	3.933
330	0.755	1.381	1.759	2.087	2.470	3.270	3.930
340	0.754	1.380	1.757	2.085	2.468	3.267	3.927
350	0.753	1.378	1.755	2.083	2.466	3.265	3.924
360	0.751	1.377	1.754	2.081	2.464	3.262	3.920
370	0.750	1.375	1.752	2.080	2.462	3.260	3.918
380	0.749	1.374	1.751	2.078	2.460	3.257	3.915
390	0.748	1.373	1.749	2.076	2.458	3.255	3.912
400	0.747	1.372	1.748	2.075	2.456	3.253	3.910
425	0.745	1.369	1.745	2.071	2.452	3.248	3.904
450	0.743	1.366	1.742	2.068	2.448	3.243	3.898
475	0.741	1.364	1.739	2.065	2.445	3.239	3.893
500	0.740	1.362	1.736	2.062	2.442	3.235	3.888
525	0.738	1.360	1.734	2.060	2.439	3.231	3.884
550	0.736	1.358	1.732	2.057	2.436	3.228	3.880
575	0.735	1.356	1.730	2.055	2.434	3.225	3.876
600	0.734	1.355	1.728	2.053	2.431	3.222	3.873
625	0.733	1.353	1.726	2.051	2.429	3.219	3.870
650	0.731	1.352	1.725	2.049	2.427	3.216	3.867
700	0.729	1.349	1.722	2.046	2.423	3.211	3.861
750	0.727	1.347	1.719	2.043	2.420	3.207	3.856
800	0.726	1.344	1.717	2.040	2.417	3.203	3.852
850	0.724	1.343	1.714	2.038	2.414	3.200	3.847
900	0.723	1.341	1.712	2.035	2.411	3.197	3.844
950	0.721	1.339	1.711	2.033	2.409	3.194	3.840
1000	0.720	1.338	1.709	2.031	2.407	3.191	3.837
1500	0.712	1.327	1.697	2.018	2.392	3.172	3.815
2000	0.707	1.321	1.690	2.010	2.383	3.161	3.802
3000	0.701	1.314	1.681	2.001	2.372	3.148	3.786
5000	0.695	1.306	1.673	1.991	2.362	3.134	3.771
10000	0.689	1.299	1.665	1.982	2.351	3.121	3.755
∞	0.674	1.282	1.645	1.960	2.326	3.090	3.719

TABLE 1.6.1. FACTORS FOR ONE-SIDED SAMPLING PLANS AND TOLERANCE LIMITS.

GAMMA = 0.750

N \P →	0.750	0.900	0.950	0.975	0.990	0.999	0.9999
2	2.225	3.992	5.122	6.112	7.267	9.672	11.650
3	1.464	2.501	3.152	3.725	4.396	5.805	6.970
4	1.255	2.134	2.681	3.162	3.726	4.911	5.891
5	1.152	1.962	2.463	2.904	3.421	4.507	5.406
6	1.088	1.859	2.336	2.754	3.244	4.273	5.125
7	1.043	1.790	2.250	2.654	3.126	4.119	4.940
8	1.010	1.740	2.188	2.581	3.042	4.008	4.807
9	0.985	1.701	2.141	2.526	2.977	3.924	4.707
10	0.964	1.671	2.104	2.483	2.927	3.858	4.628
11	0.947	1.645	2.073	2.447	2.885	3.804	4.564
12	0.932	1.624	2.048	2.418	2.851	3.760	4.511
13	0.920	1.606	2.026	2.393	2.822	3.722	4.466
14	0.909	1.591	2.007	2.371	2.797	3.690	4.427
15	0.899	1.577	1.991	2.352	2.775	3.661	4.394
16	0.891	1.565	1.976	2.336	2.756	3.636	4.364
17	0.883	1.554	1.963	2.321	2.739	3.614	4.338
18	0.876	1.545	1.952	2.308	2.723	3.595	4.314
19	0.870	1.536	1.941	2.296	2.710	3.577	4.293
20	0.864	1.528	1.932	2.285	2.697	3.560	4.274
21	0.859	1.521	1.923	2.275	2.685	3.546	4.256
22	0.854	1.514	1.915	2.266	2.675	3.532	4.240
23	0.849	1.508	1.908	2.257	2.665	3.520	4.225
24	0.845	1.502	1.901	2.249	2.656	3.508	4.212
25	0.841	1.497	1.895	2.242	2.648	3.497	4.199
26	0.838	1.492	1.889	2.235	2.640	3.487	4.187
27	0.834	1.487	1.883	2.229	2.633	3.478	4.176
28	0.831	1.483	1.878	2.223	2.626	3.469	4.166
29	0.828	1.478	1.873	2.218	2.620	3.461	4.156
30	0.825	1.475	1.869	2.212	2.614	3.453	4.147
31	0.822	1.471	1.864	2.207	2.608	3.446	4.138
32	0.820	1.467	1.860	2.203	2.602	3.439	4.130
33	0.817	1.464	1.856	2.198	2.597	3.433	4.122
34	0.815	1.461	1.853	2.194	2.593	3.427	4.115
35	0.813	1.458	1.849	2.190	2.588	3.421	4.108
36	0.810	1.455	1.846	2.186	2.584	3.415	4.102
37	0.808	1.452	1.842	2.183	2.579	3.410	4.095
38	0.806	1.450	1.839	2.179	2.575	3.405	4.089
39	0.805	1.447	1.836	2.176	2.572	3.400	4.083
40	0.803	1.445	1.834	2.173	2.568	3.395	4.078
41	0.801	1.443	1.831	2.169	2.564	3.391	4.073
42	0.799	1.440	1.828	2.167	2.561	3.386	4.068
43	0.798	1.438	1.826	2.164	2.558	3.382	4.063
44	0.796	1.436	1.824	2.161	2.555	3.378	4.058
45	0.795	1.434	1.821	2.158	2.552	3.375	4.054
46	0.793	1.432	1.819	2.156	2.549	3.371	4.049
47	0.792	1.431	1.817	2.153	2.546	3.367	4.045
48	0.791	1.429	1.815	2.151	2.543	3.364	4.041
49	0.789	1.427	1.813	2.149	2.541	3.361	4.037
50	0.788	1.425	1.811	2.147	2.538	3.357	4.033

TABLE 1.6.2. FACTORS FOR ONE-SIDED SAMPLING PLANS AND TOLERANCE LIMITS.

GAMMA = 0.750 (CONTINUED)

N \P →	0.750	0.900	0.950	0.975	0.990	0.999	0.9999
51	0.787	1.424	1.809	2.145	2.536	3.354	4.030
52	0.786	1.422	1.807	2.143	2.534	3.351	4.026
53	0.785	1.421	1.805	2.141	2.531	3.349	4.023
54	0.783	1.419	1.804	2.139	2.529	3.346	4.020
55	0.782	1.418	1.802	2.137	2.527	3.343	4.016
56	0.781	1.417	1.801	2.135	2.525	3.340	4.013
57	0.780	1.415	1.799	2.133	2.523	3.338	4.010
58	0.779	1.414	1.797	2.131	2.521	3.335	4.007
59	0.778	1.413	1.796	2.130	2.519	3.333	4.004
60	0.777	1.411	1.795	2.128	2.517	3.331	4.002
61	0.777	1.410	1.793	2.127	2.515	3.328	3.999
62	0.776	1.409	1.792	2.125	2.514	3.326	3.996
63	0.775	1.408	1.791	2.124	2.512	3.324	3.994
64	0.774	1.407	1.789	2.122	2.510	3.322	3.991
65	0.773	1.406	1.788	2.121	2.509	3.320	3.989
66	0.772	1.405	1.787	2.119	2.507	3.318	3.987
67	0.772	1.404	1.786	2.118	2.506	3.316	3.984
68	0.771	1.403	1.784	2.117	2.504	3.314	3.982
69	0.770	1.402	1.783	2.115	2.503	3.312	3.980
70	0.769	1.401	1.782	2.114	2.501	3.310	3.978
71	0.769	1.400	1.781	2.113	2.500	3.309	3.976
72	0.768	1.399	1.780	2.112	2.498	3.307	3.974
73	0.767	1.398	1.779	2.111	2.497	3.305	3.972
74	0.767	1.397	1.778	2.109	2.496	3.303	3.970
75	0.766	1.396	1.777	2.108	2.495	3.302	3.968
76	0.765	1.396	1.776	2.107	2.493	3.300	3.966
77	0.765	1.395	1.775	2.106	2.492	3.299	3.964
78	0.764	1.394	1.774	2.105	2.491	3.297	3.962
79	0.763	1.393	1.773	2.104	2.490	3.296	3.960
80	0.763	1.392	1.772	2.103	2.489	3.294	3.959
81	0.762	1.392	1.771	2.102	2.487	3.293	3.957
82	0.762	1.391	1.771	2.101	2.486	3.291	3.955
83	0.761	1.390	1.770	2.100	2.485	3.290	3.954
84	0.761	1.390	1.769	2.099	2.484	3.289	3.952
85	0.760	1.389	1.768	2.098	2.483	3.287	3.951
86	0.759	1.388	1.767	2.097	2.482	3.286	3.949
87	0.759	1.387	1.767	2.097	2.481	3.285	3.948
88	0.758	1.387	1.766	2.096	2.480	3.284	3.946
89	0.758	1.386	1.765	2.095	2.479	3.282	3.945
90	0.757	1.386	1.764	2.094	2.478	3.281	3.943
91	0.757	1.385	1.764	2.093	2.477	3.280	3.942
92	0.756	1.384	1.763	2.092	2.476	3.279	3.941
93	0.756	1.384	1.762	2.092	2.475	3.278	3.939
94	0.756	1.383	1.762	2.091	2.475	3.277	3.938
95	0.755	1.383	1.761	2.090	2.474	3.276	3.937
96	0.755	1.382	1.760	2.089	2.473	3.274	3.935
97	0.754	1.381	1.760	2.089	2.472	3.273	3.934
98	0.754	1.381	1.759	2.088	2.471	3.272	3.933
99	0.753	1.380	1.758	2.087	2.470	3.271	3.932

TABLE 1.6.3. FACTORS FOR ONE-SIDED SAMPLING PLANS AND TOLERANCE LIMITS.

GAMMA = 0.750 (CONTINUED)

N \P →	0.750	0.900	0.950	0.975	0.990	0.999	0.9999
100	0.753	1.380	1.758	2.086	2.470	3.270	3.931
102	0.752	1.379	1.756	2.085	2.468	3.268	3.928
104	0.751	1.378	1.755	2.084	2.467	3.266	3.926
106	0.751	1.377	1.754	2.082	2.465	3.265	3.924
108	0.750	1.376	1.753	2.081	2.464	3.263	3.922
110	0.749	1.375	1.752	2.080	2.462	3.261	3.920
112	0.748	1.374	1.751	2.079	2.461	3.259	3.918
114	0.748	1.373	1.750	2.078	2.460	3.258	3.916
116	0.747	1.372	1.749	2.077	2.458	3.256	3.914
118	0.746	1.371	1.748	2.075	2.457	3.255	3.912
120	0.746	1.371	1.747	2.074	2.456	3.253	3.910
122	0.745	1.370	1.746	2.073	2.455	3.252	3.908
124	0.745	1.369	1.745	2.072	2.454	3.250	3.907
126	0.744	1.368	1.744	2.071	2.453	3.249	3.905
128	0.743	1.368	1.743	2.070	2.451	3.247	3.903
130	0.743	1.367	1.743	2.070	2.450	3.246	3.902
132	0.742	1.366	1.742	2.069	2.449	3.245	3.900
134	0.742	1.365	1.741	2.068	2.448	3.243	3.899
136	0.741	1.365	1.740	2.067	2.447	3.242	3.897
138	0.741	1.364	1.740	2.066	2.446	3.241	3.896
140	0.740	1.364	1.739	2.065	2.445	3.240	3.895
142	0.740	1.363	1.738	2.064	2.445	3.239	3.893
144	0.739	1.362	1.737	2.064	2.444	3.238	3.892
146	0.739	1.362	1.737	2.063	2.443	3.236	3.891
148	0.738	1.361	1.736	2.062	2.442	3.235	3.889
150	0.738	1.361	1.735	2.061	2.441	3.234	3.888
152	0.737	1.360	1.735	2.061	2.440	3.233	3.887
154	0.737	1.359	1.734	2.060	2.439	3.232	3.886
156	0.737	1.359	1.734	2.059	2.439	3.231	3.885
158	0.736	1.358	1.733	2.059	2.438	3.230	3.883
160	0.736	1.358	1.732	2.058	2.437	3.229	3.882
162	0.735	1.357	1.732	2.057	2.436	3.228	3.881
164	0.735	1.357	1.731	2.057	2.436	3.227	3.880
166	0.735	1.356	1.731	2.056	2.435	3.227	3.879
168	0.734	1.356	1.730	2.055	2.434	3.226	3.878
170	0.734	1.355	1.730	2.055	2.434	3.225	3.877
172	0.734	1.355	1.729	2.054	2.433	3.224	3.876
174	0.733	1.355	1.728	2.054	2.432	3.223	3.875
176	0.733	1.354	1.728	2.053	2.432	3.222	3.874
178	0.733	1.354	1.727	2.052	2.431	3.221	3.873
180	0.732	1.353	1.727	2.052	2.430	3.221	3.872
185	0.731	1.352	1.726	2.051	2.429	3.219	3.870
190	0.731	1.351	1.725	2.049	2.427	3.217	3.868
195	0.730	1.350	1.724	2.048	2.426	3.215	3.866
200	0.729	1.349	1.723	2.047	2.425	3.213	3.864
205	0.728	1.348	1.721	2.046	2.423	3.212	3.862
210	0.728	1.348	1.721	2.045	2.422	3.210	3.860
215	0.727	1.347	1.720	2.044	2.421	3.209	3.858
220	0.726	1.346	1.719	2.043	2.420	3.207	3.856

TABLE 1.6.4. FACTORS FOR ONE-SIDED SAMPLING PLANS AND TOLERANCE LIMITS.

GAMMA = 0.750 (CONTINUED)

N \P →	0.750	0.900	0.950	0.975	0.990	0.999	0.9999
225	0.726	1.345	1.718	2.042	2.419	3.206	3.855
230	0.725	1.345	1.717	2.041	2.418	3.205	3.853
235	0.725	1.344	1.716	2.040	2.417	3.203	3.852
240	0.724	1.343	1.715	2.039	2.415	3.202	3.850
245	0.724	1.343	1.715	2.038	2.415	3.201	3.849
250	0.723	1.342	1.714	2.037	2.414	3.200	3.847
255	0.723	1.341	1.713	2.036	2.413	3.198	3.846
260	0.722	1.341	1.712	2.036	2.412	3.197	3.845
265	0.722	1.340	1.712	2.035	2.411	3.196	3.843
270	0.721	1.339	1.711	2.034	2.410	3.195	3.842
275	0.721	1.339	1.710	2.033	2.409	3.194	3.841
280	0.720	1.338	1.710	2.033	2.408	3.193	3.840
285	0.720	1.338	1.709	2.032	2.408	3.192	3.839
290	0.720	1.337	1.709	2.031	2.407	3.191	3.838
295	0.719	1.337	1.708	2.031	2.406	3.190	3.836
300	0.719	1.336	1.708	2.030	2.406	3.189	3.835
310	0.718	1.335	1.706	2.029	2.404	3.188	3.833
320	0.717	1.335	1.705	2.028	2.403	3.186	3.832
330	0.717	1.334	1.704	2.027	2.402	3.185	3.830
340	0.716	1.333	1.704	2.026	2.400	3.183	3.828
350	0.715	1.332	1.703	2.025	2.399	3.182	3.826
360	0.715	1.331	1.702	2.024	2.398	3.180	3.825
370	0.714	1.331	1.701	2.023	2.397	3.179	3.823
380	0.714	1.330	1.700	2.022	2.396	3.178	3.822
390	0.713	1.329	1.699	2.021	2.395	3.177	3.820
400	0.713	1.329	1.699	2.020	2.394	3.175	3.819
425	0.712	1.327	1.697	2.018	2.392	3.173	3.816
450	0.710	1.326	1.696	2.017	2.390	3.170	3.813
475	0.710	1.325	1.694	2.015	2.389	3.168	3.810
500	0.709	1.324	1.693	2.014	2.387	3.166	3.808
525	0.708	1.323	1.692	2.012	2.385	3.164	3.806
550	0.707	1.322	1.691	2.011	2.384	3.162	3.804
575	0.706	1.321	1.690	2.010	2.383	3.161	3.802
600	0.706	1.320	1.689	2.009	2.381	3.159	3.800
625	0.705	1.319	1.688	2.008	2.380	3.158	3.798
650	0.704	1.318	1.687	2.007	2.379	3.156	3.797
700	0.703	1.317	1.685	2.005	2.377	3.154	3.794
750	0.702	1.316	1.684	2.003	2.375	3.152	3.791
800	0.701	1.315	1.683	2.002	2.374	3.150	3.789
850	0.701	1.314	1.681	2.001	2.372	3.148	3.786
900	0.700	1.313	1.680	2.000	2.371	3.146	3.784
950	0.699	1.312	1.679	1.998	2.370	3.145	3.783
1000	0.698	1.311	1.678	1.997	2.369	3.143	3.781
1500	0.694	1.305	1.672	1.990	2.361	3.133	3.769
2000	0.691	1.302	1.668	1.986	2.356	3.127	3.762
3000	0.688	1.298	1.664	1.981	2.350	3.120	3.754
5000	0.685	1.295	1.660	1.976	2.345	3.113	3.746
10000	0.682	1.291	1.655	1.972	2.339	3.107	3.738
∞	0.674	1.282	1.645	1.960	2.326	3.090	3.719

TABLE 1.7.1. FACTORS FOR ONE-SIDED SAMPLING PLANS AND TOLERANCE LIMITS.

GAMMA = 0.500

N \P →	0.750	0.900	0.950	0.975	0.990	0.999	0.9999
2	0.887	1.784	2.339	2.820	3.376	4.527	5.468
3	0.773	1.498	1.938	2.321	2.764	3.688	4.447
4	0.738	1.419	1.830	2.186	2.601	3.465	4.175
5	0.722	1.382	1.779	2.124	2.526	3.362	4.050
6	0.712	1.360	1.750	2.089	2.483	3.304	3.979
7	0.705	1.347	1.732	2.066	2.455	3.266	3.933
8	0.701	1.337	1.719	2.050	2.436	3.239	3.900
9	0.698	1.330	1.709	2.038	2.421	3.220	3.877
10	0.695	1.324	1.702	2.029	2.410	3.205	3.858
11	0.693	1.320	1.696	2.022	2.402	3.193	3.844
12	0.691	1.316	1.691	2.016	2.395	3.183	3.832
13	0.690	1.313	1.687	2.011	2.389	3.175	3.822
14	0.689	1.311	1.684	2.007	2.384	3.168	3.814
15	0.687	1.309	1.681	2.004	2.379	3.163	3.807
16	0.687	1.307	1.678	2.001	2.376	3.158	3.801
17	0.686	1.305	1.676	1.998	2.373	3.153	3.796
18	0.685	1.304	1.674	1.996	2.370	3.150	3.791
19	0.685	1.302	1.673	1.994	2.367	3.146	3.787
20	0.684	1.301	1.671	1.992	2.365	3.143	3.784
21	0.684	1.300	1.670	1.990	2.363	3.141	3.780
22	0.683	1.299	1.669	1.989	2.361	3.138	3.777
23	0.683	1.299	1.668	1.988	2.360	3.136	3.775
24	0.682	1.298	1.667	1.986	2.358	3.134	3.772
25	0.682	1.297	1.666	1.985	2.357	3.132	3.770
26	0.682	1.297	1.665	1.984	2.356	3.130	3.768
27	0.681	1.296	1.664	1.983	2.355	3.129	3.766
28	0.681	1.295	1.663	1.982	2.354	3.127	3.764
29	0.681	1.295	1.663	1.982	2.353	3.126	3.763
30	0.681	1.294	1.662	1.981	2.352	3.125	3.761
31	0.681	1.294	1.661	1.980	2.351	3.124	3.760
32	0.680	1.294	1.661	1.979	2.350	3.122	3.758
33	0.680	1.293	1.660	1.979	2.349	3.121	3.757
34	0.680	1.293	1.660	1.978	2.349	3.120	3.756
35	0.680	1.293	1.659	1.978	2.348	3.120	3.755
36	0.680	1.292	1.659	1.977	2.347	3.119	3.754
37	0.679	1.292	1.659	1.977	2.347	3.118	3.753
38	0.679	1.292	1.658	1.976	2.346	3.117	3.752
39	0.679	1.291	1.658	1.976	2.346	3.116	3.751
40	0.679	1.291	1.658	1.975	2.345	3.116	3.750
41	0.679	1.291	1.657	1.975	2.345	3.115	3.749
42	0.679	1.291	1.657	1.975	2.344	3.115	3.749
43	0.679	1.290	1.657	1.974	2.344	3.114	3.748
44	0.679	1.290	1.656	1.974	2.343	3.113	3.747
45	0.679	1.290	1.656	1.974	2.343	3.113	3.747
46	0.678	1.290	1.656	1.973	2.343	3.112	3.746
47	0.678	1.290	1.656	1.973	2.342	3.112	3.745
48	0.678	1.289	1.655	1.973	2.342	3.111	3.745
49	0.678	1.289	1.655	1.973	2.342	3.111	3.744
50	0.678	1.289	1.655	1.972	2.341	3.111	3.744

TABLE 1.7.2.   FACTORS FOR ONE-SIDED SAMPLING PLANS AND TOLERANCE LIMITS.

GAMMA = 0.500 (CONTINUED)

N \P →	0.750	0.900	0.950	0.975	0.990	0.999	0.9999
51	0.678	1.289	1.655	1.972	2.341	3.110	3.743
52	0.678	1.289	1.655	1.972	2.341	3.110	3.743
53	0.678	1.289	1.654	1.972	2.340	3.109	3.742
54	0.678	1.289	1.654	1.971	2.340	3.109	3.742
55	0.678	1.288	1.654	1.971	2.340	3.109	3.741
56	0.678	1.288	1.654	1.971	2.340	3.108	3.741
57	0.678	1.288	1.654	1.971	2.339	3.108	3.741
58	0.678	1.288	1.654	1.971	2.339	3.108	3.740
59	0.678	1.288	1.653	1.970	2.339	3.107	3.740
60	0.678	1.288	1.653	1.970	2.339	3.107	3.740
61	0.677	1.288	1.653	1.970	2.338	3.107	3.739
62	0.677	1.288	1.653	1.970	2.338	3.106	3.739
63	0.677	1.288	1.653	1.970	2.338	3.106	3.739
64	0.677	1.287	1.653	1.970	2.338	3.106	3.738
65	0.677	1.287	1.653	1.969	2.338	3.106	3.738
66	0.677	1.287	1.652	1.969	2.338	3.105	3.738
67	0.677	1.287	1.652	1.969	2.337	3.105	3.737
68	0.677	1.287	1.652	1.969	2.337	3.105	3.737
69	0.677	1.287	1.652	1.969	2.337	3.105	3.737
70	0.677	1.287	1.652	1.969	2.337	3.105	3.737
71	0.677	1.287	1.652	1.969	2.337	3.104	3.736
72	0.677	1.287	1.652	1.968	2.337	3.104	3.736
73	0.677	1.287	1.652	1.968	2.336	3.104	3.736
74	0.677	1.287	1.652	1.968	2.336	3.104	3.736
75	0.677	1.287	1.652	1.968	2.336	3.104	3.735
76	0.677	1.286	1.651	1.968	2.336	3.103	3.735
77	0.677	1.286	1.651	1.968	2.336	3.103	3.735
78	0.677	1.286	1.651	1.968	2.336	3.103	3.735
79	0.677	1.286	1.651	1.968	2.336	3.103	3.734
80	0.677	1.286	1.651	1.968	2.336	3.103	3.734
81	0.677	1.286	1.651	1.967	2.335	3.103	3.734
82	0.677	1.286	1.651	1.967	2.335	3.102	3.734
83	0.677	1.286	1.651	1.967	2.335	3.102	3.734
84	0.677	1.286	1.651	1.967	2.335	3.102	3.734
85	0.677	1.286	1.651	1.967	2.335	3.102	3.733
86	0.677	1.286	1.651	1.967	2.335	3.102	3.733
87	0.677	1.286	1.651	1.967	2.335	3.102	3.733
88	0.677	1.286	1.651	1.967	2.335	3.102	3.733
89	0.677	1.286	1.650	1.967	2.335	3.101	3.733
90	0.677	1.286	1.650	1.967	2.335	3.101	3.733
91	0.676	1.286	1.650	1.967	2.334	3.101	3.732
92	0.676	1.286	1.650	1.967	2.334	3.101	3.732
93	0.676	1.286	1.650	1.966	2.334	3.101	3.732
94	0.676	1.286	1.650	1.966	2.334	3.101	3.732
95	0.676	1.285	1.650	1.966	2.334	3.101	3.732
96	0.676	1.285	1.650	1.966	2.334	3.101	3.732
97	0.676	1.285	1.650	1.966	2.334	3.101	3.732
98	0.676	1.285	1.650	1.966	2.334	3.100	3.731
99	0.676	1.285	1.650	1.966	2.334	3.100	3.731

TABLE 1.7.3. *FACTORS FOR ONE-SIDED SAMPLING PLANS AND TOLERANCE LIMITS.*

GAMMA = 0.500 (*CONTINUED*)

N \P →	0.750	0.900	0.950	0.975	0.990	0.999	0.9999
100	0.676	1.285	1.650	1.966	2.334	3.100	3.731
102	0.676	1.285	1.650	1.966	2.334	3.100	3.731
104	0.676	1.285	1.650	1.966	2.333	3.100	3.731
106	0.676	1.285	1.650	1.966	2.333	3.100	3.730
108	0.676	1.285	1.649	1.966	2.333	3.099	3.730
110	0.676	1.285	1.649	1.965	2.333	3.099	3.730
112	0.676	1.285	1.649	1.965	2.333	3.099	3.730
114	0.676	1.285	1.649	1.965	2.333	3.099	3.730
116	0.676	1.285	1.649	1.965	2.333	3.099	3.729
118	0.676	1.285	1.649	1.965	2.333	3.099	3.729
120	0.676	1.285	1.649	1.965	2.332	3.099	3.729
122	0.676	1.285	1.649	1.965	2.332	3.098	3.729
124	0.676	1.285	1.649	1.965	2.332	3.098	3.729
126	0.676	1.285	1.649	1.965	2.332	3.098	3.729
128	0.676	1.284	1.649	1.965	2.332	3.098	3.728
130	0.676	1.284	1.649	1.965	2.332	3.098	3.728
132	0.676	1.284	1.649	1.965	2.332	3.098	3.728
134	0.676	1.284	1.649	1.964	2.332	3.098	3.728
136	0.676	1.284	1.648	1.964	2.332	3.098	3.728
138	0.676	1.284	1.648	1.964	2.332	3.097	3.728
140	0.676	1.284	1.648	1.964	2.332	3.097	3.728
142	0.676	1.284	1.648	1.964	2.331	3.097	3.728
144	0.676	1.284	1.648	1.964	2.331	3.097	3.727
146	0.676	1.284	1.648	1.964	2.331	3.097	3.727
148	0.676	1.284	1.648	1.964	2.331	3.097	3.727
150	0.676	1.284	1.648	1.964	2.331	3.097	3.727
152	0.676	1.284	1.648	1.964	2.331	3.097	3.727
154	0.676	1.284	1.648	1.964	2.331	3.097	3.727
156	0.676	1.284	1.648	1.964	2.331	3.097	3.727
158	0.676	1.284	1.648	1.964	2.331	3.097	3.727
160	0.676	1.284	1.648	1.964	2.331	3.096	3.727
162	0.676	1.284	1.648	1.964	2.331	3.096	3.726
164	0.676	1.284	1.648	1.964	2.331	3.096	3.726
166	0.676	1.284	1.648	1.964	2.331	3.096	3.726
168	0.676	1.284	1.648	1.964	2.331	3.096	3.726
170	0.676	1.284	1.648	1.964	2.331	3.096	3.726
172	0.676	1.284	1.648	1.963	2.331	3.096	3.726
174	0.676	1.284	1.648	1.963	2.331	3.096	3.726
176	0.676	1.284	1.648	1.963	2.330	3.096	3.726
178	0.676	1.284	1.648	1.963	2.330	3.096	3.726
180	0.675	1.284	1.648	1.963	2.330	3.096	3.726
185	0.675	1.284	1.648	1.963	2.330	3.096	3.726
190	0.675	1.284	1.647	1.963	2.330	3.095	3.725
195	0.675	1.283	1.647	1.963	2.330	3.095	3.725
200	0.675	1.283	1.647	1.963	2.330	3.095	3.725
205	0.675	1.283	1.647	1.963	2.330	3.095	3.725
210	0.675	1.283	1.647	1.963	2.330	3.095	3.725
215	0.675	1.283	1.647	1.963	2.330	3.095	3.725
220	0.675	1.283	1.647	1.963	2.330	3.095	3.725

TABLE 1.7.4. *FACTORS FOR ONE-SIDED SAMPLING PLANS AND TOLERANCE LIMITS.*

GAMMA = 0.500 (*CONTINUED*)

N \P →	0.750	0.900	0.950	0.975	0.990	0.999	0.9999
225	0.675	1.283	1.647	1.963	2.330	3.095	3.724
230	0.675	1.283	1.647	1.963	2.330	3.095	3.724
235	0.675	1.283	1.647	1.963	2.329	3.094	3.724
240	0.675	1.283	1.647	1.962	2.329	3.094	3.724
245	0.675	1.283	1.647	1.962	2.329	3.094	3.724
250	0.675	1.283	1.647	1.962	2.329	3.094	3.724
255	0.675	1.283	1.647	1.962	2.329	3.094	3.724
260	0.675	1.283	1.647	1.962	2.329	3.094	3.724
265	0.675	1.283	1.647	1.962	2.329	3.094	3.724
270	0.675	1.283	1.647	1.962	2.329	3.094	3.723
275	0.675	1.283	1.647	1.962	2.329	3.094	3.723
280	0.675	1.283	1.647	1.962	2.329	3.094	3.723
285	0.675	1.283	1.647	1.962	2.329	3.094	3.723
290	0.675	1.283	1.647	1.962	2.329	3.094	3.723
295	0.675	1.283	1.647	1.962	2.329	3.094	3.723
300	0.675	1.283	1.646	1.962	2.329	3.094	3.723
310	0.675	1.283	1.646	1.962	2.329	3.093	3.723
320	0.675	1.283	1.646	1.962	2.329	3.093	3.723
330	0.675	1.283	1.646	1.962	2.329	3.093	3.723
340	0.675	1.283	1.646	1.962	2.328	3.093	3.723
350	0.675	1.283	1.646	1.962	2.328	3.093	3.722
360	0.675	1.283	1.646	1.962	2.328	3.093	3.722
370	0.675	1.283	1.646	1.962	2.328	3.093	3.722
380	0.675	1.283	1.646	1.962	2.328	3.093	3.722
390	0.675	1.282	1.646	1.962	2.328	3.093	3.722
400	0.675	1.282	1.646	1.961	2.328	3.093	3.722
425	0.675	1.282	1.646	1.961	2.328	3.093	3.722
450	0.675	1.282	1.646	1.961	2.328	3.092	3.722
475	0.675	1.282	1.646	1.961	2.328	3.092	3.722
500	0.675	1.282	1.646	1.961	2.328	3.092	3.721
525	0.675	1.282	1.646	1.961	2.328	3.092	3.721
550	0.675	1.282	1.646	1.961	2.328	3.092	3.721
575	0.675	1.282	1.646	1.961	2.328	3.092	3.721
600	0.675	1.282	1.646	1.961	2.328	3.092	3.721
625	0.675	1.282	1.646	1.961	2.328	3.092	3.721
650	0.675	1.282	1.646	1.961	2.327	3.092	3.721
700	0.675	1.282	1.646	1.961	2.327	3.092	3.721
750	0.675	1.282	1.646	1.961	2.327	3.092	3.721
800	0.675	1.282	1.645	1.961	2.327	3.091	3.721
850	0.675	1.282	1.645	1.961	2.327	3.091	3.720
900	0.675	1.282	1.645	1.961	2.327	3.091	3.720
950	0.675	1.282	1.645	1.961	2.327	3.091	3.720
1000	0.675	1.282	1.645	1.961	2.327	3.091	3.720
1500	0.675	1.282	1.645	1.960	2.327	3.091	3.720
2000	0.675	1.282	1.645	1.960	2.327	3.091	3.720
3000	0.675	1.282	1.645	1.960	2.327	3.091	3.719
5000	0.675	1.282	1.645	1.960	2.326	3.090	3.719
10000	0.675	1.282	1.645	1.960	2.326	3.090	3.719
∞	0.674	1.282	1.645	1.960	2.326	3.090	3.719

TABLE 1.8.1. FACTORS FOR ONE-SIDED SAMPLING PLANS AND TOLERANCE LIMITS.

GAMMA = 0.250

N \P →	0.750	0.900	0.950	0.975	0.990	0.999	0.9999
2	0.236	0.886	1.248	1.551	1.895	2.592	3.155
3	0.309	0.912	1.253	1.541	1.871	2.544	3.092
4	0.353	0.943	1.281	1.567	1.895	2.570	3.118
5	0.384	0.969	1.305	1.592	1.921	2.599	3.151
6	0.407	0.990	1.326	1.614	1.944	2.625	3.181
7	0.425	1.006	1.344	1.632	1.964	2.648	3.207
8	0.439	1.021	1.358	1.647	1.980	2.668	3.229
9	0.452	1.033	1.371	1.661	1.995	2.685	3.249
10	0.462	1.043	1.382	1.673	2.008	2.700	3.267
11	0.471	1.053	1.392	1.683	2.019	2.714	3.282
12	0.479	1.061	1.401	1.693	2.029	2.726	3.296
13	0.486	1.068	1.409	1.701	2.039	2.737	3.309
14	0.493	1.075	1.416	1.709	2.047	2.747	3.321
15	0.498	1.081	1.422	1.716	2.055	2.756	3.331
16	0.504	1.086	1.428	1.722	2.062	2.765	3.341
17	0.508	1.091	1.434	1.728	2.068	2.773	3.350
18	0.513	1.096	1.439	1.733	2.074	2.780	3.358
19	0.517	1.100	1.443	1.738	2.080	2.787	3.366
20	0.520	1.104	1.448	1.743	2.085	2.793	3.373
21	0.524	1.108	1.452	1.747	2.090	2.799	3.380
22	0.527	1.112	1.455	1.752	2.094	2.804	3.386
23	0.530	1.115	1.459	1.756	2.098	2.809	3.392
24	0.533	1.118	1.462	1.759	2.102	2.814	3.398
25	0.536	1.121	1.466	1.763	2.106	2.819	3.403
26	0.538	1.124	1.469	1.766	2.110	2.823	3.408
27	0.541	1.126	1.472	1.769	2.113	2.828	3.413
28	0.543	1.129	1.474	1.772	2.117	2.832	3.418
29	0.545	1.131	1.477	1.775	2.120	2.835	3.422
30	0.547	1.133	1.479	1.778	2.123	2.839	3.426
31	0.549	1.136	1.482	1.780	2.126	2.842	3.430
32	0.551	1.138	1.484	1.783	2.128	2.846	3.434
33	0.553	1.140	1.486	1.785	2.131	2.849	3.438
34	0.554	1.141	1.488	1.787	2.133	2.852	3.442
35	0.556	1.143	1.490	1.789	2.136	2.855	3.445
36	0.558	1.145	1.492	1.792	2.138	2.858	3.448
37	0.559	1.147	1.494	1.794	2.140	2.861	3.451
38	0.561	1.148	1.496	1.795	2.143	2.863	3.454
39	0.562	1.150	1.497	1.797	2.145	2.866	3.457
40	0.563	1.151	1.499	1.799	2.147	2.868	3.460
41	0.565	1.153	1.501	1.801	2.149	2.870	3.463
42	0.566	1.154	1.502	1.803	2.150	2.873	3.466
43	0.567	1.156	1.504	1.804	2.152	2.875	3.468
44	0.568	1.157	1.505	1.806	2.154	2.877	3.471
45	0.569	1.158	1.507	1.807	2.156	2.879	3.473
46	0.570	1.159	1.508	1.809	2.157	2.881	3.475
47	0.571	1.161	1.509	1.810	2.159	2.883	3.478
48	0.572	1.162	1.511	1.812	2.160	2.885	3.480
49	0.573	1.163	1.512	1.813	2.162	2.887	3.482
50	0.574	1.164	1.513	1.814	2.163	2.889	3.484

TABLE 1.8.2. FACTORS FOR ONE-SIDED SAMPLING PLANS AND TOLERANCE LIMITS.

GAMMA = 0.250 (CONTINUED)

N \P →	0.750	0.900	0.950	0.975	0.990	0.999	0.9999
51	0.575	1.165	1.514	1.816	2.165	2.890	3.486
52	0.576	1.166	1.515	1.817	2.166	2.892	3.488
53	0.577	1.167	1.516	1.818	2.168	2.894	3.490
54	0.578	1.168	1.518	1.819	2.169	2.895	3.492
55	0.579	1.169	1.519	1.820	2.170	2.897	3.494
56	0.580	1.170	1.520	1.822	2.171	2.899	3.496
57	0.580	1.171	1.521	1.823	2.173	2.900	3.497
58	0.581	1.172	1.522	1.824	2.174	2.902	3.499
59	0.582	1.173	1.523	1.825	2.175	2.903	3.501
60	0.583	1.173	1.523	1.826	2.176	2.904	3.502
61	0.583	1.174	1.524	1.827	2.177	2.906	3.504
62	0.584	1.175	1.525	1.828	2.178	2.907	3.505
63	0.585	1.176	1.526	1.829	2.179	2.908	3.507
64	0.586	1.177	1.527	1.830	2.180	2.910	3.508
65	0.586	1.177	1.528	1.831	2.181	2.911	3.510
66	0.587	1.178	1.529	1.831	2.182	2.912	3.511
67	0.587	1.179	1.529	1.832	2.183	2.913	3.513
68	0.588	1.180	1.530	1.833	2.184	2.914	3.514
69	0.589	1.180	1.531	1.834	2.185	2.916	3.515
70	0.589	1.181	1.532	1.835	2.186	2.917	3.517
71	0.590	1.182	1.532	1.836	2.187	2.918	3.518
72	0.590	1.182	1.533	1.836	2.188	2.919	3.519
73	0.591	1.183	1.534	1.837	2.189	2.920	3.520
74	0.592	1.183	1.535	1.838	2.190	2.921	3.522
75	0.592	1.184	1.535	1.839	2.191	2.922	3.523
76	0.593	1.185	1.536	1.839	2.191	2.923	3.524
77	0.593	1.185	1.537	1.840	2.192	2.924	3.525
78	0.594	1.186	1.537	1.841	2.193	2.925	3.526
79	0.594	1.186	1.538	1.842	2.194	2.926	3.527
80	0.595	1.187	1.538	1.842	2.194	2.927	3.528
81	0.595	1.187	1.539	1.843	2.195	2.928	3.529
82	0.596	1.188	1.540	1.844	2.196	2.929	3.530
83	0.596	1.189	1.540	1.844	2.197	2.929	3.531
84	0.596	1.189	1.541	1.845	2.197	2.930	3.532
85	0.597	1.190	1.541	1.845	2.198	2.931	3.533
86	0.597	1.190	1.542	1.846	2.199	2.932	3.534
87	0.598	1.191	1.543	1.847	2.199	2.933	3.535
88	0.598	1.191	1.543	1.847	2.200	2.934	3.536
89	0.599	1.192	1.544	1.848	2.201	2.934	3.537
90	0.599	1.192	1.544	1.848	2.201	2.935	3.538
91	0.599	1.192	1.545	1.849	2.202	2.936	3.539
92	0.600	1.193	1.545	1.850	2.203	2.937	3.540
93	0.600	1.193	1.546	1.850	2.203	2.938	3.541
94	0.601	1.194	1.546	1.851	2.204	2.938	3.542
95	0.601	1.194	1.547	1.851	2.204	2.939	3.543
96	0.601	1.195	1.547	1.852	2.205	2.940	3.543
97	0.602	1.195	1.548	1.852	2.206	2.940	3.544
98	0.602	1.196	1.548	1.853	2.206	2.941	3.545
99	0.602	1.196	1.548	1.853	2.207	2.942	3.546

TABLE 1.8.3. *FACTORS FOR ONE-SIDED SAMPLING PLANS AND TOLERANCE LIMITS.*

GAMMA = 0.250 (*CONTINUED*)

$N \backslash P \rightarrow$	0.750	0.900	0.950	0.975	0.990	0.999	0.9999
100	0.603	1.196	1.549	1.854	2.207	2.943	3.547
102	0.603	1.197	1.550	1.855	2.208	2.944	3.548
104	0.604	1.198	1.551	1.856	2.209	2.945	3.550
106	0.605	1.199	1.551	1.857	2.210	2.946	3.551
108	0.605	1.199	1.552	1.857	2.211	2.948	3.553
110	0.606	1.200	1.553	1.858	2.212	2.949	3.554
112	0.607	1.201	1.554	1.859	2.213	2.950	3.555
114	0.607	1.201	1.555	1.860	2.214	2.951	3.557
116	0.608	1.202	1.555	1.861	2.215	2.952	3.558
118	0.608	1.203	1.556	1.862	2.216	2.953	3.559
120	0.609	1.203	1.557	1.862	2.217	2.954	3.560
122	0.609	1.204	1.557	1.863	2.218	2.955	3.562
124	0.610	1.205	1.558	1.864	2.218	2.956	3.563
126	0.610	1.205	1.559	1.864	2.219	2.957	3.564
128	0.611	1.206	1.559	1.865	2.220	2.958	3.565
130	0.611	1.206	1.560	1.866	2.221	2.959	3.566
132	0.612	1.207	1.561	1.867	2.222	2.960	3.567
134	0.612	1.207	1.561	1.867	2.222	2.961	3.568
136	0.613	1.208	1.562	1.868	2.223	2.962	3.569
138	0.613	1.208	1.562	1.868	2.224	2.963	3.570
140	0.613	1.209	1.563	1.869	2.224	2.964	3.571
142	0.614	1.209	1.563	1.870	2.225	2.964	3.572
144	0.614	1.210	1.564	1.870	2.226	2.965	3.573
146	0.615	1.210	1.564	1.871	2.226	2.966	3.574
148	0.615	1.211	1.565	1.871	2.227	2.967	3.575
150	0.615	1.211	1.565	1.872	2.228	2.968	3.576
152	0.616	1.212	1.566	1.872	2.228	2.968	3.577
154	0.616	1.212	1.566	1.873	2.229	2.969	3.578
156	0.617	1.212	1.567	1.874	2.229	2.970	3.578
158	0.617	1.213	1.567	1.874	2.230	2.971	3.579
160	0.617	1.213	1.568	1.875	2.231	2.971	3.580
162	0.618	1.214	1.568	1.875	2.231	2.972	3.581
164	0.618	1.214	1.569	1.876	2.232	2.973	3.582
166	0.618	1.214	1.569	1.876	2.232	2.973	3.582
168	0.619	1.215	1.570	1.877	2.233	2.974	3.583
170	0.619	1.215	1.570	1.877	2.233	2.975	3.584
172	0.619	1.216	1.570	1.877	2.234	2.975	3.585
174	0.620	1.216	1.571	1.878	2.234	2.976	3.585
176	0.620	1.216	1.571	1.878	2.235	2.976	3.586
178	0.620	1.217	1.572	1.879	2.235	2.977	3.587
180	0.620	1.217	1.572	1.879	2.236	2.978	3.587
185	0.621	1.218	1.573	1.880	2.237	2.979	3.589
190	0.622	1.219	1.574	1.881	2.238	2.980	3.591
195	0.623	1.219	1.575	1.882	2.239	2.982	3.592
200	0.623	1.220	1.576	1.883	2.240	2.983	3.594
205	0.624	1.221	1.576	1.884	2.241	2.984	3.595
210	0.624	1.222	1.577	1.885	2.242	2.985	3.597
215	0.625	1.222	1.578	1.886	2.243	2.987	3.598
220	0.625	1.223	1.579	1.886	2.244	2.988	3.599

TABLE 1.8.4. FACTORS FOR ONE-SIDED SAMPLING PLANS AND TOLERANCE LIMITS.

GAMMA = 0.250 (CONTINUED)

N \P →	0.750	0.900	0.950	0.975	0.990	0.999	0.9999
225	0.626	1.224	1.579	1.887	2.245	2.989	3.600
230	0.627	1.224	1.580	1.888	2.246	2.990	3.602
235	0.627	1.225	1.581	1.889	2.246	2.991	3.603
240	0.628	1.225	1.581	1.889	2.247	2.992	3.604
245	0.628	1.226	1.582	1.890	2.248	2.993	3.605
250	0.628	1.226	1.583	1.891	2.249	2.994	3.606
255	0.629	1.227	1.583	1.891	2.249	2.995	3.607
260	0.629	1.227	1.584	1.892	2.250	2.995	3.608
265	0.630	1.228	1.584	1.893	2.251	2.996	3.609
270	0.630	1.228	1.585	1.893	2.251	2.997	3.610
275	0.631	1.229	1.585	1.894	2.252	2.998	3.611
280	0.631	1.229	1.586	1.894	2.253	2.999	3.612
285	0.631	1.230	1.586	1.895	2.253	2.999	3.613
290	0.632	1.230	1.587	1.896	2.254	3.000	3.614
295	0.632	1.231	1.587	1.896	2.255	3.001	3.615
300	0.632	1.231	1.588	1.897	2.255	3.002	3.615
310	0.633	1.232	1.589	1.898	2.256	3.003	3.617
320	0.634	1.233	1.589	1.898	2.257	3.004	3.619
330	0.634	1.233	1.590	1.899	2.258	3.006	3.620
340	0.635	1.234	1.591	1.900	2.259	3.007	3.621
350	0.635	1.235	1.592	1.901	2.260	3.008	3.623
360	0.636	1.235	1.593	1.902	2.261	3.009	3.624
370	0.636	1.236	1.593	1.903	2.262	3.010	3.625
380	0.637	1.236	1.594	1.903	2.263	3.011	3.626
390	0.637	1.237	1.594	1.904	2.263	3.012	3.628
400	0.638	1.238	1.595	1.905	2.264	3.013	3.629
425	0.639	1.239	1.597	1.906	2.266	3.015	3.631
450	0.640	1.240	1.598	1.908	2.268	3.017	3.634
475	0.641	1.241	1.599	1.909	2.269	3.019	3.636
500	0.642	1.242	1.600	1.910	2.271	3.021	3.638
525	0.642	1.243	1.601	1.911	2.272	3.022	3.640
550	0.643	1.244	1.602	1.913	2.273	3.024	3.641
575	0.644	1.245	1.603	1.914	2.274	3.025	3.643
600	0.645	1.245	1.604	1.914	2.275	3.027	3.645
625	0.645	1.246	1.605	1.915	2.276	3.028	3.646
650	0.646	1.247	1.605	1.916	2.277	3.029	3.647
700	0.647	1.248	1.607	1.918	2.279	3.031	3.650
750	0.648	1.249	1.608	1.919	2.280	3.033	3.652
800	0.648	1.250	1.609	1.920	2.282	3.035	3.654
850	0.649	1.251	1.610	1.922	2.283	3.036	3.656
900	0.650	1.252	1.611	1.923	2.284	3.038	3.658
950	0.651	1.253	1.612	1.924	2.285	3.039	3.659
1000	0.651	1.253	1.613	1.924	2.286	3.041	3.661
1500	0.655	1.258	1.619	1.931	2.294	3.049	3.671
2000	0.658	1.262	1.622	1.935	2.298	3.055	3.678
3000	0.661	1.265	1.626	1.939	2.303	3.061	3.685
5000	0.664	1.269	1.630	1.944	2.308	3.068	3.693
10000	0.667	1.273	1.635	1.949	2.313	3.074	3.700
∞	0.674	1.282	1.645	1.960	2.326	3.090	3.719

TABLE 1.9.1. *FACTORS FOR ONE-SIDED SAMPLING PLANS AND TOLERANCE LIMITS.*

GAMMA = 0.100

N \P →	0.750	0.900	0.950	0.975	0.990	0.999	0.9999
2	-0.333	0.403	0.717	0.961	1.225	1.741	2.148
3	-0.075	0.535	0.840	1.087	1.361	1.907	2.344
4	0.036	0.617	0.922	1.173	1.455	2.022	2.478
5	0.106	0.675	0.982	1.237	1.525	2.107	2.575
6	0.155	0.719	1.028	1.286	1.578	2.172	2.651
7	0.193	0.755	1.065	1.326	1.622	2.224	2.712
8	0.223	0.783	1.096	1.358	1.658	2.268	2.763
9	0.248	0.808	1.122	1.386	1.688	2.305	2.806
10	0.269	0.828	1.144	1.410	1.715	2.337	2.843
11	0.287	0.847	1.163	1.431	1.738	2.365	2.875
12	0.302	0.863	1.180	1.450	1.758	2.390	2.904
13	0.316	0.877	1.196	1.466	1.776	2.412	2.930
14	0.329	0.890	1.210	1.481	1.793	2.433	2.953
15	0.340	0.901	1.222	1.495	1.808	2.451	2.974
16	0.350	0.912	1.234	1.507	1.822	2.468	2.994
17	0.359	0.921	1.244	1.519	1.834	2.483	3.012
18	0.367	0.930	1.254	1.529	1.846	2.497	3.028
19	0.375	0.939	1.263	1.539	1.857	2.510	3.043
20	0.382	0.946	1.271	1.548	1.867	2.523	3.058
21	0.389	0.953	1.279	1.557	1.876	2.534	3.071
22	0.395	0.960	1.286	1.565	1.885	2.545	3.084
23	0.401	0.966	1.293	1.572	1.893	2.555	3.095
24	0.406	0.972	1.300	1.579	1.901	2.565	3.106
25	0.411	0.978	1.306	1.586	1.908	2.574	3.117
26	0.416	0.983	1.311	1.592	1.915	2.582	3.127
27	0.421	0.988	1.317	1.598	1.922	2.590	3.136
28	0.425	0.993	1.322	1.604	1.928	2.598	3.145
29	0.429	0.997	1.327	1.609	1.934	2.605	3.154
30	0.433	1.002	1.332	1.614	1.940	2.612	3.162
31	0.437	1.006	1.336	1.619	1.945	2.619	3.170
32	0.440	1.010	1.341	1.624	1.951	2.625	3.177
33	0.444	1.013	1.345	1.629	1.956	2.632	3.184
34	0.447	1.017	1.349	1.633	1.960	2.637	3.191
35	0.450	1.020	1.352	1.637	1.965	2.643	3.198
36	0.453	1.024	1.356	1.641	1.970	2.649	3.204
37	0.456	1.027	1.360	1.645	1.974	2.654	3.210
38	0.459	1.030	1.363	1.648	1.978	2.659	3.216
39	0.461	1.033	1.366	1.652	1.982	2.664	3.221
40	0.464	1.036	1.369	1.655	1.986	2.668	3.227
41	0.466	1.038	1.372	1.659	1.989	2.673	3.232
42	0.469	1.041	1.375	1.662	1.993	2.677	3.237
43	0.471	1.044	1.378	1.665	1.996	2.681	3.242
44	0.473	1.046	1.381	1.668	2.000	2.686	3.247
45	0.475	1.048	1.383	1.671	2.003	2.690	3.252
46	0.477	1.051	1.386	1.674	2.006	2.693	3.256
47	0.479	1.053	1.389	1.677	2.009	2.697	3.260
48	0.481	1.055	1.391	1.679	2.012	2.701	3.265
49	0.483	1.057	1.393	1.682	2.015	2.704	3.269
50	0.485	1.059	1.396	1.684	2.018	2.708	3.273

TABLE 1.9.2.   FACTORS FOR ONE-SIDED SAMPLING PLANS AND TOLERANCE LIMITS.

GAMMA = 0.100 (CONTINUED)

N \P →	0.750	0.900	0.950	0.975	0.990	0.999	0.9999
51	0.487	1.061	1.398	1.687	2.020	2.711	3.277
52	0.489	1.063	1.400	1.689	2.023	2.714	3.280
53	0.490	1.065	1.402	1.691	2.026	2.717	3.284
54	0.492	1.067	1.404	1.694	2.028	2.721	3.288
55	0.494	1.069	1.406	1.696	2.031	2.724	3.291
56	0.495	1.071	1.408	1.698	2.033	2.726	3.294
57	0.497	1.072	1.410	1.700	2.035	2.729	3.298
58	0.498	1.074	1.412	1.702	2.038	2.732	3.301
59	0.500	1.076	1.414	1.704	2.040	2.735	3.304
60	0.501	1.077	1.415	1.706	2.042	2.737	3.307
61	0.502	1.079	1.417	1.708	2.044	2.740	3.310
62	0.504	1.080	1.419	1.710	2.046	2.743	3.313
63	0.505	1.082	1.420	1.712	2.048	2.745	3.316
64	0.506	1.083	1.422	1.713	2.050	2.748	3.319
65	0.507	1.085	1.424	1.715	2.052	2.750	3.322
66	0.509	1.086	1.425	1.717	2.054	2.752	3.324
67	0.510	1.087	1.427	1.718	2.056	2.754	3.327
68	0.511	1.089	1.428	1.720	2.057	2.757	3.330
69	0.512	1.090	1.430	1.722	2.059	2.759	3.332
70	0.513	1.091	1.431	1.723	2.061	2.761	3.335
71	0.514	1.093	1.432	1.725	2.063	2.763	3.337
72	0.515	1.094	1.434	1.726	2.064	2.765	3.339
73	0.516	1.095	1.435	1.728	2.066	2.767	3.342
74	0.517	1.096	1.436	1.729	2.068	2.769	3.344
75	0.518	1.097	1.438	1.730	2.069	2.771	3.346
76	0.519	1.098	1.439	1.732	2.071	2.773	3.349
77	0.520	1.100	1.440	1.733	2.072	2.775	3.351
78	0.521	1.101	1.441	1.735	2.074	2.777	3.353
79	0.522	1.102	1.442	1.736	2.075	2.778	3.355
80	0.523	1.103	1.444	1.737	2.077	2.780	3.357
81	0.524	1.104	1.445	1.738	2.078	2.782	3.359
82	0.525	1.105	1.446	1.740	2.079	2.784	3.361
83	0.526	1.106	1.447	1.741	2.081	2.785	3.363
84	0.527	1.107	1.448	1.742	2.082	2.787	3.365
85	0.528	1.108	1.449	1.743	2.083	2.789	3.367
86	0.528	1.109	1.450	1.744	2.085	2.790	3.369
87	0.529	1.110	1.451	1.746	2.086	2.792	3.371
88	0.530	1.110	1.452	1.747	2.087	2.793	3.372
89	0.531	1.111	1.453	1.748	2.088	2.795	3.374
90	0.532	1.112	1.454	1.749	2.090	2.796	3.376
91	0.532	1.113	1.455	1.750	2.091	2.798	3.378
92	0.533	1.114	1.456	1.751	2.092	2.799	3.379
93	0.534	1.115	1.457	1.752	2.093	2.801	3.381
94	0.535	1.116	1.458	1.753	2.094	2.802	3.383
95	0.535	1.116	1.459	1.754	2.095	2.803	3.384
96	0.536	1.117	1.460	1.755	2.096	2.805	3.386
97	0.537	1.118	1.461	1.756	2.098	2.806	3.387
98	0.537	1.119	1.462	1.757	2.099	2.808	3.389
99	0.538	1.120	1.463	1.758	2.100	2.809	3.390

TABLE 1.9.3. FACTORS FOR ONE-SIDED SAMPLING PLANS AND TOLERANCE LIMITS.

GAMMA = 0.100 (CONTINUED)

N \P →	0.750	0.900	0.950	0.975	0.990	0.999	0.9999
100	0.539	1.120	1.463	1.759	2.101	2.810	3.392
102	0.540	1.122	1.465	1.761	2.103	2.813	3.395
104	0.541	1.123	1.467	1.762	2.105	2.815	3.398
106	0.542	1.125	1.468	1.764	2.107	2.817	3.400
108	0.544	1.126	1.470	1.766	2.109	2.820	3.403
110	0.545	1.127	1.471	1.767	2.110	2.822	3.406
112	0.546	1.129	1.473	1.769	2.112	2.824	3.408
114	0.547	1.130	1.474	1.771	2.114	2.826	3.411
116	0.548	1.131	1.475	1.772	2.116	2.829	3.413
118	0.549	1.132	1.477	1.774	2.117	2.831	3.416
120	0.550	1.134	1.478	1.775	2.119	2.833	3.418
122	0.551	1.135	1.479	1.776	2.120	2.835	3.420
124	0.552	1.136	1.481	1.778	2.122	2.836	3.423
126	0.553	1.137	1.482	1.779	2.124	2.838	3.425
128	0.554	1.138	1.483	1.781	2.125	2.840	3.427
130	0.555	1.139	1.484	1.782	2.126	2.842	3.429
132	0.556	1.140	1.485	1.783	2.128	2.844	3.431
134	0.556	1.141	1.486	1.784	2.129	2.845	3.433
136	0.557	1.142	1.488	1.786	2.131	2.847	3.435
138	0.558	1.143	1.489	1.787	2.132	2.849	3.437
140	0.559	1.144	1.490	1.788	2.133	2.850	3.439
142	0.560	1.145	1.491	1.789	2.134	2.852	3.440
144	0.561	1.146	1.492	1.790	2.136	2.853	3.442
146	0.561	1.147	1.493	1.791	2.137	2.855	3.444
148	0.562	1.148	1.494	1.792	2.138	2.856	3.446
150	0.563	1.148	1.495	1.793	2.139	2.858	3.447
152	0.563	1.149	1.496	1.794	2.140	2.859	3.449
154	0.564	1.150	1.496	1.795	2.142	2.861	3.451
156	0.565	1.151	1.497	1.796	2.143	2.862	3.452
158	0.566	1.152	1.498	1.797	2.144	2.863	3.454
160	0.566	1.152	1.499	1.798	2.145	2.865	3.455
162	0.567	1.153	1.500	1.799	2.146	2.866	3.457
164	0.567	1.154	1.501	1.800	2.147	2.867	3.458
166	0.568	1.155	1.502	1.801	2.148	2.868	3.460
168	0.569	1.155	1.502	1.802	2.149	2.870	3.461
170	0.569	1.156	1.503	1.803	2.150	2.871	3.463
172	0.570	1.157	1.504	1.804	2.151	2.872	3.464
174	0.570	1.157	1.505	1.805	2.152	2.873	3.465
176	0.571	1.158	1.506	1.805	2.153	2.874	3.467
178	0.572	1.159	1.506	1.806	2.154	2.876	3.468
180	0.572	1.159	1.507	1.807	2.155	2.877	3.469
185	0.574	1.161	1.509	1.809	2.157	2.879	3.473
190	0.575	1.163	1.510	1.811	2.159	2.882	3.476
195	0.576	1.164	1.512	1.813	2.161	2.884	3.479
200	0.577	1.165	1.514	1.814	2.163	2.887	3.481
205	0.578	1.167	1.515	1.816	2.165	2.889	3.484
210	0.580	1.168	1.517	1.818	2.167	2.892	3.487
215	0.581	1.169	1.518	1.819	2.168	2.894	3.489
220	0.582	1.171	1.519	1.821	2.170	2.896	3.492

TABLE 1.9.4. FACTORS FOR ONE-SIDED SAMPLING PLANS AND TOLERANCE LIMITS.

GAMMA = 0.100 (CONTINUED)

N \P →	0.750	0.900	0.950	0.975	0.990	0.999	0.9999
225	0.583	1.172	1.521	1.822	2.172	2.898	3.494
230	0.584	1.173	1.522	1.824	2.173	2.900	3.496
235	0.585	1.174	1.523	1.825	2.175	2.902	3.499
240	0.586	1.175	1.525	1.826	2.176	2.904	3.501
245	0.586	1.176	1.526	1.828	2.178	2.905	3.503
250	0.587	1.177	1.527	1.829	2.179	2.907	3.505
255	0.588	1.178	1.528	1.830	2.181	2.909	3.507
260	0.589	1.179	1.529	1.831	2.182	2.910	3.509
265	0.590	1.180	1.530	1.832	2.183	2.912	3.511
270	0.590	1.181	1.531	1.834	2.184	2.914	3.513
275	0.591	1.182	1.532	1.835	2.186	2.915	3.514
280	0.592	1.183	1.533	1.836	2.187	2.917	3.516
285	0.593	1.183	1.534	1.837	2.188	2.918	3.518
290	0.593	1.184	1.535	1.838	2.189	2.920	3.519
295	0.594	1.185	1.536	1.839	2.190	2.921	3.521
300	0.595	1.186	1.537	1.840	2.191	2.922	3.523
310	0.596	1.187	1.538	1.842	2.193	2.925	3.526
320	0.597	1.189	1.540	1.843	2.195	2.927	3.528
330	0.598	1.190	1.541	1.845	2.197	2.930	3.531
340	0.599	1.191	1.543	1.847	2.199	2.932	3.534
350	0.600	1.193	1.544	1.848	2.201	2.934	3.536
360	0.601	1.194	1.546	1.850	2.203	2.936	3.539
370	0.602	1.195	1.547	1.851	2.204	2.938	3.541
380	0.603	1.196	1.548	1.853	2.206	2.940	3.543
390	0.604	1.197	1.549	1.854	2.207	2.942	3.546
400	0.605	1.198	1.551	1.855	2.209	2.944	3.548
425	0.607	1.201	1.553	1.858	2.212	2.948	3.553
450	0.609	1.203	1.556	1.861	2.215	2.952	3.557
475	0.611	1.205	1.558	1.863	2.218	2.955	3.561
500	0.612	1.207	1.560	1.866	2.221	2.958	3.565
525	0.614	1.208	1.562	1.868	2.223	2.962	3.568
550	0.615	1.210	1.564	1.870	2.225	2.964	3.572
575	0.616	1.212	1.566	1.872	2.227	2.967	3.575
600	0.618	1.213	1.567	1.874	2.229	2.970	3.578
625	0.619	1.214	1.569	1.875	2.231	2.972	3.581
650	0.620	1.216	1.570	1.877	2.233	2.974	3.583
700	0.622	1.218	1.573	1.880	2.236	2.978	3.588
750	0.624	1.220	1.575	1.883	2.239	2.982	3.592
800	0.625	1.222	1.577	1.885	2.242	2.985	3.596
850	0.627	1.224	1.579	1.887	2.244	2.988	3.600
900	0.628	1.225	1.581	1.889	2.247	2.991	3.603
950	0.629	1.227	1.583	1.891	2.249	2.994	3.606
1000	0.630	1.228	1.584	1.893	2.251	2.996	3.609
1500	0.638	1.238	1.595	1.905	2.264	3.013	3.628
2000	0.643	1.244	1.602	1.912	2.272	3.023	3.640
3000	0.649	1.250	1.610	1.921	2.282	3.035	3.654
5000	0.655	1.257	1.617	1.929	2.292	3.047	3.669
10000	0.660	1.264	1.625	1.938	2.302	3.060	3.683
∞	0.674	1.282	1.645	1.960	2.326	3.090	3.719

TABLE 1.10.1. *FACTORS FOR ONE-SIDED SAMPLING PLANS AND TOLERANCE LIMITS.*

GAMMA = 0.050

N \P →	0.750	0.900	0.950	0.975	0.990	0.999	0.9999
2	-0.935	0.138	0.475	0.711	0.954	1.409	1.761
3	-0.349	0.334	0.639	0.875	1.130	1.626	2.017
4	-0.168	0.444	0.743	0.983	1.246	1.768	2.183
5	-0.066	0.519	0.818	1.061	1.331	1.871	2.302
6	0.003	0.575	0.875	1.121	1.396	1.950	2.394
7	0.054	0.619	0.920	1.169	1.449	2.014	2.468
8	0.095	0.655	0.958	1.209	1.493	2.067	2.529
9	0.128	0.686	0.990	1.243	1.530	2.112	2.581
10	0.155	0.712	1.017	1.273	1.563	2.151	2.626
11	0.179	0.734	1.041	1.299	1.591	2.185	2.666
12	0.199	0.754	1.062	1.321	1.616	2.216	2.701
13	0.217	0.772	1.081	1.342	1.638	2.243	2.733
14	0.233	0.788	1.098	1.360	1.658	2.267	2.761
15	0.247	0.802	1.114	1.377	1.677	2.290	2.787
16	0.260	0.815	1.128	1.392	1.694	2.310	2.811
17	0.272	0.827	1.141	1.406	1.709	2.329	2.833
18	0.283	0.839	1.153	1.419	1.724	2.347	2.853
19	0.293	0.849	1.164	1.431	1.737	2.363	2.872
20	0.302	0.858	1.175	1.443	1.749	2.378	2.889
21	0.310	0.867	1.184	1.453	1.761	2.392	2.906
22	0.318	0.876	1.193	1.463	1.772	2.405	2.921
23	0.326	0.884	1.202	1.472	1.782	2.418	2.936
24	0.333	0.891	1.210	1.481	1.791	2.430	2.949
25	0.339	0.898	1.217	1.489	1.801	2.441	2.962
26	0.345	0.904	1.225	1.497	1.809	2.451	2.974
27	0.351	0.911	1.231	1.504	1.817	2.461	2.986
28	0.357	0.917	1.238	1.511	1.825	2.471	2.997
29	0.362	0.922	1.244	1.518	1.833	2.480	3.008
30	0.367	0.928	1.250	1.524	1.840	2.489	3.018
31	0.372	0.933	1.255	1.530	1.846	2.497	3.027
32	0.376	0.938	1.261	1.536	1.853	2.505	3.036
33	0.381	0.942	1.266	1.542	1.859	2.512	3.045
34	0.385	0.947	1.271	1.547	1.865	2.520	3.054
35	0.389	0.951	1.276	1.552	1.871	2.527	3.062
36	0.392	0.955	1.280	1.557	1.876	2.533	3.070
37	0.396	0.959	1.284	1.562	1.882	2.540	3.077
38	0.400	0.963	1.289	1.567	1.887	2.546	3.084
39	0.403	0.967	1.293	1.571	1.892	2.552	3.091
40	0.406	0.970	1.297	1.575	1.896	2.558	3.098
41	0.409	0.974	1.300	1.580	1.901	2.564	3.105
42	0.412	0.977	1.304	1.584	1.905	2.569	3.111
43	0.415	0.980	1.308	1.587	1.910	2.574	3.117
44	0.418	0.983	1.311	1.591	1.914	2.579	3.123
45	0.421	0.986	1.314	1.595	1.918	2.584	3.129
46	0.423	0.989	1.317	1.598	1.922	2.589	3.134
47	0.426	0.992	1.321	1.602	1.925	2.594	3.140
48	0.428	0.995	1.324	1.605	1.929	2.598	3.145
49	0.431	0.997	1.327	1.608	1.933	2.603	3.150
50	0.433	1.000	1.329	1.611	1.936	2.607	3.155

TABLE 1.10.2. FACTORS FOR ONE-SIDED SAMPLING PLANS AND TOLERANCE LIMITS.

GAMMA = 0.050 (CONTINUED)

N \P →	0.750	0.900	0.950	0.975	0.990	0.999	0.9999
51	0.435	1.003	1.332	1.614	1.940	2.611	3.160
52	0.438	1.005	1.335	1.617	1.943	2.615	3.165
53	0.440	1.007	1.337	1.620	1.946	2.619	3.169
54	0.442	1.010	1.340	1.623	1.949	2.623	3.174
55	0.444	1.012	1.343	1.626	1.952	2.627	3.178
56	0.446	1.014	1.345	1.628	1.955	2.630	3.182
57	0.448	1.016	1.347	1.631	1.958	2.634	3.186
58	0.450	1.018	1.350	1.634	1.961	2.637	3.190
59	0.451	1.020	1.352	1.636	1.964	2.641	3.194
60	0.453	1.022	1.354	1.638	1.966	2.644	3.198
61	0.455	1.024	1.356	1.641	1.969	2.647	3.202
62	0.457	1.026	1.358	1.643	1.972	2.650	3.206
63	0.458	1.028	1.360	1.645	1.974	2.653	3.209
64	0.460	1.030	1.362	1.648	1.976	2.657	3.213
65	0.461	1.032	1.364	1.650	1.979	2.659	3.216
66	0.463	1.033	1.366	1.652	1.981	2.662	3.220
67	0.465	1.035	1.368	1.654	1.984	2.665	3.223
68	0.466	1.037	1.370	1.656	1.986	2.668	3.226
69	0.467	1.038	1.372	1.658	1.988	2.671	3.229
70	0.469	1.040	1.374	1.660	1.990	2.673	3.232
71	0.470	1.042	1.375	1.662	1.992	2.676	3.235
72	0.472	1.043	1.377	1.664	1.994	2.679	3.238
73	0.473	1.045	1.379	1.665	1.996	2.681	3.241
74	0.474	1.046	1.380	1.667	1.998	2.684	3.244
75	0.476	1.048	1.382	1.669	2.000	2.686	3.247
76	0.477	1.049	1.384	1.671	2.002	2.688	3.250
77	0.478	1.050	1.385	1.672	2.004	2.691	3.253
78	0.479	1.052	1.387	1.674	2.006	2.693	3.255
79	0.480	1.053	1.388	1.676	2.008	2.695	3.258
80	0.482	1.054	1.390	1.677	2.010	2.697	3.260
81	0.483	1.056	1.391	1.679	2.011	2.700	3.263
82	0.484	1.057	1.392	1.681	2.013	2.702	3.265
83	0.485	1.058	1.394	1.682	2.015	2.704	3.268
84	0.486	1.059	1.395	1.684	2.017	2.706	3.270
85	0.487	1.061	1.397	1.685	2.018	2.708	3.273
86	0.488	1.062	1.398	1.686	2.020	2.710	3.275
87	0.489	1.063	1.399	1.688	2.021	2.712	3.277
88	0.490	1.064	1.400	1.689	2.023	2.714	3.279
89	0.491	1.065	1.402	1.691	2.025	2.716	3.282
90	0.492	1.066	1.403	1.692	2.026	2.718	3.284
91	0.493	1.068	1.404	1.693	2.028	2.719	3.286
92	0.494	1.069	1.405	1.695	2.029	2.721	3.288
93	0.495	1.070	1.407	1.696	2.030	2.723	3.290
94	0.496	1.071	1.408	1.697	2.032	2.725	3.292
95	0.497	1.072	1.409	1.699	2.033	2.727	3.294
96	0.498	1.073	1.410	1.700	2.035	2.728	3.296
97	0.499	1.074	1.411	1.701	2.036	2.730	3.298
98	0.499	1.075	1.412	1.702	2.037	2.732	3.300
99	0.500	1.076	1.413	1.703	2.039	2.733	3.302

TABLE 1.10.3. FACTORS FOR ONE-SIDED SAMPLING PLANS AND TOLERANCE LIMITS.

GAMMA = 0.050

N \P →	0.750	0.900	0.950	0.975	0.990	0.999	0.9999
100	0.501	1.077	1.414	1.705	2.040	2.735	3.304
102	0.503	1.079	1.416	1.707	2.043	2.738	3.308
104	0.504	1.080	1.418	1.709	2.045	2.741	3.311
106	0.506	1.082	1.420	1.711	2.048	2.744	3.315
108	0.507	1.084	1.422	1.713	2.050	2.747	3.318
110	0.509	1.086	1.424	1.715	2.052	2.750	3.321
112	0.510	1.087	1.426	1.718	2.054	2.753	3.325
114	0.512	1.089	1.428	1.719	2.057	2.755	3.328
116	0.513	1.090	1.430	1.721	2.059	2.758	3.331
118	0.514	1.092	1.431	1.723	2.061	2.761	3.334
120	0.516	1.093	1.433	1.725	2.063	2.763	3.337
122	0.517	1.095	1.435	1.727	2.065	2.765	3.340
124	0.518	1.096	1.436	1.729	2.067	2.768	3.342
126	0.519	1.098	1.438	1.730	2.069	2.770	3.345
128	0.520	1.099	1.439	1.732	2.071	2.773	3.348
130	0.522	1.100	1.441	1.734	2.072	2.775	3.350
132	0.523	1.102	1.442	1.735	2.074	2.777	3.353
134	0.524	1.103	1.443	1.737	2.076	2.779	3.356
136	0.525	1.104	1.445	1.738	2.078	2.781	3.358
138	0.526	1.105	1.446	1.740	2.079	2.783	3.360
140	0.527	1.106	1.447	1.741	2.081	2.785	3.363
142	0.528	1.108	1.449	1.743	2.082	2.787	3.365
144	0.529	1.109	1.450	1.744	2.084	2.789	3.367
146	0.530	1.110	1.451	1.745	2.086	2.791	3.369
148	0.531	1.111	1.453	1.747	2.087	2.793	3.372
150	0.532	1.112	1.454	1.748	2.089	2.795	3.374
152	0.533	1.113	1.455	1.749	2.090	2.797	3.376
154	0.534	1.114	1.456	1.751	2.091	2.798	3.378
156	0.534	1.115	1.457	1.752	2.093	2.800	3.380
158	0.535	1.116	1.458	1.753	2.094	2.802	3.382
160	0.536	1.117	1.459	1.754	2.096	2.803	3.384
162	0.537	1.118	1.460	1.755	2.097	2.805	3.386
164	0.538	1.119	1.462	1.757	2.098	2.807	3.388
166	0.539	1.120	1.463	1.758	2.099	2.808	3.389
168	0.539	1.121	1.464	1.759	2.101	2.810	3.391
170	0.540	1.122	1.465	1.760	2.102	2.811	3.393
172	0.541	1.123	1.466	1.761	2.103	2.813	3.395
174	0.542	1.123	1.467	1.762	2.104	2.814	3.397
176	0.542	1.124	1.467	1.763	2.105	2.816	3.398
178	0.543	1.125	1.468	1.764	2.107	2.817	3.400
180	0.544	1.126	1.469	1.765	2.108	2.819	3.402
185	0.546	1.128	1.472	1.768	2.111	2.822	3.406
190	0.547	1.130	1.474	1.770	2.113	2.825	3.409
195	0.549	1.132	1.476	1.772	2.116	2.828	3.413
200	0.550	1.133	1.478	1.775	2.118	2.831	3.417
205	0.552	1.135	1.480	1.777	2.121	2.834	3.420
210	0.553	1.137	1.482	1.779	2.123	2.837	3.423
215	0.555	1.138	1.483	1.781	2.125	2.840	3.427
220	0.556	1.140	1.485	1.783	2.127	2.843	3.430

TABLE 1.10.4. *FACTORS FOR ONE-SIDED SAMPLING PLANS AND TOLERANCE LIMITS.*

GAMMA = 0.050 (*CONTINUED*)

$N \setminus P \rightarrow$	0.750	0.900	0.950	0.975	0.990	0.999	0.9999
225	0.557	1.142	1.487	1.785	2.129	2.845	3.433
230	0.558	1.143	1.488	1.786	2.131	2.848	3.436
235	0.560	1.144	1.490	1.788	2.133	2.850	3.438
240	0.561	1.146	1.492	1.790	2.135	2.853	3.441
245	0.562	1.147	1.493	1.791	2.137	2.855	3.444
250	0.563	1.148	1.494	1.793	2.139	2.857	3.446
255	0.564	1.150	1.496	1.795	2.141	2.859	3.449
260	0.565	1.151	1.497	1.796	2.142	2.861	3.451
265	0.566	1.152	1.499	1.798	2.144	2.863	3.454
270	0.567	1.153	1.500	1.799	2.145	2.865	3.456
275	0.568	1.154	1.501	1.800	2.147	2.867	3.458
280	0.569	1.155	1.502	1.802	2.149	2.869	3.460
285	0.570	1.156	1.503	1.803	2.150	2.871	3.463
290	0.571	1.157	1.505	1.804	2.151	2.873	3.465
295	0.572	1.158	1.506	1.806	2.153	2.874	3.467
300	0.572	1.159	1.507	1.807	2.154	2.876	3.469
310	0.574	1.161	1.509	1.809	2.157	2.879	3.473
320	0.576	1.163	1.511	1.811	2.159	2.883	3.476
330	0.577	1.165	1.513	1.814	2.162	2.885	3.480
340	0.578	1.167	1.515	1.816	2.164	2.888	3.483
350	0.580	1.168	1.517	1.818	2.166	2.891	3.486
360	0.581	1.170	1.518	1.819	2.168	2.894	3.489
370	0.582	1.171	1.520	1.821	2.170	2.896	3.492
380	0.584	1.172	1.522	1.823	2.172	2.899	3.495
390	0.585	1.174	1.523	1.825	2.174	2.901	3.498
400	0.586	1.175	1.525	1.826	2.176	2.903	3.500
425	0.588	1.178	1.528	1.830	2.180	2.909	3.507
450	0.591	1.181	1.531	1.834	2.184	2.913	3.512
475	0.593	1.184	1.534	1.837	2.188	2.918	3.518
500	0.595	1.186	1.537	1.840	2.191	2.922	3.522
525	0.597	1.188	1.539	1.843	2.194	2.926	3.527
550	0.599	1.190	1.542	1.845	2.197	2.930	3.531
575	0.600	1.192	1.544	1.848	2.200	2.933	3.535
600	0.602	1.194	1.546	1.850	2.203	2.936	3.539
625	0.603	1.196	1.548	1.852	2.205	2.939	3.542
650	0.605	1.197	1.550	1.854	2.207	2.942	3.545
700	0.607	1.200	1.553	1.858	2.211	2.947	3.552
750	0.609	1.203	1.556	1.861	2.215	2.952	3.557
800	0.611	1.205	1.559	1.864	2.219	2.956	3.562
850	0.613	1.208	1.561	1.867	2.222	2.960	3.566
900	0.615	1.210	1.563	1.869	2.224	2.963	3.571
950	0.616	1.212	1.566	1.872	2.227	2.967	3.574
1000	0.618	1.213	1.567	1.874	2.230	2.970	3.578
1500	0.628	1.226	1.581	1.889	2.247	2.991	3.603
2000	0.634	1.233	1.590	1.899	2.257	3.004	3.618
3000	0.642	1.242	1.600	1.910	2.270	3.020	3.636
5000	0.649	1.251	1.610	1.921	2.282	3.035	3.655
10000	0.656	1.260	1.620	1.932	2.295	3.051	3.673
∞	0.674	1.282	1.645	1.960	2.326	3.090	3.719

TABLE 1.11.1. FACTORS FOR ONE-SIDED SAMPLING PLANS AND TOLERANCE LIMITS.

GAMMA = 0.025

N \P →	0.750	0.900	0.950	0.975	0.990	0.999	0.9999
2	-2.006	-0.143	0.273	0.523	0.761	1.184	1.501
3	-0.662	0.159	0.478	0.713	0.958	1.423	1.784
4	-0.373	0.298	0.601	0.835	1.088	1.579	1.965
5	-0.229	0.389	0.687	0.923	1.182	1.693	2.096
6	-0.136	0.455	0.752	0.991	1.256	1.781	2.198
7	-0.070	0.507	0.804	1.046	1.315	1.852	2.280
8	-0.019	0.550	0.847	1.091	1.364	1.911	2.349
9	0.022	0.585	0.884	1.130	1.406	1.961	2.407
10	0.056	0.615	0.915	1.163	1.442	2.005	2.457
11	0.084	0.642	0.943	1.192	1.474	2.043	2.502
12	0.109	0.665	0.967	1.218	1.502	2.078	2.541
13	0.131	0.685	0.989	1.242	1.528	2.108	2.577
14	0.150	0.704	1.008	1.262	1.551	2.136	2.609
15	0.168	0.721	1.026	1.282	1.572	2.161	2.638
16	0.183	0.736	1.042	1.299	1.591	2.185	2.665
17	0.197	0.750	1.057	1.315	1.608	2.206	2.690
18	0.210	0.763	1.071	1.330	1.625	2.226	2.713
19	0.222	0.775	1.084	1.344	1.640	2.244	2.734
20	0.233	0.786	1.095	1.356	1.654	2.261	2.754
21	0.243	0.796	1.107	1.368	1.667	2.278	2.773
22	0.253	0.806	1.117	1.380	1.680	2.293	2.790
23	0.262	0.815	1.127	1.390	1.691	2.307	2.806
24	0.270	0.823	1.136	1.400	1.702	2.320	2.822
25	0.278	0.831	1.145	1.410	1.713	2.333	2.837
26	0.285	0.839	1.153	1.419	1.723	2.345	2.851
27	0.292	0.846	1.161	1.427	1.732	2.356	2.864
28	0.298	0.853	1.168	1.435	1.741	2.367	2.877
29	0.305	0.860	1.175	1.443	1.749	2.378	2.889
30	0.310	0.866	1.182	1.450	1.757	2.388	2.900
31	0.316	0.872	1.189	1.457	1.765	2.397	2.911
32	0.321	0.878	1.195	1.464	1.773	2.406	2.922
33	0.327	0.883	1.201	1.471	1.780	2.415	2.932
34	0.331	0.888	1.206	1.477	1.787	2.423	2.941
35	0.336	0.893	1.212	1.483	1.793	2.431	2.951
36	0.341	0.898	1.217	1.488	1.799	2.439	2.960
37	0.345	0.903	1.222	1.494	1.806	2.446	2.968
38	0.349	0.907	1.227	1.499	1.811	2.454	2.977
39	0.353	0.911	1.232	1.504	1.817	2.460	2.985
40	0.357	0.916	1.236	1.509	1.823	2.467	2.993
41	0.361	0.920	1.241	1.514	1.828	2.474	3.000
42	0.364	0.923	1.245	1.519	1.833	2.480	3.007
43	0.368	0.927	1.249	1.523	1.838	2.486	3.014
44	0.371	0.931	1.253	1.527	1.843	2.492	3.021
45	0.374	0.934	1.257	1.532	1.847	2.498	3.028
46	0.377	0.938	1.260	1.536	1.852	2.503	3.034
47	0.380	0.941	1.264	1.540	1.856	2.508	3.040
48	0.383	0.944	1.267	1.543	1.860	2.514	3.046
49	0.386	0.947	1.271	1.547	1.865	2.519	3.052
50	0.389	0.950	1.274	1.551	1.869	2.524	3.058

TABLE 1.11.2. FACTORS FOR ONE-SIDED SAMPLING PLANS AND TOLERANCE LIMITS.

GAMMA = 0.025 (CONTINUED)

N \P →	0.750	0.900	0.950	0.975	0.990	0.999	0.9999
51	0.391	0.953	1.277	1.554	1.873	2.528	3.064
52	0.394	0.956	1.281	1.558	1.876	2.533	3.069
53	0.397	0.959	1.284	1.561	1.880	2.538	3.074
54	0.399	0.961	1.287	1.564	1.884	2.542	3.079
55	0.401	0.964	1.289	1.567	1.887	2.546	3.084
56	0.404	0.967	1.292	1.570	1.891	2.551	3.089
57	0.406	0.969	1.295	1.573	1.894	2.555	3.094
58	0.408	0.972	1.298	1.576	1.897	2.559	3.099
59	0.410	0.974	1.300	1.579	1.900	2.563	3.103
60	0.412	0.976	1.303	1.582	1.903	2.566	3.108
61	0.414	0.979	1.305	1.585	1.907	2.570	3.112
62	0.416	0.981	1.308	1.588	1.910	2.574	3.116
63	0.418	0.983	1.310	1.590	1.912	2.577	3.121
64	0.420	0.985	1.313	1.593	1.915	2.581	3.125
65	0.422	0.987	1.315	1.595	1.918	2.584	3.129
66	0.424	0.989	1.317	1.598	1.921	2.588	3.133
67	0.426	0.991	1.319	1.600	1.923	2.591	3.136
68	0.428	0.993	1.321	1.602	1.926	2.594	3.140
69	0.429	0.995	1.324	1.605	1.929	2.597	3.144
70	0.431	0.997	1.326	1.607	1.931	2.600	3.147
71	0.433	0.999	1.328	1.609	1.934	2.603	3.151
72	0.434	1.001	1.330	1.611	1.936	2.606	3.154
73	0.436	1.002	1.332	1.614	1.938	2.609	3.158
74	0.437	1.004	1.333	1.616	1.941	2.612	3.161
75	0.439	1.006	1.335	1.618	1.943	2.615	3.164
76	0.440	1.007	1.337	1.620	1.945	2.618	3.168
77	0.442	1.009	1.339	1.622	1.947	2.620	3.171
78	0.443	1.011	1.341	1.624	1.950	2.623	3.174
79	0.445	1.012	1.342	1.625	1.952	2.626	3.177
80	0.446	1.014	1.344	1.627	1.954	2.628	3.180
81	0.447	1.015	1.346	1.629	1.956	2.631	3.183
82	0.449	1.017	1.348	1.631	1.958	2.633	3.186
83	0.450	1.018	1.349	1.633	1.960	2.636	3.188
84	0.451	1.020	1.351	1.635	1.962	2.638	3.191
85	0.452	1.021	1.352	1.636	1.964	2.641	3.194
86	0.454	1.022	1.354	1.638	1.966	2.643	3.197
87	0.455	1.024	1.355	1.640	1.967	2.645	3.199
88	0.456	1.025	1.357	1.641	1.969	2.647	3.202
89	0.457	1.026	1.358	1.643	1.971	2.650	3.205
90	0.458	1.028	1.360	1.644	1.973	2.652	3.207
91	0.460	1.029	1.361	1.646	1.975	2.654	3.210
92	0.461	1.030	1.363	1.648	1.976	2.656	3.212
93	0.462	1.032	1.364	1.649	1.978	2.658	3.214
94	0.463	1.033	1.365	1.651	1.980	2.660	3.217
95	0.464	1.034	1.367	1.652	1.981	2.662	3.219
96	0.465	1.035	1.368	1.653	1.983	2.664	3.222
97	0.466	1.036	1.369	1.655	1.984	2.666	3.224
98	0.467	1.037	1.371	1.656	1.986	2.668	3.226
99	0.468	1.039	1.372	1.658	1.988	2.670	3.228

TABLE 1.11.3. FACTORS FOR ONE-SIDED SAMPLING PLANS AND TOLERANCE LIMITS.

GAMMA = 0.025 (CONTINUED)

N \P →	0.750	0.900	0.950	0.975	0.990	0.999	0.9999
100	0.469	1.040	1.373	1.659	1.989	2.672	3.230
102	0.471	1.042	1.375	1.662	1.992	2.676	3.235
104	0.473	1.044	1.378	1.664	1.995	2.679	3.239
106	0.475	1.046	1.380	1.667	1.998	2.683	3.243
108	0.476	1.048	1.382	1.669	2.001	2.686	3.247
110	0.478	1.050	1.385	1.672	2.003	2.689	3.251
112	0.480	1.052	1.387	1.674	2.006	2.692	3.255
114	0.481	1.054	1.389	1.676	2.008	2.696	3.258
116	0.483	1.056	1.391	1.679	2.011	2.699	3.262
118	0.485	1.057	1.393	1.681	2.013	2.702	3.265
120	0.486	1.059	1.395	1.683	2.016	2.705	3.269
122	0.488	1.061	1.397	1.685	2.018	2.707	3.272
124	0.489	1.063	1.398	1.687	2.020	2.710	3.275
126	0.490	1.064	1.400	1.689	2.023	2.713	3.278
128	0.492	1.066	1.402	1.691	2.025	2.716	3.282
130	0.493	1.067	1.404	1.693	2.027	2.718	3.285
132	0.495	1.069	1.405	1.695	2.029	2.721	3.288
134	0.496	1.070	1.407	1.697	2.031	2.723	3.290
136	0.497	1.072	1.409	1.698	2.033	2.726	3.293
138	0.498	1.073	1.410	1.700	2.035	2.728	3.296
140	0.500	1.075	1.412	1.702	2.037	2.731	3.299
142	0.501	1.076	1.413	1.703	2.039	2.733	3.302
144	0.502	1.077	1.415	1.705	2.040	2.735	3.304
146	0.503	1.079	1.416	1.707	2.042	2.737	3.307
148	0.504	1.080	1.418	1.708	2.044	2.740	3.309
150	0.505	1.081	1.419	1.710	2.046	2.742	3.312
152	0.506	1.082	1.420	1.711	2.047	2.744	3.314
154	0.507	1.084	1.422	1.713	2.049	2.746	3.317
156	0.508	1.085	1.423	1.714	2.051	2.748	3.319
158	0.509	1.086	1.424	1.716	2.052	2.750	3.321
160	0.510	1.087	1.426	1.717	2.054	2.752	3.324
162	0.511	1.088	1.427	1.718	2.055	2.754	3.326
164	0.512	1.089	1.428	1.720	2.057	2.756	3.328
166	0.513	1.090	1.429	1.721	2.058	2.757	3.330
168	0.514	1.091	1.431	1.722	2.060	2.759	3.332
170	0.515	1.092	1.432	1.724	2.061	2.761	3.334
172	0.516	1.094	1.433	1.725	2.063	2.763	3.336
174	0.517	1.095	1.434	1.726	2.064	2.764	3.338
176	0.518	1.096	1.435	1.728	2.066	2.766	3.340
178	0.519	1.097	1.436	1.729	2.067	2.768	3.342
180	0.519	1.097	1.437	1.730	2.068	2.769	3.344
185	0.521	1.100	1.440	1.733	2.071	2.773	3.349
190	0.523	1.102	1.443	1.736	2.075	2.777	3.353
195	0.525	1.104	1.445	1.738	2.078	2.781	3.358
200	0.527	1.106	1.447	1.741	2.080	2.785	3.362
205	0.529	1.108	1.450	1.743	2.083	2.788	3.366
210	0.531	1.110	1.452	1.746	2.086	2.791	3.370
215	0.532	1.112	1.454	1.748	2.089	2.795	3.374
220	0.534	1.114	1.456	1.750	2.091	2.798	3.377

TABLE 1.11.4. *FACTORS FOR ONE-SIDED SAMPLING PLANS AND TOLERANCE LIMITS.*

GAMMA = 0.025 (CONTINUED)

N \P →	0.750	0.900	0.950	0.975	0.990	0.999	0.9999
225	0.535	1.116	1.458	1.753	2.093	2.801	3.381
230	0.537	1.118	1.460	1.755	2.096	2.804	3.384
235	0.538	1.119	1.462	1.757	2.098	2.806	3.387
240	0.540	1.121	1.463	1.759	2.100	2.809	3.391
245	0.541	1.122	1.465	1.761	2.103	2.812	3.394
250	0.542	1.124	1.467	1.762	2.105	2.815	3.397
255	0.543	1.125	1.469	1.764	2.107	2.817	3.400
260	0.545	1.127	1.470	1.766	2.109	2.819	3.403
265	0.546	1.128	1.472	1.768	2.111	2.822	3.405
270	0.547	1.129	1.473	1.769	2.112	2.824	3.408
275	0.548	1.131	1.475	1.771	2.114	2.826	3.411
280	0.549	1.132	1.476	1.773	2.116	2.829	3.413
285	0.550	1.133	1.477	1.774	2.118	2.831	3.416
290	0.551	1.135	1.479	1.776	2.119	2.833	3.418
295	0.552	1.136	1.480	1.777	2.121	2.835	3.421
300	0.553	1.137	1.481	1.779	2.123	2.837	3.423
310	0.555	1.139	1.484	1.781	2.126	2.841	3.427
320	0.557	1.141	1.486	1.784	2.129	2.844	3.432
330	0.559	1.143	1.489	1.787	2.132	2.848	3.436
340	0.560	1.145	1.491	1.789	2.134	2.851	3.440
350	0.562	1.147	1.493	1.791	2.137	2.855	3.443
360	0.564	1.149	1.495	1.794	2.139	2.858	3.447
370	0.565	1.151	1.497	1.796	2.142	2.861	3.451
380	0.566	1.152	1.499	1.798	2.144	2.863	3.454
390	0.568	1.154	1.501	1.800	2.146	2.866	3.457
400	0.569	1.155	1.502	1.802	2.148	2.869	3.460
425	0.572	1.159	1.506	1.806	2.153	2.875	3.467
450	0.575	1.162	1.510	1.810	2.158	2.881	3.474
475	0.578	1.165	1.513	1.814	2.162	2.886	3.480
500	0.580	1.168	1.517	1.818	2.166	2.891	3.486
525	0.582	1.171	1.520	1.821	2.170	2.896	3.491
550	0.584	1.173	1.522	1.824	2.173	2.900	3.496
575	0.586	1.176	1.525	1.827	2.177	2.904	3.501
600	0.588	1.178	1.527	1.829	2.180	2.907	3.505
625	0.590	1.180	1.530	1.832	2.182	2.911	3.509
650	0.591	1.182	1.532	1.834	2.185	2.914	3.513
700	0.594	1.185	1.536	1.839	2.190	2.920	3.520
750	0.597	1.188	1.539	1.843	2.194	2.926	3.527
800	0.599	1.191	1.543	1.846	2.198	2.931	3.533
850	0.602	1.194	1.545	1.849	2.202	2.935	3.538
900	0.604	1.196	1.548	1.852	2.205	2.940	3.543
950	0.605	1.198	1.551	1.855	2.209	2.944	3.547
1000	0.607	1.200	1.553	1.858	2.211	2.947	3.552
1500	0.619	1.215	1.569	1.876	2.232	2.973	3.581
2000	0.627	1.224	1.579	1.887	2.244	2.988	3.599
3000	0.635	1.234	1.591	1.900	2.259	3.006	3.621
5000	0.644	1.245	1.603	1.913	2.274	3.025	3.642
10000	0.653	1.255	1.615	1.927	2.289	3.044	3.665
∞	0.674	1.282	1.645	1.960	2.326	3.090	3.719

TABLE 1.12.1. FACTORS FOR ONE-SIDED SAMPLING PLANS AND TOLERANCE LIMITS.

GAMMA = 0.010

N \P →	0.750	0.900	0.950	0.975	0.990	0.999	0.9999
2	-5.110	-0.707	-0.000	0.311	0.564	0.969	1.259
3	-1.204	-0.072	0.295	0.539	0.782	1.222	1.555
4	-0.670	0.123	0.443	0.678	0.924	1.388	1.747
5	-0.447	0.238	0.543	0.777	1.027	1.510	1.888
6	-0.316	0.319	0.618	0.853	1.108	1.605	1.997
7	-0.226	0.381	0.678	0.914	1.173	1.683	2.086
8	-0.159	0.431	0.727	0.965	1.227	1.747	2.160
9	-0.107	0.472	0.768	1.008	1.273	1.803	2.224
10	-0.064	0.508	0.804	1.045	1.314	1.851	2.279
11	-0.028	0.538	0.835	1.078	1.349	1.893	2.328
12	0.003	0.565	0.862	1.107	1.381	1.931	2.372
13	0.030	0.589	0.887	1.133	1.409	1.965	2.411
14	0.053	0.610	0.909	1.156	1.434	1.996	2.446
15	0.074	0.629	0.929	1.178	1.458	2.024	2.479
16	0.093	0.647	0.948	1.197	1.479	2.049	2.509
17	0.110	0.663	0.965	1.215	1.499	2.073	2.536
18	0.126	0.678	0.980	1.232	1.517	2.095	2.562
19	0.140	0.692	0.995	1.248	1.534	2.116	2.585
20	0.153	0.705	1.008	1.262	1.550	2.135	2.608
21	0.165	0.716	1.021	1.276	1.565	2.153	2.628
22	0.177	0.728	1.033	1.289	1.579	2.170	2.648
23	0.187	0.738	1.044	1.301	1.592	2.186	2.666
24	0.197	0.748	1.054	1.312	1.605	2.201	2.684
25	0.206	0.757	1.064	1.323	1.616	2.215	2.700
26	0.215	0.766	1.074	1.333	1.628	2.229	2.716
27	0.223	0.774	1.083	1.342	1.638	2.242	2.731
28	0.231	0.782	1.091	1.351	1.648	2.254	2.745
29	0.238	0.790	1.099	1.360	1.658	2.266	2.759
30	0.245	0.797	1.107	1.369	1.667	2.277	2.772
31	0.252	0.804	1.114	1.377	1.676	2.288	2.784
32	0.258	0.810	1.121	1.384	1.684	2.298	2.796
33	0.264	0.817	1.128	1.392	1.692	2.308	2.807
34	0.270	0.823	1.135	1.399	1.700	2.317	2.818
35	0.276	0.828	1.141	1.405	1.708	2.326	2.829
36	0.281	0.834	1.147	1.412	1.715	2.335	2.839
37	0.286	0.839	1.153	1.418	1.722	2.343	2.849
38	0.291	0.844	1.158	1.424	1.728	2.352	2.858
39	0.296	0.849	1.164	1.430	1.735	2.359	2.867
40	0.300	0.854	1.169	1.436	1.741	2.367	2.876
41	0.304	0.859	1.174	1.441	1.747	2.374	2.885
42	0.309	0.863	1.179	1.446	1.753	2.381	2.893
43	0.313	0.867	1.183	1.451	1.758	2.388	2.901
44	0.317	0.872	1.188	1.456	1.764	2.395	2.909
45	0.320	0.876	1.192	1.461	1.769	2.402	2.916
46	0.324	0.880	1.197	1.466	1.774	2.408	2.923
47	0.328	0.883	1.201	1.470	1.779	2.414	2.931
48	0.331	0.887	1.205	1.475	1.784	2.420	2.937
49	0.334	0.891	1.209	1.479	1.789	2.426	2.944
50	0.338	0.894	1.212	1.483	1.793	2.431	2.951

TABLE 1.12.2. FACTORS FOR ONE-SIDED SAMPLING PLANS AND TOLERANCE LIMITS.

GAMMA = 0.010 (CONTINUED)

$N \backslash P \rightarrow$	0.750	0.900	0.950	0.975	0.990	0.999	0.9999
51	0.341	0.897	1.216	1.487	1.798	2.437	2.957
52	0.344	0.901	1.220	1.491	1.802	2.442	2.963
53	0.347	0.904	1.223	1.495	1.806	2.447	2.969
54	0.350	0.907	1.227	1.499	1.811	2.452	2.975
55	0.353	0.910	1.230	1.502	1.815	2.457	2.981
56	0.355	0.913	1.233	1.506	1.819	2.462	2.986
57	0.358	0.916	1.236	1.509	1.822	2.467	2.992
58	0.361	0.919	1.239	1.513	1.826	2.471	2.997
59	0.363	0.922	1.242	1.516	1.830	2.476	3.002
60	0.366	0.924	1.245	1.519	1.833	2.480	3.007
61	0.368	0.927	1.248	1.522	1.837	2.484	3.012
62	0.370	0.929	1.251	1.525	1.840	2.489	3.017
63	0.373	0.932	1.254	1.528	1.844	2.493	3.022
64	0.375	0.934	1.257	1.531	1.847	2.497	3.027
65	0.377	0.937	1.259	1.534	1.850	2.501	3.031
66	0.379	0.939	1.262	1.537	1.853	2.504	3.036
67	0.381	0.941	1.264	1.540	1.856	2.508	3.040
68	0.383	0.944	1.267	1.542	1.859	2.512	3.044
69	0.385	0.946	1.269	1.545	1.862	2.515	3.048
70	0.387	0.948	1.272	1.548	1.865	2.519	3.053
71	0.389	0.950	1.274	1.550	1.868	2.522	3.057
72	0.391	0.952	1.276	1.553	1.871	2.526	3.061
73	0.393	0.954	1.278	1.555	1.873	2.529	3.064
74	0.395	0.956	1.281	1.558	1.876	2.533	3.068
75	0.397	0.958	1.283	1.560	1.879	2.536	3.072
76	0.398	0.960	1.285	1.562	1.881	2.539	3.076
77	0.400	0.962	1.287	1.565	1.884	2.542	3.079
78	0.402	0.964	1.289	1.567	1.886	2.545	3.083
79	0.403	0.966	1.291	1.569	1.889	2.548	3.086
80	0.405	0.968	1.293	1.571	1.891	2.551	3.090
81	0.407	0.969	1.295	1.573	1.893	2.554	3.093
82	0.408	0.971	1.297	1.575	1.896	2.557	3.096
83	0.410	0.973	1.299	1.577	1.898	2.560	3.100
84	0.411	0.974	1.301	1.579	1.900	2.562	3.103
85	0.413	0.976	1.302	1.581	1.903	2.565	3.106
86	0.414	0.978	1.304	1.583	1.905	2.568	3.109
87	0.416	0.979	1.306	1.585	1.907	2.570	3.112
88	0.417	0.981	1.308	1.587	1.909	2.573	3.115
89	0.418	0.982	1.309	1.589	1.911	2.575	3.118
90	0.420	0.984	1.311	1.591	1.913	2.578	3.121
91	0.421	0.985	1.313	1.593	1.915	2.580	3.124
92	0.422	0.987	1.314	1.594	1.917	2.583	3.127
93	0.424	0.988	1.316	1.596	1.919	2.585	3.130
94	0.425	0.990	1.317	1.598	1.921	2.588	3.132
95	0.426	0.991	1.319	1.600	1.923	2.590	3.135
96	0.427	0.992	1.321	1.601	1.925	2.592	3.138
97	0.429	0.994	1.322	1.603	1.926	2.594	3.140
98	0.430	0.995	1.323	1.605	1.928	2.597	3.143
99	0.431	0.996	1.325	1.606	1.930	2.599	3.145

TABLE 1.12.3. FACTORS FOR ONE-SIDED SAMPLING PLANS AND TOLERANCE LIMITS.

GAMMA = 0.010 (CONTINUED)

N \P →	0.750	0.900	0.950	0.975	0.990	0.999	0.9999
100	0.432	0.998	1.326	1.608	1.932	2.601	3.148
102	0.434	1.000	1.329	1.611	1.935	2.605	3.153
104	0.437	1.003	1.332	1.614	1.939	2.609	3.158
106	0.439	1.005	1.335	1.617	1.942	2.613	3.162
108	0.441	1.008	1.337	1.620	1.945	2.617	3.167
110	0.443	1.010	1.340	1.622	1.948	2.621	3.171
112	0.445	1.012	1.342	1.625	1.951	2.625	3.176
114	0.447	1.014	1.345	1.628	1.954	2.628	3.180
116	0.449	1.016	1.347	1.630	1.957	2.632	3.184
118	0.450	1.018	1.349	1.633	1.960	2.635	3.188
120	0.452	1.020	1.352	1.635	1.963	2.639	3.192
122	0.454	1.022	1.354	1.638	1.965	2.642	3.196
124	0.456	1.024	1.356	1.640	1.968	2.645	3.200
126	0.457	1.026	1.358	1.642	1.970	2.649	3.203
128	0.459	1.028	1.360	1.645	1.973	2.652	3.207
130	0.461	1.030	1.362	1.647	1.975	2.655	3.210
132	0.462	1.032	1.364	1.649	1.978	2.658	3.214
134	0.464	1.033	1.366	1.651	1.980	2.660	3.217
136	0.465	1.035	1.368	1.653	1.982	2.663	3.220
138	0.467	1.037	1.370	1.655	1.985	2.666	3.224
140	0.468	1.038	1.371	1.657	1.987	2.669	3.227
142	0.469	1.040	1.373	1.659	1.989	2.671	3.230
144	0.471	1.041	1.375	1.661	1.991	2.674	3.233
146	0.472	1.043	1.377	1.663	1.993	2.677	3.236
148	0.473	1.044	1.378	1.665	1.995	2.679	3.239
150	0.475	1.046	1.380	1.666	1.997	2.682	3.242
152	0.476	1.047	1.381	1.668	1.999	2.684	3.245
154	0.477	1.049	1.383	1.670	2.001	2.686	3.247
156	0.478	1.050	1.385	1.672	2.003	2.689	3.250
158	0.480	1.052	1.386	1.673	2.005	2.691	3.253
160	0.481	1.053	1.388	1.675	2.007	2.693	3.255
162	0.482	1.054	1.389	1.676	2.008	2.696	3.258
164	0.483	1.055	1.390	1.678	2.010	2.698	3.261
166	0.484	1.057	1.392	1.680	2.012	2.700	3.263
168	0.485	1.058	1.393	1.681	2.014	2.702	3.265
170	0.486	1.059	1.395	1.683	2.015	2.704	3.268
172	0.487	1.060	1.396	1.684	2.017	2.706	3.270
174	0.488	1.062	1.397	1.686	2.019	2.708	3.273
176	0.489	1.063	1.399	1.687	2.020	2.710	3.275
178	0.490	1.064	1.400	1.688	2.022	2.712	3.277
180	0.491	1.065	1.401	1.690	2.023	2.714	3.279
185	0.494	1.068	1.404	1.693	2.027	2.719	3.285
190	0.496	1.070	1.407	1.696	2.031	2.723	3.290
195	0.498	1.073	1.410	1.700	2.034	2.727	3.295
200	0.500	1.075	1.413	1.703	2.038	2.731	3.300
205	0.503	1.078	1.415	1.705	2.041	2.735	3.305
210	0.505	1.080	1.418	1.708	2.044	2.739	3.309
215	0.506	1.082	1.420	1.711	2.047	2.743	3.313
220	0.508	1.084	1.423	1.714	2.050	2.747	3.318

TABLE 1.12.4. FACTORS FOR ONE-SIDED SAMPLING PLANS AND TOLERANCE LIMITS.

GAMMA = 0.010 (CONTINUED)

N \P →	0.750	0.900	0.950	0.975	0.990	0.999	0.9999
225	0.510	1.086	1.425	1.716	2.053	2.750	3.322
230	0.512	1.088	1.427	1.719	2.055	2.754	3.326
235	0.513	1.090	1.429	1.721	2.058	2.757	3.330
240	0.515	1.092	1.431	1.723	2.061	2.760	3.333
245	0.517	1.094	1.433	1.726	2.063	2.763	3.337
250	0.518	1.096	1.435	1.728	2.066	2.766	3.340
255	0.520	1.098	1.437	1.730	2.068	2.769	3.344
260	0.521	1.099	1.439	1.732	2.070	2.772	3.347
265	0.523	1.101	1.441	1.734	2.073	2.775	3.350
270	0.524	1.102	1.443	1.736	2.075	2.777	3.354
275	0.525	1.104	1.445	1.738	2.077	2.780	3.357
280	0.527	1.105	1.446	1.740	2.079	2.783	3.360
285	0.528	1.107	1.448	1.741	2.081	2.785	3.362
290	0.529	1.108	1.449	1.743	2.083	2.788	3.365
295	0.530	1.110	1.451	1.745	2.085	2.790	3.368
300	0.531	1.111	1.452	1.747	2.087	2.792	3.371
310	0.534	1.114	1.455	1.750	2.090	2.797	3.376
320	0.536	1.116	1.458	1.753	2.094	2.801	3.381
330	0.538	1.119	1.461	1.756	2.097	2.805	3.386
340	0.540	1.121	1.463	1.759	2.100	2.809	3.390
350	0.542	1.123	1.466	1.761	2.103	2.813	3.395
360	0.543	1.125	1.468	1.764	2.106	2.816	3.399
370	0.545	1.127	1.471	1.766	2.109	2.820	3.403
380	0.547	1.129	1.473	1.769	2.112	2.823	3.407
390	0.548	1.131	1.475	1.771	2.114	2.826	3.411
400	0.550	1.133	1.477	1.773	2.117	2.830	3.414
425	0.554	1.137	1.482	1.779	2.123	2.837	3.423
450	0.557	1.141	1.486	1.783	2.128	2.844	3.431
475	0.560	1.144	1.490	1.788	2.133	2.850	3.438
500	0.563	1.148	1.494	1.792	2.138	2.855	3.444
525	0.565	1.151	1.497	1.796	2.142	2.861	3.451
550	0.568	1.154	1.500	1.799	2.146	2.866	3.457
575	0.570	1.156	1.503	1.803	2.150	2.870	3.462
600	0.572	1.159	1.506	1.806	2.153	2.875	3.467
625	0.574	1.161	1.509	1.809	2.157	2.879	3.472
650	0.576	1.164	1.511	1.812	2.160	2.883	3.476
700	0.580	1.168	1.516	1.817	2.165	2.890	3.485
750	0.583	1.171	1.520	1.821	2.171	2.896	3.492
800	0.586	1.175	1.524	1.826	2.175	2.902	3.499
850	0.588	1.178	1.527	1.830	2.180	2.908	3.505
900	0.591	1.181	1.531	1.833	2.184	2.912	3.511
950	0.593	1.183	1.534	1.836	2.187	2.917	3.516
1000	0.595	1.186	1.536	1.839	2.191	2.921	3.521
1500	0.609	1.203	1.556	1.861	2.215	2.951	3.556
2000	0.618	1.213	1.567	1.874	2.229	2.969	3.577
3000	0.628	1.225	1.581	1.889	2.247	2.991	3.603
5000	0.638	1.238	1.595	1.905	2.264	3.013	3.628
10000	0.649	1.251	1.610	1.921	2.282	3.035	3.655
∞	0.674	1.282	1.645	1.960	2.326	3.090	3.719

TABLE 1.13.1. FACTORS FOR ONE-SIDED SAMPLING PLANS AND TOLERANCE LIMITS.

GAMMA = 0.005

N \P →	0.750	0.900	0.950	0.975	0.990	0.999	0.9999
2	-10.247	-1.524	-0.273	0.153	0.438	0.844	1.120
3	-1.778	-0.266	0.163	0.425	0.671	1.102	1.420
4	-0.934	-0.007	0.336	0.577	0.821	1.273	1.616
5	-0.624	0.132	0.448	0.683	0.930	1.399	1.761
6	-0.454	0.225	0.531	0.764	1.015	1.497	1.874
7	-0.343	0.295	0.595	0.830	1.083	1.578	1.966
8	-0.262	0.351	0.648	0.884	1.141	1.645	2.044
9	-0.199	0.397	0.693	0.929	1.190	1.703	2.110
10	-0.149	0.436	0.731	0.969	1.232	1.754	2.168
11	-0.107	0.469	0.765	1.004	1.270	1.798	2.219
12	-0.072	0.499	0.794	1.035	1.303	1.838	2.265
13	-0.041	0.525	0.821	1.063	1.333	1.874	2.306
14	-0.014	0.548	0.845	1.088	1.360	1.906	2.344
15	0.010	0.569	0.866	1.111	1.385	1.936	2.378
16	0.031	0.588	0.886	1.132	1.408	1.963	2.409
17	0.050	0.606	0.904	1.151	1.429	1.989	2.439
18	0.068	0.622	0.921	1.169	1.448	2.012	2.466
19	0.084	0.637	0.937	1.186	1.466	2.034	2.491
20	0.098	0.651	0.951	1.201	1.483	2.054	2.514
21	0.112	0.664	0.965	1.216	1.499	2.073	2.536
22	0.125	0.676	0.978	1.229	1.514	2.091	2.557
23	0.136	0.687	0.990	1.242	1.528	2.108	2.577
24	0.147	0.698	1.001	1.254	1.541	2.124	2.595
25	0.158	0.708	1.012	1.266	1.554	2.140	2.613
26	0.167	0.718	1.022	1.277	1.566	2.154	2.630
27	0.176	0.727	1.032	1.287	1.577	2.168	2.645
28	0.185	0.735	1.041	1.297	1.588	2.181	2.661
29	0.193	0.743	1.049	1.306	1.598	2.193	2.675
30	0.201	0.751	1.058	1.315	1.608	2.205	2.689
31	0.208	0.759	1.066	1.324	1.618	2.217	2.702
32	0.215	0.766	1.073	1.332	1.627	2.228	2.715
33	0.222	0.773	1.081	1.340	1.635	2.238	2.727
34	0.229	0.779	1.088	1.348	1.644	2.248	2.739
35	0.235	0.785	1.094	1.355	1.652	2.258	2.750
36	0.241	0.791	1.101	1.362	1.659	2.268	2.761
37	0.246	0.797	1.107	1.369	1.667	2.277	2.771
38	0.252	0.803	1.113	1.375	1.674	2.285	2.781
39	0.257	0.808	1.119	1.381	1.681	2.294	2.791
40	0.262	0.813	1.125	1.387	1.688	2.302	2.801
41	0.267	0.818	1.130	1.393	1.694	2.310	2.810
42	0.271	0.823	1.135	1.399	1.700	2.317	2.819
43	0.276	0.828	1.140	1.404	1.706	2.325	2.827
44	0.280	0.832	1.145	1.410	1.712	2.332	2.835
45	0.284	0.837	1.150	1.415	1.718	2.339	2.844
46	0.288	0.841	1.155	1.420	1.724	2.346	2.851
47	0.292	0.845	1.159	1.425	1.729	2.352	2.859
48	0.296	0.849	1.163	1.430	1.734	2.359	2.866
49	0.300	0.853	1.168	1.434	1.739	2.365	2.874
50	0.303	0.857	1.172	1.439	1.744	2.371	2.881

TABLE 1.13.2. FACTORS FOR ONE-SIDED SAMPLING PLANS AND TOLERANCE LIMITS.

GAMMA = 0.005 (CONTINUED)

N \P →	0.750	0.900	0.950	0.975	0.990	0.999	0.9999
51	0.307	0.861	1.176	1.443	1.749	2.377	2.887
52	0.310	0.864	1.180	1.447	1.754	2.383	2.894
53	0.313	0.868	1.183	1.451	1.758	2.388	2.900
54	0.317	0.871	1.187	1.455	1.763	2.394	2.907
55	0.320	0.874	1.191	1.459	1.767	2.399	2.913
56	0.323	0.878	1.194	1.463	1.771	2.404	2.919
57	0.326	0.881	1.198	1.467	1.775	2.409	2.925
58	0.328	0.884	1.201	1.471	1.779	2.414	2.931
59	0.331	0.887	1.204	1.474	1.783	2.419	2.936
60	0.334	0.890	1.207	1.478	1.787	2.424	2.942
61	0.337	0.893	1.211	1.481	1.791	2.428	2.947
62	0.339	0.895	1.214	1.484	1.795	2.433	2.952
63	0.342	0.898	1.217	1.488	1.798	2.437	2.957
64	0.344	0.901	1.220	1.491	1.802	2.441	2.962
65	0.347	0.903	1.222	1.494	1.805	2.446	2.967
66	0.349	0.906	1.225	1.497	1.809	2.450	2.972
67	0.351	0.908	1.228	1.500	1.812	2.454	2.977
68	0.354	0.911	1.231	1.503	1.815	2.458	2.981
69	0.356	0.913	1.233	1.506	1.818	2.462	2.986
70	0.358	0.916	1.236	1.509	1.822	2.466	2.990
71	0.360	0.918	1.238	1.511	1.825	2.469	2.995
72	0.362	0.920	1.241	1.514	1.828	2.473	2.999
73	0.364	0.922	1.243	1.517	1.831	2.477	3.003
74	0.366	0.925	1.246	1.519	1.833	2.480	3.007
75	0.368	0.927	1.248	1.522	1.836	2.484	3.011
76	0.370	0.929	1.250	1.524	1.839	2.487	3.015
77	0.372	0.931	1.253	1.527	1.842	2.490	3.019
78	0.374	0.933	1.255	1.529	1.845	2.494	3.023
79	0.376	0.935	1.257	1.532	1.847	2.497	3.027
80	0.377	0.937	1.259	1.534	1.850	2.500	3.031
81	0.379	0.939	1.261	1.536	1.852	2.503	3.034
82	0.381	0.941	1.263	1.539	1.855	2.506	3.038
83	0.383	0.943	1.265	1.541	1.857	2.509	3.041
84	0.384	0.944	1.267	1.543	1.860	2.512	3.045
85	0.386	0.946	1.269	1.545	1.862	2.515	3.048
86	0.387	0.948	1.271	1.547	1.864	2.518	3.052
87	0.389	0.950	1.273	1.549	1.867	2.521	3.055
88	0.391	0.951	1.275	1.551	1.869	2.524	3.058
89	0.392	0.953	1.277	1.553	1.871	2.527	3.061
90	0.394	0.955	1.279	1.555	1.874	2.529	3.064
91	0.395	0.956	1.280	1.557	1.876	2.532	3.068
92	0.396	0.958	1.282	1.559	1.878	2.535	3.071
93	0.398	0.959	1.284	1.561	1.880	2.537	3.074
94	0.399	0.961	1.286	1.563	1.882	2.540	3.077
95	0.401	0.963	1.287	1.565	1.884	2.542	3.080
96	0.402	0.964	1.289	1.567	1.886	2.545	3.082
97	0.403	0.965	1.291	1.568	1.888	2.547	3.085
98	0.405	0.967	1.292	1.570	1.890	2.550	3.088
99	0.406	0.968	1.294	1.572	1.892	2.552	3.091

TABLE 1.13.3. FACTORS FOR ONE-SIDED SAMPLING PLANS AND TOLERANCE LIMITS.

GAMMA = 0.005 (CONTINUED)

N \P →	0.750	0.900	0.950	0.975	0.990	0.999	0.9999
100	0.407	0.970	1.295	1.574	1.894	2.554	3.094
102	0.410	0.973	1.298	1.577	1.898	2.559	3.099
104	0.412	0.975	1.301	1.580	1.901	2.563	3.104
106	0.415	0.978	1.304	1.583	1.905	2.568	3.109
108	0.417	0.980	1.307	1.587	1.908	2.572	3.114
110	0.419	0.983	1.310	1.590	1.912	2.576	3.119
112	0.421	0.985	1.313	1.593	1.915	2.580	3.124
114	0.423	0.988	1.315	1.595	1.918	2.584	3.128
116	0.425	0.990	1.318	1.598	1.921	2.588	3.133
118	0.427	0.992	1.320	1.601	1.924	2.592	3.137
120	0.429	0.995	1.323	1.604	1.927	2.595	3.141
122	0.431	0.997	1.325	1.606	1.930	2.599	3.145
124	0.433	0.999	1.327	1.609	1.933	2.602	3.150
126	0.435	1.001	1.330	1.611	1.936	2.606	3.154
128	0.437	1.003	1.332	1.614	1.939	2.609	3.157
130	0.439	1.005	1.334	1.616	1.941	2.612	3.161
132	0.440	1.007	1.336	1.619	1.944	2.616	3.165
134	0.442	1.009	1.338	1.621	1.946	2.619	3.169
136	0.444	1.011	1.340	1.623	1.949	2.622	3.172
138	0.445	1.012	1.342	1.625	1.951	2.625	3.176
140	0.447	1.014	1.344	1.627	1.954	2.628	3.179
142	0.448	1.016	1.346	1.630	1.956	2.631	3.183
144	0.450	1.018	1.348	1.632	1.958	2.634	3.186
146	0.451	1.019	1.350	1.634	1.961	2.636	3.189
148	0.453	1.021	1.352	1.636	1.963	2.639	3.192
150	0.454	1.022	1.354	1.638	1.965	2.642	3.195
152	0.456	1.024	1.355	1.639	1.967	2.644	3.198
154	0.457	1.026	1.357	1.641	1.969	2.647	3.201
156	0.458	1.027	1.359	1.643	1.971	2.650	3.204
158	0.460	1.029	1.360	1.645	1.973	2.652	3.207
160	0.461	1.030	1.362	1.647	1.975	2.655	3.210
162	0.462	1.031	1.364	1.649	1.977	2.657	3.213
164	0.463	1.033	1.365	1.650	1.979	2.659	3.216
166	0.465	1.034	1.367	1.652	1.981	2.662	3.218
168	0.466	1.036	1.368	1.654	1.983	2.664	3.221
170	0.467	1.037	1.370	1.655	1.985	2.666	3.224
172	0.468	1.038	1.371	1.657	1.987	2.668	3.226
174	0.469	1.040	1.373	1.658	1.988	2.671	3.229
176	0.470	1.041	1.374	1.660	1.990	2.673	3.231
178	0.471	1.042	1.375	1.662	1.992	2.675	3.234
180	0.473	1.043	1.377	1.663	1.993	2.677	3.236
185	0.475	1.046	1.380	1.667	1.998	2.682	3.242
190	0.478	1.049	1.383	1.670	2.001	2.687	3.248
195	0.480	1.052	1.386	1.674	2.005	2.692	3.253
200	0.482	1.055	1.389	1.677	2.009	2.696	3.259
205	0.485	1.057	1.392	1.680	2.012	2.700	3.264
210	0.487	1.060	1.395	1.683	2.016	2.705	3.269
215	0.489	1.062	1.398	1.686	2.019	2.709	3.273
220	0.491	1.064	1.400	1.689	2.022	2.713	3.278

TABLE 1.13.4. FACTORS FOR ONE-SIDED SAMPLING PLANS AND TOLERANCE LIMITS.

GAMMA = 0.005 (CONTINUED)

N \P →	0.750	0.900	0.950	0.975	0.990	0.999	0.9999
225	0.493	1.067	1.403	1.692	2.025	2.716	3.282
230	0.495	1.069	1.405	1.694	2.028	2.720	3.287
235	0.497	1.071	1.408	1.697	2.031	2.724	3.291
240	0.499	1.073	1.410	1.700	2.034	2.727	3.295
245	0.500	1.075	1.412	1.702	2.037	2.731	3.299
250	0.502	1.077	1.414	1.704	2.040	2.734	3.303
255	0.504	1.079	1.416	1.707	2.042	2.737	3.307
260	0.505	1.081	1.419	1.709	2.045	2.740	3.310
265	0.507	1.082	1.421	1.711	2.047	2.743	3.314
270	0.508	1.084	1.422	1.713	2.050	2.746	3.317
275	0.510	1.086	1.424	1.715	2.052	2.749	3.320
280	0.511	1.088	1.426	1.717	2.054	2.752	3.324
285	0.513	1.089	1.428	1.719	2.056	2.755	3.327
290	0.514	1.091	1.430	1.721	2.059	2.757	3.330
295	0.515	1.092	1.431	1.723	2.061	2.760	3.333
300	0.516	1.094	1.433	1.725	2.063	2.762	3.336
310	0.519	1.097	1.436	1.729	2.067	2.767	3.342
320	0.521	1.099	1.439	1.732	2.070	2.772	3.347
330	0.524	1.102	1.442	1.735	2.074	2.777	3.352
340	0.526	1.104	1.445	1.738	2.078	2.781	3.357
350	0.528	1.107	1.448	1.741	2.081	2.785	3.362
360	0.530	1.109	1.450	1.744	2.084	2.789	3.367
370	0.532	1.111	1.453	1.747	2.087	2.793	3.371
380	0.534	1.114	1.455	1.749	2.090	2.796	3.376
390	0.535	1.116	1.457	1.752	2.093	2.800	3.380
400	0.537	1.118	1.460	1.754	2.096	2.803	3.384
425	0.541	1.122	1.465	1.760	2.102	2.811	3.393
450	0.545	1.126	1.470	1.766	2.108	2.819	3.401
475	0.548	1.130	1.474	1.770	2.113	2.825	3.409
500	0.551	1.134	1.478	1.775	2.118	2.832	3.417
525	0.554	1.137	1.482	1.779	2.123	2.837	3.423
550	0.557	1.141	1.486	1.783	2.128	2.843	3.430
575	0.559	1.144	1.489	1.787	2.132	2.848	3.436
600	0.562	1.146	1.492	1.790	2.136	2.853	3.441
625	0.564	1.149	1.495	1.793	2.139	2.857	3.447
650	0.566	1.151	1.498	1.796	2.143	2.862	3.452
700	0.570	1.156	1.503	1.802	2.149	2.869	3.461
750	0.573	1.160	1.507	1.807	2.155	2.876	3.469
800	0.576	1.164	1.512	1.812	2.160	2.883	3.477
850	0.579	1.167	1.515	1.816	2.165	2.889	3.483
900	0.582	1.170	1.519	1.820	2.169	2.894	3.490
950	0.584	1.173	1.522	1.823	2.173	2.899	3.496
1000	0.586	1.176	1.525	1.827	2.177	2.904	3.501
1500	0.602	1.195	1.546	1.850	2.203	2.937	3.539
2000	0.612	1.206	1.559	1.865	2.219	2.957	3.563
3000	0.623	1.219	1.574	1.882	2.238	2.980	3.590
5000	0.635	1.233	1.590	1.899	2.258	3.005	3.619
10000	0.646	1.247	1.606	1.917	2.277	3.029	3.648
∞	0.674	1.282	1.645	1.960	2.326	3.090	3.719

TABLE 2

Sample size requirements for one-sided sampling plans.

TABLE 2.1. SAMPLE SIZE REQUIREMENTS FOR ONE-SIDED SAMPLING PLANS.

ALPHA = 0.010   BETA = 0.050

P1 = 0.005

P2	N	K
0.010	1006	2.430
0.015	361	2.339
0.020	209	2.271
0.025	146	2.216
0.030	112	2.170
0.035	90	2.130
0.040	76	2.095
0.045	65	2.063
0.050	57	2.034
0.055	51	2.007
0.060	46	1.982
0.065	42	1.958
0.070	39	1.936
0.075	36	1.915
0.080	34	1.895
0.085	31	1.877
0.090	30	1.858
0.095	28	1.841
0.100	26	1.825
0.150	17	1.685
0.200	13	1.572
0.300	8	1.397

P1 = 0.010

P2	N	K
0.020	714	2.167
0.025	382	2.112
0.030	251	2.066
0.035	184	2.026
0.040	144	1.990
0.045	117	1.958
0.050	99	1.929
0.055	85	1.902
0.060	75	1.877
0.065	67	1.853
0.070	60	1.831
0.075	54	1.810
0.080	50	1.791
0.085	46	1.772
0.090	43	1.753
0.095	40	1.736
0.100	37	1.719
0.150	23	1.578
0.200	16	1.467
0.300	10	1.285

P1 = 0.015

P2	N	K
0.030	569	2.001
0.035	362	1.961
0.040	258	1.925
0.045	197	1.893
0.050	158	1.864
0.055	131	1.837
0.060	112	1.811
0.065	97	1.788
0.070	85	1.766
0.075	76	1.745
0.080	69	1.725
0.085	62	1.706
0.090	57	1.688
0.095	53	1.670
0.100	49	1.654
0.150	28	1.513
0.200	19	1.401
0.300	11	1.221

P1 = 0.020

P2	N	K
0.040	477	1.877
0.045	334	1.844
0.050	252	1.815
0.055	199	1.788
0.060	164	1.763
0.065	138	1.739
0.070	119	1.717
0.075	104	1.696
0.080	92	1.676
0.085	82	1.657
0.090	74	1.639
0.095	68	1.621
0.100	62	1.605
0.150	33	1.464
0.200	22	1.352
0.300	12	1.173

P1 = 0.025

P2	N	K
0.050	412	1.776
0.055	307	1.749
0.060	241	1.723
0.065	196	1.700
0.070	164	1.678
0.075	140	1.656
0.080	121	1.636
0.085	107	1.617
0.090	95	1.599
0.095	86	1.582
0.100	78	1.565
0.150	38	1.424
0.200	25	1.312
0.300	14	1.130

P1 = 0.030

P2	N	K
0.060	363	1.690
0.065	282	1.667
0.070	228	1.644
0.075	190	1.623
0.080	161	1.603
0.085	139	1.584
0.090	122	1.566
0.095	108	1.549
0.100	97	1.532
0.150	45	1.390
0.200	28	1.279
0.300	15	1.097

P1 = 0.035

P2	N	K
0.070	324	1.616
0.075	260	1.595
0.080	215	1.574
0.085	182	1.555
0.090	157	1.537
0.095	137	1.520
0.100	121	1.503
0.150	52	1.361
0.200	31	1.250
0.300	16	1.068

P1 = 0.040

P2	N	K
0.080	292	1.549
0.085	241	1.530
0.090	203	1.512
0.095	174	1.494
0.100	151	1.477
0.150	60	1.335
0.200	34	1.224
0.300	17	1.043

P1 = 0.045

P2	N	K
0.090	266	1.488
0.095	223	1.471
0.100	191	1.454
0.150	69	1.312
0.200	38	1.200
0.300	18	1.019

P1 = 0.050

P2	N	K
0.100	244	1.433
0.150	79	1.291
0.200	43	1.179
0.300	20	0.997

P1 = 0.075

P2	N	K
0.150	169	1.205
0.200	71	1.092
0.300	27	0.910

P1 = 0.100

P2	N	K
0.200	125	1.025
0.300	38	0.842

TABLE 2.2. SAMPLE SIZE REQUIREMENTS FOR ONE-SIDED SAMPLING PLANS.

ALPHA = 0.010   BETA = 0.100

P2	N	K	P2	N	K	P2	N	K
P1 = 0.005			**P1 = 0.015**			**P1 = 0.030**		
0.010	825	2.415	0.070	69	1.726	0.095	88	1.516
0.015	295	2.315	0.075	62	1.704	0.100	79	1.498
0.020	171	2.241	0.080	55	1.682	0.150	36	1.343
0.025	119	2.181	0.085	50	1.661	0.200	22	1.221
0.030	90	2.131	0.090	46	1.641	0.300	12	1.025
0.035	73	2.087	0.095	42	1.622			
0.040	61	2.049	0.100	39	1.604	**P1 = 0.035**		
0.045	53	2.014	0.150	22	1.451			
0.050	46	1.982	0.200	15	1.330	0.070	265	1.596
0.055	41	1.952	0.300	9	1.134	0.075	212	1.573
0.060	37	1.925				0.080	175	1.551
0.065	34	1.900	**P1 = 0.020**			0.085	148	1.530
0.070	31	1.876				0.090	127	1.510
0.075	29	1.853	0.040	390	1.859	0.095	111	1.491
0.080	27	1.832	0.045	273	1.824	0.100	98	1.472
0.085	25	1.811	0.050	205	1.791	0.150	42	1.318
0.090	24	1.792	0.055	162	1.762	0.200	25	1.196
0.095	22	1.773	0.060	133	1.734	0.300	13	0.999
0.100	21	1.755	0.065	112	1.708			
0.150	14	1.603	0.070	96	1.684	**P1 = 0.040**		
0.200	10	1.483	0.075	84	1.661			
0.300	7	1.284	0.080	74	1.639	0.080	239	1.529
			0.085	66	1.618	0.085	196	1.508
P1 = 0.010			0.090	60	1.599	0.090	165	1.488
			0.095	55	1.580	0.095	142	1.469
0.020	585	2.151	0.100	50	1.561	0.100	123	1.450
0.025	312	2.091	0.150	26	1.407	0.150	48	1.295
0.030	205	2.041	0.200	17	1.286	0.200	28	1.173
0.035	150	1.997	0.300	10	1.090	0.300	14	0.976
0.040	117	1.958						
0.045	95	1.923	**P1 = 0.025**			**P1 = 0.045**		
0.050	80	1.891						
0.055	69	1.861	0.050	337	1.758	0.090	217	1.468
0.060	61	1.834	0.055	251	1.728	0.095	182	1.449
0.065	54	1.808	0.060	196	1.700	0.100	156	1.430
0.070	48	1.784	0.065	160	1.674	0.150	56	1.275
0.075	44	1.761	0.070	133	1.650	0.200	31	1.153
0.080	40	1.739	0.075	114	1.627	0.300	15	0.955
0.085	37	1.719	0.080	99	1.605			
0.090	34	1.699	0.085	87	1.584	**P1 = 0.050**		
0.095	32	1.680	0.090	77	1.564			
0.100	30	1.662	0.095	69	1.545	0.100	199	1.412
0.150	18	1.509	0.100	63	1.527	0.150	64	1.256
0.200	13	1.389	0.150	31	1.373	0.200	34	1.133
0.300	8	1.193	0.200	20	1.251	0.300	16	0.936
			0.300	11	1.055			
P1 = 0.015						**P1 = 0.075**		
			P1 = 0.030					
0.030	466	1.984				0.150	138	1.181
0.035	296	1.940	0.060	296	1.671	0.200	58	1.058
0.040	211	1.901	0.065	230	1.645	0.300	22	0.858
0.045	161	1.866	0.070	186	1.621			
0.050	129	1.834	0.075	154	1.598	**P1 = 0.100**		
0.055	107	1.804	0.080	131	1.576			
0.060	91	1.777	0.085	113	1.555	0.200	102	1.000
0.065	79	1.751	0.090	99	1.535	0.300	31	0.800

TABLE 2.3. SAMPLE SIZE REQUIREMENTS FOR ONE-SIDED SAMPLING PLANS.

ALPHA = 0.010   BETA = 0.200

P2	N	K	P2	N	K	P2	N	K
P1 = 0.005			**P1 = 0.015**			**P1 = 0.030**		
0.010	630	2.393	0.070	52	1.668	0.095	66	1.467
0.015	224	2.280	0.075	46	1.642	0.100	59	1.446
0.020	129	2.196	0.080	41	1.617	0.150	27	1.273
0.025	89	2.128	0.085	38	1.595	0.200	16	1.134
0.030	68	2.072	0.090	34	1.571	0.300	9	0.918
0.035	55	2.023	0.095	32	1.551			
0.040	46	1.980	0.100	29	1.530	**P1 = 0.035**		
0.045	39	1.940	0.150	16	1.357			
0.050	35	1.905	0.200	11	1.223	0.070	201	1.567
0.055	31	1.872	0.300	7	1.010	0.075	161	1.541
0.060	28	1.842				0.080	133	1.516
0.065	25	1.812	**P1 = 0.020**			0.085	112	1.492
0.070	23	1.785				0.090	96	1.469
0.075	22	1.762	0.040	297	1.833	0.095	84	1.448
0.080	20	1.737	0.045	207	1.792	0.100	74	1.427
0.085	19	1.715	0.050	156	1.756	0.150	31	1.252
0.090	18	1.693	0.055	123	1.722	0.200	18	1.114
0.095	17	1.672	0.060	100	1.691	0.300	9	0.892
0.100	16	1.652	0.065	84	1.662			
0.150	10	1.481	0.070	72	1.634	**P1 = 0.040**		
0.200	8	1.352	0.075	63	1.608			
0.300	5	1.135	0.080	56	1.584	0.080	181	1.499
			0.085	50	1.561	0.085	149	1.475
P1 = 0.010			0.090	45	1.538	0.090	125	1.452
			0.095	41	1.517	0.095	107	1.431
0.020	446	2.127	0.100	37	1.496	0.100	93	1.410
0.025	237	2.059	0.150	20	1.325	0.150	36	1.234
0.030	155	2.002	0.200	13	1.189	0.200	21	1.098
0.035	113	1.952	0.300	7	0.968	0.300	10	0.875
0.040	88	1.908						
0.045	72	1.869	**P1 = 0.025**			**P1 = 0.045**		
0.050	60	1.832						
0.055	52	1.800	0.050	256	1.730	0.090	165	1.437
0.060	45	1.768	0.055	190	1.696	0.095	138	1.415
0.065	40	1.739	0.060	149	1.665	0.100	118	1.394
0.070	36	1.712	0.065	121	1.636	0.150	42	1.219
0.075	33	1.688	0.070	101	1.608	0.200	23	1.081
0.080	30	1.663	0.075	86	1.582	0.300	11	0.859
0.085	28	1.641	0.080	74	1.557			
0.090	26	1.619	0.085	65	1.534	**P1 = 0.050**		
0.095	24	1.597	0.090	58	1.511			
0.100	22	1.576	0.095	52	1.490	0.100	151	1.380
0.150	13	1.405	0.100	47	1.469	0.150	48	1.204
0.200	9	1.270	0.150	23	1.296	0.200	26	1.067
0.300	6	1.057	0.200	15	1.161	0.300	12	0.845
			0.300	8	0.941			
P1 = 0.015						**P1 = 0.075**		
			P1 = 0.030					
0.030	354	1.959				0.150	104	1.146
0.035	224	1.909	0.060	225	1.643	0.200	43	1.006
0.040	159	1.864	0.065	175	1.614	0.300	16	0.779
0.045	122	1.825	0.070	141	1.586			
0.050	97	1.788	0.075	117	1.560	**P1 = 0.100**		
0.055	80	1.755	0.080	99	1.535			
0.060	68	1.724	0.085	85	1.511	0.200	77	0.962
0.065	59	1.695	0.090	75	1.489	0.300	23	0.735

74

TABLE 2.4. SAMPLE SIZE REQUIREMENTS FOR ONE-SIDED SAMPLING PLANS.

ALPHA = 0.025   BETA = 0.050

P2	N	K	P2	N	K	P2	N	K
P1 = 0.005			**P1 = 0.015**			**P1 = 0.030**		
0.010	833	2.441	0.070	72	1.796	0.095	91	1.573
0.015	300	2.356	0.075	64	1.777	0.100	81	1.558
0.020	175	2.293	0.080	58	1.758	0.150	38	1.427
0.025	122	2.243	0.085	53	1.741	0.200	24	1.324
0.030	93	2.201	0.090	48	1.724	0.300	13	1.156
0.035	76	2.164	0.095	44	1.708			
0.040	64	2.131	0.100	41	1.693	**P1 = 0.035**		
0.045	55	2.102	0.150	24	1.562			
0.050	49	2.075	0.200	16	1.461	0.070	269	1.630
0.055	43	2.050	0.300	10	1.290	0.075	217	1.611
0.060	39	2.027				0.080	179	1.592
0.065	36	2.005	**P1 = 0.020**			0.085	152	1.574
0.070	33	1.985				0.090	131	1.558
0.075	31	1.965	0.040	396	1.890	0.095	114	1.541
0.080	29	1.947	0.045	278	1.860	0.100	101	1.526
0.085	27	1.930	0.050	210	1.833	0.150	44	1.395
0.090	25	1.914	0.055	166	1.808	0.200	26	1.293
0.095	24	1.897	0.060	137	1.784	0.300	14	1.124
0.100	23	1.881	0.065	115	1.762			
0.150	15	1.753	0.070	99	1.742	**P1 = 0.040**		
0.200	11	1.653	0.075	87	1.723			
0.300	8	1.469	0.080	77	1.704	0.080	243	1.564
			0.085	69	1.687	0.085	200	1.546
P1 = 0.010			0.090	62	1.670	0.090	169	1.529
			0.095	57	1.654	0.095	145	1.513
0.020	592	2.179	0.100	52	1.638	0.100	126	1.497
0.025	318	2.128	0.150	28	1.508	0.150	50	1.367
0.030	209	2.085	0.200	19	1.405	0.200	29	1.264
0.035	154	2.048	0.300	11	1.237	0.300	15	1.095
0.040	120	2.015						
0.045	98	1.986	**P1 = 0.025**			**P1 = 0.045**		
0.050	83	1.959						
0.055	72	1.934	0.050	342	1.789	0.090	221	1.504
0.060	63	1.911	0.055	255	1.764	0.095	186	1.487
0.065	56	1.889	0.060	201	1.741	0.100	159	1.472
0.070	51	1.869	0.065	163	1.719	0.150	58	1.341
0.075	46	1.849	0.070	137	1.698	0.200	32	1.238
0.080	42	1.831	0.075	117	1.679	0.300	16	1.070
0.085	39	1.814	0.080	102	1.661			
0.090	36	1.797	0.085	90	1.643	**P1 = 0.050**		
0.095	34	1.781	0.090	80	1.626			
0.100	32	1.765	0.095	72	1.610	0.100	203	1.448
0.150	19	1.637	0.100	65	1.595	0.150	67	1.317
0.200	14	1.533	0.150	33	1.464	0.200	36	1.214
0.300	9	1.361	0.200	21	1.362	0.300	17	1.046
			0.300	12	1.193			
P1 = 0.015						**P1 = 0.075**		
			P1 = 0.030					
0.030	472	2.013				0.150	141	1.222
0.035	301	1.976	0.060	301	1.704	0.200	60	1.118
0.040	215	1.943	0.065	235	1.682	0.300	23	0.950
0.045	165	1.914	0.070	190	1.662			
0.050	132	1.886	0.075	158	1.642	**P1 = 0.100**		
0.055	110	1.861	0.080	134	1.624			
0.060	94	1.838	0.085	116	1.606	0.200	105	1.044
0.065	81	1.816	0.090	102	1.589	0.300	32	0.875

TABLE 2.5. SAMPLE SIZE REQUIREMENTS FOR ONE-SIDED SAMPLING PLANS.

ALPHA = 0.025   BETA = 0.100

P1 = 0.005

P2	N	K
0.010	669	2.426
0.015	240	2.332
0.020	139	2.263
0.025	97	2.207
0.030	74	2.160
0.035	60	2.120
0.040	51	2.084
0.045	44	2.052
0.050	38	2.022
0.055	34	1.995
0.060	31	1.969
0.065	28	1.946
0.070	26	1.924
0.075	24	1.903
0.080	23	1.882
0.085	21	1.864
0.090	20	1.845
0.095	19	1.828
0.100	18	1.811
0.150	12	1.670
0.200	9	1.558
0.300	6	1.374

P1 = 0.010

P2	N	K
0.020	475	2.162
0.025	254	2.106
0.030	167	2.059
0.035	123	2.018
0.040	96	1.982
0.045	78	1.949
0.050	66	1.919
0.055	57	1.892
0.060	50	1.866
0.065	44	1.843
0.070	40	1.820
0.075	36	1.799
0.080	33	1.779
0.085	31	1.760
0.090	29	1.741
0.095	27	1.724
0.100	25	1.707
0.150	15	1.566
0.200	11	1.453
0.300	7	1.270

P1 = 0.015

P2	N	K
0.030	378	1.996
0.035	241	1.955
0.040	172	1.918
0.045	131	1.886
0.050	105	1.856
0.055	88	1.828
0.060	74	1.802
0.065	65	1.778
0.070	57	1.756
0.075	51	1.735
0.080	46	1.714
0.085	42	1.695
0.090	38	1.677
0.095	35	1.659
0.100	33	1.642
0.150	19	1.500
0.200	13	1.387
0.300	8	1.203

P1 = 0.020

P2	N	K
0.040	317	1.872
0.045	222	1.839
0.050	168	1.808
0.055	133	1.781
0.060	109	1.755
0.065	92	1.731
0.070	79	1.708
0.075	69	1.687
0.080	61	1.667
0.085	55	1.647
0.090	49	1.629
0.095	45	1.611
0.100	41	1.594
0.150	22	1.451
0.200	15	1.339
0.300	8	1.160

P1 = 0.025

P2	N	K
0.050	274	1.771
0.055	204	1.743
0.060	160	1.717
0.065	130	1.693
0.070	109	1.670
0.075	93	1.649
0.080	81	1.628
0.085	71	1.609
0.090	64	1.590
0.095	57	1.572
0.100	52	1.555
0.150	26	1.412
0.200	17	1.299
0.300	9	1.118

P1 = 0.030

P2	N	K
0.060	241	1.685
0.065	188	1.661
0.070	152	1.638
0.075	126	1.616
0.080	107	1.596
0.085	93	1.577
0.090	81	1.558
0.095	72	1.540
0.100	65	1.523
0.150	30	1.379
0.200	19	1.266
0.300	10	1.084

P1 = 0.035

P2	N	K
0.070	215	1.610
0.075	173	1.588
0.080	143	1.568
0.085	121	1.548
0.090	104	1.530
0.095	91	1.512
0.100	81	1.495
0.150	35	1.351
0.200	21	1.237
0.300	11	1.054

P1 = 0.040

P2	N	K
0.080	195	1.543
0.085	160	1.524
0.090	135	1.505
0.095	116	1.487
0.100	101	1.470
0.150	40	1.326
0.200	23	1.212
0.300	12	1.028

P1 = 0.045

P2	N	K
0.090	177	1.483
0.095	149	1.465
0.100	127	1.447
0.150	46	1.303
0.200	26	1.189
0.300	12	1.006

P1 = 0.050

P2	N	K
0.100	162	1.427
0.150	53	1.282
0.200	28	1.168
0.300	13	0.984

P1 = 0.075

P2	N	K
0.150	112	1.198
0.200	48	1.083
0.300	18	0.897

P1 = 0.100

P2	N	K
0.200	83	1.018
0.300	25	0.831

TABLE 2.6. SAMPLE SIZE REQUIREMENTS FOR ONE-SIDED SAMPLING PLANS.

ALPHA = 0.025   BETA = 0.200

P2	N	K	P2	N	K	P2	N	K
P1 = 0.005			**P1 = 0.015**			**P1 = 0.030**		
0.010	495	2.402	0.070	41	1.694	0.095	53	1.489
0.015	176	2.295	0.075	37	1.670	0.100	47	1.469
0.020	102	2.215	0.080	33	1.647	0.150	22	1.306
0.025	71	2.152	0.085	30	1.625	0.200	13	1.176
0.030	54	2.098	0.090	28	1.605	0.300	7	0.971
0.035	44	2.053	0.095	25	1.583			
0.040	37	2.012	0.100	24	1.566	**P1 = 0.035**		
0.045	32	1.975	0.150	13	1.403			
0.050	28	1.941	0.200	9	1.278	0.070	158	1.579
0.055	25	1.911	0.300	6	1.075	0.075	127	1.554
0.060	22	1.881				0.080	105	1.531
0.065	21	1.856	**P1 = 0.020**			0.085	89	1.508
0.070	19	1.830				0.090	76	1.487
0.075	17	1.805	0.040	234	1.844	0.095	66	1.466
0.080	16	1.783	0.045	163	1.805	0.100	59	1.447
0.085	15	1.762	0.050	123	1.771	0.150	25	1.282
0.090	14	1.741	0.055	97	1.739	0.200	15	1.154
0.095	14	1.723	0.060	80	1.710	0.300	8	0.948
0.100	13	1.704	0.065	67	1.682			
0.150	9	1.547	0.070	58	1.657	**P1 = 0.040**		
0.200	6	1.424	0.075	50	1.632			
0.300	4	1.226	0.080	44	1.608	0.080	143	1.511
			0.085	40	1.587	0.085	117	1.489
P1 = 0.010			0.090	36	1.566	0.090	99	1.467
			0.095	33	1.546	0.095	85	1.447
0.020	351	2.137	0.100	30	1.526	0.100	74	1.427
0.025	187	2.072	0.150	16	1.364	0.150	29	1.262
0.030	123	2.018	0.200	11	1.239	0.200	17	1.134
0.035	90	1.972	0.300	6	1.033	0.300	8	0.924
0.040	70	1.930						
0.045	57	1.893	**P1 = 0.025**			**P1 = 0.045**		
0.050	48	1.859						
0.055	41	1.827	0.050	202	1.742	0.090	130	1.450
0.060	36	1.798	0.055	150	1.709	0.095	109	1.429
0.065	32	1.771	0.060	117	1.680	0.100	93	1.409
0.070	29	1.746	0.065	95	1.652	0.150	33	1.243
0.075	26	1.721	0.070	80	1.626	0.200	19	1.115
0.080	24	1.699	0.075	68	1.601	0.300	9	0.906
0.085	22	1.677	0.080	59	1.578			
0.090	21	1.657	0.085	52	1.556	**P1 = 0.050**		
0.095	19	1.636	0.090	46	1.535			
0.100	18	1.618	0.095	42	1.515	0.100	119	1.393
0.150	11	1.459	0.100	38	1.495	0.150	38	1.227
0.200	8	1.334	0.150	19	1.333	0.200	21	1.098
0.300	5	1.131	0.200	12	1.205	0.300	10	0.890
			0.300	7	1.001			
P1 = 0.015						**P1 = 0.075**		
			P1 = 0.030					
0.030	279	1.969				0.150	82	1.161
0.035	177	1.922	0.060	177	1.655	0.200	35	1.029
0.040	126	1.880	0.065	138	1.627	0.300	13	0.816
0.045	96	1.842	0.070	111	1.601			
0.050	77	1.808	0.075	92	1.576	**P1 = 0.100**		
0.055	64	1.777	0.080	78	1.553			
0.060	54	1.747	0.085	68	1.531	0.200	61	0.978
0.065	47	1.720	0.090	59	1.509	0.300	18	0.762

TABLE 2.7. SAMPLE SIZE REQUIREMENTS FOR ONE-SIDED SAMPLING PLANS.

ALPHA = 0.050   BETA = 0.050

P2	N	K	P2	N	K	P2	N	K
P1 = 0.005			**P1 = 0.015**			**P1 = 0.030**		
0.010	697	2.452	0.070	61	1.828	0.095	77	1.599
0.015	252	2.374	0.075	54	1.810	0.100	69	1.585
0.020	147	2.317	0.080	49	1.793	0.150	32	1.467
0.025	103	2.271	0.085	45	1.777	0.200	20	1.374
0.030	79	2.232	0.090	41	1.762	0.300	11	1.221
0.035	64	2.199	0.095	38	1.747			
0.040	54	2.169	0.100	35	1.734	**P1 = 0.035**		
0.045	47	2.142	0.150	20	1.617			
0.050	42	2.117	0.200	14	1.523	0.070	226	1.645
0.055	37	2.095	0.300	9	1.363	0.075	182	1.627
0.060	34	2.074				0.080	151	1.610
0.065	31	2.054	**P1 = 0.020**			0.085	128	1.594
0.070	29	2.035				0.090	110	1.579
0.075	26	2.020	0.040	332	1.903	0.095	97	1.564
0.080	25	2.001	0.045	234	1.876	0.100	86	1.550
0.085	23	1.987	0.050	177	1.851	0.150	37	1.431
0.090	22	1.971	0.055	140	1.828	0.200	23	1.336
0.095	21	1.955	0.060	115	1.807	0.300	12	1.184
0.100	20	1.941	0.065	97	1.787			
0.150	13	1.826	0.070	84	1.768	**P1 = 0.040**		
0.200	10	1.730	0.075	74	1.750			
0.300	7	1.565	0.080	65	1.734	0.080	204	1.579
			0.085	59	1.718	0.085	169	1.563
P1 = 0.010			0.090	53	1.702	0.090	142	1.548
			0.095	48	1.688	0.095	122	1.533
0.020	496	2.191	0.100	45	1.673	0.100	107	1.519
0.025	267	2.144	0.150	24	1.555	0.150	43	1.399
0.030	176	2.105	0.200	16	1.463	0.200	25	1.305
0.035	130	2.072	0.300	9	1.317	0.300	13	1.152
0.040	102	2.041						
0.045	83	2.015	**P1 = 0.025**			**P1 = 0.045**		
0.050	70	1.990						
0.055	61	1.967	0.050	287	1.803	0.090	186	1.519
0.060	54	1.946	0.055	215	1.780	0.095	157	1.505
0.065	48	1.926	0.060	169	1.759	0.100	134	1.490
0.070	43	1.908	0.065	138	1.739	0.150	49	1.371
0.075	39	1.891	0.070	115	1.720	0.200	28	1.276
0.080	36	1.874	0.075	99	1.703	0.300	14	1.122
0.085	33	1.858	0.080	86	1.686			
0.090	31	1.843	0.085	76	1.670	**P1 = 0.050**		
0.095	29	1.828	0.090	68	1.654			
0.100	27	1.814	0.095	61	1.640	0.100	171	1.465
0.150	17	1.695	0.100	56	1.625	0.150	57	1.345
0.200	12	1.604	0.150	28	1.507	0.200	31	1.250
0.300	8	1.443	0.200	18	1.414	0.300	15	1.096
			0.300	10	1.265			
P1 = 0.015						**P1 = 0.075**		
			P1 = 0.030					
0.030	396	2.026				0.150	119	1.240
0.035	253	1.992	0.060	253	1.719	0.200	51	1.145
0.040	181	1.962	0.065	198	1.699	0.300	20	0.991
0.045	139	1.935	0.070	160	1.680			
0.050	112	1.910	0.075	133	1.662	**P1 = 0.100**		
0.055	93	1.887	0.080	113	1.645			
0.060	79	1.866	0.085	98	1.629	0.200	88	1.064
0.065	69	1.846	0.090	86	1.614	0.300	27	0.910

TABLE 2.8.1. *SAMPLE SIZE REQUIREMENTS FOR ONE-SIDED SAMPLING PLANS.*

ALPHA = 0.050   BETA = 0.100

P2	N	K	P2	N	K	P2	N	K
	P1 = 0.0025			P1 = 0.0050			P1 = 0.0075	
0.0050	737	2.678	0.0100	548	2.436	0.0150	452	2.286
0.0075	269	2.598	0.0125	297	2.389	0.0175	290	2.252
0.0100	158	2.540	0.0150	197	2.350	0.0200	209	2.222
0.0125	112	2.493	0.0175	146	2.316	0.0225	161	2.195
0.0150	86	2.454	0.0200	115	2.286	0.0250	130	2.170
0.0175	70	2.421	0.0225	95	2.259			
0.0200	60	2.391				0.0275	109	2.147
0.0225	52	2.364	0.0250	80	2.235	0.0300	93	2.127
0.0250	46	2.340	0.0275	70	2.213	0.0325	81	2.107
0.0275	41	2.318	0.0300	62	2.192	0.0350	72	2.089
			0.0325	55	2.173	0.0375	65	2.072
0.0300	38	2.298	0.0350	50	2.155			
0.0325	35	2.278	0.0375	46	2.138	0.0400	59	2.056
0.0350	32	2.261				0.0425	54	2.041
0.0375	30	2.244	0.0400	42	2.122	0.0450	49	2.026
0.0400	28	2.228	0.0425	39	2.106	0.0475	46	2.012
0.0425	26	2.213	0.0450	37	2.092	0.0500	43	1.999
0.0450	25	2.199	0.0475	34	2.078			
0.0475	24	2.184	0.0500	32	2.065	0.0525	40	1.986
			0.0525	30	2.052	0.0550	38	1.973
0.0500	23	2.171				0.0575	35	1.962
0.0525	22	2.158	0.0550	29	2.040	0.0600	34	1.950
0.0550	21	2.146	0.0575	27	2.028	0.0625	32	1.939
0.0575	20	2.135	0.0600	26	2.017			
0.0600	19	2.124	0.0625	25	2.006	0.0650	30	1.929
0.0625	18	2.114	0.0650	24	1.995	0.0675	29	1.918
0.0650	18	2.102	0.0675	23	1.985	0.0700	28	1.908
0.0675	17	2.093				0.0725	26	1.899
			0.0700	22	1.975	0.0750	25	1.889
0.0700	17	2.081	0.0725	21	1.965	0.0775	24	1.880
0.0725	16	2.073	0.0750	21	1.954			
0.0750	16	2.061	0.0775	20	1.946	0.0800	24	1.870
0.0775	15	2.054	0.0800	19	1.937	0.0825	23	1.861
0.0800	15	2.044	0.0825	19	1.927	0.0850	22	1.853
0.0825	14	2.038				0.0875	21	1.845
0.0850	14	2.027	0.0850	18	1.920	0.0900	20	1.837
0.0875	14	2.017	0.0875	17	1.913	0.0925	20	1.828
			0.0900	17	1.903			
0.0900	13	2.013	0.0925	17	1.893	0.0950	19	1.821
0.0925	13	2.003	0.0950	16	1.887	0.0975	19	1.812
0.0950	13	1.993	0.0975	16	1.878	0.1000	18	1.805
0.0975	12	1.990				0.1100	16	1.777
0.1000	12	1.981	0.1000	15	1.873	0.1200	15	1.748
0.1100	11	1.952	0.1100	14	1.842			
0.1200	10	1.928	0.1200	13	1.814	0.1300	14	1.721
0.1300	10	1.895	0.1300	12	1.788	0.1400	13	1.696
0.1400	9	1.875	0.1400	11	1.765	0.1500	12	1.673
			0.1500	10	1.745	0.1600	11	1.653
0.1500	9	1.845	0.1600	10	1.717	0.1700	10	1.634
0.1600	8	1.831						
0.1700	8	1.802	0.1700	9	1.700	0.1800	10	1.608
0.1800	7	1.795	0.1800	9	1.674	0.1900	9	1.593
0.1900	7	1.768	0.1900	8	1.662	0.2000	9	1.569
0.2000	7	1.741	0.2000	8	1.636	0.2500	7	1.482
0.2500	6	1.644	0.2500	6	1.560	0.3000	6	1.396
0.3000	5	1.565	0.3000	5	1.483			

TABLE 2.8.2. SAMPLE SIZE REQUIREMENTS FOR ONE-SIDED SAMPLING PLANS.

ALPHA = 0.050   BETA = 0.100 (CONTINUED)

P2	N	K
P1 = 0.0100		
0.0200	390	2.174
0.0225	275	2.147
0.0250	209	2.122
0.0275	167	2.100
0.0300	138	2.079
0.0325	117	2.059
0.0350	101	2.041
0.0375	89	2.024
0.0400	79	2.008
0.0425	71	1.992
0.0450	65	1.978
0.0475	59	1.964
0.0500	55	1.950
0.0525	51	1.937
0.0550	47	1.925
0.0575	44	1.913
0.0600	42	1.901
0.0625	39	1.890
0.0650	37	1.880
0.0675	35	1.869
0.0700	34	1.859
0.0725	32	1.849
0.0750	31	1.839
0.0775	29	1.830
0.0800	28	1.821
0.0825	27	1.812
0.0850	26	1.804
0.0875	25	1.795
0.0900	24	1.787
0.0925	23	1.779
0.0950	22	1.771
0.0975	22	1.763
0.1000	21	1.756
0.1100	19	1.726
0.1200	17	1.699
0.1300	15	1.675
0.1400	14	1.649
0.1500	13	1.626
0.1600	12	1.604
0.1700	11	1.585
0.1800	11	1.559
0.1900	10	1.542
0.2000	9	1.528
0.2500	7	1.442
0.3000	6	1.357
P1 = 0.0125		
0.0250	345	2.084
0.0275	259	2.062
0.0300	205	2.041
0.0325	168	2.021
0.0350	141	2.003

P2	N	K
P1 = 0.0125		
0.0375	121	1.985
0.0400	106	1.969
0.0425	94	1.954
0.0450	84	1.939
0.0475	76	1.925
0.0500	69	1.911
0.0525	64	1.898
0.0550	59	1.886
0.0575	55	1.874
0.0600	51	1.863
0.0625	48	1.851
0.0650	45	1.841
0.0675	42	1.830
0.0700	40	1.820
0.0725	38	1.810
0.0750	36	1.801
0.0775	34	1.791
0.0800	33	1.782
0.0825	31	1.773
0.0850	30	1.764
0.0875	29	1.756
0.0900	28	1.747
0.0925	27	1.739
0.0950	26	1.731
0.0975	25	1.724
0.1000	24	1.716
0.1100	21	1.687
0.1200	19	1.659
0.1300	17	1.634
0.1400	16	1.608
0.1500	14	1.587
0.1600	13	1.565
0.1700	12	1.545
0.1800	12	1.520
0.1900	11	1.502
0.2000	10	1.485
0.2500	8	1.394
0.3000	6	1.325
P1 = 0.0150		
0.0300	311	2.009
0.0325	243	1.989
0.0350	198	1.971
0.0375	166	1.953
0.0400	142	1.937
0.0425	123	1.921
0.0450	109	1.907
0.0475	97	1.893
0.0500	87	1.879
0.0525	79	1.866
0.0550	73	1.854
0.0575	67	1.842
0.0600	62	1.830

P2	N	K
P1 = 0.0150		
0.0625	57	1.819
0.0650	54	1.808
0.0675	50	1.797
0.0700	47	1.787
0.0725	45	1.777
0.0750	42	1.768
0.0775	40	1.758
0.0800	38	1.749
0.0825	36	1.740
0.0850	35	1.731
0.0875	33	1.723
0.0900	32	1.715
0.0925	31	1.706
0.0950	29	1.699
0.0975	28	1.691
0.1000	27	1.683
0.1100	24	1.653
0.1200	21	1.626
0.1300	19	1.600
0.1400	17	1.576
0.1500	16	1.552
0.1600	15	1.529
0.1700	13	1.511
0.1800	13	1.487
0.1900	12	1.467
0.2000	11	1.450
0.2500	8	1.366
0.3000	7	1.278
P1 = 0.0175		
0.0350	283	1.943
0.0375	229	1.926
0.0400	191	1.909
0.0425	162	1.894
0.0450	140	1.879
0.0475	123	1.865
0.0500	110	1.851
0.0525	98	1.838
0.0550	89	1.826
0.0575	81	1.814
0.0600	75	1.802
0.0625	69	1.791
0.0650	64	1.780
0.0675	60	1.769
0.0700	56	1.759
0.0725	52	1.749
0.0750	49	1.739
0.0775	47	1.730
0.0800	44	1.721
0.0825	42	1.712
0.0850	40	1.703
0.0875	38	1.695
0.0900	36	1.686

TABLE 2.8.3. SAMPLE SIZE REQUIREMENTS FOR ONE-SIDED SAMPLING PLANS.

ALPHA = 0.050   BETA = 0.100 (CONTINUED)

P1 = 0.0175

P2	N	K
0.0925	35	1.678
0.0950	33	1.670
0.0975	32	1.662
0.1000	31	1.654
0.1100	27	1.625
0.1200	23	1.598
0.1300	21	1.571
0.1400	19	1.547
0.1500	17	1.524
0.1600	16	1.501
0.1700	15	1.479
0.1800	13	1.462
0.1900	13	1.438
0.2000	12	1.420
0.2500	9	1.331
0.3000	7	1.254

P1 = 0.0200

P2	N	K
0.0400	261	1.885
0.0425	216	1.869
0.0450	183	1.854
0.0475	158	1.840
0.0500	138	1.826
0.0525	122	1.813
0.0550	110	1.801
0.0575	99	1.789
0.0600	90	1.777
0.0625	83	1.766
0.0650	76	1.755
0.0675	70	1.744
0.0700	65	1.734
0.0725	61	1.724
0.0750	57	1.714
0.0775	54	1.705
0.0800	51	1.696
0.0825	48	1.687
0.0850	46	1.678
0.0875	43	1.669
0.0900	41	1.661
0.0925	39	1.653
0.0950	38	1.645
0.0975	36	1.637
0.1000	35	1.629
0.1100	30	1.599
0.1200	26	1.572
0.1300	23	1.546
0.1400	21	1.521
0.1500	19	1.498
0.1600	17	1.476
0.1700	16	1.454
0.1800	14	1.436

P1 = 0.0200

P2	N	K
0.1900	13	1.416
0.2000	13	1.393
0.2500	9	1.310
0.3000	7	1.233

P1 = 0.0225

P2	N	K
0.0450	241	1.832
0.0475	204	1.818
0.0500	175	1.804
0.0525	153	1.791
0.0550	135	1.779
0.0575	121	1.767
0.0600	109	1.755
0.0625	99	1.744
0.0650	90	1.733
0.0675	83	1.722
0.0700	77	1.712
0.0725	71	1.702
0.0750	66	1.692
0.0775	62	1.683
0.0800	58	1.673
0.0825	55	1.664
0.0850	52	1.656
0.0875	49	1.647
0.0900	47	1.638
0.0925	44	1.630
0.0950	42	1.622
0.0975	40	1.614
0.1000	39	1.607
0.1100	33	1.577
0.1200	28	1.550
0.1300	25	1.523
0.1400	22	1.499
0.1500	20	1.476
0.1600	18	1.454
0.1700	17	1.431
0.1800	15	1.412
0.1900	14	1.392
0.2000	13	1.373
0.2500	10	1.283
0.3000	8	1.201

P1 = 0.0250

P2	N	K
0.0500	225	1.784
0.0525	193	1.771
0.0550	168	1.759
0.0575	148	1.747
0.0600	132	1.735
0.0625	119	1.723
0.0650	108	1.713
0.0675	98	1.702

P1 = 0.0250

P2	N	K
0.0700	90	1.692
0.0725	83	1.682
0.0750	77	1.672
0.0775	72	1.662
0.0800	67	1.653
0.0825	63	1.644
0.0850	59	1.635
0.0875	56	1.627
0.0900	53	1.618
0.0925	50	1.610
0.0950	48	1.602
0.0975	45	1.594
0.1000	43	1.586
0.1100	36	1.557
0.1200	31	1.529
0.1300	27	1.503
0.1400	24	1.478
0.1500	22	1.454
0.1600	20	1.432
0.1700	18	1.411
0.1800	16	1.391
0.1900	15	1.371
0.2000	14	1.352
0.2500	10	1.265
0.3000	8	1.184

P1 = 0.0275

P2	N	K
0.0550	211	1.740
0.0575	183	1.728
0.0600	161	1.716
0.0625	143	1.705
0.0650	129	1.694
0.0675	116	1.683
0.0700	106	1.673
0.0725	97	1.663
0.0750	89	1.653
0.0775	83	1.644
0.0800	77	1.634
0.0825	72	1.625
0.0850	67	1.616
0.0875	63	1.608
0.0900	60	1.599
0.0925	56	1.591
0.0950	53	1.583
0.0975	51	1.575
0.1000	48	1.567
0.1100	40	1.538
0.1200	34	1.510
0.1300	30	1.484
0.1400	26	1.459
0.1500	23	1.436
0.1600	21	1.413

TABLE 2.8.4. SAMPLE SIZE REQUIREMENTS FOR ONE-SIDED SAMPLING PLANS.

ALPHA = 0.050   BETA = 0.100 (CONTINUED)

P2	N	K	P2	N	K	P2	N	K
P1 = 0.0275			**P1 = 0.0325**			**P1 = 0.0375**		
0.1700	19	1.392	0.0950	67	1.549	0.0950	85	1.520
0.1800	18	1.370	0.0975	63	1.541	0.0975	79	1.512
0.1900	16	1.351	0.1000	60	1.534	0.1000	75	1.504
0.2000	15	1.332	0.1100	49	1.504	0.1100	59	1.474
0.2500	11	1.242	0.1200	41	1.476	0.1200	49	1.446
0.3000	8	1.168	0.1300	35	1.450	0.1300	41	1.420
			0.1400	31	1.425	0.1400	36	1.395
P1 = 0.0300			0.1500	27	1.401	0.1500	31	1.372
0.0600	198	1.699	0.1600	24	1.379	0.1600	27	1.349
0.0625	174	1.688	0.1700	22	1.357	0.1700	25	1.327
0.0650	155	1.677	0.1800	20	1.336	0.1800	22	1.307
0.0675	139	1.666	0.1900	18	1.317	0.1900	20	1.287
0.0700	125	1.656	0.2000	17	1.297	0.2000	18	1.268
0.0725	114	1.646	0.2500	12	1.208	0.2500	13	1.178
0.0750	104	1.636	0.3000	9	1.129	0.3000	10	1.097
0.0775	96	1.626						
0.0800	89	1.617	**P1 = 0.0350**			**P1 = 0.0400**		
0.0825	82	1.608	0.0700	177	1.625	0.0800	160	1.558
0.0850	77	1.599	0.0725	158	1.615	0.0825	145	1.549
0.0875	72	1.590	0.0750	143	1.605	0.0850	132	1.540
0.0900	67	1.582	0.0775	129	1.595	0.0875	121	1.532
0.0925	63	1.574	0.0800	118	1.586	0.0900	111	1.523
0.0950	60	1.566	0.0825	108	1.577	0.0925	103	1.515
0.0975	57	1.558	0.0850	100	1.568	0.0950	96	1.507
0.1000	54	1.550	0.0875	93	1.559	0.0975	89	1.499
0.1100	44	1.520	0.0900	86	1.551	0.1000	83	1.491
0.1200	37	1.492	0.0925	80	1.542	0.1100	66	1.461
0.1300	32	1.466	0.0950	75	1.534			
			0.0975	71	1.526	0.1200	53	1.433
0.1400	28	1.442				0.1300	45	1.406
0.1500	25	1.418	0.1000	67	1.518	0.1400	38	1.381
0.1600	23	1.395	0.1100	54	1.489	0.1500	33	1.358
0.1700	20	1.374	0.1200	45	1.461	0.1600	29	1.335
0.1800	19	1.353	0.1300	38	1.434	0.1700	26	1.314
0.1900	17	1.334	0.1400	33	1.410	0.1800	24	1.293
0.2000	16	1.314	0.1500	29	1.386	0.1900	21	1.273
0.2500	11	1.227	0.1600	26	1.363	0.2000	19	1.254
0.3000	9	1.144	0.1700	23	1.342	0.2500	13	1.166
			0.1800	21	1.321	0.3000	10	1.085
P1 = 0.0325			0.1900	19	1.301			
			0.2000	18	1.281	**P1 = 0.0425**		
0.0650	187	1.661	0.2500	12	1.194			
0.0675	166	1.650	0.3000	9	1.116	0.0850	153	1.528
0.0700	148	1.640				0.0875	139	1.519
0.0725	134	1.630	**P1 = 0.0375**			0.0900	127	1.510
0.0750	122	1.620				0.0925	117	1.502
0.0775	111	1.610	0.0750	168	1.591	0.0950	108	1.494
			0.0775	151	1.581	0.0975	100	1.486
0.0800	102	1.601	0.0800	137	1.572	0.1000	93	1.478
0.0825	94	1.592	0.0825	125	1.563	0.1100	72	1.448
0.0850	87	1.583	0.0850	115	1.554	0.1200	59	1.420
0.0875	81	1.574	0.0875	106	1.545	0.1300	49	1.393
0.0900	76	1.566	0.0900	98	1.537	0.1400	41	1.368
0.0925	71	1.558	0.0925	91	1.528	0.1500	36	1.345

TABLE 2.8.5. SAMPLE SIZE REQUIREMENTS FOR ONE-SIDED SAMPLING PLANS.

ALPHA = 0.050   BETA = 0.100 (CONTINUED)

P2	N	K	P2	N	K	P2	N	K
P1 = 0.0425			**P1 = 0.0500**			**P1 = 0.0700**		
0.1600	31	1.322	0.1600	38	1.286	0.1800	47	1.166
0.1700	28	1.300	0.1700	33	1.265	0.1900	41	1.146
0.1800	25	1.280	0.1800	30	1.244	0.2000	36	1.126
0.1900	23	1.259	0.1900	26	1.224	0.2500	21	1.037
0.2000	20	1.241	0.2000	24	1.204	0.3000	15	0.955
0.2500	14	1.151	0.2500	16	1.114			
0.3000	10	1.073	0.3000	11	1.037	**P1 = 0.0750**		
P1 = 0.0450			**P1 = 0.0550**			0.1500	93	1.216
						0.1600	75	1.193
0.0900	146	1.498	0.1100	123	1.392	0.1700	62	1.171
0.0925	133	1.490	0.1200	93	1.363	0.1800	53	1.150
0.0950	123	1.482	0.1300	74	1.337	0.1900	45	1.129
0.0975	113	1.474	0.1400	60	1.312	0.2000	40	1.110
0.1000	105	1.466	0.1500	51	1.288	0.2500	23	1.019
0.1100	80	1.436	0.1600	43	1.265	0.3000	16	0.938
0.1200	64	1.407	0.1700	38	1.243			
0.1300	53	1.381	0.1800	33	1.222	**P1 = 0.0800**		
0.1400	45	1.356	0.1900	29	1.202			
			0.2000	26	1.183	0.1600	87	1.177
0.1500	38	1.332	0.2500	17	1.093	0.1700	71	1.155
0.1600	33	1.310	0.3000	12	1.014	0.1800	60	1.134
0.1700	30	1.288				0.1900	51	1.114
0.1800	26	1.267	**P1 = 0.0600**			0.2000	44	1.094
0.1900	24	1.247				0.2500	25	1.003
0.2000	22	1.227	0.1200	114	1.344	0.3000	17	0.922
0.2500	14	1.140	0.1300	88	1.317			
0.3000	11	1.057	0.1400	71	1.292	**P1 = 0.0850**		
			0.1500	59	1.268			
P1 = 0.0475			0.1600	49	1.245	0.1700	82	1.140
			0.1700	42	1.223	0.1800	68	1.119
0.0950	139	1.470	0.1800	37	1.202	0.1900	57	1.099
0.0975	128	1.462	0.1900	33	1.182	0.2000	49	1.079
0.1000	118	1.454	0.2000	29	1.163	0.2500	27	0.988
0.1100	89	1.424	0.2500	18	1.073	0.3000	18	0.907
0.1200	70	1.396	0.3000	13	0.993			
0.1300	57	1.369				**P1 = 0.0900**		
0.1400	48	1.344	**P1 = 0.0650**					
0.1500	41	1.321				0.1800	77	1.105
			0.1300	106	1.299	0.1900	64	1.084
0.1600	36	1.298	0.1400	83	1.274	0.2000	55	1.065
0.1700	31	1.276	0.1500	68	1.250	0.2500	29	0.974
0.1800	28	1.255	0.1600	57	1.227	0.3000	19	0.893
0.1900	25	1.235	0.1700	48	1.205			
0.2000	23	1.216	0.1800	42	1.184	**P1 = 0.0950**		
0.2500	15	1.127	0.1900	36	1.164			
0.3000	11	1.047	0.2000	32	1.144	0.1900	73	1.071
			0.2500	20	1.054	0.2000	61	1.051
P1 = 0.0500			0.3000	14	0.973	0.2500	32	0.960
						0.3000	20	0.879
0.1000	134	1.443	**P1 = 0.0700**					
0.1100	99	1.413				**P1 = 0.1000**		
0.1200	77	1.384	0.1400	99	1.256			
0.1300	62	1.358	0.1500	79	1.232	0.2000	69	1.038
0.1400	52	1.333	0.1600	65	1.209	0.2500	34	0.947
0.1500	44	1.309	0.1700	55	1.187	0.3000	21	0.866

TABLE 2.9. SAMPLE SIZE REQUIREMENTS FOR ONE-SIDED SAMPLING PLANS.

ALPHA = 0.050   BETA = 0.200

P1 = 0.005

P2	N	K
0.010	392	2.412
0.015	140	2.311
0.020	82	2.237
0.025	57	2.178
0.030	44	2.128
0.035	35	2.085
0.040	30	2.047
0.045	26	2.013
0.050	23	1.982
0.055	20	1.953
0.060	18	1.927
0.065	17	1.902
0.070	16	1.879
0.075	14	1.857
0.080	13	1.836
0.085	13	1.817
0.090	12	1.798
0.095	11	1.780
0.100	11	1.762
0.150	7	1.619
0.200	6	1.499
0.300	4	1.309

P1 = 0.010

P2	N	K
0.020	278	2.148
0.025	149	2.087
0.030	98	2.037
0.035	72	1.993
0.040	56	1.954
0.045	46	1.919
0.050	39	1.888
0.055	33	1.858
0.060	29	1.831
0.065	26	1.806
0.070	24	1.783
0.075	21	1.760
0.080	20	1.740
0.085	18	1.719
0.090	17	1.700
0.095	16	1.682
0.100	15	1.664
0.150	9	1.518
0.200	7	1.400
0.300	4	1.223

P1 = 0.015

P2	N	K
0.030	221	1.981
0.035	141	1.936
0.040	100	1.897
0.045	77	1.862
0.050	62	1.830
0.055	51	1.800
0.060	44	1.773
0.065	38	1.748
0.070	33	1.723
0.075	30	1.701
0.080	27	1.680
0.085	24	1.659
0.090	22	1.640
0.095	21	1.622
0.100	19	1.603
0.150	11	1.455
0.200	8	1.338
0.300	5	1.148

P1 = 0.020

P2	N	K
0.040	185	1.856
0.045	130	1.820
0.050	98	1.788
0.055	78	1.758
0.060	64	1.730
0.065	54	1.705
0.070	46	1.680
0.075	40	1.657
0.080	36	1.636
0.085	32	1.615
0.090	29	1.596
0.095	26	1.577
0.100	24	1.559
0.150	13	1.409
0.200	9	1.292
0.300	5	1.105

P1 = 0.025

P2	N	K
0.050	160	1.754
0.055	119	1.724
0.060	94	1.696
0.065	76	1.670
0.070	64	1.646
0.075	54	1.623
0.080	47	1.601
0.085	42	1.581
0.090	37	1.561
0.095	33	1.542
0.100	30	1.524
0.150	15	1.372
0.200	10	1.255
0.300	6	1.064

P1 = 0.030

P2	N	K
0.060	141	1.668
0.065	110	1.642
0.070	89	1.617
0.075	74	1.594
0.080	63	1.572
0.085	54	1.551
0.090	47	1.531
0.095	42	1.512
0.100	38	1.494
0.150	18	1.342
0.200	11	1.223
0.300	6	1.033

P1 = 0.035

P2	N	K
0.070	126	1.593
0.075	101	1.569
0.080	84	1.547
0.085	71	1.526
0.090	61	1.506
0.095	53	1.487
0.100	47	1.469
0.150	20	1.315
0.200	12	1.196
0.300	6	1.006

P1 = 0.040

P2	N	K
0.080	114	1.525
0.085	94	1.504
0.090	79	1.484
0.095	68	1.465
0.100	59	1.446
0.150	23	1.292
0.200	14	1.173
0.300	7	0.980

P1 = 0.045

P2	N	K
0.090	103	1.464
0.095	87	1.445
0.100	74	1.426
0.150	27	1.272
0.200	15	1.151
0.300	7	0.958

P1 = 0.050

P2	N	K
0.100	95	1.408
0.150	31	1.253
0.200	17	1.132
0.300	8	0.938

P1 = 0.075

P2	N	K
0.150	66	1.177
0.200	28	1.054
0.300	11	0.857

P1 = 0.100

P2	N	K
0.200	49	0.996
0.300	15	0.796

TABLE 3

Two-sided tolerance limit factors for a normal distribution (control center) for values of P = 0.75, 0.90, 0.95, 0.975, 0.99, 0.995, 0.999 and N = 2(1)100(2)180(5)300(10)400(25)650(50)1000, 1500, 2000, 3000, 5000, 10000, ∞.

TABLE 3.1.1. *TWO-SIDED TOLERANCE LIMIT FACTORS (CONTROL CENTER).*

GAMMA = 0.995

N \P →	0.750	0.900	0.950	0.975	0.990	0.995	0.999
2	223.997	311.145	365.448	413.370	469.765	508.471	588.832
3	19.027	26.599	31.340	35.534	40.480	43.879	50.946
4	8.491	11.919	14.072	15.979	18.231	19.780	23.004
5	5.666	7.978	9.433	10.724	12.249	13.298	15.483
6	4.428	6.251	7.400	8.419	9.625	10.455	12.183
7	3.744	5.296	6.275	7.145	8.173	8.882	10.358
8	3.312	4.692	5.564	6.339	7.255	7.886	9.203
9	3.014	4.276	5.074	5.783	6.622	7.200	8.406
10	2.797	3.972	4.715	5.376	6.158	6.697	7.822
11	2.630	3.739	4.440	5.064	5.803	6.313	7.375
12	2.499	3.554	4.223	4.818	5.522	6.008	7.022
13	2.392	3.405	4.046	4.618	5.294	5.761	6.734
14	2.304	3.281	3.900	4.451	5.105	5.556	6.496
15	2.229	3.176	3.777	4.311	4.945	5.382	6.294
16	2.166	3.087	3.671	4.191	4.808	5.234	6.122
17	2.110	3.009	3.579	4.087	4.689	5.105	5.972
18	2.062	2.941	3.499	3.996	4.585	4.992	5.841
19	2.019	2.881	3.428	3.915	4.493	4.892	5.725
20	1.981	2.827	3.364	3.843	4.411	4.803	5.621
21	1.947	2.779	3.308	3.779	4.337	4.723	5.528
22	1.917	2.736	3.256	3.720	4.271	4.651	5.444
23	1.889	2.696	3.210	3.667	4.210	4.585	5.367
24	1.863	2.661	3.167	3.619	4.155	4.525	5.298
25	1.840	2.628	3.128	3.575	4.104	4.470	5.234
26	1.819	2.597	3.092	3.534	4.058	4.419	5.175
27	1.799	2.569	3.059	3.496	4.014	4.372	5.120
28	1.781	2.543	3.028	3.461	3.974	4.329	5.070
29	1.764	2.519	3.000	3.428	3.937	4.289	5.022
30	1.748	2.497	2.973	3.398	3.902	4.251	4.978
31	1.733	2.476	2.948	3.370	3.870	4.215	4.937
32	1.719	2.456	2.925	3.343	3.839	4.182	4.899
33	1.706	2.437	2.903	3.318	3.811	4.151	4.862
34	1.694	2.420	2.882	3.294	3.784	4.122	4.828
35	1.682	2.403	2.862	3.272	3.758	4.094	4.796
36	1.671	2.388	2.844	3.251	3.734	4.068	4.765
37	1.660	2.373	2.826	3.231	3.711	4.043	4.736
38	1.651	2.359	2.810	3.212	3.690	4.019	4.709
39	1.641	2.346	2.794	3.194	3.669	3.997	4.683
40	1.632	2.333	2.779	3.177	3.649	3.976	4.658
41	1.624	2.321	2.764	3.160	3.631	3.955	4.634
42	1.616	2.309	2.751	3.145	3.613	3.936	4.611
43	1.608	2.298	2.738	3.130	3.596	3.917	4.590
44	1.601	2.288	2.725	3.116	3.579	3.900	4.569
45	1.593	2.278	2.713	3.102	3.564	3.883	4.549
46	1.587	2.268	2.702	3.089	3.549	3.866	4.530
47	1.580	2.259	2.691	3.076	3.534	3.851	4.512
48	1.574	2.250	2.680	3.064	3.520	3.835	4.494
49	1.568	2.241	2.670	3.052	3.507	3.821	4.477
50	1.562	2.233	2.660	3.041	3.494	3.807	4.461

TABLE 3.1.2. TWO-SIDED TOLERANCE LIMIT FACTORS (CONTROL CENTER).

GAMMA = 0.995 (CONTINUED)

N \P →	0.750	0.900	0.950	0.975	0.990	0.995	0.999
51	1.556	2.225	2.650	3.030	3.482	3.794	4.445
52	1.551	2.217	2.641	3.020	3.470	3.781	4.430
53	1.546	2.210	2.633	3.010	3.458	3.768	4.416
54	1.541	2.203	2.624	3.000	3.447	3.756	4.401
55	1.536	2.196	2.616	2.991	3.436	3.744	4.388
56	1.531	2.189	2.608	2.982	3.426	3.733	4.375
57	1.527	2.183	2.600	2.973	3.416	3.722	4.362
58	1.522	2.176	2.593	2.965	3.406	3.711	4.349
59	1.518	2.170	2.586	2.956	3.397	3.701	4.337
60	1.514	2.164	2.579	2.948	3.388	3.691	4.326
61	1.510	2.159	2.572	2.941	3.379	3.682	4.315
62	1.506	2.153	2.565	2.933	3.370	3.672	4.304
63	1.502	2.148	2.559	2.926	3.362	3.663	4.293
64	1.499	2.142	2.553	2.919	3.354	3.654	4.283
65	1.495	2.137	2.547	2.912	3.346	3.646	4.273
66	1.492	2.132	2.541	2.905	3.338	3.637	4.263
67	1.488	2.128	2.535	2.899	3.331	3.629	4.253
68	1.485	2.123	2.529	2.892	3.323	3.621	4.244
69	1.482	2.118	2.524	2.886	3.316	3.614	4.235
70	1.479	2.114	2.519	2.880	3.309	3.606	4.226
71	1.476	2.110	2.514	2.874	3.303	3.599	4.218
72	1.473	2.105	2.509	2.868	3.296	3.592	4.209
73	1.470	2.101	2.504	2.863	3.290	3.585	4.201
74	1.467	2.097	2.499	2.857	3.283	3.578	4.193
75	1.464	2.093	2.494	2.852	3.277	3.571	4.185
76	1.462	2.090	2.490	2.847	3.271	3.565	4.178
77	1.459	2.086	2.485	2.842	3.266	3.558	4.170
78	1.456	2.082	2.481	2.837	3.260	3.552	4.163
79	1.454	2.079	2.477	2.832	3.254	3.546	4.156
80	1.451	2.075	2.473	2.827	3.249	3.540	4.149
81	1.449	2.072	2.468	2.823	3.244	3.534	4.142
82	1.447	2.068	2.464	2.818	3.238	3.529	4.136
83	1.444	2.065	2.461	2.814	3.233	3.523	4.129
84	1.442	2.062	2.457	2.809	3.228	3.518	4.123
85	1.440	2.059	2.453	2.805	3.223	3.512	4.117
86	1.438	2.056	2.449	2.801	3.219	3.507	4.111
87	1.436	2.053	2.446	2.797	3.214	3.502	4.105
88	1.434	2.050	2.442	2.793	3.209	3.497	4.099
89	1.432	2.047	2.439	2.789	3.205	3.492	4.093
90	1.430	2.044	2.436	2.785	3.200	3.487	4.088
91	1.428	2.041	2.432	2.781	3.196	3.483	4.082
92	1.426	2.039	2.429	2.778	3.192	3.478	4.077
93	1.424	2.036	2.426	2.774	3.188	3.474	4.071
94	1.422	2.033	2.423	2.770	3.184	3.469	4.066
95	1.420	2.031	2.420	2.767	3.180	3.465	4.061
96	1.419	2.028	2.417	2.764	3.176	3.460	4.056
97	1.417	2.026	2.414	2.760	3.172	3.456	4.051
98	1.415	2.023	2.411	2.757	3.168	3.452	4.046
99	1.413	2.021	2.408	2.754	3.164	3.448	4.042

TABLE 3.1.3. TWO-SIDED TOLERANCE LIMIT FACTORS (CONTROL CENTER).

GAMMA = 0.995 (CONTINUED)

N \P →	0.750	0.900	0.950	0.975	0.990	0.995	0.999
100	1.412	2.019	2.405	2.750	3.161	3.444	4.037
102	1.409	2.014	2.400	2.744	3.153	3.436	4.028
104	1.406	2.010	2.394	2.738	3.147	3.429	4.019
106	1.403	2.005	2.389	2.732	3.140	3.422	4.010
108	1.400	2.001	2.384	2.727	3.133	3.414	4.002
110	1.397	1.997	2.380	2.721	3.127	3.408	3.994
112	1.394	1.993	2.375	2.716	3.121	3.401	3.987
114	1.391	1.989	2.371	2.711	3.115	3.395	3.979
116	1.389	1.986	2.366	2.706	3.109	3.388	3.972
118	1.386	1.982	2.362	2.701	3.104	3.382	3.965
120	1.384	1.979	2.358	2.696	3.099	3.377	3.958
122	1.382	1.975	2.354	2.692	3.093	3.371	3.951
124	1.379	1.972	2.350	2.687	3.088	3.365	3.945
126	1.377	1.969	2.346	2.683	3.083	3.360	3.938
128	1.375	1.966	2.343	2.679	3.078	3.355	3.932
130	1.373	1.963	2.339	2.675	3.074	3.350	3.926
132	1.371	1.960	2.335	2.671	3.069	3.345	3.920
134	1.369	1.957	2.332	2.667	3.065	3.340	3.915
136	1.367	1.954	2.329	2.663	3.060	3.335	3.909
138	1.365	1.952	2.326	2.659	3.056	3.330	3.904
140	1.363	1.949	2.322	2.656	3.052	3.326	3.899
142	1.361	1.947	2.319	2.652	3.048	3.321	3.893
144	1.360	1.944	2.316	2.649	3.044	3.317	3.888
146	1.358	1.942	2.313	2.646	3.040	3.313	3.883
148	1.356	1.939	2.311	2.642	3.036	3.309	3.879
150	1.355	1.937	2.308	2.639	3.033	3.305	3.874
152	1.353	1.934	2.305	2.636	3.029	3.301	3.869
154	1.351	1.932	2.302	2.633	3.026	3.297	3.865
156	1.350	1.930	2.300	2.630	3.022	3.293	3.861
158	1.348	1.928	2.297	2.627	3.019	3.290	3.856
160	1.347	1.926	2.295	2.624	3.016	3.286	3.852
162	1.345	1.924	2.292	2.621	3.012	3.283	3.848
164	1.344	1.922	2.290	2.619	3.009	3.279	3.844
166	1.343	1.920	2.287	2.616	3.006	3.276	3.840
168	1.341	1.918	2.285	2.613	3.003	3.273	3.836
170	1.340	1.916	2.283	2.611	3.000	3.269	3.832
172	1.339	1.914	2.281	2.608	2.997	3.266	3.829
174	1.337	1.912	2.278	2.606	2.994	3.263	3.825
176	1.336	1.910	2.276	2.603	2.992	3.260	3.821
178	1.335	1.909	2.274	2.601	2.989	3.257	3.818
180	1.334	1.907	2.272	2.598	2.986	3.254	3.814
185	1.331	1.903	2.267	2.593	2.980	3.247	3.806
190	1.328	1.899	2.262	2.587	2.973	3.240	3.798
195	1.325	1.895	2.258	2.582	2.967	3.234	3.791
200	1.323	1.891	2.254	2.577	2.962	3.227	3.783
205	1.320	1.888	2.249	2.572	2.956	3.221	3.776
210	1.318	1.884	2.245	2.568	2.951	3.216	3.769
215	1.316	1.881	2.242	2.563	2.946	3.210	3.763
220	1.313	1.878	2.238	2.559	2.941	3.205	3.757

TABLE 3.1.4. TWO-SIDED TOLERANCE LIMIT FACTORS (CONTROL CENTER).

GAMMA = 0.995 (CONTINUED)

N \P →	0.750	0.900	0.950	0.975	0.990	0.995	0.999
225	1.311	1.875	2.234	2.555	2.936	3.200	3.751
230	1.309	1.872	2.231	2.551	2.932	3.195	3.745
235	1.307	1.869	2.227	2.547	2.927	3.190	3.739
240	1.305	1.867	2.224	2.544	2.923	3.185	3.734
245	1.304	1.864	2.221	2.540	2.919	3.181	3.729
250	1.302	1.862	2.218	2.537	2.915	3.177	3.724
255	1.300	1.859	2.215	2.533	2.911	3.173	3.719
260	1.299	1.857	2.212	2.530	2.908	3.169	3.714
265	1.297	1.854	2.210	2.527	2.904	3.165	3.710
270	1.295	1.852	2.207	2.524	2.901	3.161	3.705
275	1.294	1.850	2.204	2.521	2.897	3.157	3.701
280	1.292	1.848	2.202	2.518	2.894	3.154	3.697
285	1.291	1.846	2.200	2.515	2.891	3.150	3.693
290	1.290	1.844	2.197	2.513	2.888	3.147	3.689
295	1.288	1.842	2.195	2.510	2.885	3.144	3.685
300	1.287	1.840	2.193	2.508	2.882	3.140	3.681
310	1.284	1.837	2.188	2.503	2.876	3.134	3.674
320	1.282	1.833	2.184	2.498	2.871	3.128	3.667
330	1.280	1.830	2.181	2.494	2.866	3.123	3.661
340	1.278	1.827	2.177	2.489	2.861	3.118	3.655
350	1.276	1.824	2.173	2.485	2.856	3.113	3.649
360	1.274	1.821	2.170	2.482	2.852	3.108	3.643
370	1.272	1.819	2.167	2.478	2.848	3.103	3.638
380	1.270	1.816	2.164	2.475	2.844	3.099	3.633
390	1.268	1.814	2.161	2.471	2.840	3.095	3.628
400	1.267	1.811	2.158	2.468	2.836	3.091	3.623
425	1.263	1.806	2.152	2.461	2.828	3.081	3.612
450	1.259	1.801	2.146	2.454	2.820	3.073	3.602
475	1.256	1.796	2.140	2.447	2.813	3.065	3.593
500	1.253	1.792	2.135	2.442	2.806	3.058	3.585
525	1.250	1.788	2.131	2.436	2.800	3.051	3.577
550	1.248	1.784	2.126	2.432	2.794	3.045	3.570
575	1.246	1.781	2.122	2.427	2.789	3.039	3.563
600	1.243	1.778	2.119	2.423	2.784	3.034	3.557
625	1.241	1.775	2.115	2.419	2.780	3.029	3.551
650	1.239	1.772	2.112	2.415	2.775	3.024	3.545
700	1.236	1.767	2.106	2.408	2.768	3.016	3.535
750	1.233	1.763	2.100	2.402	2.761	3.008	3.526
800	1.230	1.759	2.096	2.397	2.754	3.001	3.518
850	1.227	1.755	2.091	2.392	2.748	2.995	3.511
900	1.225	1.752	2.087	2.387	2.743	2.990	3.504
950	1.223	1.749	2.084	2.383	2.739	2.984	3.498
1000	1.221	1.746	2.080	2.379	2.734	2.980	3.493
1500	1.207	1.726	2.057	2.352	2.703	2.946	3.454
2000	1.199	1.715	2.043	2.337	2.686	2.927	3.431
3000	1.190	1.702	2.028	2.319	2.665	2.904	3.404
5000	1.181	1.688	2.012	2.301	2.644	2.881	3.378
10000	1.172	1.675	1.996	2.283	2.624	2.859	3.352
∞	1.150	1.645	1.960	2.241	2.576	2.807	3.291

TABLE 3.2.1. *TWO-SIDED TOLERANCE LIMIT FACTORS (CONTROL CENTER).*

GAMMA = 0.990

N \P →	0.750	0.900	0.950	0.975	0.990	0.995	0.999
2	111.996	155.569	182.720	206.681	234.877	254.230	294.410
3	13.435	18.782	22.131	25.093	28.586	30.986	35.977
4	6.706	9.416	11.118	12.626	14.405	15.630	18.177
5	4.724	6.655	7.870	8.947	10.220	11.096	12.920
6	3.812	5.383	6.373	7.252	8.292	9.007	10.497
7	3.291	4.658	5.520	6.285	7.191	7.814	9.114
8	2.955	4.189	4.968	5.660	6.479	7.043	8.220
9	2.720	3.860	4.581	5.222	5.980	6.503	7.593
10	2.546	3.617	4.294	4.897	5.610	6.102	7.127
11	2.411	3.429	4.073	4.646	5.324	5.792	6.768
12	2.304	3.279	3.896	4.445	5.096	5.545	6.481
13	2.216	3.156	3.751	4.281	4.909	5.342	6.246
14	2.144	3.054	3.631	4.145	4.753	5.173	6.050
15	2.082	2.967	3.529	4.029	4.621	5.030	5.883
16	2.029	2.893	3.441	3.929	4.507	4.907	5.740
17	1.983	2.828	3.364	3.842	4.408	4.799	5.615
18	1.942	2.771	3.297	3.765	4.321	4.705	5.505
19	1.907	2.720	3.237	3.698	4.244	4.621	5.408
20	1.875	2.675	3.184	3.637	4.175	4.546	5.321
21	1.846	2.635	3.136	3.583	4.113	4.478	5.242
22	1.820	2.598	3.092	3.533	4.056	4.417	5.171
23	1.796	2.564	3.053	3.488	4.005	4.362	5.106
24	1.775	2.534	3.017	3.447	3.958	4.311	5.047
25	1.755	2.506	2.984	3.409	3.915	4.264	4.993
26	1.736	2.480	2.953	3.375	3.875	4.221	4.942
27	1.720	2.456	2.925	3.342	3.838	4.181	4.896
28	1.704	2.434	2.898	3.313	3.804	4.144	4.853
29	1.689	2.413	2.874	3.285	3.772	4.109	4.812
30	1.676	2.394	2.851	3.259	3.742	4.077	4.775
31	1.663	2.376	2.829	3.234	3.715	4.046	4.739
32	1.651	2.359	2.809	3.211	3.688	4.018	4.706
33	1.640	2.343	2.790	3.190	3.664	3.991	4.675
34	1.629	2.328	2.773	3.169	3.640	3.966	4.646
35	1.619	2.314	2.756	3.150	3.618	3.942	4.618
36	1.610	2.300	2.740	3.132	3.598	3.919	4.592
37	1.601	2.287	2.725	3.115	3.578	3.898	4.567
38	1.592	2.275	2.710	3.098	3.559	3.878	4.543
39	1.584	2.264	2.697	3.083	3.541	3.858	4.520
40	1.576	2.253	2.684	3.068	3.524	3.840	4.499
41	1.569	2.242	2.671	3.054	3.508	3.822	4.478
42	1.562	2.232	2.659	3.040	3.493	3.805	4.459
43	1.555	2.223	2.648	3.027	3.478	3.789	4.440
44	1.549	2.214	2.637	3.015	3.464	3.774	4.422
45	1.543	2.205	2.627	3.003	3.450	3.759	4.405
46	1.537	2.196	2.617	2.992	3.437	3.745	4.388
47	1.531	2.188	2.607	2.981	3.425	3.731	4.372
48	1.526	2.181	2.598	2.970	3.412	3.718	4.357
49	1.520	2.173	2.589	2.960	3.401	3.705	4.342
50	1.515	2.166	2.580	2.950	3.390	3.693	4.328

TABLE 3.2.2. *TWO-SIDED TOLERANCE LIMIT FACTORS (CONTROL CENTER).*

GAMMA = 0.990 (CONTINUED)

N \P →	0.750	0.900	0.950	0.975	0.990	0.995	0.999
51	1.510	2.159	2.572	2.941	3.379	3.682	4.314
52	1.506	2.152	2.564	2.932	3.369	3.670	4.301
53	1.501	2.146	2.557	2.923	3.359	3.659	4.288
54	1.497	2.140	2.549	2.915	3.349	3.649	4.276
55	1.493	2.134	2.542	2.907	3.339	3.639	4.264
56	1.488	2.128	2.535	2.899	3.330	3.629	4.253
57	1.484	2.122	2.528	2.891	3.322	3.619	4.241
58	1.481	2.117	2.522	2.884	3.313	3.610	4.231
59	1.477	2.111	2.516	2.876	3.305	3.601	4.220
60	1.473	2.106	2.509	2.869	3.297	3.592	4.210
61	1.470	2.101	2.503	2.863	3.289	3.584	4.200
62	1.466	2.096	2.498	2.856	3.282	3.576	4.191
63	1.463	2.092	2.492	2.850	3.274	3.568	4.181
64	1.460	2.087	2.487	2.843	3.267	3.560	4.172
65	1.457	2.083	2.481	2.837	3.260	3.553	4.164
66	1.454	2.078	2.476	2.832	3.254	3.545	4.155
67	1.451	2.074	2.471	2.826	3.247	3.538	4.147
68	1.448	2.070	2.466	2.820	3.241	3.531	4.139
69	1.445	2.066	2.462	2.815	3.234	3.524	4.131
70	1.442	2.062	2.457	2.810	3.228	3.518	4.123
71	1.440	2.058	2.453	2.804	3.222	3.511	4.115
72	1.437	2.055	2.448	2.799	3.217	3.505	4.108
73	1.435	2.051	2.444	2.795	3.211	3.499	4.101
74	1.432	2.048	2.440	2.790	3.206	3.493	4.094
75	1.430	2.044	2.436	2.785	3.200	3.487	4.087
76	1.427	2.041	2.432	2.781	3.195	3.482	4.081
77	1.425	2.038	2.428	2.776	3.190	3.476	4.074
78	1.423	2.034	2.424	2.772	3.185	3.471	4.068
79	1.421	2.031	2.420	2.767	3.180	3.465	4.061
80	1.419	2.028	2.416	2.763	3.175	3.460	4.055
81	1.416	2.025	2.413	2.759	3.171	3.455	4.049
82	1.414	2.022	2.409	2.755	3.166	3.450	4.044
83	1.412	2.019	2.406	2.751	3.162	3.445	4.038
84	1.410	2.017	2.403	2.747	3.157	3.440	4.032
85	1.408	2.014	2.399	2.744	3.153	3.436	4.027
86	1.407	2.011	2.396	2.740	3.149	3.431	4.022
87	1.405	2.008	2.393	2.737	3.145	3.427	4.016
88	1.403	2.006	2.390	2.733	3.141	3.422	4.011
89	1.401	2.003	2.387	2.730	3.137	3.418	4.006
90	1.399	2.001	2.384	2.726	3.133	3.414	4.001
91	1.398	1.998	2.381	2.723	3.129	3.410	3.996
92	1.396	1.996	2.378	2.720	3.125	3.405	3.992
93	1.394	1.994	2.375	2.716	3.121	3.401	3.987
94	1.393	1.991	2.373	2.713	3.118	3.398	3.982
95	1.391	1.989	2.370	2.710	3.114	3.394	3.978
96	1.390	1.987	2.367	2.707	3.111	3.390	3.973
97	1.388	1.985	2.365	2.704	3.107	3.386	3.969
98	1.387	1.983	2.362	2.701	3.104	3.383	3.965
99	1.385	1.980	2.360	2.698	3.101	3.379	3.961

TABLE 3.2.3. TWO-SIDED TOLERANCE LIMIT FACTORS (CONTROL CENTER).

GAMMA = 0.990 (CONTINUED)

N \P →	0.750	0.900	0.950	0.975	0.990	0.995	0.999
100	1.384	1.978	2.357	2.696	3.098	3.375	3.956
102	1.381	1.974	2.352	2.690	3.091	3.369	3.948
104	1.378	1.970	2.348	2.685	3.085	3.362	3.941
106	1.375	1.967	2.343	2.680	3.079	3.356	3.933
108	1.373	1.963	2.339	2.675	3.074	3.349	3.926
110	1.370	1.959	2.335	2.670	3.068	3.343	3.919
112	1.368	1.956	2.331	2.665	3.063	3.337	3.912
114	1.366	1.953	2.327	2.661	3.057	3.332	3.905
116	1.363	1.949	2.323	2.656	3.052	3.326	3.899
118	1.361	1.946	2.319	2.652	3.048	3.321	3.893
120	1.359	1.943	2.315	2.648	3.043	3.316	3.887
122	1.357	1.940	2.312	2.644	3.038	3.311	3.881
124	1.355	1.937	2.309	2.640	3.034	3.306	3.875
126	1.353	1.935	2.305	2.636	3.029	3.301	3.869
128	1.351	1.932	2.302	2.632	3.025	3.296	3.864
130	1.349	1.929	2.299	2.629	3.021	3.292	3.859
132	1.347	1.927	2.296	2.625	3.017	3.288	3.854
134	1.346	1.924	2.293	2.622	3.013	3.283	3.849
136	1.344	1.922	2.290	2.618	3.009	3.279	3.844
138	1.342	1.919	2.287	2.615	3.005	3.275	3.839
140	1.341	1.917	2.284	2.612	3.002	3.271	3.834
142	1.339	1.915	2.281	2.609	2.998	3.267	3.830
144	1.338	1.912	2.279	2.606	2.995	3.263	3.825
146	1.336	1.910	2.276	2.603	2.991	3.260	3.821
148	1.334	1.908	2.274	2.600	2.988	3.256	3.817
150	1.333	1.906	2.271	2.597	2.985	3.252	3.812
152	1.332	1.904	2.269	2.594	2.981	3.249	3.808
154	1.330	1.902	2.266	2.592	2.978	3.246	3.804
156	1.329	1.900	2.264	2.589	2.975	3.242	3.801
158	1.328	1.898	2.262	2.586	2.972	3.239	3.797
160	1.326	1.896	2.260	2.584	2.969	3.236	3.793
162	1.325	1.894	2.257	2.581	2.967	3.233	3.789
164	1.324	1.893	2.255	2.579	2.964	3.230	3.786
166	1.322	1.891	2.253	2.577	2.961	3.227	3.782
168	1.321	1.889	2.251	2.574	2.958	3.224	3.779
170	1.320	1.888	2.249	2.572	2.956	3.221	3.776
172	1.319	1.886	2.247	2.570	2.953	3.218	3.772
174	1.318	1.884	2.245	2.568	2.951	3.215	3.769
176	1.317	1.883	2.243	2.565	2.948	3.213	3.766
178	1.316	1.881	2.241	2.563	2.946	3.210	3.763
180	1.315	1.880	2.240	2.561	2.943	3.207	3.760
185	1.312	1.876	2.235	2.556	2.937	3.201	3.752
190	1.309	1.872	2.231	2.551	2.932	3.195	3.745
195	1.307	1.869	2.227	2.547	2.927	3.189	3.738
200	1.305	1.866	2.223	2.542	2.921	3.184	3.732
205	1.303	1.863	2.219	2.538	2.917	3.178	3.726
210	1.301	1.860	2.216	2.534	2.912	3.173	3.720
215	1.298	1.857	2.212	2.530	2.907	3.168	3.714
220	1.297	1.854	2.209	2.526	2.903	3.164	3.708

TABLE 3.2.4. TWO-SIDED TOLERANCE LIMIT FACTORS (CONTROL CENTER).

GAMMA = 0.990 (CONTINUED)

N \P →	0.750	0.900	0.950	0.975	0.990	0.995	0.999
225	1.295	1.851	2.206	2.523	2.899	3.159	3.703
230	1.293	1.849	2.203	2.519	2.895	3.155	3.698
235	1.291	1.846	2.200	2.516	2.891	3.150	3.693
240	1.289	1.844	2.197	2.512	2.887	3.146	3.688
245	1.288	1.841	2.194	2.509	2.884	3.142	3.684
250	1.286	1.839	2.191	2.506	2.880	3.139	3.679
255	1.285	1.837	2.189	2.503	2.877	3.135	3.675
260	1.283	1.835	2.186	2.500	2.873	3.131	3.671
265	1.282	1.833	2.184	2.498	2.870	3.128	3.666
270	1.280	1.831	2.182	2.495	2.867	3.124	3.663
275	1.279	1.829	2.179	2.492	2.864	3.121	3.659
280	1.278	1.827	2.177	2.490	2.861	3.118	3.655
285	1.277	1.825	2.175	2.487	2.858	3.115	3.651
290	1.275	1.824	2.173	2.485	2.856	3.112	3.648
295	1.274	1.822	2.171	2.483	2.853	3.109	3.644
300	1.273	1.820	2.169	2.480	2.850	3.106	3.641
310	1.271	1.817	2.165	2.476	2.845	3.101	3.635
320	1.269	1.814	2.161	2.472	2.841	3.096	3.629
330	1.267	1.811	2.158	2.468	2.836	3.091	3.623
340	1.265	1.808	2.155	2.464	2.832	3.086	3.617
350	1.263	1.806	2.152	2.461	2.828	3.081	3.612
360	1.261	1.803	2.149	2.457	2.824	3.077	3.607
370	1.259	1.801	2.146	2.454	2.820	3.073	3.602
380	1.258	1.799	2.143	2.451	2.816	3.069	3.598
390	1.256	1.796	2.140	2.448	2.813	3.066	3.593
400	1.255	1.794	2.138	2.445	2.810	3.062	3.589
425	1.251	1.789	2.132	2.438	2.802	3.054	3.579
450	1.248	1.785	2.127	2.432	2.795	3.046	3.571
475	1.245	1.781	2.122	2.427	2.789	3.039	3.562
500	1.243	1.777	2.117	2.421	2.783	3.033	3.555
525	1.240	1.774	2.113	2.417	2.777	3.027	3.548
550	1.238	1.770	2.109	2.412	2.772	3.021	3.541
575	1.236	1.767	2.106	2.408	2.768	3.016	3.535
600	1.234	1.765	2.103	2.404	2.763	3.011	3.530
625	1.232	1.762	2.099	2.401	2.759	3.007	3.525
650	1.231	1.759	2.097	2.398	2.755	3.003	3.520
700	1.227	1.755	2.091	2.391	2.748	2.995	3.511
750	1.225	1.751	2.086	2.386	2.742	2.988	3.503
800	1.222	1.747	2.082	2.381	2.736	2.982	3.496
850	1.220	1.744	2.078	2.377	2.731	2.976	3.489
900	1.218	1.741	2.075	2.373	2.727	2.971	3.483
950	1.216	1.738	2.071	2.369	2.722	2.967	3.478
1000	1.214	1.736	2.068	2.365	2.718	2.962	3.473
1500	1.202	1.718	2.047	2.341	2.691	2.932	3.437
2000	1.195	1.708	2.035	2.327	2.675	2.915	3.417
3000	1.186	1.696	2.021	2.311	2.656	2.894	3.393
5000	1.178	1.684	2.007	2.295	2.637	2.874	3.369
10000	1.170	1.672	1.993	2.279	2.619	2.854	3.346
∞	1.150	1.645	1.960	2.241	2.576	2.807	3.291

TABLE 3.3.1.   TWO-SIDED TOLERANCE LIMIT FACTORS (CONTROL CENTER).

GAMMA = 0.975

N \P →	0.750	0.900	0.950	0.975	0.990	0.995	0.999
2	44.791	62.218	73.077	82.660	93.937	101.678	117.747
3	8.460	11.830	13.940	15.806	18.007	19.520	22.665
4	4.885	6.863	8.104	9.205	10.503	11.397	13.256
5	3.692	5.205	6.157	7.001	7.998	8.685	10.114
6	3.105	4.389	5.198	5.916	6.765	7.350	8.567
7	2.756	3.904	4.628	5.271	6.031	6.555	7.647
8	2.524	3.580	4.248	4.841	5.543	6.026	7.034
9	2.358	3.349	3.976	4.533	5.192	5.647	6.594
10	2.233	3.175	3.771	4.301	4.928	5.361	6.263
11	2.135	3.038	3.610	4.119	4.721	5.137	6.003
12	2.057	2.928	3.480	3.972	4.554	4.956	5.793
13	1.992	2.837	3.374	3.851	4.416	4.806	5.620
14	1.937	2.761	3.284	3.749	4.300	4.681	5.474
15	1.891	2.696	3.207	3.662	4.201	4.573	5.350
16	1.851	2.640	3.140	3.587	4.115	4.480	5.242
17	1.816	2.591	3.082	3.521	4.040	4.399	5.147
18	1.785	2.547	3.031	3.462	3.974	4.327	5.064
19	1.758	2.508	2.985	3.411	3.915	4.263	4.990
20	1.733	2.474	2.944	3.364	3.862	4.205	4.923
21	1.711	2.442	2.907	3.322	3.814	4.153	4.862
22	1.691	2.414	2.874	3.284	3.770	4.106	4.807
23	1.672	2.388	2.843	3.249	3.731	4.063	4.757
24	1.656	2.364	2.815	3.217	3.694	4.024	4.711
25	1.640	2.343	2.789	3.188	3.661	3.987	4.669
26	1.626	2.322	2.765	3.161	3.630	3.953	4.630
27	1.613	2.304	2.743	3.135	3.601	3.922	4.594
28	1.600	2.286	2.723	3.112	3.574	3.893	4.560
29	1.589	2.270	2.703	3.090	3.549	3.866	4.528
30	1.578	2.255	2.685	3.070	3.526	3.841	4.499
31	1.568	2.241	2.669	3.050	3.504	3.817	4.471
32	1.559	2.227	2.653	3.032	3.483	3.794	4.445
33	1.550	2.215	2.638	3.015	3.463	3.773	4.420
34	1.541	2.203	2.624	2.999	3.445	3.753	4.397
35	1.533	2.191	2.610	2.984	3.428	3.734	4.375
36	1.526	2.181	2.598	2.970	3.411	3.716	4.354
37	1.519	2.171	2.586	2.956	3.396	3.699	4.334
38	1.512	2.161	2.574	2.943	3.381	3.683	4.315
39	1.505	2.152	2.563	2.930	3.366	3.668	4.297
40	1.499	2.143	2.553	2.919	3.353	3.653	4.280
41	1.493	2.135	2.543	2.907	3.340	3.639	4.264
42	1.488	2.127	2.533	2.897	3.328	3.626	4.248
43	1.483	2.119	2.524	2.886	3.316	3.613	4.233
44	1.477	2.112	2.516	2.876	3.305	3.600	4.219
45	1.472	2.105	2.507	2.867	3.294	3.589	4.205
46	1.468	2.098	2.499	2.858	3.283	3.577	4.192
47	1.463	2.092	2.492	2.849	3.273	3.566	4.179
48	1.459	2.085	2.484	2.840	3.264	3.556	4.167
49	1.455	2.079	2.477	2.832	3.254	3.546	4.155
50	1.451	2.074	2.470	2.825	3.245	3.536	4.144

TABLE 3.3.2. TWO-SIDED TOLERANCE LIMIT FACTORS (CONTROL CENTER).

GAMMA = 0.975 (CONTINUED)

N \P →	0.750	0.900	0.950	0.975	0.990	0.995	0.999
51	1.447	2.068	2.464	2.817	3.237	3.527	4.133
52	1.443	2.063	2.457	2.810	3.228	3.518	4.122
53	1.439	2.057	2.451	2.803	3.220	3.509	4.112
54	1.436	2.052	2.445	2.796	3.212	3.500	4.102
55	1.432	2.048	2.439	2.789	3.205	3.492	4.092
56	1.429	2.043	2.434	2.783	3.198	3.484	4.083
57	1.426	2.038	2.428	2.777	3.191	3.476	4.074
58	1.423	2.034	2.423	2.771	3.184	3.469	4.065
59	1.420	2.030	2.418	2.765	3.177	3.462	4.057
60	1.417	2.025	2.413	2.759	3.171	3.455	4.049
61	1.414	2.021	2.408	2.754	3.164	3.448	4.041
62	1.411	2.018	2.404	2.749	3.158	3.441	4.033
63	1.409	2.014	2.399	2.743	3.152	3.435	4.026
64	1.406	2.010	2.395	2.738	3.147	3.429	4.018
65	1.403	2.006	2.391	2.734	3.141	3.423	4.011
66	1.401	2.003	2.386	2.729	3.136	3.417	4.004
67	1.399	2.000	2.382	2.724	3.130	3.411	3.998
68	1.396	1.996	2.378	2.720	3.125	3.405	3.991
69	1.394	1.993	2.375	2.715	3.120	3.400	3.985
70	1.392	1.990	2.371	2.711	3.115	3.394	3.978
71	1.390	1.987	2.367	2.707	3.110	3.389	3.972
72	1.388	1.984	2.364	2.703	3.106	3.384	3.966
73	1.385	1.981	2.360	2.699	3.101	3.379	3.961
74	1.383	1.978	2.357	2.695	3.097	3.374	3.955
75	1.382	1.975	2.353	2.691	3.092	3.370	3.949
76	1.380	1.972	2.350	2.687	3.088	3.365	3.944
77	1.378	1.970	2.347	2.684	3.084	3.360	3.939
78	1.376	1.967	2.344	2.680	3.080	3.356	3.934
79	1.374	1.965	2.341	2.677	3.076	3.352	3.929
80	1.372	1.962	2.338	2.673	3.072	3.348	3.924
81	1.371	1.960	2.335	2.670	3.068	3.343	3.919
82	1.369	1.957	2.332	2.667	3.064	3.339	3.914
83	1.367	1.955	2.329	2.664	3.061	3.335	3.909
84	1.366	1.953	2.327	2.661	3.057	3.332	3.905
85	1.364	1.950	2.324	2.658	3.054	3.328	3.900
86	1.363	1.948	2.321	2.655	3.050	3.324	3.896
87	1.361	1.946	2.319	2.652	3.047	3.320	3.892
88	1.360	1.944	2.316	2.649	3.044	3.317	3.888
89	1.358	1.942	2.314	2.646	3.041	3.313	3.883
90	1.357	1.940	2.311	2.643	3.037	3.310	3.879
91	1.355	1.938	2.309	2.640	3.034	3.306	3.875
92	1.354	1.936	2.307	2.638	3.031	3.303	3.872
93	1.353	1.934	2.304	2.635	3.028	3.300	3.868
94	1.351	1.932	2.302	2.633	3.025	3.297	3.864
95	1.350	1.930	2.300	2.630	3.022	3.294	3.860
96	1.349	1.928	2.298	2.628	3.020	3.290	3.857
97	1.348	1.927	2.296	2.625	3.017	3.287	3.853
98	1.346	1.925	2.294	2.623	3.014	3.284	3.850
99	1.345	1.923	2.292	2.621	3.011	3.282	3.846

TABLE 3.3.3. TWO-SIDED TOLERANCE LIMIT FACTORS (CONTROL CENTER).

GAMMA = 0.975 (CONTINUED)

N \P →	0.750	0.900	0.950	0.975	0.990	0.995	0.999
100	1.344	1.922	2.290	2.618	3.009	3.279	3.843
102	1.342	1.918	2.286	2.614	3.004	3.273	3.836
104	1.339	1.915	2.282	2.609	2.999	3.268	3.830
106	1.337	1.912	2.278	2.605	2.994	3.262	3.824
108	1.335	1.909	2.275	2.601	2.989	3.257	3.818
110	1.333	1.906	2.271	2.597	2.985	3.252	3.812
112	1.331	1.903	2.268	2.593	2.980	3.248	3.807
114	1.329	1.901	2.265	2.590	2.976	3.243	3.801
116	1.327	1.898	2.261	2.586	2.972	3.238	3.796
118	1.326	1.895	2.258	2.583	2.968	3.234	3.791
120	1.324	1.893	2.255	2.579	2.964	3.230	3.786
122	1.322	1.890	2.253	2.576	2.960	3.226	3.781
124	1.320	1.888	2.250	2.573	2.956	3.222	3.776
126	1.319	1.886	2.247	2.570	2.953	3.218	3.772
128	1.317	1.884	2.244	2.567	2.949	3.214	3.767
130	1.316	1.881	2.242	2.564	2.946	3.210	3.763
132	1.314	1.879	2.239	2.561	2.943	3.207	3.759
134	1.313	1.877	2.237	2.558	2.939	3.203	3.755
136	1.311	1.875	2.234	2.555	2.936	3.200	3.751
138	1.310	1.873	2.232	2.552	2.933	3.196	3.747
140	1.309	1.871	2.230	2.550	2.930	3.193	3.743
142	1.307	1.869	2.227	2.547	2.927	3.190	3.739
144	1.306	1.868	2.225	2.545	2.924	3.187	3.736
146	1.305	1.866	2.223	2.542	2.922	3.184	3.732
148	1.304	1.864	2.221	2.540	2.919	3.181	3.728
150	1.302	1.862	2.219	2.538	2.916	3.178	3.725
152	1.301	1.861	2.217	2.535	2.914	3.175	3.722
154	1.300	1.859	2.215	2.533	2.911	3.172	3.718
156	1.299	1.857	2.213	2.531	2.908	3.169	3.715
158	1.298	1.856	2.211	2.529	2.906	3.167	3.712
160	1.297	1.854	2.209	2.527	2.904	3.164	3.709
162	1.296	1.853	2.208	2.525	2.901	3.162	3.706
164	1.295	1.851	2.206	2.523	2.899	3.159	3.703
166	1.294	1.850	2.204	2.521	2.897	3.157	3.700
168	1.293	1.848	2.202	2.519	2.894	3.154	3.697
170	1.292	1.847	2.201	2.517	2.892	3.152	3.695
172	1.291	1.846	2.199	2.515	2.890	3.150	3.692
174	1.290	1.844	2.198	2.513	2.888	3.147	3.689
176	1.289	1.843	2.196	2.511	2.886	3.145	3.687
178	1.288	1.842	2.195	2.510	2.884	3.143	3.684
180	1.287	1.840	2.193	2.508	2.882	3.141	3.682
185	1.285	1.837	2.189	2.504	2.877	3.135	3.675
190	1.283	1.834	2.186	2.500	2.873	3.131	3.670
195	1.281	1.832	2.183	2.496	2.868	3.126	3.664
200	1.279	1.829	2.179	2.492	2.864	3.121	3.659
205	1.277	1.826	2.176	2.489	2.860	3.117	3.654
210	1.276	1.824	2.173	2.485	2.856	3.113	3.649
215	1.274	1.822	2.171	2.482	2.852	3.108	3.644
220	1.272	1.819	2.168	2.479	2.849	3.105	3.639

TABLE 3.3.4.   TWO-SIDED TOLERANCE LIMIT FACTORS (CONTROL CENTER).

GAMMA = 0.975 (CONTINUED)

N \P →	0.750	0.900	0.950	0.975	0.990	0.995	0.999
225	1.271	1.817	2.165	2.476	2.845	3.101	3.635
230	1.269	1.815	2.163	2.473	2.842	3.097	3.631
235	1.268	1.813	2.160	2.470	2.839	3.094	3.627
240	1.266	1.811	2.158	2.468	2.836	3.090	3.623
245	1.265	1.809	2.156	2.465	2.833	3.087	3.619
250	1.264	1.807	2.153	2.462	2.830	3.084	3.615
255	1.263	1.805	2.151	2.460	2.827	3.081	3.611
260	1.261	1.804	2.149	2.458	2.824	3.078	3.608
265	1.260	1.802	2.147	2.455	2.822	3.075	3.605
270	1.259	1.800	2.145	2.453	2.819	3.072	3.601
275	1.258	1.799	2.143	2.451	2.817	3.069	3.598
280	1.257	1.797	2.141	2.449	2.814	3.067	3.595
285	1.256	1.796	2.140	2.447	2.812	3.064	3.592
290	1.255	1.794	2.138	2.445	2.810	3.062	3.589
295	1.254	1.793	2.136	2.443	2.807	3.059	3.586
300	1.253	1.791	2.135	2.441	2.805	3.057	3.584
310	1.251	1.789	2.131	2.437	2.801	3.053	3.578
320	1.249	1.786	2.128	2.434	2.797	3.048	3.573
330	1.248	1.784	2.126	2.431	2.793	3.044	3.568
340	1.246	1.782	2.123	2.428	2.790	3.040	3.564
350	1.244	1.779	2.120	2.425	2.786	3.037	3.560
360	1.243	1.777	2.118	2.422	2.783	3.033	3.555
370	1.242	1.775	2.115	2.419	2.780	3.030	3.551
380	1.240	1.773	2.113	2.417	2.777	3.026	3.548
390	1.239	1.772	2.111	2.414	2.774	3.023	3.544
400	1.238	1.770	2.109	2.412	2.772	3.020	3.541
425	1.235	1.766	2.104	2.406	2.765	3.013	3.532
450	1.232	1.762	2.100	2.401	2.759	3.007	3.525
475	1.230	1.759	2.096	2.396	2.754	3.001	3.518
500	1.228	1.756	2.092	2.392	2.749	2.996	3.512
525	1.226	1.753	2.088	2.388	2.745	2.991	3.506
550	1.224	1.750	2.085	2.385	2.740	2.986	3.501
575	1.222	1.747	2.082	2.381	2.737	2.982	3.496
600	1.220	1.745	2.079	2.378	2.733	2.978	3.491
625	1.219	1.743	2.077	2.375	2.729	2.974	3.487
650	1.218	1.741	2.074	2.372	2.726	2.971	3.483
700	1.215	1.737	2.070	2.367	2.720	2.965	3.475
750	1.213	1.734	2.066	2.363	2.715	2.959	3.468
800	1.210	1.731	2.062	2.359	2.710	2.954	3.462
850	1.209	1.728	2.059	2.355	2.706	2.949	3.457
900	1.207	1.726	2.056	2.351	2.702	2.945	3.452
950	1.205	1.723	2.053	2.348	2.699	2.941	3.447
1000	1.204	1.721	2.051	2.345	2.695	2.937	3.443
1500	1.193	1.707	2.033	2.325	2.672	2.912	3.414
2000	1.187	1.698	2.023	2.314	2.659	2.898	3.397
3000	1.180	1.688	2.011	2.300	2.643	2.880	3.377
5000	1.173	1.678	1.999	2.286	2.628	2.863	3.357
10000	1.167	1.668	1.988	2.273	2.612	2.847	3.337
∞	1.150	1.645	1.960	2.241	2.576	2.807	3.291

TABLE 3.4.1. *TWO-SIDED TOLERANCE LIMIT FACTORS (CONTROL CENTER).*

GAMMA = 0.950

$N \backslash P \rightarrow$	0.750	0.900	0.950	0.975	0.990	0.995	0.999
2	22.383	31.092	36.519	41.308	46.944	50.813	58.844
3	5.937	8.306	9.789	11.101	12.647	13.710	15.920
4	3.818	5.368	6.341	7.203	8.221	8.921	10.377
5	3.041	4.291	5.077	5.774	6.598	7.165	8.345
6	2.638	3.733	4.422	5.034	5.758	6.256	7.294
7	2.391	3.390	4.020	4.579	5.241	5.697	6.647
8	2.223	3.156	3.746	4.269	4.889	5.316	6.206
9	2.101	2.986	3.546	4.044	4.633	5.039	5.885
10	2.008	2.856	3.393	3.871	4.437	4.827	5.640
11	1.934	2.754	3.273	3.735	4.282	4.659	5.446
12	1.874	2.670	3.175	3.624	4.156	4.522	5.287
13	1.825	2.601	3.093	3.531	4.051	4.409	5.156
14	1.783	2.542	3.024	3.453	3.962	4.312	5.044
15	1.747	2.492	2.965	3.386	3.885	4.230	4.949
16	1.716	2.449	2.913	3.328	3.819	4.158	4.865
17	1.689	2.410	2.868	3.277	3.761	4.095	4.792
18	1.665	2.376	2.828	3.231	3.709	4.039	4.727
19	1.643	2.346	2.793	3.191	3.663	3.988	4.669
20	1.624	2.319	2.760	3.154	3.621	3.943	4.616
21	1.607	2.294	2.731	3.121	3.583	3.903	4.569
22	1.591	2.272	2.705	3.091	3.549	3.865	4.526
23	1.576	2.251	2.681	3.063	3.518	3.831	4.486
24	1.563	2.232	2.658	3.038	3.489	3.800	4.450
25	1.551	2.215	2.638	3.015	3.462	3.771	4.416
26	1.539	2.199	2.619	2.993	3.437	3.744	4.385
27	1.529	2.184	2.601	2.973	3.415	3.720	4.356
28	1.519	2.170	2.585	2.954	3.393	3.696	4.330
29	1.510	2.157	2.569	2.937	3.373	3.675	4.304
30	1.501	2.145	2.555	2.921	3.355	3.654	4.281
31	1.493	2.134	2.541	2.905	3.337	3.635	4.259
32	1.486	2.123	2.529	2.891	3.320	3.617	4.238
33	1.478	2.113	2.517	2.877	3.305	3.600	4.218
34	1.472	2.103	2.505	2.864	3.290	3.584	4.199
35	1.465	2.094	2.495	2.852	3.276	3.569	4.182
36	1.459	2.086	2.484	2.840	3.263	3.555	4.165
37	1.454	2.077	2.475	2.829	3.250	3.541	4.149
38	1.448	2.070	2.466	2.819	3.238	3.528	4.134
39	1.443	2.062	2.457	2.809	3.227	3.516	4.119
40	1.438	2.055	2.448	2.799	3.216	3.504	4.105
41	1.433	2.049	2.440	2.790	3.205	3.492	4.092
42	1.429	2.042	2.433	2.781	3.196	3.482	4.080
43	1.424	2.036	2.425	2.773	3.186	3.471	4.068
44	1.420	2.030	2.418	2.765	3.177	3.461	4.056
45	1.416	2.024	2.412	2.757	3.168	3.452	4.045
46	1.412	2.019	2.405	2.750	3.160	3.443	4.034
47	1.409	2.014	2.399	2.743	3.151	3.434	4.024
48	1.405	2.009	2.393	2.736	3.144	3.425	4.014
49	1.402	2.004	2.387	2.729	3.136	3.417	4.004
50	1.398	1.999	2.382	2.723	3.129	3.409	3.995

TABLE 3.4.2. TWO-SIDED TOLERANCE LIMIT FACTORS (CONTROL CENTER).

GAMMA = 0.950 (CONTINUED)

$N \backslash P \rightarrow$	0.750	0.900	0.950	0.975	0.990	0.995	0.999
51	1.395	1.994	2.376	2.717	3.122	3.401	3.986
52	1.392	1.990	2.371	2.711	3.115	3.394	3.978
53	1.389	1.986	2.366	2.705	3.108	3.387	3.969
54	1.386	1.982	2.361	2.700	3.102	3.380	3.961
55	1.383	1.978	2.356	2.694	3.096	3.373	3.953
56	1.381	1.974	2.352	2.689	3.090	3.367	3.946
57	1.378	1.970	2.347	2.684	3.084	3.361	3.939
58	1.376	1.967	2.343	2.679	3.079	3.355	3.932
59	1.373	1.963	2.339	2.675	3.073	3.349	3.925
60	1.371	1.960	2.335	2.670	3.068	3.343	3.918
61	1.369	1.957	2.331	2.666	3.063	3.338	3.912
62	1.366	1.953	2.327	2.661	3.058	3.332	3.905
63	1.364	1.950	2.324	2.657	3.053	3.327	3.899
64	1.362	1.947	2.320	2.653	3.048	3.322	3.893
65	1.360	1.944	2.317	2.649	3.044	3.317	3.887
66	1.358	1.941	2.313	2.645	3.039	3.312	3.882
67	1.356	1.939	2.310	2.641	3.035	3.307	3.876
68	1.354	1.936	2.307	2.638	3.031	3.303	3.871
69	1.352	1.933	2.304	2.634	3.027	3.298	3.866
70	1.350	1.931	2.300	2.631	3.023	3.294	3.861
71	1.349	1.928	2.297	2.627	3.019	3.290	3.856
72	1.347	1.926	2.295	2.624	3.015	3.285	3.851
73	1.345	1.923	2.292	2.621	3.011	3.281	3.846
74	1.344	1.921	2.289	2.617	3.008	3.277	3.841
75	1.342	1.919	2.286	2.614	3.004	3.274	3.837
76	1.341	1.917	2.284	2.611	3.001	3.270	3.832
77	1.339	1.914	2.281	2.608	2.997	3.266	3.828
78	1.337	1.912	2.278	2.605	2.994	3.262	3.824
79	1.336	1.910	2.276	2.603	2.991	3.259	3.820
80	1.335	1.908	2.274	2.600	2.988	3.255	3.816
81	1.333	1.906	2.271	2.597	2.984	3.252	3.812
82	1.332	1.904	2.269	2.594	2.981	3.249	3.808
83	1.330	1.902	2.267	2.592	2.978	3.246	3.804
84	1.329	1.900	2.264	2.589	2.975	3.242	3.800
85	1.328	1.899	2.262	2.587	2.973	3.239	3.797
86	1.327	1.897	2.260	2.584	2.970	3.236	3.793
87	1.325	1.895	2.258	2.582	2.967	3.233	3.790
88	1.324	1.893	2.256	2.580	2.964	3.230	3.786
89	1.323	1.892	2.254	2.577	2.962	3.227	3.783
90	1.322	1.890	2.252	2.575	2.959	3.225	3.780
91	1.321	1.888	2.250	2.573	2.957	3.222	3.776
92	1.320	1.887	2.248	2.571	2.954	3.219	3.773
93	1.318	1.885	2.246	2.569	2.952	3.216	3.770
94	1.317	1.884	2.244	2.566	2.949	3.214	3.767
95	1.316	1.882	2.242	2.564	2.947	3.211	3.764
96	1.315	1.881	2.241	2.562	2.944	3.209	3.761
97	1.314	1.879	2.239	2.560	2.942	3.206	3.758
98	1.313	1.878	2.237	2.558	2.940	3.204	3.755
99	1.312	1.876	2.236	2.556	2.938	3.201	3.752

TABLE 3.4.3.   TWO-SIDED TOLERANCE LIMIT FACTORS (CONTROL CENTER).

GAMMA = 0.950 (CONTINUED)

N \P →	0.750	0.900	0.950	0.975	0.990	0.995	0.999
100	1.311	1.875	2.234	2.555	2.936	3.199	3.750
102	1.309	1.872	2.231	2.551	2.931	3.194	3.744
104	1.308	1.869	2.228	2.547	2.927	3.190	3.739
106	1.306	1.867	2.225	2.544	2.923	3.186	3.734
108	1.304	1.864	2.222	2.541	2.919	3.181	3.729
110	1.302	1.862	2.219	2.537	2.916	3.177	3.724
112	1.301	1.860	2.216	2.534	2.912	3.173	3.720
114	1.299	1.858	2.213	2.531	2.909	3.170	3.715
116	1.298	1.855	2.211	2.528	2.905	3.166	3.711
118	1.296	1.853	2.208	2.525	2.902	3.162	3.707
120	1.295	1.851	2.206	2.522	2.899	3.159	3.703
122	1.293	1.849	2.203	2.520	2.896	3.155	3.699
124	1.292	1.847	2.201	2.517	2.893	3.152	3.695
126	1.291	1.845	2.199	2.514	2.890	3.149	3.691
128	1.289	1.843	2.197	2.512	2.887	3.146	3.687
130	1.288	1.842	2.194	2.510	2.884	3.143	3.684
132	1.287	1.840	2.192	2.507	2.881	3.140	3.680
134	1.286	1.838	2.190	2.505	2.878	3.137	3.677
136	1.284	1.837	2.188	2.503	2.876	3.134	3.674
138	1.283	1.835	2.186	2.500	2.873	3.131	3.670
140	1.282	1.833	2.185	2.498	2.871	3.128	3.667
142	1.281	1.832	2.183	2.496	2.868	3.126	3.664
144	1.280	1.830	2.181	2.494	2.866	3.123	3.661
146	1.279	1.829	2.179	2.492	2.864	3.121	3.658
148	1.278	1.827	2.177	2.490	2.861	3.118	3.655
150	1.277	1.826	2.176	2.488	2.859	3.116	3.652
152	1.276	1.825	2.174	2.486	2.857	3.114	3.650
154	1.275	1.823	2.172	2.484	2.855	3.111	3.647
156	1.274	1.822	2.171	2.483	2.853	3.109	3.644
158	1.273	1.821	2.169	2.481	2.851	3.107	3.642
160	1.272	1.819	2.168	2.479	2.849	3.105	3.639
162	1.272	1.818	2.166	2.477	2.847	3.102	3.637
164	1.271	1.817	2.165	2.476	2.845	3.100	3.634
166	1.270	1.816	2.163	2.474	2.843	3.098	3.632
168	1.269	1.815	2.162	2.473	2.841	3.096	3.630
170	1.268	1.813	2.161	2.471	2.840	3.094	3.627
172	1.267	1.812	2.159	2.469	2.838	3.092	3.625
174	1.267	1.811	2.158	2.468	2.836	3.091	3.623
176	1.266	1.810	2.157	2.466	2.834	3.089	3.621
178	1.265	1.809	2.155	2.465	2.833	3.087	3.619
180	1.264	1.808	2.154	2.464	2.831	3.085	3.616
185	1.263	1.805	2.151	2.460	2.827	3.081	3.611
190	1.261	1.803	2.148	2.457	2.823	3.077	3.607
195	1.259	1.801	2.146	2.454	2.820	3.073	3.602
200	1.258	1.798	2.143	2.451	2.816	3.069	3.598
205	1.256	1.796	2.140	2.448	2.813	3.065	3.593
210	1.255	1.794	2.138	2.445	2.810	3.062	3.589
215	1.253	1.792	2.136	2.442	2.807	3.059	3.585
220	1.252	1.790	2.133	2.440	2.804	3.055	3.581

TABLE 3.4.4. *TWO-SIDED TOLERANCE LIMIT FACTORS* (*CONTROL CENTER*).

GAMMA = 0.950 (*CONTINUED*)

$N \backslash P \rightarrow$	0.750	0.900	0.950	0.975	0.990	0.995	0.999
225	1.251	1.789	2.131	2.437	2.801	3.052	3.578
230	1.250	1.787	2.129	2.435	2.798	3.049	3.574
235	1.248	1.785	2.127	2.432	2.795	3.046	3.571
240	1.247	1.783	2.125	2.430	2.793	3.043	3.568
245	1.246	1.782	2.123	2.428	2.790	3.041	3.564
250	1.245	1.780	2.121	2.426	2.788	3.038	3.561
255	1.244	1.779	2.120	2.424	2.786	3.036	3.558
260	1.243	1.777	2.118	2.422	2.783	3.033	3.555
265	1.242	1.776	2.116	2.420	2.781	3.031	3.553
270	1.241	1.775	2.115	2.418	2.779	3.028	3.550
275	1.240	1.773	2.113	2.416	2.777	3.026	3.547
280	1.239	1.772	2.111	2.415	2.775	3.024	3.545
285	1.238	1.771	2.110	2.413	2.773	3.022	3.542
290	1.238	1.770	2.109	2.411	2.771	3.020	3.540
295	1.237	1.768	2.107	2.410	2.769	3.018	3.538
300	1.236	1.767	2.106	2.408	2.767	3.016	3.535
310	1.234	1.765	2.103	2.405	2.764	3.012	3.531
320	1.233	1.763	2.101	2.402	2.761	3.008	3.527
330	1.232	1.761	2.098	2.400	2.758	3.005	3.523
340	1.230	1.759	2.096	2.397	2.755	3.002	3.519
350	1.229	1.757	2.094	2.395	2.752	2.999	3.515
360	1.228	1.756	2.092	2.392	2.749	2.996	3.512
370	1.227	1.754	2.090	2.390	2.747	2.993	3.509
380	1.225	1.752	2.088	2.388	2.744	2.990	3.505
390	1.224	1.751	2.086	2.386	2.742	2.988	3.502
400	1.223	1.749	2.084	2.384	2.739	2.985	3.499
425	1.221	1.746	2.080	2.379	2.734	2.979	3.493
450	1.219	1.743	2.077	2.375	2.729	2.974	3.486
475	1.217	1.740	2.073	2.371	2.725	2.969	3.481
500	1.215	1.737	2.070	2.368	2.721	2.965	3.476
525	1.213	1.735	2.067	2.364	2.717	2.961	3.471
550	1.212	1.733	2.065	2.361	2.713	2.957	3.466
575	1.210	1.731	2.062	2.358	2.710	2.953	3.462
600	1.209	1.729	2.060	2.356	2.707	2.950	3.458
625	1.208	1.727	2.058	2.353	2.704	2.947	3.455
650	1.207	1.725	2.056	2.351	2.702	2.944	3.451
700	1.204	1.722	2.052	2.347	2.697	2.939	3.445
750	1.202	1.719	2.049	2.343	2.692	2.934	3.439
800	1.201	1.717	2.046	2.339	2.688	2.930	3.434
850	1.199	1.715	2.043	2.336	2.685	2.926	3.430
900	1.198	1.712	2.040	2.333	2.682	2.922	3.426
950	1.196	1.711	2.038	2.331	2.679	2.919	3.422
1000	1.195	1.709	2.036	2.328	2.676	2.916	3.418
1500	1.186	1.697	2.022	2.312	2.657	2.895	3.394
2000	1.181	1.689	2.013	2.302	2.645	2.883	3.379
3000	1.176	1.681	2.003	2.291	2.632	2.869	3.363
5000	1.170	1.673	1.993	2.279	2.619	2.854	3.346
10000	1.164	1.664	1.983	2.268	2.606	2.840	3.329
∞	1.150	1.645	1.960	2.241	2.576	2.807	3.291

TABLE 3.5.1. TWO-SIDED TOLERANCE LIMIT FACTORS (CONTROL CENTER).

GAMMA = 0.900

N \P →	0.750	0.900	0.950	0.975	0.990	0.995	0.999
2	11.166	15.512	18.221	20.611	23.423	25.354	29.362
3	4.134	5.788	6.823	7.739	8.819	9.561	11.103
4	2.954	4.157	4.913	5.582	6.372	6.916	8.046
5	2.477	3.499	4.142	4.713	5.387	5.851	6.816
6	2.217	3.141	3.723	4.239	4.850	5.270	6.146
7	2.053	2.913	3.456	3.938	4.508	4.901	5.720
8	1.938	2.754	3.270	3.728	4.271	4.644	5.423
9	1.854	2.637	3.132	3.573	4.094	4.454	5.203
10	1.788	2.546	3.026	3.453	3.958	4.306	5.033
11	1.736	2.473	2.941	3.356	3.849	4.188	4.897
12	1.694	2.414	2.871	3.277	3.759	4.091	4.785
13	1.658	2.364	2.812	3.211	3.684	4.010	4.691
14	1.628	2.322	2.762	3.155	3.620	3.941	4.611
15	1.602	2.285	2.720	3.106	3.565	3.881	4.541
16	1.579	2.254	2.682	3.064	3.517	3.829	4.481
17	1.559	2.226	2.649	3.026	3.474	3.783	4.428
18	1.541	2.201	2.620	2.993	3.436	3.742	4.380
19	1.526	2.178	2.593	2.963	3.402	3.705	4.338
20	1.511	2.158	2.570	2.936	3.372	3.672	4.299
21	1.498	2.140	2.548	2.912	3.344	3.642	4.264
22	1.487	2.123	2.528	2.890	3.318	3.614	4.232
23	1.476	2.108	2.510	2.869	3.295	3.589	4.203
24	1.466	2.094	2.494	2.850	3.274	3.566	4.176
25	1.457	2.081	2.479	2.833	3.254	3.544	4.151
26	1.448	2.069	2.464	2.817	3.235	3.524	4.128
27	1.440	2.058	2.451	2.802	3.218	3.506	4.106
28	1.433	2.048	2.439	2.788	3.202	3.488	4.086
29	1.426	2.038	2.427	2.775	3.187	3.472	4.068
30	1.420	2.029	2.417	2.763	3.173	3.457	4.050
31	1.414	2.020	2.406	2.751	3.160	3.443	4.033
32	1.408	2.012	2.397	2.740	3.148	3.429	4.018
33	1.403	2.005	2.388	2.730	3.136	3.416	4.003
34	1.397	1.997	2.379	2.720	3.125	3.404	3.989
35	1.393	1.991	2.371	2.711	3.114	3.393	3.975
36	1.388	1.984	2.363	2.702	3.104	3.382	3.963
37	1.384	1.978	2.356	2.694	3.095	3.372	3.951
38	1.380	1.972	2.349	2.686	3.086	3.362	3.939
39	1.376	1.966	2.343	2.678	3.077	3.353	3.928
40	1.372	1.961	2.336	2.671	3.069	3.344	3.918
41	1.368	1.956	2.330	2.664	3.061	3.335	3.908
42	1.365	1.951	2.324	2.658	3.053	3.327	3.898
43	1.362	1.946	2.319	2.651	3.046	3.319	3.889
44	1.358	1.942	2.313	2.645	3.039	3.311	3.880
45	1.355	1.938	2.308	2.639	3.032	3.304	3.872
46	1.352	1.933	2.303	2.634	3.026	3.297	3.864
47	1.350	1.929	2.299	2.628	3.020	3.290	3.856
48	1.347	1.926	2.294	2.623	3.014	3.284	3.848
49	1.344	1.922	2.290	2.618	3.008	3.278	3.841
50	1.342	1.918	2.285	2.613	3.003	3.272	3.834

TABLE 3.5.2. TWO-SIDED TOLERANCE LIMIT FACTORS (CONTROL CENTER).

GAMMA = 0.900 (CONTINUED)

N \P →	0.750	0.900	0.950	0.975	0.990	0.995	0.999
51	1.339	1.915	2.281	2.609	2.997	3.266	3.827
52	1.337	1.912	2.277	2.604	2.992	3.260	3.821
53	1.335	1.908	2.274	2.600	2.987	3.255	3.815
54	1.333	1.905	2.270	2.596	2.982	3.250	3.808
55	1.331	1.902	2.266	2.591	2.978	3.245	3.803
56	1.328	1.899	2.263	2.587	2.973	3.240	3.797
57	1.326	1.896	2.259	2.584	2.969	3.235	3.791
58	1.325	1.894	2.256	2.580	2.964	3.230	3.786
59	1.323	1.891	2.253	2.576	2.960	3.226	3.781
60	1.321	1.888	2.250	2.573	2.956	3.221	3.775
61	1.319	1.886	2.247	2.569	2.952	3.217	3.771
62	1.317	1.884	2.244	2.566	2.949	3.213	3.766
63	1.316	1.881	2.241	2.563	2.945	3.209	3.761
64	1.314	1.879	2.239	2.560	2.941	3.205	3.756
65	1.313	1.877	2.236	2.557	2.938	3.201	3.752
66	1.311	1.874	2.233	2.554	2.934	3.198	3.748
67	1.310	1.872	2.231	2.551	2.931	3.194	3.744
68	1.308	1.870	2.228	2.548	2.928	3.191	3.739
69	1.307	1.868	2.226	2.545	2.925	3.187	3.735
70	1.305	1.866	2.224	2.543	2.922	3.184	3.732
71	1.304	1.864	2.221	2.540	2.919	3.180	3.728
72	1.303	1.862	2.219	2.537	2.916	3.177	3.724
73	1.301	1.861	2.217	2.535	2.913	3.174	3.720
74	1.300	1.859	2.215	2.533	2.910	3.171	3.717
75	1.299	1.857	2.213	2.530	2.907	3.168	3.713
76	1.298	1.855	2.211	2.528	2.905	3.165	3.710
77	1.296	1.854	2.209	2.526	2.902	3.162	3.707
78	1.295	1.852	2.207	2.523	2.900	3.160	3.703
79	1.294	1.850	2.205	2.521	2.897	3.157	3.700
80	1.293	1.849	2.203	2.519	2.895	3.154	3.697
81	1.292	1.847	2.201	2.517	2.892	3.152	3.694
82	1.291	1.846	2.199	2.515	2.890	3.149	3.691
83	1.290	1.844	2.198	2.513	2.888	3.147	3.688
84	1.289	1.843	2.196	2.511	2.885	3.144	3.685
85	1.288	1.841	2.194	2.509	2.883	3.142	3.683
86	1.287	1.840	2.192	2.507	2.881	3.140	3.680
87	1.286	1.839	2.191	2.505	2.879	3.137	3.677
88	1.285	1.837	2.189	2.504	2.877	3.135	3.675
89	1.284	1.836	2.188	2.502	2.875	3.133	3.672
90	1.283	1.835	2.186	2.500	2.873	3.131	3.669
91	1.282	1.834	2.185	2.498	2.871	3.128	3.667
92	1.282	1.832	2.183	2.497	2.869	3.126	3.664
93	1.281	1.831	2.182	2.495	2.867	3.124	3.662
94	1.280	1.830	2.180	2.493	2.865	3.122	3.660
95	1.279	1.829	2.179	2.492	2.863	3.120	3.657
96	1.278	1.828	2.178	2.490	2.862	3.118	3.655
97	1.277	1.826	2.176	2.489	2.860	3.116	3.653
98	1.277	1.825	2.175	2.487	2.858	3.115	3.651
99	1.276	1.824	2.174	2.486	2.856	3.113	3.649

TABLE 3.5.3. TWO-SIDED TOLERANCE LIMIT FACTORS (CONTROL CENTER).

GAMMA = 0.900 (CONTINUED)

N \P →	0.750	0.900	0.950	0.975	0.990	0.995	0.999
100	1.275	1.823	2.172	2.484	2.855	3.111	3.646
102	1.274	1.821	2.170	2.481	2.851	3.107	3.642
104	1.272	1.819	2.167	2.479	2.848	3.104	3.638
106	1.271	1.817	2.165	2.476	2.845	3.101	3.634
108	1.270	1.815	2.163	2.473	2.842	3.097	3.631
110	1.268	1.813	2.161	2.471	2.839	3.094	3.627
112	1.267	1.812	2.159	2.468	2.837	3.091	3.623
114	1.266	1.810	2.157	2.466	2.834	3.088	3.620
116	1.265	1.808	2.155	2.464	2.831	3.085	3.617
118	1.263	1.807	2.153	2.462	2.829	3.083	3.613
120	1.262	1.805	2.151	2.459	2.826	3.080	3.610
122	1.261	1.803	2.149	2.457	2.824	3.077	3.607
124	1.260	1.802	2.147	2.455	2.822	3.075	3.604
126	1.259	1.800	2.145	2.453	2.819	3.072	3.601
128	1.258	1.799	2.144	2.451	2.817	3.070	3.598
130	1.257	1.798	2.142	2.450	2.815	3.068	3.596
132	1.256	1.796	2.140	2.448	2.813	3.065	3.593
134	1.255	1.795	2.139	2.446	2.811	3.063	3.590
136	1.254	1.794	2.137	2.444	2.809	3.061	3.588
138	1.254	1.792	2.136	2.442	2.807	3.059	3.585
140	1.253	1.791	2.134	2.441	2.805	3.057	3.583
142	1.252	1.790	2.133	2.439	2.803	3.055	3.581
144	1.251	1.789	2.132	2.438	2.801	3.053	3.578
146	1.250	1.788	2.130	2.436	2.799	3.051	3.576
148	1.250	1.787	2.129	2.434	2.798	3.049	3.574
150	1.249	1.786	2.128	2.433	2.796	3.047	3.572
152	1.248	1.784	2.126	2.432	2.794	3.045	3.569
154	1.247	1.783	2.125	2.430	2.793	3.043	3.567
156	1.247	1.782	2.124	2.429	2.791	3.042	3.565
158	1.246	1.781	2.123	2.427	2.789	3.040	3.563
160	1.245	1.780	2.121	2.426	2.788	3.038	3.561
162	1.244	1.779	2.120	2.425	2.786	3.037	3.559
164	1.244	1.778	2.119	2.423	2.785	3.035	3.558
166	1.243	1.778	2.118	2.422	2.784	3.033	3.556
168	1.243	1.777	2.117	2.421	2.782	3.032	3.554
170	1.242	1.776	2.116	2.420	2.781	3.030	3.552
172	1.241	1.775	2.115	2.419	2.779	3.029	3.550
174	1.241	1.774	2.114	2.417	2.778	3.027	3.549
176	1.240	1.773	2.113	2.416	2.777	3.026	3.547
178	1.240	1.772	2.112	2.415	2.775	3.025	3.545
180	1.239	1.772	2.111	2.414	2.774	3.023	3.544
185	1.238	1.770	2.109	2.411	2.771	3.020	3.540
190	1.236	1.768	2.106	2.409	2.768	3.017	3.536
195	1.235	1.766	2.104	2.406	2.765	3.014	3.533
200	1.234	1.764	2.102	2.404	2.763	3.011	3.529
205	1.233	1.763	2.100	2.402	2.760	3.008	3.526
210	1.232	1.761	2.098	2.400	2.758	3.005	3.523
215	1.231	1.759	2.096	2.398	2.755	3.002	3.520
220	1.229	1.758	2.095	2.396	2.753	3.000	3.517

TABLE 3.5.4. TWO-SIDED TOLERANCE LIMIT FACTORS (CONTROL CENTER).

GAMMA = 0.900 (CONTINUED)

N \P →	0.750	0.900	0.950	0.975	0.990	0.995	0.999
225	1.228	1.757	2.093	2.394	2.751	2.998	3.514
230	1.228	1.755	2.091	2.392	2.749	2.995	3.511
235	1.227	1.754	2.090	2.390	2.746	2.993	3.508
240	1.226	1.753	2.088	2.388	2.744	2.991	3.506
245	1.225	1.751	2.087	2.387	2.743	2.989	3.503
250	1.224	1.750	2.085	2.385	2.741	2.987	3.501
255	1.223	1.749	2.084	2.383	2.739	2.985	3.499
260	1.222	1.748	2.083	2.382	2.737	2.983	3.497
265	1.222	1.747	2.081	2.380	2.735	2.981	3.494
270	1.221	1.746	2.080	2.379	2.734	2.979	3.492
275	1.220	1.745	2.079	2.377	2.732	2.977	3.490
280	1.220	1.744	2.078	2.376	2.731	2.976	3.488
285	1.219	1.743	2.077	2.375	2.729	2.974	3.486
290	1.218	1.742	2.076	2.374	2.728	2.972	3.484
295	1.218	1.741	2.074	2.372	2.726	2.971	3.483
300	1.217	1.740	2.073	2.371	2.725	2.969	3.481
310	1.216	1.738	2.071	2.369	2.722	2.966	3.477
320	1.215	1.737	2.069	2.367	2.720	2.964	3.474
330	1.214	1.735	2.068	2.364	2.717	2.961	3.471
340	1.212	1.734	2.066	2.362	2.715	2.959	3.468
350	1.212	1.732	2.064	2.361	2.713	2.956	3.465
360	1.211	1.731	2.063	2.359	2.711	2.954	3.463
370	1.210	1.730	2.061	2.357	2.709	2.952	3.460
380	1.209	1.728	2.060	2.355	2.707	2.950	3.458
390	1.208	1.727	2.058	2.354	2.705	2.948	3.455
400	1.207	1.726	2.057	2.352	2.703	2.946	3.453
425	1.205	1.724	2.054	2.349	2.699	2.941	3.448
450	1.204	1.721	2.051	2.345	2.695	2.937	3.443
475	1.202	1.719	2.048	2.342	2.692	2.933	3.439
500	1.201	1.717	2.046	2.340	2.689	2.930	3.435
525	1.199	1.715	2.044	2.337	2.686	2.927	3.431
550	1.198	1.713	2.041	2.335	2.683	2.924	3.427
575	1.197	1.712	2.040	2.332	2.680	2.921	3.424
600	1.196	1.710	2.038	2.330	2.678	2.918	3.421
625	1.195	1.709	2.036	2.328	2.676	2.916	3.418
650	1.194	1.707	2.034	2.327	2.674	2.914	3.416
700	1.192	1.705	2.032	2.323	2.670	2.910	3.411
750	1.191	1.703	2.029	2.320	2.667	2.906	3.406
800	1.190	1.701	2.027	2.318	2.664	2.903	3.403
850	1.188	1.699	2.025	2.315	2.661	2.900	3.399
900	1.187	1.697	2.023	2.313	2.658	2.897	3.396
950	1.186	1.696	2.021	2.311	2.656	2.894	3.393
1000	1.185	1.695	2.019	2.309	2.654	2.892	3.390
1500	1.178	1.685	2.008	2.296	2.639	2.876	3.371
2000	1.175	1.679	2.001	2.289	2.630	2.866	3.360
3000	1.170	1.673	1.993	2.280	2.620	2.855	3.347
5000	1.165	1.666	1.986	2.271	2.610	2.844	3.334
10000	1.161	1.660	1.978	2.262	2.600	2.833	3.321
∞	1.150	1.645	1.960	2.241	2.576	2.807	3.291

TABLE 3.6.1.   *TWO-SIDED TOLERANCE LIMIT FACTORS (CONTROL CENTER).*

GAMMA = 0.750

N \P →	0.750	0.900	0.950	0.975	0.990	0.995	0.999
2	4.393	6.109	7.178	8.121	9.231	9.992	11.574
3	2.486	3.489	4.116	4.671	5.325	5.775	6.710
4	2.035	2.871	3.397	3.862	4.412	4.789	5.575
5	1.829	2.589	3.068	3.493	3.995	4.341	5.060
6	1.709	2.425	2.877	3.278	3.752	4.079	4.759
7	1.629	2.316	2.750	3.135	3.591	3.905	4.560
8	1.572	2.238	2.659	3.033	3.476	3.781	4.417
9	1.530	2.179	2.590	2.955	3.388	3.687	4.309
10	1.496	2.132	2.535	2.894	3.319	3.612	4.223
11	1.469	2.095	2.491	2.845	3.263	3.552	4.154
12	1.447	2.064	2.455	2.804	3.217	3.502	4.097
13	1.428	2.037	2.424	2.769	3.178	3.460	4.048
14	1.412	2.015	2.398	2.739	3.144	3.424	4.006
15	1.398	1.996	2.375	2.714	3.115	3.392	3.970
16	1.386	1.978	2.355	2.691	3.089	3.364	3.938
17	1.375	1.963	2.338	2.671	3.067	3.340	3.910
18	1.365	1.950	2.322	2.653	3.046	3.318	3.885
19	1.357	1.938	2.307	2.637	3.028	3.298	3.862
20	1.349	1.927	2.295	2.623	3.012	3.280	3.841
21	1.342	1.917	2.283	2.609	2.997	3.264	3.823
22	1.335	1.908	2.272	2.597	2.983	3.249	3.805
23	1.329	1.900	2.262	2.586	2.970	3.235	3.790
24	1.324	1.892	2.253	2.576	2.958	3.223	3.775
25	1.319	1.885	2.245	2.566	2.948	3.211	3.761
26	1.314	1.878	2.237	2.557	2.938	3.200	3.749
27	1.310	1.872	2.230	2.549	2.928	3.190	3.737
28	1.306	1.866	2.223	2.542	2.920	3.181	3.726
29	1.302	1.861	2.217	2.534	2.911	3.172	3.716
30	1.299	1.856	2.211	2.528	2.904	3.163	3.706
31	1.295	1.851	2.205	2.521	2.896	3.156	3.697
32	1.292	1.847	2.200	2.515	2.890	3.148	3.689
33	1.289	1.843	2.195	2.510	2.883	3.141	3.681
34	1.286	1.839	2.190	2.504	2.877	3.135	3.673
35	1.284	1.835	2.186	2.499	2.871	3.128	3.666
36	1.281	1.831	2.182	2.494	2.866	3.122	3.659
37	1.279	1.828	2.178	2.490	2.860	3.117	3.652
38	1.276	1.825	2.174	2.485	2.855	3.111	3.646
39	1.274	1.822	2.170	2.481	2.851	3.106	3.640
40	1.272	1.819	2.167	2.477	2.846	3.101	3.634
41	1.270	1.816	2.163	2.473	2.842	3.096	3.629
42	1.268	1.813	2.160	2.470	2.838	3.092	3.623
43	1.266	1.810	2.157	2.466	2.834	3.088	3.618
44	1.265	1.808	2.154	2.463	2.830	3.083	3.613
45	1.263	1.806	2.151	2.460	2.826	3.079	3.609
46	1.261	1.803	2.148	2.456	2.823	3.075	3.604
47	1.260	1.801	2.146	2.454	2.819	3.072	3.600
48	1.258	1.799	2.143	2.451	2.816	3.068	3.596
49	1.257	1.797	2.141	2.448	2.813	3.065	3.592
50	1.255	1.795	2.138	2.445	2.810	3.061	3.588

TABLE 3.6.2. TWO-SIDED TOLERANCE LIMIT FACTORS (CONTROL CENTER).

GAMMA = 0.750 (CONTINUED)

N \P →	0.750	0.900	0.950	0.975	0.990	0.995	0.999
51	1.254	1.793	2.136	2.443	2.807	3.058	3.584
52	1.253	1.791	2.134	2.440	2.804	3.055	3.581
53	1.252	1.789	2.132	2.438	2.801	3.052	3.577
54	1.250	1.788	2.130	2.435	2.798	3.049	3.574
55	1.249	1.786	2.128	2.433	2.796	3.046	3.570
56	1.248	1.784	2.126	2.431	2.793	3.044	3.567
57	1.247	1.783	2.124	2.429	2.791	3.041	3.564
58	1.246	1.781	2.122	2.427	2.788	3.038	3.561
59	1.245	1.780	2.120	2.425	2.786	3.036	3.558
60	1.244	1.778	2.119	2.423	2.784	3.034	3.555
61	1.243	1.777	2.117	2.421	2.782	3.031	3.553
62	1.242	1.775	2.115	2.419	2.780	3.029	3.550
63	1.241	1.774	2.114	2.417	2.778	3.027	3.547
64	1.240	1.773	2.112	2.415	2.776	3.025	3.545
65	1.239	1.772	2.111	2.414	2.774	3.022	3.543
66	1.238	1.770	2.109	2.412	2.772	3.020	3.540
67	1.237	1.769	2.108	2.411	2.770	3.018	3.538
68	1.237	1.768	2.107	2.409	2.768	3.016	3.536
69	1.236	1.767	2.105	2.407	2.766	3.015	3.533
70	1.235	1.766	2.104	2.406	2.765	3.013	3.531
71	1.234	1.765	2.103	2.405	2.763	3.011	3.529
72	1.234	1.764	2.102	2.403	2.761	3.009	3.527
73	1.233	1.763	2.100	2.402	2.760	3.007	3.525
74	1.232	1.762	2.099	2.400	2.758	3.006	3.523
75	1.231	1.761	2.098	2.399	2.757	3.004	3.521
76	1.231	1.760	2.097	2.398	2.755	3.003	3.519
77	1.230	1.759	2.096	2.397	2.754	3.001	3.517
78	1.230	1.758	2.095	2.395	2.752	2.999	3.516
79	1.229	1.757	2.094	2.394	2.751	2.998	3.514
80	1.228	1.756	2.093	2.393	2.750	2.996	3.512
81	1.228	1.755	2.092	2.392	2.748	2.995	3.511
82	1.227	1.755	2.091	2.391	2.747	2.994	3.509
83	1.227	1.754	2.090	2.390	2.746	2.992	3.507
84	1.226	1.753	2.089	2.388	2.745	2.991	3.506
85	1.225	1.752	2.088	2.387	2.743	2.990	3.504
86	1.225	1.751	2.087	2.386	2.742	2.988	3.503
87	1.224	1.751	2.086	2.385	2.741	2.987	3.501
88	1.224	1.750	2.085	2.384	2.740	2.986	3.500
89	1.223	1.749	2.084	2.383	2.739	2.984	3.498
90	1.223	1.748	2.083	2.382	2.738	2.983	3.497
91	1.222	1.748	2.082	2.381	2.737	2.982	3.495
92	1.222	1.747	2.082	2.380	2.736	2.981	3.494
93	1.221	1.746	2.081	2.380	2.734	2.980	3.493
94	1.221	1.746	2.080	2.379	2.733	2.979	3.491
95	1.220	1.745	2.079	2.378	2.732	2.978	3.490
96	1.220	1.744	2.079	2.377	2.731	2.977	3.489
97	1.220	1.744	2.078	2.376	2.730	2.975	3.488
98	1.219	1.743	2.077	2.375	2.730	2.974	3.486
99	1.219	1.743	2.076	2.374	2.729	2.973	3.485

TABLE 3.6.3. TWO-SIDED TOLERANCE LIMIT FACTORS (CONTROL CENTER).

GAMMA = 0.750 (CONTINUED)

N \P →	0.750	0.900	0.950	0.975	0.990	0.995	0.999
100	1.218	1.742	2.076	2.374	2.728	2.972	3.484
102	1.218	1.741	2.074	2.372	2.726	2.970	3.482
104	1.217	1.740	2.073	2.371	2.724	2.969	3.480
106	1.216	1.739	2.072	2.369	2.722	2.967	3.477
108	1.215	1.738	2.070	2.368	2.721	2.965	3.475
110	1.215	1.737	2.069	2.366	2.719	2.963	3.473
112	1.214	1.736	2.068	2.365	2.718	2.962	3.471
114	1.213	1.735	2.067	2.364	2.716	2.960	3.470
116	1.213	1.734	2.066	2.362	2.715	2.958	3.468
118	1.212	1.733	2.065	2.361	2.713	2.957	3.466
120	1.211	1.732	2.064	2.360	2.712	2.955	3.464
122	1.211	1.731	2.063	2.359	2.711	2.954	3.462
124	1.210	1.730	2.062	2.358	2.709	2.952	3.461
126	1.210	1.729	2.061	2.357	2.708	2.951	3.459
128	1.209	1.729	2.060	2.355	2.707	2.950	3.458
130	1.208	1.728	2.059	2.354	2.706	2.948	3.456
132	1.208	1.727	2.058	2.353	2.704	2.947	3.455
134	1.207	1.726	2.057	2.352	2.703	2.946	3.453
136	1.207	1.726	2.056	2.351	2.702	2.945	3.452
138	1.206	1.725	2.055	2.350	2.701	2.944	3.450
140	1.206	1.724	2.055	2.350	2.700	2.942	3.449
142	1.205	1.724	2.054	2.349	2.699	2.941	3.448
144	1.205	1.723	2.053	2.348	2.698	2.940	3.446
146	1.205	1.722	2.052	2.347	2.697	2.939	3.445
148	1.204	1.722	2.052	2.346	2.696	2.938	3.444
150	1.204	1.721	2.051	2.345	2.695	2.937	3.443
152	1.203	1.721	2.050	2.344	2.694	2.936	3.442
154	1.203	1.720	2.049	2.344	2.693	2.935	3.440
156	1.202	1.719	2.049	2.343	2.692	2.934	3.439
158	1.202	1.719	2.048	2.342	2.692	2.933	3.438
160	1.202	1.718	2.047	2.341	2.691	2.932	3.437
162	1.201	1.718	2.047	2.341	2.690	2.931	3.436
164	1.201	1.717	2.046	2.340	2.689	2.930	3.435
166	1.201	1.717	2.046	2.339	2.688	2.930	3.434
168	1.200	1.716	2.045	2.339	2.688	2.929	3.433
170	1.200	1.716	2.044	2.338	2.687	2.928	3.432
172	1.200	1.715	2.044	2.337	2.686	2.927	3.431
174	1.199	1.715	2.043	2.337	2.685	2.926	3.430
176	1.199	1.714	2.043	2.336	2.685	2.925	3.429
178	1.199	1.714	2.042	2.335	2.684	2.925	3.428
180	1.198	1.713	2.042	2.335	2.683	2.924	3.427
185	1.198	1.712	2.040	2.333	2.681	2.922	3.425
190	1.197	1.711	2.039	2.332	2.680	2.920	3.423
195	1.196	1.710	2.038	2.331	2.678	2.919	3.421
200	1.195	1.709	2.037	2.329	2.677	2.917	3.419
205	1.195	1.708	2.036	2.328	2.675	2.916	3.418
210	1.194	1.708	2.035	2.327	2.674	2.914	3.416
215	1.194	1.707	2.034	2.326	2.673	2.913	3.414
220	1.193	1.706	2.033	2.325	2.671	2.911	3.413

TABLE 3.6.4. TWO-SIDED TOLERANCE LIMIT FACTORS (CONTROL CENTER).

GAMMA = 0.750 (CONTINUED)

N \P →	0.750	0.900	0.950	0.975	0.990	0.995	0.999
225	1.193	1.705	2.032	2.324	2.670	2.910	3.411
230	1.192	1.704	2.031	2.323	2.669	2.909	3.410
235	1.192	1.704	2.030	2.322	2.668	2.907	3.408
240	1.191	1.703	2.029	2.321	2.667	2.906	3.407
245	1.191	1.702	2.028	2.320	2.666	2.905	3.405
250	1.190	1.702	2.028	2.319	2.665	2.904	3.404
255	1.190	1.701	2.027	2.318	2.664	2.903	3.403
260	1.189	1.700	2.026	2.317	2.663	2.902	3.402
265	1.189	1.700	2.025	2.316	2.662	2.901	3.400
270	1.188	1.699	2.025	2.315	2.661	2.900	3.399
275	1.188	1.699	2.024	2.315	2.660	2.899	3.398
280	1.188	1.698	2.023	2.314	2.659	2.898	3.397
285	1.187	1.698	2.023	2.313	2.658	2.897	3.396
290	1.187	1.697	2.022	2.313	2.658	2.896	3.395
295	1.187	1.697	2.022	2.312	2.657	2.895	3.394
300	1.186	1.696	2.021	2.311	2.656	2.894	3.393
310	1.186	1.695	2.020	2.310	2.655	2.893	3.391
320	1.185	1.694	2.019	2.309	2.653	2.891	3.389
330	1.184	1.693	2.018	2.308	2.652	2.890	3.388
340	1.184	1.693	2.017	2.307	2.651	2.889	3.386
350	1.183	1.692	2.016	2.305	2.649	2.887	3.385
360	1.183	1.691	2.015	2.304	2.648	2.886	3.383
370	1.182	1.690	2.014	2.304	2.647	2.885	3.382
380	1.182	1.690	2.013	2.303	2.646	2.884	3.380
390	1.181	1.689	2.013	2.302	2.645	2.883	3.379
400	1.181	1.689	2.012	2.301	2.644	2.882	3.378
425	1.180	1.687	2.010	2.299	2.642	2.879	3.375
450	1.179	1.686	2.009	2.297	2.640	2.877	3.372
475	1.178	1.685	2.007	2.296	2.638	2.875	3.370
500	1.177	1.683	2.006	2.294	2.636	2.873	3.368
525	1.177	1.682	2.005	2.293	2.635	2.871	3.366
550	1.176	1.682	2.004	2.291	2.633	2.870	3.364
575	1.175	1.681	2.003	2.290	2.632	2.868	3.362
600	1.175	1.680	2.002	2.289	2.631	2.867	3.360
625	1.174	1.679	2.001	2.288	2.629	2.865	3.359
650	1.174	1.678	2.000	2.287	2.628	2.864	3.357
700	1.173	1.677	1.998	2.285	2.626	2.862	3.355
750	1.172	1.676	1.997	2.284	2.624	2.860	3.352
800	1.171	1.675	1.996	2.282	2.623	2.858	3.350
850	1.171	1.674	1.994	2.281	2.621	2.856	3.348
900	1.170	1.673	1.993	2.280	2.620	2.855	3.347
950	1.169	1.672	1.992	2.279	2.619	2.854	3.345
1000	1.169	1.671	1.992	2.278	2.617	2.852	3.344
1500	1.165	1.666	1.985	2.271	2.609	2.844	3.333
2000	1.163	1.663	1.982	2.266	2.605	2.838	3.327
3000	1.161	1.660	1.978	2.262	2.599	2.832	3.320
5000	1.158	1.656	1.974	2.257	2.594	2.827	3.313
10000	1.156	1.653	1.970	2.252	2.588	2.821	3.307
∞	1.150	1.645	1.960	2.241	2.576	2.807	3.291

TABLE 3.7.1. TWO-SIDED TOLERANCE LIMIT FACTORS (CONTROL CENTER).

GAMMA = 0.500

N \P →	0.750	0.900	0.950	0.975	0.990	0.995	0.999
2	2.057	2.869	3.376	3.822	4.348	4.708	5.456
3	1.582	2.229	2.634	2.993	3.415	3.706	4.309
4	1.440	2.039	2.416	2.750	3.144	3.415	3.978
5	1.370	1.945	2.308	2.630	3.010	3.272	3.817
6	1.328	1.888	2.243	2.558	2.930	3.186	3.720
7	1.299	1.850	2.199	2.509	2.876	3.128	3.655
8	1.279	1.823	2.167	2.473	2.836	3.086	3.608
9	1.263	1.802	2.143	2.447	2.806	3.055	3.572
10	1.251	1.785	2.124	2.425	2.783	3.029	3.543
11	1.242	1.772	2.109	2.408	2.764	3.009	3.521
12	1.234	1.761	2.096	2.394	2.748	2.992	3.502
13	1.227	1.752	2.085	2.382	2.735	2.978	3.486
14	1.221	1.744	2.076	2.372	2.723	2.966	3.472
15	1.216	1.737	2.068	2.363	2.714	2.955	3.460
16	1.212	1.731	2.061	2.356	2.705	2.946	3.450
17	1.208	1.726	2.055	2.349	2.698	2.938	3.441
18	1.205	1.721	2.050	2.343	2.691	2.931	3.433
19	1.202	1.717	2.045	2.338	2.685	2.925	3.425
20	1.199	1.714	2.041	2.333	2.680	2.919	3.419
21	1.197	1.710	2.037	2.329	2.675	2.914	3.413
22	1.195	1.707	2.034	2.325	2.670	2.909	3.407
23	1.193	1.705	2.030	2.321	2.666	2.905	3.403
24	1.191	1.702	2.027	2.318	2.662	2.901	3.398
25	1.189	1.700	2.025	2.315	2.659	2.897	3.394
26	1.188	1.698	2.022	2.312	2.656	2.894	3.390
27	1.186	1.696	2.020	2.309	2.653	2.890	3.386
28	1.185	1.694	2.018	2.307	2.650	2.888	3.383
29	1.184	1.692	2.016	2.305	2.648	2.885	3.380
30	1.183	1.691	2.014	2.303	2.645	2.882	3.377
31	1.182	1.689	2.012	2.301	2.643	2.880	3.374
32	1.181	1.688	2.011	2.299	2.641	2.878	3.372
33	1.180	1.686	2.009	2.297	2.639	2.876	3.370
34	1.179	1.685	2.008	2.295	2.637	2.874	3.367
35	1.178	1.684	2.006	2.294	2.636	2.872	3.365
36	1.177	1.683	2.005	2.293	2.634	2.870	3.363
37	1.177	1.682	2.004	2.291	2.632	2.868	3.361
38	1.176	1.681	2.003	2.290	2.631	2.867	3.359
39	1.175	1.680	2.002	2.289	2.630	2.865	3.358
40	1.175	1.679	2.001	2.287	2.628	2.864	3.356
41	1.174	1.678	2.000	2.286	2.627	2.862	3.355
42	1.173	1.677	1.999	2.285	2.626	2.861	3.353
43	1.173	1.677	1.998	2.284	2.625	2.860	3.352
44	1.172	1.676	1.997	2.283	2.624	2.859	3.350
45	1.172	1.675	1.996	2.282	2.623	2.858	3.349
46	1.171	1.675	1.995	2.281	2.622	2.857	3.348
47	1.171	1.674	1.994	2.281	2.621	2.856	3.347
48	1.170	1.673	1.994	2.280	2.620	2.855	3.345
49	1.170	1.673	1.993	2.279	2.619	2.854	3.344
50	1.170	1.672	1.992	2.278	2.618	2.853	3.343

TABLE 3.7.2. TWO-SIDED TOLERANCE LIMIT FACTORS (CONTROL CENTER).

GAMMA = 0.500 (CONTINUED)

N \P →	0.750	0.900	0.950	0.975	0.990	0.995	0.999
51	1.169	1.672	1.992	2.278	2.617	2.852	3.342
52	1.169	1.671	1.991	2.277	2.616	2.851	3.341
53	1.169	1.671	1.991	2.276	2.616	2.850	3.340
54	1.168	1.670	1.990	2.276	2.615	2.849	3.340
55	1.168	1.670	1.989	2.275	2.614	2.849	3.339
56	1.168	1.669	1.989	2.274	2.613	2.848	3.338
57	1.167	1.669	1.988	2.274	2.613	2.847	3.337
58	1.167	1.668	1.988	2.273	2.612	2.846	3.336
59	1.167	1.668	1.987	2.273	2.612	2.846	3.335
60	1.166	1.668	1.987	2.272	2.611	2.845	3.335
61	1.166	1.667	1.987	2.272	2.610	2.845	3.334
62	1.166	1.667	1.986	2.271	2.610	2.844	3.333
63	1.166	1.667	1.986	2.271	2.609	2.843	3.333
64	1.165	1.666	1.985	2.270	2.609	2.843	3.332
65	1.165	1.666	1.985	2.270	2.608	2.842	3.331
66	1.165	1.666	1.985	2.269	2.608	2.842	3.331
67	1.165	1.665	1.984	2.269	2.607	2.841	3.330
68	1.165	1.665	1.984	2.269	2.607	2.841	3.330
69	1.164	1.665	1.984	2.268	2.606	2.840	3.329
70	1.164	1.664	1.983	2.268	2.606	2.840	3.329
71	1.164	1.664	1.983	2.267	2.606	2.839	3.328
72	1.164	1.664	1.983	2.267	2.605	2.839	3.328
73	1.164	1.664	1.982	2.267	2.605	2.838	3.327
74	1.163	1.663	1.982	2.266	2.604	2.838	3.327
75	1.163	1.663	1.982	2.266	2.604	2.838	3.326
76	1.163	1.663	1.981	2.266	2.604	2.837	3.326
77	1.163	1.663	1.981	2.265	2.603	2.837	3.325
78	1.163	1.662	1.981	2.265	2.603	2.836	3.325
79	1.163	1.662	1.981	2.265	2.603	2.836	3.324
80	1.162	1.662	1.980	2.265	2.602	2.836	3.324
81	1.162	1.662	1.980	2.264	2.602	2.835	3.324
82	1.162	1.662	1.980	2.264	2.602	2.835	3.323
83	1.162	1.661	1.980	2.264	2.601	2.835	3.323
84	1.162	1.661	1.979	2.263	2.601	2.834	3.322
85	1.162	1.661	1.979	2.263	2.601	2.834	3.322
86	1.162	1.661	1.979	2.263	2.600	2.834	3.322
87	1.161	1.661	1.979	2.263	2.600	2.833	3.321
88	1.161	1.660	1.978	2.262	2.600	2.833	3.321
89	1.161	1.660	1.978	2.262	2.600	2.833	3.321
90	1.161	1.660	1.978	2.262	2.599	2.833	3.320
91	1.161	1.660	1.978	2.262	2.599	2.832	3.320
92	1.161	1.660	1.978	2.262	2.599	2.832	3.320
93	1.161	1.660	1.977	2.261	2.599	2.832	3.319
94	1.161	1.659	1.977	2.261	2.598	2.832	3.319
95	1.160	1.659	1.977	2.261	2.598	2.831	3.319
96	1.160	1.659	1.977	2.261	2.598	2.831	3.318
97	1.160	1.659	1.977	2.261	2.598	2.831	3.318
98	1.160	1.659	1.977	2.260	2.597	2.831	3.318
99	1.160	1.659	1.976	2.260	2.597	2.830	3.318

TABLE 3.7.3. *TWO-SIDED TOLERANCE LIMIT FACTORS (CONTROL CENTER).*

GAMMA = 0.500 (*CONTINUED*)

N \P →	0.750	0.900	0.950	0.975	0.990	0.995	0.999
100	1.160	1.659	1.976	2.260	2.597	2.830	3.317
102	1.160	1.658	1.976	2.260	2.597	2.830	3.317
104	1.160	1.658	1.976	2.259	2.596	2.829	3.316
106	1.159	1.658	1.975	2.259	2.596	2.829	3.316
108	1.159	1.658	1.975	2.259	2.595	2.828	3.315
110	1.159	1.657	1.975	2.258	2.595	2.828	3.315
112	1.159	1.657	1.974	2.258	2.595	2.828	3.315
114	1.159	1.657	1.974	2.258	2.594	2.827	3.314
116	1.159	1.657	1.974	2.257	2.594	2.827	3.314
118	1.158	1.656	1.974	2.257	2.594	2.827	3.313
120	1.158	1.656	1.974	2.257	2.594	2.826	3.313
122	1.158	1.656	1.973	2.257	2.593	2.826	3.313
124	1.158	1.656	1.973	2.256	2.593	2.826	3.312
126	1.158	1.656	1.973	2.256	2.593	2.825	3.312
128	1.158	1.656	1.973	2.256	2.592	2.825	3.312
130	1.158	1.655	1.972	2.256	2.592	2.825	3.311
132	1.158	1.655	1.972	2.255	2.592	2.825	3.311
134	1.158	1.655	1.972	2.255	2.592	2.824	3.311
136	1.157	1.655	1.972	2.255	2.591	2.824	3.310
138	1.157	1.655	1.972	2.255	2.591	2.824	3.310
140	1.157	1.655	1.972	2.255	2.591	2.824	3.310
142	1.157	1.655	1.971	2.254	2.591	2.823	3.310
144	1.157	1.654	1.971	2.254	2.591	2.823	3.309
146	1.157	1.654	1.971	2.254	2.590	2.823	3.309
148	1.157	1.654	1.971	2.254	2.590	2.823	3.309
150	1.157	1.654	1.971	2.254	2.590	2.822	3.309
152	1.157	1.654	1.971	2.254	2.590	2.822	3.308
154	1.157	1.654	1.971	2.253	2.590	2.822	3.308
156	1.157	1.654	1.970	2.253	2.589	2.822	3.308
158	1.156	1.654	1.970	2.253	2.589	2.822	3.308
160	1.156	1.653	1.970	2.253	2.589	2.822	3.307
162	1.156	1.653	1.970	2.253	2.589	2.821	3.307
164	1.156	1.653	1.970	2.253	2.589	2.821	3.307
166	1.156	1.653	1.970	2.253	2.589	2.821	3.307
168	1.156	1.653	1.970	2.252	2.589	2.821	3.307
170	1.156	1.653	1.970	2.252	2.588	2.821	3.306
172	1.156	1.653	1.969	2.252	2.588	2.821	3.306
174	1.156	1.653	1.969	2.252	2.588	2.820	3.306
176	1.156	1.653	1.969	2.252	2.588	2.820	3.306
178	1.156	1.653	1.969	2.252	2.588	2.820	3.306
180	1.156	1.652	1.969	2.252	2.588	2.820	3.306
185	1.156	1.652	1.969	2.251	2.587	2.820	3.305
190	1.155	1.652	1.969	2.251	2.587	2.819	3.305
195	1.155	1.652	1.968	2.251	2.587	2.819	3.304
200	1.155	1.652	1.968	2.251	2.586	2.819	3.304
205	1.155	1.652	1.968	2.250	2.586	2.818	3.304
210	1.155	1.651	1.968	2.250	2.586	2.818	3.303
215	1.155	1.651	1.968	2.250	2.586	2.818	3.303
220	1.155	1.651	1.967	2.250	2.586	2.818	3.303

TABLE 3.7.4.   TWO-SIDED TOLERANCE LIMIT FACTORS (CONTROL CENTER).

GAMMA = 0.500 (CONTINUED)

N \P →	0.750	0.900	0.950	0.975	0.990	0.995	0.999
225	1.155	1.651	1.967	2.250	2.585	2.817	3.303
230	1.155	1.651	1.967	2.249	2.585	2.817	3.302
235	1.154	1.651	1.967	2.249	2.585	2.817	3.302
240	1.154	1.651	1.967	2.249	2.585	2.817	3.302
245	1.154	1.650	1.967	2.249	2.585	2.817	3.302
250	1.154	1.650	1.966	2.249	2.584	2.816	3.301
255	1.154	1.650	1.966	2.249	2.584	2.816	3.301
260	1.154	1.650	1.966	2.249	2.584	2.816	3.301
265	1.154	1.650	1.966	2.248	2.584	2.816	3.301
270	1.154	1.650	1.966	2.248	2.584	2.816	3.301
275	1.154	1.650	1.966	2.248	2.584	2.815	3.300
280	1.154	1.650	1.966	2.248	2.583	2.815	3.300
285	1.154	1.650	1.966	2.248	2.583	2.815	3.300
290	1.154	1.650	1.966	2.248	2.583	2.815	3.300
295	1.154	1.649	1.965	2.248	2.583	2.815	3.300
300	1.154	1.649	1.965	2.248	2.583	2.815	3.300
310	1.153	1.649	1.965	2.247	2.583	2.815	3.299
320	1.153	1.649	1.965	2.247	2.583	2.814	3.299
330	1.153	1.649	1.965	2.247	2.582	2.814	3.299
340	1.153	1.649	1.965	2.247	2.582	2.814	3.299
350	1.153	1.649	1.965	2.247	2.582	2.814	3.298
360	1.153	1.649	1.964	2.247	2.582	2.814	3.298
370	1.153	1.649	1.964	2.246	2.582	2.813	3.298
380	1.153	1.648	1.964	2.246	2.581	2.813	3.298
390	1.153	1.648	1.964	2.246	2.581	2.813	3.298
400	1.153	1.648	1.964	2.246	2.581	2.813	3.297
425	1.153	1.648	1.964	2.246	2.581	2.813	3.297
450	1.152	1.648	1.964	2.246	2.581	2.812	3.297
475	1.152	1.648	1.963	2.245	2.580	2.812	3.296
500	1.152	1.648	1.963	2.245	2.580	2.812	3.296
525	1.152	1.647	1.963	2.245	2.580	2.811	3.296
550	1.152	1.647	1.963	2.245	2.580	2.811	3.295
575	1.152	1.647	1.963	2.245	2.580	2.811	3.295
600	1.152	1.647	1.963	2.245	2.579	2.811	3.295
625	1.152	1.647	1.963	2.244	2.579	2.811	3.295
650	1.152	1.647	1.962	2.244	2.579	2.811	3.295
700	1.152	1.647	1.962	2.244	2.579	2.810	3.294
750	1.152	1.647	1.962	2.244	2.579	2.810	3.294
800	1.152	1.647	1.962	2.244	2.579	2.810	3.294
850	1.151	1.646	1.962	2.244	2.578	2.810	3.294
900	1.151	1.646	1.962	2.243	2.578	2.810	3.294
950	1.151	1.646	1.962	2.243	2.578	2.809	3.293
1000	1.151	1.646	1.962	2.243	2.578	2.809	3.293
1500	1.151	1.646	1.961	2.243	2.577	2.809	3.292
2000	1.151	1.646	1.961	2.242	2.577	2.808	3.292
3000	1.151	1.645	1.961	2.242	2.577	2.808	3.291
5000	1.151	1.645	1.960	2.242	2.576	2.808	3.291
10000	1.150	1.645	1.960	2.242	2.576	2.807	3.291
∞	1.150	1.645	1.960	2.241	2.576	2.807	3.291

TABLE 4

Two-sided tolerance limit factors for a normal distribution (control both tails) for values of P = 0.125, 0.10, 0.05, 0.025, 0.01, 0.005, 0.0005 and N = 2(1)100(2)180(5)300(10)400(25)650(50)1000, 1500, 2000, 3000, 5000, 10000, ∞.

TABLE 4.1.1. *TWÓ-SIDED TOLERANCE LIMIT FACTORS (CONTROL BOTH TAILS).*

GAMMA = 0.995

N \P →	0.125	0.100	0.050	0.025	0.010	0.005	0.0005
2	273.599	294.535	352.510	402.794	461.260	501.071	615.120
3	23.273	25.088	30.138	34.538	39.667	43.166	53.208
4	10.392	11.212	13.501	15.501	17.838	19.433	24.018
5	6.933	7.486	9.031	10.385	11.968	13.049	16.160
6	5.415	5.850	7.070	8.140	9.392	10.248	12.711
7	4.573	4.944	5.984	6.897	7.966	8.697	10.802
8	4.039	4.369	5.296	6.109	7.063	7.715	9.593
9	3.669	3.972	4.820	5.566	6.440	7.038	8.759
10	3.398	3.680	4.471	5.167	5.982	6.540	8.147
11	3.189	3.455	4.203	4.861	5.632	6.160	7.679
12	3.024	3.277	3.991	4.618	5.354	5.858	7.308
13	2.888	3.132	3.818	4.421	5.128	5.612	7.006
14	2.776	3.012	3.674	4.257	4.941	5.408	6.755
15	2.681	2.909	3.553	4.118	4.782	5.236	6.544
16	2.599	2.822	3.448	3.999	4.645	5.088	6.362
17	2.527	2.745	3.357	3.896	4.527	4.959	6.204
18	2.465	2.678	3.278	3.805	4.423	4.847	6.065
19	2.409	2.618	3.207	3.724	4.331	4.747	5.943
20	2.359	2.565	3.144	3.652	4.249	4.657	5.833
21	2.314	2.517	3.087	3.588	4.175	4.577	5.735
22	2.274	2.474	3.035	3.529	4.108	4.505	5.646
23	2.237	2.434	2.989	3.476	4.048	4.439	5.565
24	2.203	2.398	2.946	3.427	3.992	4.378	5.491
25	2.172	2.365	2.906	3.383	3.941	4.323	5.423
26	2.143	2.334	2.870	3.341	3.894	4.272	5.361
27	2.117	2.306	2.837	3.303	3.850	4.225	5.303
28	2.092	2.279	2.805	3.268	3.810	4.181	5.249
29	2.069	2.254	2.776	3.235	3.772	4.140	5.199
30	2.047	2.231	2.749	3.204	3.737	4.102	5.152
31	2.027	2.210	2.723	3.175	3.704	4.066	5.108
32	2.008	2.189	2.699	3.148	3.673	4.032	5.067
33	1.990	2.170	2.677	3.122	3.644	4.001	5.028
34	1.973	2.152	2.656	3.098	3.616	3.971	4.991
35	1.957	2.135	2.635	3.075	3.590	3.943	4.957
36	1.942	2.119	2.616	3.053	3.566	3.916	4.924
37	1.927	2.104	2.598	3.033	3.542	3.891	4.893
38	1.913	2.089	2.581	3.013	3.520	3.866	4.864
39	1.900	2.075	2.565	2.995	3.499	3.844	4.836
40	1.888	2.062	2.549	2.977	3.479	3.822	4.809
41	1.876	2.049	2.534	2.960	3.459	3.801	4.784
42	1.865	2.037	2.520	2.944	3.441	3.781	4.759
43	1.854	2.025	2.506	2.929	3.424	3.762	4.736
44	1.843	2.014	2.493	2.914	3.407	3.744	4.714
45	1.833	2.004	2.481	2.900	3.391	3.726	4.692
46	1.824	1.993	2.469	2.886	3.375	3.710	4.672
47	1.814	1.983	2.457	2.873	3.360	3.693	4.652
48	1.806	1.974	2.446	2.861	3.346	3.678	4.633
49	1.797	1.965	2.435	2.848	3.332	3.663	4.615
50	1.789	1.956	2.425	2.837	3.319	3.649	4.597

TABLE 4.1.2. TWO-SIDED TOLERANCE LIMIT FACTORS (CONTROL BOTH TAILS).

GAMMA = 0.995 (CONTINUED)

N \P →	0.125	0.100	0.050	0.025	0.010	0.005	0.0005
51	1.781	1.948	2.415	2.826	3.306	3.635	4.580
52	1.773	1.939	2.406	2.815	3.294	3.621	4.564
53	1.766	1.931	2.396	2.804	3.282	3.608	4.548
54	1.758	1.924	2.387	2.794	3.270	3.596	4.533
55	1.751	1.916	2.379	2.784	3.259	3.584	4.518
56	1.745	1.909	2.370	2.775	3.248	3.572	4.504
57	1.738	1.902	2.362	2.766	3.238	3.561	4.490
58	1.732	1.896	2.354	2.757	3.228	3.550	4.477
59	1.726	1.889	2.347	2.748	3.218	3.539	4.464
60	1.720	1.883	2.339	2.740	3.208	3.529	4.451
61	1.714	1.877	2.332	2.732	3.199	3.519	4.439
62	1.708	1.871	2.325	2.724	3.190	3.509	4.427
63	1.703	1.865	2.318	2.716	3.182	3.500	4.415
64	1.697	1.859	2.312	2.708	3.173	3.491	4.404
65	1.692	1.854	2.305	2.701	3.165	3.482	4.393
66	1.687	1.848	2.299	2.694	3.157	3.473	4.383
67	1.682	1.843	2.293	2.687	3.149	3.465	4.372
68	1.678	1.838	2.287	2.681	3.141	3.456	4.362
69	1.673	1.833	2.281	2.674	3.134	3.448	4.352
70	1.668	1.828	2.276	2.668	3.127	3.440	4.343
71	1.664	1.823	2.270	2.661	3.120	3.433	4.333
72	1.660	1.819	2.265	2.655	3.113	3.425	4.324
73	1.655	1.814	2.259	2.649	3.106	3.418	4.315
74	1.651	1.810	2.254	2.644	3.099	3.411	4.307
75	1.647	1.806	2.249	2.638	3.093	3.404	4.298
76	1.643	1.801	2.244	2.633	3.087	3.397	4.290
77	1.639	1.797	2.240	2.627	3.081	3.391	4.282
78	1.636	1.793	2.235	2.622	3.075	3.384	4.274
79	1.632	1.789	2.230	2.617	3.069	3.378	4.266
80	1.628	1.786	2.226	2.612	3.063	3.372	4.259
81	1.625	1.782	2.222	2.607	3.057	3.366	4.251
82	1.621	1.778	2.217	2.602	3.052	3.360	4.244
83	1.618	1.775	2.213	2.597	3.047	3.354	4.237
84	1.615	1.771	2.209	2.593	3.041	3.348	4.230
85	1.611	1.768	2.205	2.588	3.036	3.343	4.223
86	1.608	1.764	2.201	2.584	3.031	3.337	4.217
87	1.605	1.761	2.197	2.579	3.026	3.332	4.210
88	1.602	1.758	2.193	2.575	3.021	3.326	4.204
89	1.599	1.754	2.190	2.571	3.017	3.321	4.197
90	1.596	1.751	2.186	2.567	3.012	3.316	4.191
91	1.593	1.748	2.182	2.563	3.007	3.311	4.185
92	1.590	1.745	2.179	2.559	3.003	3.306	4.179
93	1.587	1.742	2.175	2.555	2.998	3.302	4.173
94	1.585	1.739	2.172	2.551	2.994	3.297	4.168
95	1.582	1.737	2.169	2.547	2.990	3.292	4.162
96	1.579	1.734	2.165	2.544	2.986	3.288	4.157
97	1.577	1.731	2.162	2.540	2.982	3.283	4.151
98	1.574	1.728	2.159	2.536	2.978	3.279	4.146
99	1.572	1.726	2.156	2.533	2.974	3.275	4.141

TABLE 4.1.3. TWO-SIDED TOLERANCE LIMIT FACTORS (CONTROL BOTH TAILS).

GAMMA = 0.995 (CONTINUED)

N \P →	0.125	0.100	0.050	0.025	0.010	0.005	0.0005
100	1.569	1.723	2.153	2.529	2.970	3.271	4.136
102	1.564	1.718	2.147	2.523	2.962	3.262	4.125
104	1.560	1.713	2.141	2.516	2.955	3.254	4.116
106	1.555	1.708	2.136	2.510	2.948	3.247	4.106
108	1.551	1.704	2.130	2.504	2.941	3.239	4.097
110	1.547	1.699	2.125	2.498	2.934	3.232	4.089
112	1.543	1.695	2.120	2.492	2.928	3.225	4.080
114	1.539	1.690	2.115	2.487	2.921	3.218	4.072
116	1.535	1.686	2.110	2.481	2.915	3.212	4.064
118	1.531	1.682	2.106	2.476	2.909	3.205	4.056
120	1.527	1.678	2.101	2.471	2.904	3.199	4.048
122	1.524	1.675	2.097	2.466	2.898	3.193	4.041
124	1.520	1.671	2.093	2.461	2.893	3.187	4.034
126	1.517	1.667	2.088	2.457	2.887	3.181	4.027
128	1.514	1.664	2.084	2.452	2.882	3.176	4.020
130	1.510	1.661	2.080	2.448	2.877	3.170	4.013
132	1.507	1.657	2.077	2.443	2.872	3.165	4.007
134	1.504	1.654	2.073	2.439	2.868	3.160	4.001
136	1.501	1.651	2.069	2.435	2.863	3.155	3.995
138	1.498	1.648	2.066	2.431	2.858	3.150	3.989
140	1.495	1.645	2.062	2.427	2.854	3.145	3.983
142	1.493	1.642	2.059	2.423	2.850	3.141	3.977
144	1.490	1.639	2.056	2.420	2.845	3.136	3.971
146	1.487	1.636	2.052	2.416	2.841	3.132	3.966
148	1.485	1.634	2.049	2.412	2.837	3.127	3.961
150	1.482	1.631	2.046	2.409	2.833	3.123	3.955
152	1.480	1.628	2.043	2.406	2.829	3.119	3.950
154	1.477	1.626	2.040	2.402	2.826	3.115	3.945
156	1.475	1.623	2.037	2.399	2.822	3.111	3.941
158	1.473	1.621	2.034	2.396	2.818	3.107	3.936
160	1.470	1.618	2.031	2.393	2.815	3.103	3.931
162	1.468	1.616	2.029	2.390	2.811	3.099	3.926
164	1.466	1.614	2.026	2.387	2.808	3.095	3.922
166	1.464	1.611	2.023	2.384	2.804	3.092	3.918
168	1.462	1.609	2.021	2.381	2.801	3.088	3.913
170	1.460	1.607	2.018	2.378	2.798	3.085	3.909
172	1.458	1.605	2.016	2.375	2.795	3.081	3.905
174	1.456	1.603	2.013	2.372	2.792	3.078	3.901
176	1.454	1.601	2.011	2.370	2.789	3.075	3.897
178	1.452	1.599	2.009	2.367	2.786	3.072	3.893
180	1.450	1.597	2.006	2.365	2.783	3.069	3.889
185	1.445	1.592	2.001	2.358	2.776	3.061	3.880
190	1.441	1.587	1.996	2.352	2.769	3.054	3.871
195	1.437	1.583	1.990	2.347	2.763	3.047	3.862
200	1.433	1.579	1.986	2.341	2.756	3.040	3.854
205	1.429	1.575	1.981	2.336	2.750	3.033	3.846
210	1.425	1.571	1.976	2.331	2.745	3.027	3.839
215	1.422	1.567	1.972	2.326	2.739	3.021	3.832
220	1.418	1.563	1.968	2.321	2.734	3.016	3.825

TABLE 4.1.4. *TWO-SIDED TOLERANCE LIMIT FACTORS (CONTROL BOTH TAILS).*

GAMMA = 0.995 (CONTINUED)

N \P →	0.125	0.100	0.050	0.025	0.010	0.005	0.0005
225	1.415	1.560	1.964	2.317	2.729	3.010	3.818
230	1.412	1.557	1.960	2.312	2.724	3.005	3.811
235	1.409	1.553	1.956	2.308	2.719	3.000	3.805
240	1.406	1.550	1.953	2.304	2.714	2.995	3.799
245	1.403	1.547	1.949	2.300	2.710	2.990	3.793
250	1.400	1.544	1.946	2.296	2.706	2.985	3.788
255	1.397	1.541	1.942	2.293	2.702	2.981	3.782
260	1.395	1.539	1.939	2.289	2.697	2.976	3.777
265	1.392	1.536	1.936	2.286	2.694	2.972	3.772
270	1.390	1.533	1.933	2.282	2.690	2.968	3.767
275	1.387	1.531	1.930	2.279	2.686	2.964	3.762
280	1.385	1.528	1.927	2.276	2.683	2.960	3.757
285	1.383	1.526	1.925	2.273	2.679	2.956	3.753
290	1.381	1.524	1.922	2.270	2.676	2.953	3.748
295	1.379	1.521	1.919	2.267	2.672	2.949	3.744
300	1.376	1.519	1.917	2.264	2.669	2.946	3.740
310	1.372	1.515	1.912	2.259	2.663	2.939	3.732
320	1.369	1.511	1.907	2.253	2.657	2.933	3.724
330	1.365	1.507	1.903	2.249	2.652	2.927	3.717
340	1.362	1.504	1.899	2.244	2.646	2.921	3.710
350	1.358	1.500	1.895	2.239	2.641	2.916	3.703
360	1.355	1.497	1.891	2.235	2.636	2.910	3.697
370	1.352	1.494	1.888	2.231	2.632	2.905	3.691
380	1.349	1.491	1.884	2.227	2.627	2.901	3.685
390	1.347	1.488	1.881	2.223	2.623	2.896	3.679
400	1.344	1.485	1.877	2.220	2.619	2.892	3.674
425	1.338	1.478	1.870	2.211	2.610	2.881	3.661
450	1.332	1.472	1.863	2.204	2.601	2.872	3.650
475	1.327	1.467	1.857	2.197	2.593	2.864	3.640
500	1.322	1.462	1.851	2.190	2.586	2.856	3.630
525	1.318	1.457	1.846	2.184	2.579	2.848	3.621
550	1.314	1.453	1.841	2.179	2.573	2.842	3.613
575	1.310	1.449	1.836	2.174	2.567	2.835	3.605
600	1.306	1.445	1.832	2.169	2.562	2.830	3.598
625	1.303	1.442	1.828	2.164	2.557	2.824	3.592
650	1.300	1.439	1.824	2.160	2.552	2.819	3.585
700	1.294	1.433	1.817	2.152	2.543	2.809	3.574
750	1.289	1.427	1.811	2.145	2.535	2.801	3.564
800	1.284	1.422	1.805	2.139	2.528	2.793	3.555
850	1.280	1.418	1.800	2.134	2.522	2.787	3.546
900	1.276	1.414	1.796	2.128	2.516	2.780	3.539
950	1.273	1.410	1.792	2.124	2.511	2.775	3.532
1000	1.270	1.407	1.788	2.119	2.506	2.769	3.525
1500	1.247	1.383	1.760	2.089	2.471	2.732	3.480
2000	1.234	1.369	1.744	2.071	2.451	2.710	3.454
3000	1.218	1.352	1.726	2.050	2.428	2.685	3.423
5000	1.202	1.336	1.707	2.029	2.404	2.660	3.392
10000	1.187	1.320	1.689	2.009	2.381	2.635	3.362
∞	1.150	1.282	1.645	1.960	2.326	2.576	3.291

TABLE 4.2.1. *TWO-SIDED TOLERANCE LIMIT FACTORS (CONTROL BOTH TAILS).*

GAMMA = 0.990

N \P →	0.125	0.100	0.050	0.025	0.010	0.005	0.0005
2	136.796	147.264	176.251	201.393	230.625	250.530	307.554
3	16.432	17.714	21.281	24.388	28.011	30.483	37.575
4	8.205	8.853	10.664	12.246	14.093	15.354	18.978
5	5.777	6.239	7.531	8.661	9.984	10.887	13.484
6	4.655	5.032	6.085	7.008	8.089	8.827	10.951
7	4.013	4.340	5.258	6.063	7.006	7.650	9.504
8	3.596	3.892	4.723	5.451	6.304	6.888	8.567
9	3.303	3.577	4.346	5.021	5.812	6.353	7.910
10	3.084	3.342	4.066	4.702	5.447	5.956	7.422
11	2.914	3.160	3.849	4.454	5.163	5.649	7.045
12	2.778	3.014	3.675	4.256	4.937	5.403	6.743
13	2.667	2.894	3.533	4.094	4.752	5.201	6.496
14	2.573	2.794	3.414	3.958	4.597	5.033	6.290
15	2.493	2.708	3.312	3.843	4.465	4.890	6.114
16	2.424	2.635	3.225	3.743	4.351	4.767	5.963
17	2.364	2.570	3.149	3.657	4.252	4.659	5.831
18	2.311	2.513	3.081	3.580	4.165	4.564	5.715
19	2.264	2.463	3.021	3.512	4.087	4.480	5.612
20	2.221	2.417	2.968	3.451	4.017	4.405	5.520
21	2.183	2.377	2.919	3.396	3.955	4.337	5.436
22	2.148	2.339	2.876	3.346	3.898	4.275	5.361
23	2.116	2.305	2.835	3.301	3.846	4.219	5.292
24	2.087	2.274	2.799	3.259	3.799	4.168	5.229
25	2.060	2.246	2.765	3.221	3.755	4.120	5.171
26	2.036	2.219	2.734	3.185	3.715	4.076	5.118
27	2.013	2.194	2.705	3.153	3.677	4.036	5.068
28	1.991	2.172	2.678	3.122	3.642	3.998	5.022
29	1.971	2.150	2.653	3.093	3.610	3.963	4.979
30	1.952	2.130	2.629	3.067	3.579	3.930	4.939
31	1.935	2.111	2.607	3.042	3.551	3.899	4.901
32	1.918	2.094	2.586	3.018	3.524	3.870	4.865
33	1.902	2.077	2.567	2.996	3.499	3.843	4.832
34	1.888	2.061	2.548	2.975	3.475	3.817	4.800
35	1.874	2.046	2.530	2.955	3.452	3.792	4.771
36	1.860	2.032	2.514	2.936	3.431	3.769	4.742
37	1.848	2.019	2.498	2.919	3.411	3.747	4.715
38	1.836	2.006	2.483	2.902	3.391	3.726	4.690
39	1.824	1.994	2.469	2.885	3.373	3.706	4.666
40	1.813	1.982	2.455	2.870	3.355	3.687	4.642
41	1.803	1.971	2.442	2.855	3.339	3.669	4.620
42	1.793	1.960	2.430	2.841	3.323	3.652	4.599
43	1.783	1.950	2.418	2.828	3.307	3.635	4.579
44	1.774	1.940	2.406	2.815	3.293	3.619	4.559
45	1.765	1.931	2.395	2.802	3.279	3.604	4.541
46	1.757	1.922	2.385	2.790	3.265	3.590	4.523
47	1.748	1.913	2.375	2.779	3.252	3.575	4.506
48	1.741	1.905	2.365	2.768	3.239	3.562	4.489
49	1.733	1.897	2.355	2.757	3.227	3.549	4.473
50	1.726	1.889	2.346	2.747	3.216	3.536	4.458

TABLE 4.2.2. *TWO-SIDED TOLERANCE LIMIT FACTORS (CONTROL BOTH TAILS).*

GAMMA = 0.990 *(CONTINUED)*

N \P →	0.125	0.100	0.050	0.025	0.010	0.005	0.0005
51	1.719	1.882	2.337	2.737	3.205	3.524	4.443
52	1.712	1.874	2.329	2.727	3.194	3.512	4.429
53	1.705	1.867	2.321	2.718	3.183	3.501	4.415
54	1.699	1.861	2.313	2.709	3.173	3.490	4.401
55	1.693	1.854	2.305	2.701	3.163	3.479	4.388
56	1.687	1.848	2.298	2.692	3.154	3.469	4.376
57	1.681	1.841	2.291	2.684	3.145	3.459	4.364
58	1.675	1.836	2.284	2.676	3.136	3.450	4.352
59	1.670	1.830	2.277	2.669	3.127	3.440	4.341
60	1.665	1.824	2.270	2.661	3.119	3.431	4.330
61	1.659	1.819	2.264	2.654	3.110	3.422	4.319
62	1.654	1.813	2.258	2.647	3.103	3.414	4.308
63	1.650	1.808	2.252	2.640	3.095	3.405	4.298
64	1.645	1.803	2.246	2.634	3.087	3.397	4.288
65	1.640	1.798	2.240	2.627	3.080	3.389	4.279
66	1.636	1.793	2.235	2.621	3.073	3.382	4.269
67	1.631	1.789	2.229	2.615	3.066	3.374	4.260
68	1.627	1.784	2.224	2.609	3.059	3.367	4.251
69	1.623	1.780	2.219	2.603	3.053	3.360	4.243
70	1.619	1.776	2.214	2.598	3.046	3.353	4.234
71	1.615	1.771	2.209	2.592	3.040	3.346	4.226
72	1.611	1.767	2.204	2.587	3.034	3.340	4.218
73	1.607	1.763	2.200	2.582	3.028	3.333	4.210
74	1.604	1.759	2.195	2.576	3.022	3.327	4.203
75	1.600	1.756	2.191	2.571	3.017	3.321	4.195
76	1.597	1.752	2.186	2.567	3.011	3.315	4.188
77	1.593	1.748	2.182	2.562	3.006	3.309	4.181
78	1.590	1.745	2.178	2.557	3.000	3.303	4.174
79	1.586	1.741	2.174	2.552	2.995	3.298	4.167
80	1.583	1.738	2.170	2.548	2.990	3.292	4.160
81	1.580	1.734	2.166	2.544	2.985	3.287	4.154
82	1.577	1.731	2.162	2.539	2.980	3.282	4.147
83	1.574	1.728	2.158	2.535	2.976	3.276	4.141
84	1.571	1.725	2.155	2.531	2.971	3.271	4.135
85	1.568	1.722	2.151	2.527	2.966	3.266	4.129
86	1.565	1.719	2.148	2.523	2.962	3.262	4.123
87	1.562	1.716	2.144	2.519	2.957	3.257	4.117
88	1.560	1.713	2.141	2.515	2.953	3.252	4.111
89	1.557	1.710	2.138	2.512	2.949	3.248	4.106
90	1.554	1.707	2.134	2.508	2.945	3.243	4.100
91	1.552	1.704	2.131	2.504	2.941	3.239	4.095
92	1.549	1.702	2.128	2.501	2.937	3.234	4.090
93	1.547	1.699	2.125	2.497	2.933	3.230	4.085
94	1.544	1.696	2.122	2.494	2.929	3.226	4.080
95	1.542	1.694	2.119	2.491	2.925	3.222	4.075
96	1.539	1.691	2.116	2.487	2.921	3.218	4.070
97	1.537	1.689	2.113	2.484	2.918	3.214	4.065
98	1.535	1.687	2.110	2.481	2.914	3.210	4.060
99	1.533	1.684	2.108	2.478	2.911	3.206	4.056

TABLE 4.2.3. TWO-SIDED TOLERANCE LIMIT FACTORS (CONTROL BOTH TAILS).

GAMMA = 0.990 (CONTINUED)

N \P →	0.125	0.100	0.050	0.025	0.010	0.005	0.0005
100	1.530	1.682	2.105	2.475	2.907	3.203	4.051
102	1.526	1.677	2.100	2.469	2.900	3.195	4.042
104	1.522	1.673	2.094	2.463	2.894	3.188	4.034
106	1.518	1.669	2.089	2.457	2.888	3.181	4.025
108	1.514	1.664	2.085	2.452	2.881	3.175	4.017
110	1.510	1.660	2.080	2.447	2.875	3.168	4.009
112	1.506	1.656	2.075	2.442	2.870	3.162	4.002
114	1.503	1.653	2.071	2.437	2.864	3.156	3.994
116	1.499	1.649	2.067	2.432	2.859	3.150	3.987
118	1.496	1.645	2.063	2.427	2.853	3.144	3.980
120	1.493	1.642	2.059	2.423	2.848	3.139	3.974
122	1.489	1.639	2.055	2.418	2.843	3.133	3.967
124	1.486	1.635	2.051	2.414	2.838	3.128	3.961
126	1.483	1.632	2.047	2.410	2.834	3.123	3.954
128	1.480	1.629	2.044	2.406	2.829	3.118	3.948
130	1.477	1.626	2.040	2.402	2.825	3.113	3.943
132	1.475	1.623	2.037	2.398	2.820	3.109	3.937
134	1.472	1.620	2.033	2.394	2.816	3.104	3.931
136	1.469	1.617	2.030	2.391	2.812	3.099	3.926
138	1.467	1.614	2.027	2.387	2.808	3.095	3.920
140	1.464	1.612	2.024	2.383	2.804	3.091	3.915
142	1.462	1.609	2.021	2.380	2.800	3.087	3.910
144	1.459	1.606	2.018	2.377	2.796	3.083	3.905
146	1.457	1.604	2.015	2.373	2.792	3.079	3.900
148	1.454	1.601	2.012	2.370	2.789	3.075	3.895
150	1.452	1.599	2.009	2.367	2.785	3.071	3.891
152	1.450	1.597	2.006	2.364	2.782	3.067	3.886
154	1.448	1.594	2.004	2.361	2.778	3.063	3.882
156	1.446	1.592	2.001	2.358	2.775	3.060	3.877
158	1.444	1.590	1.998	2.355	2.772	3.056	3.873
160	1.441	1.588	1.996	2.352	2.769	3.053	3.869
162	1.439	1.586	1.993	2.350	2.766	3.050	3.865
164	1.437	1.584	1.991	2.347	2.763	3.046	3.861
166	1.436	1.582	1.989	2.344	2.760	3.043	3.857
168	1.434	1.580	1.986	2.342	2.757	3.040	3.853
170	1.432	1.578	1.984	2.339	2.754	3.037	3.849
172	1.430	1.576	1.982	2.337	2.751	3.034	3.846
174	1.428	1.574	1.980	2.334	2.748	3.031	3.842
176	1.426	1.572	1.978	2.332	2.745	3.028	3.839
178	1.425	1.570	1.976	2.329	2.743	3.025	3.835
180	1.423	1.568	1.973	2.327	2.740	3.022	3.832
185	1.419	1.564	1.968	2.322	2.734	3.015	3.823
190	1.415	1.560	1.964	2.316	2.728	3.009	3.815
195	1.411	1.556	1.959	2.311	2.722	3.002	3.808
200	1.408	1.552	1.955	2.306	2.716	2.996	3.800
205	1.404	1.548	1.950	2.301	2.711	2.991	3.793
210	1.401	1.545	1.946	2.297	2.706	2.985	3.787
215	1.398	1.541	1.942	2.292	2.701	2.980	3.780
220	1.394	1.538	1.939	2.288	2.696	2.975	3.774

TABLE 4.2.4. *TWO-SIDED TOLERANCE LIMIT FACTORS (CONTROL BOTH TAILS).*

GAMMA = 0.990 *(CONTINUED)*

N \P →	0.125	0.100	0.050	0.025	0.010	0.005	0.0005
225	1.392	1.535	1.935	2.284	2.692	2.970	3.768
230	1.389	1.532	1.932	2.280	2.687	2.965	3.762
235	1.386	1.529	1.928	2.276	2.683	2.960	3.756
240	1.383	1.526	1.925	2.273	2.679	2.956	3.751
245	1.381	1.523	1.922	2.269	2.675	2.951	3.746
250	1.378	1.521	1.919	2.266	2.671	2.947	3.741
255	1.376	1.518	1.916	2.262	2.667	2.943	3.736
260	1.373	1.516	1.913	2.259	2.663	2.939	3.731
265	1.371	1.513	1.910	2.256	2.660	2.935	3.726
270	1.369	1.511	1.907	2.253	2.656	2.932	3.722
275	1.367	1.509	1.905	2.250	2.653	2.928	3.717
280	1.364	1.507	1.902	2.247	2.650	2.925	3.713
285	1.362	1.504	1.900	2.245	2.647	2.921	3.709
290	1.360	1.502	1.897	2.242	2.644	2.918	3.705
295	1.358	1.500	1.895	2.239	2.641	2.915	3.701
300	1.357	1.498	1.893	2.237	2.638	2.912	3.697
310	1.353	1.494	1.888	2.232	2.632	2.906	3.690
320	1.350	1.491	1.884	2.227	2.627	2.900	3.683
330	1.346	1.487	1.880	2.223	2.622	2.895	3.677
340	1.343	1.484	1.876	2.218	2.617	2.889	3.671
350	1.340	1.481	1.873	2.214	2.613	2.885	3.665
360	1.337	1.478	1.869	2.210	2.608	2.880	3.659
370	1.335	1.475	1.866	2.207	2.604	2.875	3.653
380	1.332	1.472	1.863	2.203	2.600	2.871	3.648
390	1.330	1.470	1.860	2.200	2.596	2.867	3.643
400	1.327	1.467	1.857	2.197	2.593	2.863	3.638
425	1.321	1.461	1.850	2.189	2.584	2.854	3.627
450	1.316	1.456	1.844	2.182	2.576	2.845	3.617
475	1.312	1.451	1.838	2.176	2.569	2.838	3.608
500	1.307	1.446	1.833	2.170	2.563	2.830	3.599
525	1.303	1.442	1.828	2.164	2.556	2.824	3.591
550	1.300	1.438	1.824	2.159	2.551	2.818	3.584
575	1.296	1.435	1.820	2.155	2.546	2.812	3.577
600	1.293	1.431	1.816	2.150	2.541	2.807	3.570
625	1.290	1.428	1.812	2.146	2.536	2.802	3.564
650	1.287	1.425	1.809	2.142	2.532	2.797	3.559
700	1.282	1.420	1.802	2.135	2.524	2.789	3.548
750	1.277	1.415	1.797	2.129	2.517	2.781	3.539
800	1.273	1.410	1.792	2.124	2.510	2.774	3.531
850	1.269	1.406	1.787	2.118	2.505	2.768	3.523
900	1.266	1.402	1.783	2.114	2.499	2.762	3.516
950	1.262	1.399	1.779	2.109	2.494	2.757	3.510
1000	1.259	1.396	1.775	2.105	2.490	2.752	3.504
1500	1.239	1.374	1.751	2.078	2.459	2.718	3.463
2000	1.227	1.362	1.736	2.061	2.440	2.699	3.439
3000	1.212	1.346	1.719	2.042	2.419	2.675	3.411
5000	1.198	1.332	1.702	2.023	2.398	2.653	3.383
10000	1.184	1.317	1.685	2.004	2.376	2.630	3.356
∞	1.150	1.282	1.645	1.960	2.326	2.576	3.291

TABLE 4.3.1. *TWO-SIDED TOLERANCE LIMIT FACTORS (CONTROL BOTH TAILS).*

GAMMA = 0.975

N \P →	0.125	0.100	0.050	0.025	0.010	0.005	0.0005
2	54.709	58.896	70.490	80.545	92.236	100.198	123.004
3	10.344	11.153	13.402	15.360	17.644	19.201	23.671
4	5.972	6.446	7.769	8.924	10.273	11.194	13.839
5	4.508	4.871	5.886	6.773	7.810	8.519	10.554
6	3.784	4.092	4.956	5.712	6.596	7.200	8.936
7	3.350	3.627	4.401	5.079	5.872	6.414	7.972
8	3.060	3.315	4.030	4.656	5.389	5.889	7.329
9	2.851	3.091	3.763	4.352	5.042	5.513	6.868
10	2.692	2.921	3.561	4.122	4.779	5.228	6.520
11	2.567	2.787	3.402	3.942	4.573	5.005	6.246
12	2.466	2.678	3.274	3.796	4.407	4.824	6.025
13	2.382	2.588	3.167	3.675	4.269	4.675	5.843
14	2.310	2.512	3.077	3.573	4.153	4.549	5.689
15	2.249	2.447	3.000	3.485	4.053	4.441	5.557
16	2.196	2.390	2.934	3.410	3.967	4.347	5.443
17	2.150	2.340	2.875	3.343	3.891	4.265	5.342
18	2.108	2.296	2.823	3.284	3.824	4.193	5.254
19	2.071	2.257	2.776	3.231	3.764	4.128	5.175
20	2.038	2.221	2.734	3.184	3.710	4.070	5.104
21	2.008	2.189	2.697	3.141	3.661	4.017	5.039
22	1.980	2.160	2.662	3.102	3.617	3.969	4.981
23	1.955	2.133	2.631	3.066	3.576	3.925	4.927
24	1.932	2.108	2.602	3.034	3.539	3.885	4.878
25	1.910	2.085	2.575	3.003	3.505	3.847	4.833
26	1.891	2.064	2.550	2.975	3.473	3.813	4.791
27	1.872	2.044	2.527	2.949	3.443	3.781	4.752
28	1.855	2.026	2.505	2.925	3.416	3.751	4.715
29	1.839	2.009	2.485	2.902	3.390	3.723	4.681
30	1.823	1.993	2.466	2.881	3.366	3.697	4.650
31	1.809	1.978	2.449	2.861	3.343	3.673	4.620
32	1.796	1.963	2.432	2.842	3.322	3.649	4.592
33	1.783	1.950	2.416	2.824	3.302	3.628	4.565
34	1.771	1.937	2.401	2.807	3.282	3.607	4.540
35	1.760	1.925	2.387	2.791	3.264	3.587	4.516
36	1.749	1.913	2.374	2.776	3.247	3.569	4.493
37	1.739	1.902	2.361	2.762	3.231	3.551	4.472
38	1.729	1.892	2.349	2.748	3.215	3.534	4.452
39	1.719	1.882	2.337	2.735	3.200	3.518	4.432
40	1.710	1.873	2.326	2.723	3.186	3.503	4.413
41	1.702	1.863	2.315	2.711	3.173	3.488	4.396
42	1.694	1.855	2.305	2.699	3.160	3.474	4.379
43	1.686	1.846	2.295	2.688	3.147	3.461	4.362
44	1.678	1.838	2.286	2.678	3.135	3.448	4.347
45	1.671	1.831	2.277	2.668	3.124	3.436	4.332
46	1.664	1.823	2.268	2.658	3.113	3.424	4.317
47	1.657	1.816	2.260	2.649	3.102	3.412	4.303
48	1.651	1.809	2.252	2.640	3.092	3.401	4.290
49	1.645	1.803	2.244	2.631	3.082	3.391	4.277
50	1.638	1.796	2.237	2.622	3.073	3.381	4.265

TABLE 4.3.2. TWO-SIDED TOLERANCE LIMIT FACTORS (CONTROL BOTH TAILS).

GAMMA = 0.975 (CONTINUED)

N \P →	0.125	0.100	0.050	0.025	0.010	0.005	0.0005
51	1.633	1.790	2.230	2.614	3.064	3.371	4.253
52	1.627	1.784	2.223	2.607	3.055	3.361	4.241
53	1.622	1.778	2.216	2.599	3.046	3.352	4.230
54	1.616	1.773	2.210	2.592	3.038	3.343	4.219
55	1.611	1.767	2.203	2.585	3.030	3.334	4.208
56	1.606	1.762	2.197	2.578	3.022	3.326	4.198
57	1.601	1.757	2.191	2.571	3.015	3.318	4.188
58	1.597	1.752	2.186	2.565	3.008	3.310	4.179
59	1.592	1.747	2.180	2.558	3.000	3.302	4.170
60	1.588	1.743	2.175	2.552	2.994	3.295	4.160
61	1.584	1.738	2.169	2.546	2.987	3.288	4.152
62	1.579	1.734	2.164	2.541	2.980	3.281	4.143
63	1.575	1.729	2.159	2.535	2.974	3.274	4.135
64	1.571	1.725	2.154	2.530	2.968	3.267	4.127
65	1.568	1.721	2.150	2.524	2.962	3.261	4.119
66	1.564	1.717	2.145	2.519	2.956	3.254	4.111
67	1.560	1.713	2.141	2.514	2.950	3.248	4.104
68	1.557	1.710	2.136	2.509	2.945	3.242	4.097
69	1.553	1.706	2.132	2.504	2.939	3.236	4.090
70	1.550	1.702	2.128	2.500	2.934	3.231	4.083
71	1.547	1.699	2.124	2.495	2.929	3.225	4.076
72	1.543	1.695	2.120	2.491	2.924	3.220	4.069
73	1.540	1.692	2.116	2.487	2.919	3.215	4.063
74	1.537	1.689	2.112	2.482	2.914	3.209	4.057
75	1.534	1.686	2.109	2.478	2.910	3.204	4.051
76	1.531	1.683	2.105	2.474	2.905	3.199	4.045
77	1.528	1.680	2.101	2.470	2.901	3.195	4.039
78	1.525	1.677	2.098	2.466	2.896	3.190	4.033
79	1.523	1.674	2.095	2.462	2.892	3.185	4.027
80	1.520	1.671	2.091	2.459	2.888	3.181	4.022
81	1.517	1.668	2.088	2.455	2.884	3.176	4.017
82	1.515	1.665	2.085	2.452	2.880	3.172	4.011
83	1.512	1.663	2.082	2.448	2.876	3.168	4.006
84	1.510	1.660	2.079	2.445	2.872	3.163	4.001
85	1.507	1.657	2.076	2.441	2.868	3.159	3.996
86	1.505	1.655	2.073	2.438	2.864	3.155	3.991
87	1.503	1.652	2.070	2.435	2.861	3.151	3.986
88	1.500	1.650	2.067	2.432	2.857	3.148	3.982
89	1.498	1.647	2.064	2.429	2.854	3.144	3.977
90	1.496	1.645	2.062	2.425	2.850	3.140	3.973
91	1.494	1.643	2.059	2.422	2.847	3.136	3.968
92	1.492	1.641	2.056	2.420	2.843	3.133	3.964
93	1.489	1.638	2.054	2.417	2.840	3.129	3.960
94	1.487	1.636	2.051	2.414	2.837	3.126	3.956
95	1.485	1.634	2.049	2.411	2.834	3.123	3.951
96	1.483	1.632	2.046	2.408	2.831	3.119	3.947
97	1.481	1.630	2.044	2.406	2.828	3.116	3.943
98	1.479	1.628	2.042	2.403	2.825	3.113	3.940
99	1.478	1.626	2.039	2.400	2.822	3.110	3.936

TABLE 4.3.3. TWO-SIDED TOLERANCE LIMIT FACTORS (CONTROL BOTH TAILS).

GAMMA = 0.975 (CONTINUED)

N \P →	0.125	0.100	0.050	0.025	0.010	0.005	0.0005
100	1.476	1.624	2.037	2.398	2.819	3.106	3.932
102	1.472	1.620	2.033	2.393	2.813	3.100	3.925
104	1.469	1.616	2.028	2.388	2.808	3.095	3.917
106	1.465	1.613	2.024	2.383	2.803	3.089	3.911
108	1.462	1.609	2.020	2.379	2.798	3.083	3.904
110	1.459	1.606	2.016	2.374	2.793	3.078	3.897
112	1.455	1.603	2.012	2.370	2.788	3.073	3.891
114	1.452	1.599	2.009	2.366	2.783	3.068	3.885
116	1.450	1.596	2.005	2.362	2.779	3.063	3.879
118	1.447	1.593	2.002	2.358	2.774	3.058	3.873
120	1.444	1.590	1.998	2.354	2.770	3.054	3.868
122	1.441	1.587	1.995	2.351	2.766	3.049	3.862
124	1.439	1.585	1.992	2.347	2.762	3.045	3.857
126	1.436	1.582	1.989	2.344	2.758	3.040	3.852
128	1.433	1.579	1.986	2.340	2.754	3.036	3.847
130	1.431	1.577	1.983	2.337	2.750	3.032	3.842
132	1.429	1.574	1.980	2.334	2.747	3.028	3.837
134	1.426	1.572	1.977	2.330	2.743	3.024	3.833
136	1.424	1.569	1.974	2.327	2.739	3.021	3.828
138	1.422	1.567	1.972	2.324	2.736	3.017	3.824
140	1.420	1.565	1.969	2.321	2.733	3.013	3.819
142	1.418	1.563	1.966	2.319	2.729	3.010	3.815
144	1.416	1.560	1.964	2.316	2.726	3.007	3.811
146	1.414	1.558	1.961	2.313	2.723	3.003	3.807
148	1.412	1.556	1.959	2.310	2.720	3.000	3.803
150	1.410	1.554	1.957	2.308	2.717	2.997	3.799
152	1.408	1.552	1.954	2.305	2.714	2.994	3.795
154	1.406	1.550	1.952	2.303	2.711	2.991	3.791
156	1.404	1.548	1.950	2.300	2.709	2.988	3.788
158	1.402	1.546	1.948	2.298	2.706	2.985	3.784
160	1.401	1.545	1.946	2.295	2.703	2.982	3.781
162	1.399	1.543	1.944	2.293	2.701	2.979	3.777
164	1.397	1.541	1.941	2.291	2.698	2.976	3.774
166	1.396	1.539	1.939	2.288	2.696	2.973	3.771
168	1.394	1.538	1.938	2.286	2.693	2.971	3.768
170	1.392	1.536	1.936	2.284	2.691	2.968	3.764
172	1.391	1.534	1.934	2.282	2.688	2.966	3.761
174	1.389	1.533	1.932	2.280	2.686	2.963	3.758
176	1.388	1.531	1.930	2.278	2.684	2.961	3.755
178	1.386	1.530	1.928	2.276	2.682	2.958	3.752
180	1.385	1.528	1.927	2.274	2.679	2.956	3.750
185	1.381	1.524	1.922	2.269	2.674	2.950	3.743
190	1.378	1.521	1.918	2.265	2.669	2.945	3.736
195	1.375	1.517	1.914	2.260	2.664	2.939	3.730
200	1.372	1.514	1.911	2.256	2.659	2.934	3.723
205	1.369	1.511	1.907	2.252	2.655	2.929	3.717
210	1.366	1.508	1.904	2.248	2.650	2.925	3.712
215	1.363	1.505	1.900	2.245	2.646	2.920	3.706
220	1.361	1.502	1.897	2.241	2.642	2.916	3.701

TABLE 4.3.4. *TWO-SIDED TOLERANCE LIMIT FACTORS (CONTROL BOTH TAILS).*

GAMMA = 0.975 *(CONTINUED)*

N \P →	0.125	0.100	0.050	0.025	0.010	0.005	0.0005
225	1.358	1.500	1.894	2.238	2.638	2.912	3.696
230	1.356	1.497	1.891	2.234	2.635	2.908	3.691
235	1.353	1.495	1.888	2.231	2.631	2.904	3.686
240	1.351	1.492	1.885	2.228	2.628	2.900	3.682
245	1.349	1.490	1.883	2.225	2.624	2.896	3.678
250	1.347	1.488	1.880	2.222	2.621	2.893	3.673
255	1.345	1.485	1.878	2.219	2.618	2.889	3.669
260	1.342	1.483	1.875	2.217	2.615	2.886	3.665
265	1.341	1.481	1.873	2.214	2.612	2.883	3.661
270	1.339	1.479	1.871	2.211	2.609	2.880	3.657
275	1.337	1.477	1.868	2.209	2.606	2.877	3.654
280	1.335	1.475	1.866	2.206	2.603	2.874	3.650
285	1.333	1.474	1.864	2.204	2.601	2.871	3.647
290	1.332	1.472	1.862	2.202	2.598	2.868	3.643
295	1.330	1.470	1.860	2.200	2.595	2.866	3.640
300	1.328	1.468	1.858	2.197	2.593	2.863	3.637
310	1.325	1.465	1.854	2.193	2.588	2.858	3.631
320	1.322	1.462	1.851	2.189	2.584	2.853	3.625
330	1.319	1.459	1.847	2.186	2.580	2.848	3.620
340	1.317	1.456	1.844	2.182	2.576	2.844	3.614
350	1.314	1.454	1.841	2.178	2.572	2.840	3.609
360	1.312	1.451	1.838	2.175	2.568	2.836	3.604
370	1.309	1.449	1.835	2.172	2.565	2.832	3.600
380	1.307	1.446	1.833	2.169	2.561	2.828	3.595
390	1.305	1.444	1.830	2.166	2.558	2.825	3.591
400	1.303	1.442	1.828	2.163	2.555	2.822	3.587
425	1.298	1.437	1.822	2.157	2.547	2.814	3.578
450	1.294	1.432	1.816	2.151	2.541	2.807	3.569
475	1.290	1.428	1.812	2.146	2.535	2.800	3.561
500	1.286	1.424	1.807	2.141	2.529	2.794	3.554
525	1.283	1.420	1.803	2.136	2.524	2.788	3.547
550	1.279	1.417	1.799	2.132	2.519	2.783	3.541
575	1.276	1.414	1.796	2.128	2.515	2.778	3.535
600	1.274	1.411	1.792	2.124	2.510	2.774	3.529
625	1.271	1.408	1.789	2.120	2.506	2.770	3.524
650	1.269	1.406	1.786	2.117	2.503	2.766	3.520
700	1.264	1.401	1.781	2.111	2.496	2.758	3.511
750	1.260	1.397	1.776	2.106	2.490	2.752	3.503
800	1.256	1.393	1.772	2.101	2.485	2.746	3.496
850	1.253	1.389	1.768	2.097	2.480	2.741	3.489
900	1.250	1.386	1.764	2.093	2.475	2.736	3.484
950	1.247	1.383	1.761	2.089	2.471	2.731	3.478
1000	1.245	1.381	1.758	2.085	2.467	2.727	3.473
1500	1.227	1.362	1.736	2.062	2.440	2.698	3.438
2000	1.217	1.351	1.724	2.048	2.425	2.682	3.418
3000	1.204	1.338	1.709	2.031	2.406	2.662	3.394
5000	1.192	1.325	1.694	2.015	2.388	2.642	3.370
10000	1.180	1.312	1.680	1.999	2.370	2.622	3.347
∞	1.150	1.282	1.645	1.960	2.326	2.576	3.291

TABLE 4.4.1. *TWO-SIDED TOLERANCE LIMIT FACTORS (CONTROL BOTH TAILS).*

GAMMA = 0.950

N \P →	0.125	0.100	0.050	0.025	0.010	0.005	0.0005
2	27.339	29.431	35.225	40.251	46.094	50.073	61.471
3	7.258	7.826	9.408	10.785	12.390	13.485	16.626
4	4.663	5.035	6.074	6.980	8.038	8.760	10.833
5	3.705	4.007	4.847	5.582	6.439	7.025	8.707
6	3.206	3.471	4.209	4.855	5.610	6.125	7.606
7	2.896	3.139	3.815	4.407	5.098	5.571	6.928
8	2.684	2.911	3.546	4.100	4.749	5.192	6.465
9	2.528	2.744	3.348	3.876	4.494	4.915	6.128
10	2.408	2.616	3.197	3.704	4.298	4.704	5.869
11	2.313	2.514	3.076	3.568	4.143	4.535	5.664
12	2.234	2.430	2.978	3.456	4.016	4.398	5.497
13	2.169	2.360	2.895	3.363	3.910	4.284	5.357
14	2.113	2.301	2.825	3.284	3.820	4.187	5.239
15	2.065	2.250	2.765	3.216	3.743	4.103	5.137
16	2.023	2.205	2.713	3.157	3.676	4.030	5.049
17	1.986	2.165	2.666	3.104	3.616	3.966	4.971
18	1.953	2.130	2.625	3.058	3.563	3.909	4.901
19	1.923	2.099	2.588	3.016	3.516	3.858	4.839
20	1.897	2.070	2.555	2.978	3.473	3.812	4.783
21	1.872	2.044	2.524	2.944	3.435	3.770	4.732
22	1.850	2.020	2.497	2.913	3.399	3.731	4.686
23	1.829	1.999	2.471	2.884	3.367	3.696	4.644
24	1.811	1.979	2.448	2.858	3.337	3.664	4.604
25	1.793	1.960	2.426	2.833	3.309	3.634	4.568
26	1.777	1.943	2.406	2.811	3.284	3.607	4.535
27	1.762	1.927	2.387	2.790	3.260	3.581	4.504
28	1.748	1.912	2.370	2.770	3.238	3.557	4.474
29	1.734	1.898	2.353	2.752	3.217	3.535	4.447
30	1.722	1.884	2.338	2.734	3.197	3.513	4.422
31	1.710	1.872	2.323	2.718	3.179	3.494	4.397
32	1.699	1.860	2.310	2.703	3.161	3.475	4.375
33	1.689	1.849	2.297	2.688	3.145	3.457	4.353
34	1.679	1.838	2.285	2.674	3.130	3.440	4.333
35	1.669	1.828	2.273	2.661	3.115	3.424	4.314
36	1.660	1.819	2.262	2.649	3.101	3.409	4.295
37	1.652	1.810	2.251	2.637	3.087	3.395	4.278
38	1.644	1.801	2.241	2.626	3.075	3.381	4.261
39	1.636	1.793	2.232	2.615	3.063	3.368	4.246
40	1.628	1.785	2.223	2.605	3.051	3.356	4.230
41	1.621	1.778	2.214	2.595	3.040	3.344	4.216
42	1.614	1.770	2.206	2.586	3.029	3.332	4.202
43	1.608	1.763	2.197	2.577	3.019	3.321	4.189
44	1.602	1.757	2.190	2.568	3.009	3.311	4.176
45	1.595	1.750	2:182	2.560	3.000	3.300	4.164
46	1.590	1.744	2.175	2.551	2.991	3.291	4.152
47	1.584	1.738	2.168	2.544	2.982	3.281	4.141
48	1.579	1.732	2.162	2.536	2.974	3.272	4.130
49	1.573	1.727	2.155	2.529	2.966	3.264	4.119
50	1.568	1.722	2.149	2.522	2.958	3.255	4.109

TABLE 4.4.2. *TWO-SIDED TOLERANCE LIMIT FACTORS (CONTROL BOTH TAILS).*

GAMMA = 0.950 (CONTINUED)

N \P →	0.125	0.100	0.050	0.025	0.010	0.005	0.0005
51	1.563	1.716	2.143	2.516	2.950	3.247	4.099
52	1.559	1.711	2.137	2.509	2.943	3.239	4.089
53	1.554	1.707	2.132	2.503	2.936	3.231	4.080
54	1.550	1.702	2.126	2.497	2.929	3.224	4.071
55	1.545	1.697	2.121	2.491	2.922	3.217	4.063
56	1.541	1.693	2.116	2.485	2.916	3.210	4.054
57	1.537	1.689	2.111	2.480	2.910	3.203	4.046
58	1.533	1.685	2.106	2.474	2.904	3.197	4.038
59	1.530	1.681	2.102	2.469	2.898	3.190	4.031
60	1.526	1.677	2.097	2.464	2.892	3.184	4.023
61	1.522	1.673	2.093	2.459	2.887	3.178	4.016
62	1.519	1.669	2.088	2.454	2.881	3.173	4.009
63	1.515	1.666	2.084	2.450	2.876	3.167	4.002
64	1.512	1.662	2.080	2.445	2.871	3.161	3.996
65	1.509	1.659	2.076	2.441	2.866	3.156	3.989
66	1.506	1.655	2.072	2.436	2.861	3.151	3.983
67	1.503	1.652	2.069	2.432	2.856	3.146	3.977
68	1.500	1.649	2.065	2.428	2.852	3.141	3.971
69	1.497	1.646	2.061	2.424	2.847	3.136	3.965
70	1.494	1.643	2.058	2.420	2.843	3.131	3.959
71	1.491	1.640	2.055	2.416	2.839	3.127	3.953
72	1.488	1.637	2.051	2.413	2.834	3.122	3.948
73	1.486	1.634	2.048	2.409	2.830	3.118	3.943
74	1.483	1.631	2.045	2.405	2.826	3.113	3.938
75	1.480	1.629	2.042	2.402	2.822	3.109	3.932
76	1.478	1.626	2.039	2.399	2.819	3.105	3.927
77	1.476	1.624	2.036	2.395	2.815	3.101	3.923
78	1.473	1.621	2.033	2.392	2.811	3.097	3.918
79	1.471	1.619	2.030	2.389	2.808	3.093	3.913
80	1.469	1.616	2.027	2.386	2.804	3.089	3.909
81	1.466	1.614	2.025	2.383	2.801	3.086	3.904
82	1.464	1.611	2.022	2.380	2.797	3.082	3.900
83	1.462	1.609	2.019	2.377	2.794	3.079	3.896
84	1.460	1.607	2.017	2.374	2.791	3.075	3.891
85	1.458	1.605	2.014	2.371	2.788	3.072	3.887
86	1.456	1.603	2.012	2.368	2.784	3.068	3.883
87	1.454	1.601	2.009	2.366	2.781	3.065	3.879
88	1.452	1.598	2.007	2.363	2.778	3.062	3.875
89	1.450	1.596	2.005	2.360	2.776	3.059	3.872
90	1.448	1.594	2.002	2.358	2.773	3.056	3.868
91	1.446	1.593	2.000	2.355	2.770	3.053	3.864
92	1.444	1.591	1.998	2.353	2.767	3.050	3.861
93	1.443	1.589	1.996	2.350	2.764	3.047	3.857
94	1.441	1.587	1.994	2.348	2.762	3.044	3.854
95	1.439	1.585	1.991	2.346	2.759	3.041	3.850
96	1.437	1.583	1.989	2.343	2.756	3.038	3.847
97	1.436	1.582	1.987	2.341	2.754	3.035	3.843
98	1.434	1.580	1.985	2.339	2.751	3.033	3.840
99	1.433	1.578	1.983	2.337	2.749	3.030	3.837

TABLE 4.4.3.   TWO-SIDED TOLERANCE LIMIT FACTORS (CONTROL BOTH TAILS).

GAMMA = 0.950 (CONTINUED)

N \P →	0.125	0.100	0.050	0.025	0.010	0.005	0.0005
100	1.431	1.576	1.982	2.335	2.747	3.027	3.834
102	1.428	1.573	1.978	2.330	2.742	3.022	3.828
104	1.425	1.570	1.974	2.326	2.737	3.018	3.822
106	1.422	1.567	1.971	2.322	2.733	3.013	3.816
108	1.419	1.564	1.967	2.319	2.729	3.008	3.811
110	1.416	1.561	1.964	2.315	2.724	3.004	3.805
112	1.414	1.558	1.961	2.311	2.720	2.999	3.800
114	1.411	1.556	1.958	2.308	2.716	2.995	3.795
116	1.409	1.553	1.955	2.304	2.713	2.991	3.790
118	1.406	1.550	1.952	2.301	2.709	2.987	3.785
120	1.404	1.548	1.949	2.298	2.705	2.983	3.780
122	1.402	1.545	1.946	2.295	2.702	2.979	3.776
124	1.399	1.543	1.943	2.292	2.698	2.976	3.771
126	1.397	1.541	1.941	2.289	2.695	2.972	3.767
128	1.395	1.539	1.938	2.286	2.692	2.969	3.763
130	1.393	1.536	1.935	2.283	2.689	2.965	3.759
132	1.391	1.534	1.933	2.280	2.686	2.962	3.755
134	1.389	1.532	1.931	2.278	2.683	2.959	3.751
136	1.387	1.530	1.928	2.275	2.680	2.955	3.747
138	1.385	1.528	1.926	2.273	2.677	2.952	3.743
140	1.383	1.526	1.924	2.270	2.674	2.949	3.740
142	1.381	1.524	1.922	2.268	2.671	2.946	3.736
144	1.380	1.522	1.919	2.265	2.669	2.944	3.733
146	1.378	1.521	1.917	2.263	2.666	2.941	3.729
148	1.376	1.519	1.915	2.261	2.663	2.938	3.726
150	1.375	1.517	1.913	2.258	2.661	2.935	3.723
152	1.373	1.515	1.911	2.256	2.658	2.933	3.720
154	1.371	1.514	1.909	2.254	2.656	2.930	3.716
156	1.370	1.512	1.908	2.252	2.654	2.928	3.713
158	1.368	1.510	1.906	2.250	2.651	2.925	3.710
160	1.367	1.509	1.904	2.248	2.649	2.923	3.708
162	1.365	1.507	1.902	2.246	2.647	2.920	3.705
164	1.364	1.506	1.900	2.244	2.645	2.918	3.702
166	1.363	1.504	1.899	2.242	2.643	2.916	3.699
168	1.361	1.503	1.897	2.240	2.641	2.913	3.696
170	1.360	1.501	1.895	2.239	2.638	2.911	3.694
172	1.359	1.500	1.894	2.237	2.636	2.909	3.691
174	1.357	1.499	1.892	2.235	2.634	2.907	3.689
176	1.356	1.497	1.891	2.233	2.633	2.905	3.686
178	1.355	1.496	1.889	2.232	2.631	2.903	3.684
180	1.353	1.495	1.888	2.230	2.629	2.901	3.681
185	1.350	1.492	1.884	2.226	2.624	2.896	3.675
190	1.348	1.489	1.881	2.222	2.620	2.891	3.670
195	1.345	1.486	1.877	2.218	2.616	2.887	3.664
200	1.342	1.483	1.874	2.215	2.612	2.883	3.659
205	1.340	1.480	1.871	2.211	2.608	2.878	3.654
210	1.337	1.478	1.868	2.208	2.604	2.874	3.649
215	1.335	1.475	1.865	2.205	2.601	2.871	3.645
220	1.333	1.473	1.863	2.202	2.597	2.867	3.640

TABLE 4.4.4.   TWO-SIDED TOLERANCE LIMIT FACTORS (CONTROL BOTH TAILS).

GAMMA = 0.950 (CONTINUED)

N \P →	0.125	0.100	0.050	0.025	0.010	0.005	0.0005
225	1.330	1.471	1.860	2.199	2.594	2.863	3.636
230	1.328	1.468	1.857	2.196	2.591	2.860	3.632
235	1.326	1.466	1.855	2.193	2.588	2.857	3.628
240	1.324	1.464	1.853	2.191	2.585	2.854	3.624
245	1.322	1.462	1.850	2.188	2.582	2.850	3.621
250	1.321	1.460	1.848	2.186	2.579	2.847	3.617
255	1.319	1.458	1.846	2.183	2.577	2.845	3.613
260	1.317	1.456	1.844	2.181	2.574	2.842	3.610
265	1.315	1.455	1.842	2.179	2.571	2.839	3.607
270	1.314	1.453	1.840	2.177	2.569	2.836	3.604
275	1.312	1.451	1.838	2.174	2.567	2.834	3.601
280	1.311	1.450	1.836	2.172	2.564	2.831	3.598
285	1.309	1.448	1.834	2.170	2.562	2.829	3.595
290	1.308	1.446	1.832	2.168	2.560	2.827	3.592
295	1.306	1.445	1.831	2.166	2.558	2.824	3.589
300	1.305	1.444	1.829	2.165	2.556	2.822	3.586
310	1.302	1.441	1.826	2.161	2.552	2.818	3.581
320	1.300	1.438	1.823	2.158	2.548	2.814	3.576
330	1.297	1.436	1.820	2.154	2.544	2.810	3.572
340	1.295	1.433	1.817	2.151	2.541	2.806	3.567
350	1.293	1.431	1.815	2.148	2.537	2.802	3.563
360	1.291	1.429	1.812	2.146	2.534	2.799	3.559
370	1.289	1.426	1.810	2.143	2.531	2.796	3.555
380	1.287	1.424	1.807	2.140	2.528	2.793	3.551
390	1.285	1.423	1.805	2.138	2.526	2.790	3.548
400	1.283	1.421	1.803	2.136	2.523	2.787	3.544
425	1.279	1.416	1.798	2.130	2.517	2.780	3.536
450	1.275	1.412	1.793	2.125	2.511	2.774	3.529
475	1.272	1.409	1.789	2.120	2.506	2.769	3.522
500	1.268	1.405	1.785	2.116	2.501	2.763	3.516
525	1.265	1.402	1.782	2.112	2.497	2.759	3.510
550	1.263	1.399	1.779	2.108	2.492	2.754	3.505
575	1.260	1.397	1.775	2.105	2.489	2.750	3.500
600	1.258	1.394	1.773	2.102	2.485	2.746	3.495
625	1.255	1.392	1.770	2.099	2.482	2.743	3.491
650	1.253	1.389	1.767	2.096	2.478	2.739	3.487
700	1.249	1.385	1.763	2.091	2.473	2.733	3.479
750	1.246	1.382	1.759	2.086	2.467	2.727	3.472
800	1.243	1.378	1.755	2.082	2.463	2.722	3.466
850	1.240	1.375	1.751	2.078	2.459	2.718	3.461
900	1.237	1.373	1.748	2.075	2.455	2.714	3.456
950	1.235	1.370	1.745	2.071	2.451	2.710	3.451
1000	1.233	1.368	1.743	2.069	2.448	2.706	3.447
1500	1.217	1.352	1.724	2.048	2.425	2.682	3.417
2000	1.208	1.342	1.713	2.036	2.411	2.667	3.400
3000	1.197	1.331	1.701	2.022	2.395	2.650	3.379
5000	1.187	1.319	1.688	2.008	2.380	2.633	3.359
10000	1.176	1.308	1.675	1.993	2.364	2.616	3.339
∞	1.150	1.282	1.645	1.960	2.326	2.576	3.291

TABLE 4.5.1. *TWO-SIDED TOLERANCE LIMIT FACTORS (CONTROL BOTH TAILS).*

GAMMA = 0.900

N \P →	0.125	0.100	0.050	0.025	0.010	0.005	0.0005
2	13.637	14.682	17.574	20.082	22.998	24.984	30.672
3	5.050	5.448	6.554	7.516	8.637	9.402	11.595
4	3.600	3.890	4.700	5.405	6.227	6.788	8.398
5	3.009	3.257	3.948	4.550	5.253	5.733	7.110
6	2.683	2.908	3.535	4.082	4.721	5.156	6.407
7	2.474	2.685	3.271	3.783	4.381	4.789	5.960
8	2.327	2.527	3.086	3.574	4.143	4.531	5.647
9	2.217	2.410	2.948	3.418	3.966	4.340	5.415
10	2.131	2.318	2.840	3.296	3.828	4.192	5.235
11	2.061	2.244	2.754	3.199	3.718	4.072	5.090
12	2.004	2.183	2.682	3.118	3.627	3.974	4.971
13	1.955	2.132	2.622	3.050	3.550	3.891	4.871
14	1.914	2.087	2.571	2.993	3.485	3.821	4.786
15	1.878	2.049	2.526	2.942	3.428	3.760	4.712
16	1.846	2.015	2.487	2.898	3.378	3.706	4.647
17	1.818	1.986	2.452	2.859	3.334	3.659	4.590
18	1.793	1.959	2.421	2.825	3.295	3.616	4.539
19	1.770	1.935	2.393	2.793	3.260	3.578	4.493
20	1.750	1.913	2.368	2.765	3.228	3.544	4.451
21	1.731	1.893	2.345	2.739	3.199	3.513	4.413
22	1.714	1.875	2.324	2.715	3.172	3.484	4.379
23	1.698	1.858	2.304	2.694	3.148	3.458	4.347
24	1.683	1.843	2.287	2.674	3.125	3.433	4.318
25	1.670	1.828	2.270	2.655	3.104	3.411	4.291
26	1.657	1.815	2.254	2.638	3.085	3.390	4.265
27	1.645	1.802	2.240	2.621	3.066	3.370	4.242
28	1.634	1.791	2.226	2.606	3.049	3.352	4.220
29	1.624	1.780	2.214	2.592	3.034	3.335	4.199
30	1.614	1.770	2.202	2.579	3.019	3.319	4.180
31	1.605	1.760	2.191	2.566	3.004	3.303	4.161
32	1.596	1.751	2.180	2.554	2.991	3.289	4.144
33	1.588	1.742	2.170	2.543	2.978	3.275	4.128
34	1.580	1.734	2.160	2.533	2.966	3.263	4.112
35	1.573	1.726	2.151	2.522	2.955	3.250	4.098
36	1.566	1.718	2.143	2.513	2.944	3.239	4.084
37	1.559	1.711	2.135	2.504	2.934	3.228	4.070
38	1.553	1.704	2.127	2.495	2.924	3.217	4.058
39	1.546	1.698	2.119	2.486	2.915	3.207	4.045
40	1.540	1.692	2.112	2.478	2.906	3.197	4.034
41	1.535	1.686	2.105	2.471	2.897	3.188	4.023
42	1.529	1.680	2.099	2.463	2.889	3.179	4.012
43	1.524	1.674	2.092	2.456	2.881	3.171	4.002
44	1.519	1.669	2.086	2.450	2.873	3.162	3.992
45	1.514	1.664	2.080	2.443	2.866	3.155	3.982
46	1.510	1.659	2.075	2.437	2.859	3.147	3.973
47	1.505	1.654	2.069	2.431	2.852	3.140	3.965
48	1.501	1.650	2.064	2.425	2.846	3.133	3.956
49	1.497	1.645	2.059	2.419	2.839	3.126	3.948
50	1.493	1.641	2.054	2.414	2.833	3.119	3.940

TABLE 4.5.2. TWO-SIDED TOLERANCE LIMIT FACTORS (CONTROL BOTH TAILS).

GAMMA = 0.900 (CONTINUED)

N \P →	0.125	0.100	0.050	0.025	0.010	0.005	0.0005
51	1.489	1.637	2.049	2.408	2.827	3.113	3.933
52	1.485	1.633	2.045	2.403	2.822	3.107	3.925
53	1.481	1.629	2.040	2.398	2.816	3.101	3.918
54	1.478	1.625	2.036	2.394	2.811	3.095	3.911
55	1.474	1.622	2.032	2.389	2.805	3.089	3.904
56	1.471	1.618	2.028	2.385	2.800	3.084	3.898
57	1.468	1.615	2.024	2.380	2.796	3.079	3.892
58	1.465	1.612	2.020	2.376	2.791	3.074	3.886
59	1.462	1.608	2.016	2.372	2.786	3.069	3.880
60	1.459	1.605	2.013	2.368	2.782	3.064	3.874
61	1.456	1.602	2.009	2.364	2.777	3.059	3.868
62	1.453	1.599	2.006	2.360	2.773	3.055	3.863
63	1.450	1.596	2.003	2.356	2.769	3.050	3.857
64	1.448	1.593	1.999	2.353	2.765	3.046	3.852
65	1.445	1.591	1.996	2.349	2.761	3.042	3.847
66	1.442	1.588	1.993	2.346	2.757	3.038	3.842
67	1.440	1.585	1.990	2.343	2.754	3.034	3.837
68	1.438	1.583	1.987	2.339	2.750	3.030	3.833
69	1.435	1.580	1.984	2.336	2.746	3.026	3.828
70	1.433	1.578	1.982	2.333	2.743	3.022	3.824
71	1.431	1.576	1.979	2.330	2.739	3.019	3.819
72	1.428	1.573	1.976	2.327	2.736	3.015	3.815
73	1.426	1.571	1.974	2.324	2.733	3.012	3.811
74	1.424	1.569	1.971	2.322	2.730	3.008	3.807
75	1.422	1.567	1.969	2.319	2.727	3.005	3.803
76	1.420	1.565	1.966	2.316	2.724	3.002	3.799
77	1.418	1.562	1.964	2.313	2.721	2.998	3.795
78	1.416	1.560	1.962	2.311	2.718	2.995	3.792
79	1.414	1.558	1.959	2.308	2.715	2.992	3.788
80	1.412	1.557	1.957	2.306	2.712	2.989	3.784
81	1.411	1.555	1.955	2.303	2.710	2.986	3.781
82	1.409	1.553	1.953	2.301	2.707	2.984	3.777
83	1.407	1.551	1.951	2.299	2.704	2.981	3.774
84	1.405	1.549	1.949	2.296	2.702	2.978	3.771
85	1.404	1.547	1.947	2.294	2.699	2.975	3.768
86	1.402	1.546	1.945	2.292	2.697	2.973	3.764
87	1.401	1.544	1.943	2.290	2.694	2.970	3.761
88	1.399	1.542	1.941	2.288	2.692	2.968	3.758
89	1.397	1.541	1.939	2.286	2.690	2.965	3.755
90	1.396	1.539	1.937	2.284	2.687	2.963	3.752
91	1.394	1.537	1.935	2.282	2.685	2.960	3.749
92	1.393	1.536	1.934	2.280	2.683	2.958	3.747
93	1.391	1.534	1.932	2.278	2.681	2.956	3.744
94	1.390	1.533	1.930	2.276	2.679	2.953	3.741
95	1.389	1.531	1.928	2.274	2.677	2.951	3.738
96	1.387	1.530	1.927	2.272	2.674	2.949	3.736
97	1.386	1.529	1.925	2.270	2.672	2.947	3.733
98	1.385	1.527	1.924	2.268	2.670	2.945	3.731
99	1.383	1.526	1.922	2.267	2.669	2.942	3.728

TABLE 4.5.3. TWO-SIDED TOLERANCE LIMIT FACTORS (CONTROL BOTH TAILS).

GAMMA = 0.900 (CONTINUED)

N \P →	0.125	0.100	0.050	0.025	0.010	0.005	0.0005
100	1.382	1.524	1.920	2.265	2.667	2.940	3.726
102	1.379	1.522	1.917	2.262	2.663	2.936	3.721
104	1.377	1.519	1.914	2.258	2.659	2.932	3.716
106	1.375	1.517	1.912	2.255	2.656	2.929	3.712
108	1.372	1.514	1.909	2.252	2.652	2.925	3.707
110	1.370	1.512	1.906	2.249	2.649	2.921	3.703
112	1.368	1.510	1.904	2.246	2.646	2.918	3.699
114	1.366	1.508	1.901	2.244	2.643	2.915	3.695
116	1.364	1.505	1.899	2.241	2.640	2.911	3.691
118	1.362	1.503	1.896	2.238	2.637	2.908	3.687
120	1.360	1.501	1.894	2.236	2.634	2.905	3.683
122	1.358	1.499	1.892	2.233	2.631	2.902	3.680
124	1.356	1.497	1.890	2.231	2.628	2.899	3.676
126	1.354	1.495	1.887	2.228	2.626	2.896	3.673
128	1.353	1.494	1.885	2.226	2.623	2.894	3.670
130	1.351	1.492	1.883	2.224	2.620	2.891	3.666
132	1.349	1.490	1.881	2.222	2.618	2.888	3.663
134	1.348	1.488	1.879	2.219	2.616	2.886	3.660
136	1.346	1.487	1.877	2.217	2.613	2.883	3.657
138	1.345	1.485	1.876	2.215	2.611	2.881	3.654
140	1.343	1.484	1.874	2.213	2.609	2.878	3.651
142	1.342	1.482	1.872	2.211	2.606	2.876	3.649
144	1.340	1.480	1.870	2.209	2.604	2.874	3.646
146	1.339	1.479	1.869	2.207	2.602	2.871	3.643
148	1.337	1.478	1.867	2.206	2.600	2.869	3.640
150	1.336	1.476	1.865	2.204	2.598	2.867	3.638
152	1.335	1.475	1.864	2.202	2.596	2.865	3.635
154	1.333	1.473	1.862	2.200	2.594	2.863	3.633
156	1.332	1.472	1.861	2.199	2.592	2.861	3.631
158	1.331	1.471	1.859	2.197	2.591	2.859	3.628
160	1.330	1.469	1.858	2.195	2.589	2.857	3.626
162	1.328	1.468	1.856	2.194	2.587	2.855	3.624
164	1.327	1.467	1.855	2.192	2.585	2.853	3.621
166	1.326	1.466	1.854	2.191	2.584	2.851	3.619
168	1.325	1.465	1.852	2.189	2.582	2.850	3.617
170	1.324	1.463	1.851	2.188	2.580	2.848	3.615
172	1.323	1.462	1.850	2.186	2.579	2.846	3.613
174	1.322	1.461	1.848	2.185	2.577	2.844	3.611
176	1.321	1.460	1.847	2.184	2.576	2.843	3.609
178	1.320	1.459	1.846	2.182	2.574	2.841	3.607
180	1.319	1.458	1.845	2.181	2.572	2.839	3.605
185	1.316	1.455	1.842	2.178	2.569	2.836	3.600
190	1.314	1.453	1.839	2.174	2.565	2.832	3.596
195	1.312	1.451	1.836	2.171	2.562	2.828	3.592
200	1.309	1.448	1.834	2.169	2.559	2.825	3.587
205	1.307	1.446	1.831	2.166	2.556	2.822	3.584
210	1.305	1.444	1.829	2.163	2.553	2.818	3.580
215	1.303	1.442	1.826	2.161	2.550	2.815	3.576
220	1.302	1.440	1.824	2.158	2.547	2.812	3.573

TABLE 4.5.4. *TWO-SIDED TOLERANCE LIMIT FACTORS (CONTROL BOTH TAILS).*

GAMMA = 0.900 *(CONTINUED)*

$N \backslash P \rightarrow$	0.125	0.100	0.050	0.025	0.010	0.005	0.0005
225	1.300	1.438	1.822	2.156	2.545	2.809	3.569
230	1.298	1.436	1.820	2.154	2.542	2.807	3.566
235	1.296	1.434	1.818	2.151	2.540	2.804	3.563
240	1.295	1.433	1.816	2.149	2.537	2.802	3.560
245	1.293	1.431	1.814	2.147	2.535	2.799	3.557
250	1.292	1.430	1.812	2.145	2.533	2.797	3.554
255	1.290	1.428	1.811	2.143	2.530	2.794	3.551
260	1.289	1.427	1.809	2.141	2.528	2.792	3.548
265	1.287	1.425	1.807	2.139	2.526	2.790	3.546
270	1.286	1.424	1.806	2.138	2.524	2.788	3.543
275	1.285	1.422	1.804	2.136	2.522	2.786	3.541
280	1.283	1.421	1.803	2.134	2.520	2.784	3.538
285	1.282	1.420	1.801	2.133	2.519	2.782	3.536
290	1.281	1.418	1.800	2.131	2.517	2.780	3.534
295	1.280	1.417	1.798	2.129	2.515	2.778	3.531
300	1.279	1.416	1.797	2.128	2.513	2.776	3.529
310	1.277	1.414	1.794	2.125	2.510	2.773	3.525
320	1.274	1.411	1.792	2.122	2.507	2.769	3.521
330	1.272	1.409	1.789	2.120	2.504	2.766	3.517
340	1.271	1.407	1.787	2.117	2.501	2.763	3.514
350	1.269	1.405	1.785	2.115	2.499	2.760	3.511
360	1.267	1.404	1.783	2.112	2.496	2.758	3.507
370	1.265	1.402	1.781	2.110	2.494	2.755	3.504
380	1.264	1.400	1.779	2.108	2.491	2.753	3.501
390	1.262	1.399	1.777	2.106	2.489	2.750	3.498
400	1.261	1.397	1.775	2.104	2.487	2.748	3.495
425	1.257	1.393	1.771	2.100	2.482	2.742	3.489
450	1.254	1.390	1.768	2.096	2.477	2.737	3.483
475	1.251	1.387	1.764	2.092	2.473	2.733	3.478
500	1.249	1.384	1.761	2.088	2.469	2.729	3.473
525	1.246	1.382	1.758	2.085	2.466	2.725	3.468
550	1.244	1.379	1.755	2.082	2.462	2.721	3.464
575	1.242	1.377	1.753	2.079	2.459	2.718	3.460
600	1.240	1.375	1.750	2.077	2.456	2.715	3.456
625	1.238	1.373	1.748	2.074	2.453	2.712	3.453
650	1.236	1.371	1.746	2.072	2.451	2.709	3.449
700	1.233	1.368	1.742	2.068	2.446	2.704	3.443
750	1.230	1.365	1.739	2.064	2.442	2.700	3.438
800	1.227	1.362	1.736	2.060	2.438	2.696	3.433
850	1.225	1.360	1.733	2.057	2.435	2.692	3.429
900	1.223	1.357	1.730	2.054	2.432	2.688	3.425
950	1.221	1.355	1.728	2.052	2.429	2.685	3.421
1000	1.219	1.353	1.726	2.049	2.426	2.682	3.418
1500	1.206	1.340	1.711	2.033	2.407	2.662	3.394
2000	1.199	1.332	1.702	2.023	2.396	2.650	3.379
3000	1.190	1.323	1.691	2.011	2.383	2.637	3.363
5000	1.181	1.313	1.681	1.999	2.370	2.623	3.346
10000	1.172	1.304	1.670	1.988	2.357	2.609	3.330
∞	1.150	1.282	1.645	1.960	2.326	2.576	3.291

TABLE 4.6.1.   TWO-SIDED TOLERANCE LIMIT FACTORS (CONTROL BOTH TAILS).

GAMMA = 0.750

N \P →	0.125	0.100	0.050	0.025	0.010	0.005	0.0005
2	5.363	5.776	6.919	7.910	9.061	9.845	12.090
3	3.027	3.271	3.945	4.530	5.211	5.675	7.005
4	2.466	2.671	3.238	3.731	4.305	4.696	5.817
5	2.204	2.391	2.911	3.363	3.889	4.247	5.275
6	2.048	2.225	2.718	3.146	3.644	3.984	4.958
7	1.942	2.113	2.588	3.001	3.481	3.808	4.747
8	1.865	2.032	2.494	2.895	3.363	3.682	4.595
9	1.806	1.969	2.421	2.815	3.273	3.585	4.480
10	1.759	1.919	2.364	2.751	3.201	3.508	4.388
11	1.720	1.878	2.317	2.699	3.143	3.446	4.314
12	1.687	1.844	2.278	2.655	3.094	3.393	4.252
13	1.660	1.815	2.244	2.618	3.053	3.349	4.199
14	1.636	1.789	2.215	2.586	3.017	3.311	4.153
15	1.615	1.767	2.190	2.558	2.986	3.277	4.114
16	1.596	1.747	2.168	2.533	2.958	3.248	4.079
17	1.579	1.730	2.148	2.511	2.934	3.222	4.048
18	1.564	1.714	2.130	2.491	2.912	3.198	4.020
19	1.551	1.700	2.114	2.473	2.892	3.177	3.995
20	1.538	1.687	2.099	2.457	2.873	3.158	3.972
21	1.527	1.675	2.085	2.442	2.857	3.140	3.951
22	1.516	1.664	2.073	2.428	2.842	3.124	3.932
23	1.507	1.654	2.061	2.415	2.828	3.109	3.914
24	1.498	1.644	2.051	2.404	2.815	3.095	3.898
25	1.489	1.636	2.041	2.393	2.803	3.082	3.882
26	1.482	1.627	2.031	2.383	2.791	3.070	3.868
27	1.474	1.620	2.023	2.373	2.781	3.058	3.855
28	1.468	1.612	2.015	2.364	2.771	3.048	3.842
29	1.461	1.606	2.007	2.356	2.761	3.038	3.831
30	1.455	1.599	2.000	2.348	2.753	3.029	3.820
31	1.449	1.593	1.993	2.340	2.744	3.020	3.809
32	1.444	1.588	1.987	2.333	2.737	3.011	3.799
33	1.439	1.582	1.980	2.326	2.729	3.004	3.790
34	1.434	1.577	1.975	2.320	2.722	2.996	3.781
35	1.429	1.572	1.969	2.314	2.715	2.989	3.773
36	1.425	1.567	1.964	2.308	2.709	2.982	3.765
37	1.420	1.563	1.959	2.303	2.703	2.976	3.757
38	1.416	1.559	1.954	2.297	2.697	2.969	3.750
39	1.412	1.555	1.949	2.292	2.691	2.963	3.743
40	1.409	1.551	1.945	2.287	2.686	2.958	3.736
41	1.405	1.547	1.941	2.283	2.681	2.952	3.730
42	1.402	1.543	1.937	2.278	2.676	2.947	3.724
43	1.398	1.540	1.933	2.274	2.671	2.942	3.718
44	1.395	1.537	1.929	2.270	2.667	2.937	3.712
45	1.392	1.533	1.925	2.266	2.662	2.932	3.706
46	1.389	1.530	1.922	2.262	2.658	2.928	3.701
47	1.386	1.527	1.919	2.258	2.654	2.924	3.696
48	1.383	1.524	1.915	2.255	2.650	2.919	3.691
49	1.381	1.522	1.912	2.251	2.646	2.915	3.686
50	1.378	1.519	1.909	2.248	2.643	2.911	3.682

TABLE 4.6.2. TWO-SIDED TOLERANCE LIMIT FACTORS (CONTROL BOTH TAILS).

GAMMA = 0.750 (CONTINUED)

N \P →	0.125	0.100	0.050	0.025	0.010	0.005	0.0005
51	1.376	1.516	1.906	2.245	2.639	2.908	3.677
52	1.373	1.514	1.903	2.242	2.636	2.904	3.673
53	1.371	1.511	1.901	2.239	2.632	2.900	3.669
54	1.369	1.509	1.898	2.236	2.629	2.897	3.665
55	1.366	1.507	1.895	2.233	2.626	2.894	3.661
56	1.364	1.504	1.893	2.230	2.623	2.890	3.657
57	1.362	1.502	1.890	2.228	2.620	2.887	3.654
58	1.360	1.500	1.888	2.225	2.617	2.884	3.650
59	1.358	1.498	1.886	2.222	2.614	2.881	3.646
60	1.356	1.496	1.883	2.220	2.612	2.878	3.643
61	1.354	1.494	1.881	2.218	2.609	2.876	3.640
62	1.353	1.492	1.879	2.215	2.606	2.873	3.637
63	1.351	1.490	1.877	2.213	2.604	2.870	3.633
64	1.349	1.488	1.875	2.211	2.601	2.868	3.630
65	1.347	1.487	1.873	2.209	2.599	2.865	3.627
66	1.346	1.485	1.871	2.206	2.597	2.863	3.625
67	1.344	1.483	1.869	2.204	2.594	2.860	3.622
68	1.343	1.482	1.867	2.202	2.592	2.858	3.619
69	1.341	1.480	1.866	2.200	2.590	2.856	3.616
70	1.339	1.478	1.864	2.199	2.588	2.853	3.614
71	1.338	1.477	1.862	2.197	2.586	2.851	3.611
72	1.337	1.475	1.860	2.195	2.584	2.849	3.609
73	1.335	1.474	1.859	2.193	2.582	2.847	3.606
74	1.334	1.473	1.857	2.191	2.580	2.845	3.604
75	1.332	1.471	1.856	2.190	2.578	2.843	3.601
76	1.331	1.470	1.854	2.188	2.576	2.841	3.599
77	1.330	1.468	1.853	2.186	2.574	2.839	3.597
78	1.329	1.467	1.851	2.185	2.573	2.837	3.595
79	1.327	1.466	1.850	2.183	2.571	2.835	3.593
80	1.326	1.465	1.848	2.182	2.569	2.833	3.590
81	1.325	1.463	1.847	2.180	2.568	2.832	3.588
82	1.324	1.462	1.846	2.179	2.566	2.830	3.586
83	1.323	1.461	1.844	2.177	2.564	2.828	3.584
84	1.322	1.460	1.843	2.176	2.563	2.827	3.582
85	1.321	1.459	1.842	2.174	2.561	2.825	3.580
86	1.319	1.458	1.840	2.173	2.560	2.823	3.579
87	1.318	1.456	1.839	2.172	2.558	2.822	3.577
88	1.317	1.455	1.838	2.170	2.557	2.820	3.575
89	1.316	1.454	1.837	2.169	2.555	2.819	3.573
90	1.315	1.453	1.836	2.168	2.554	2.817	3.571
91	1.314	1.452	1.835	2.166	2.553	2.816	3.570
92	1.313	1.451	1.833	2.165	2.551	2.814	3.568
93	1.312	1.450	1.832	2.164	2.550	2.813	3.566
94	1.312	1.449	1.831	2.163	2.549	2.811	3.565
95	1.311	1.448	1.830	2.162	2.547	2.810	3.563
96	1.310	1.447	1.829	2.160	2.546	2.809	3.562
97	1.309	1.446	1.828	2.159	2.545	2.807	3.560
98	1.308	1.446	1.827	2.158	2.544	2.806	3.559
99	1.307	1.445	1.826	2.157	2.542	2.805	3.557

TABLE 4.6.3. TWO-SIDED TOLERANCE LIMIT FACTORS (CONTROL BOTH TAILS).

GAMMA = 0.750 (CONTINUED)

N \P →	0.125	0.100	0.050	0.025	0.010	0.005	0.0005
100	1.306	1.444	1.825	2.156	2.541	2.804	3.556
102	1.305	1.442	1.823	2.154	2.539	2.801	3.553
104	1.303	1.440	1.821	2.152	2.537	2.799	3.550
106	1.301	1.439	1.819	2.150	2.534	2.796	3.547
108	1.300	1.437	1.818	2.148	2.532	2.794	3.545
110	1.298	1.436	1.816	2.146	2.530	2.792	3.542
112	1.297	1.434	1.814	2.144	2.528	2.790	3.540
114	1.296	1.433	1.813	2.143	2.526	2.788	3.537
116	1.294	1.431	1.811	2.141	2.525	2.786	3.535
118	1.293	1.430	1.810	2.139	2.523	2.784	3.533
120	1.292	1.429	1.808	2.138	2.521	2.782	3.530
122	1.290	1.427	1.807	2.136	2.519	2.780	3.528
124	1.289	1.426	1.805	2.134	2.517	2.778	3.526
126	1.288	1.425	1.804	2.133	2.516	2.777	3.524
128	1.287	1.424	1.803	2.132	2.514	2.775	3.522
130	1.286	1.422	1.801	2.130	2.513	2.773	3.520
132	1.285	1.421	1.800	2.129	2.511	2.772	3.518
134	1.284	1.420	1.799	2.127	2.510	2.770	3.516
136	1.283	1.419	1.798	2.126	2.508	2.769	3.515
138	1.282	1.418	1.796	2.125	2.507	2.767	3.513
140	1.281	1.417	1.795	2.123	2.505	2.766	3.511
142	1.280	1.416	1.794	2.122	2.504	2.764	3.509
144	1.279	1.415	1.793	2.121	2.503	2.763	3.508
146	1.278	1.414	1.792	2.120	2.501	2.761	3.506
148	1.277	1.413	1.791	2.119	2.500	2.760	3.505
150	1.276	1.412	1.790	2.118	2.499	2.759	3.503
152	1.275	1.411	1.789	2.116	2.498	2.757	3.502
154	1.274	1.410	1.788	2.115	2.496	2.756	3.500
156	1.273	1.409	1.787	2.114	2.495	2.755	3.499
158	1.272	1.409	1.786	2.113	2.494	2.754	3.497
160	1.272	1.408	1.785	2.112	2.493	2.752	3.496
162	1.271	1.407	1.784	2.111	2.492	2.751	3.494
164	1.270	1.406	1.783	2.110	2.491	2.750	3.493
166	1.269	1.405	1.782	2.109	2.490	2.749	3.492
168	1.269	1.405	1.781	2.108	2.489	2.748	3.490
170	1.268	1.404	1.780	2.107	2.488	2.747	3.489
172	1.267	1.403	1.780	2.106	2.487	2.746	3.488
174	1.266	1.402	1.779	2.106	2.486	2.745	3.487
176	1.266	1.402	1.778	2.105	2.485	2.744	3.486
178	1.265	1.401	1.777	2.104	2.484	2.743	3.484
180	1.264	1.400	1.776	2.103	2.483	2.742	3.483
185	1.263	1.398	1.774	2.101	2.481	2.739	3.480
190	1.261	1.397	1.773	2.099	2.478	2.737	3.478
195	1.260	1.395	1.771	2.097	2.476	2.735	3.475
200	1.258	1.394	1.769	2.095	2.474	2.733	3.473
205	1.257	1.392	1.768	2.093	2.472	2.730	3.470
210	1.256	1.391	1.766	2.092	2.470	2.729	3.468
215	1.254	1.390	1.765	2.090	2.469	2.727	3.466
220	1.253	1.388	1.763	2.088	2.467	2.725	3.464

TABLE 4.6.4. TWO-SIDED TOLERANCE LIMIT FACTORS (CONTROL BOTH TAILS).

GAMMA = 0.750 (CONTINUED)

N \P →	0.125	0.100	0.050	0.025	0.010	0.005	0.0005
225	1.252	1.387	1.762	2.087	2.465	2.723	3.462
230	1.251	1.386	1.760	2.086	2.464	2.721	3.460
235	1.249	1.385	1.759	2.084	2.462	2.720	3.458
240	1.248	1.383	1.758	2.083	2.461	2.718	3.456
245	1.247	1.382	1.757	2.081	2.459	2.717	3.454
250	1.246	1.381	1.755	2.080	2.458	2.715	3.452
255	1.245	1.380	1.754	2.079	2.456	2.714	3.451
260	1.244	1.379	1.753	2.078	2.455	2.712	3.449
265	1.243	1.378	1.752	2.076	2.454	2.711	3.447
270	1.243	1.377	1.751	2.075	2.453	2.710	3.446
275	1.242	1.376	1.750	2.074	2.451	2.708	3.444
280	1.241	1.376	1.749	2.073	2.450	2.707	3.443
285	1.240	1.375	1.748	2.072	2.449	2.706	3.441
290	1.239	1.374	1.747	2.071	2.448	2.705	3.440
295	1.238	1.373	1.746	2.070	2.447	2.703	3.439
300	1.238	1.372	1.745	2.069	2.446	2.702	3.437
310	1.236	1.371	1.744	2.067	2.444	2.700	3.435
320	1.235	1.369	1.742	2.066	2.442	2.698	3.432
330	1.233	1.368	1.740	2.064	2.440	2.696	3.430
340	1.232	1.367	1.739	2.062	2.438	2.694	3.428
350	1.231	1.365	1.738	2.061	2.437	2.692	3.426
360	1.230	1.364	1.736	2.059	2.435	2.691	3.424
370	1.229	1.363	1.735	2.058	2.433	2.689	3.422
380	1.228	1.362	1.734	2.056	2.432	2.688	3.420
390	1.227	1.361	1.733	2.055	2.430	2.686	3.418
400	1.226	1.360	1.731	2.054	2.429	2.685	3.417
425	1.223	1.357	1.729	2.051	2.426	2.681	3.413
450	1.221	1.355	1.726	2.048	2.423	2.678	3.409
475	1.219	1.353	1.724	2.046	2.420	2.675	3.406
500	1.217	1.351	1.722	2.044	2.418	2.673	3.403
525	1.216	1.349	1.720	2.042	2.416	2.670	3.400
550	1.214	1.348	1.718	2.040	2.413	2.668	3.397
575	1.213	1.346	1.717	2.038	2.411	2.666	3.395
600	1.211	1.345	1.715	2.036	2.410	2.664	3.393
625	1.210	1.344	1.714	2.035	2.408	2.662	3.391
650	1.209	1.342	1.712	2.033	2.406	2.660	3.389
700	1.207	1.340	1.710	2.030	2.403	2.657	3.385
750	1.205	1.338	1.707	2.028	2.401	2.654	3.382
800	1.203	1.336	1.705	2.026	2.398	2.652	3.379
850	1.201	1.335	1.703	2.024	2.396	2.649	3.376
900	1.200	1.333	1.702	2.022	2.394	2.647	3.373
950	1.199	1.332	1.700	2.020	2.392	2.645	3.371
1000	1.197	1.330	1.699	2.019	2.390	2.644	3.369
1500	1.189	1.321	1.689	2.008	2.378	2.631	3.354
2000	1.183	1.316	1.683	2.001	2.371	2.623	3.346
3000	1.177	1.309	1.676	1.993	2.363	2.614	3.335
5000	1.171	1.303	1.669	1.986	2.355	2.606	3.325
10000	1.165	1.297	1.662	1.978	2.346	2.597	3.315
∞	1.150	1.282	1.645	1.960	2.326	2.576	3.291

TABLE 4.7.1. TWO-SIDED TOLERANCE LIMIT FACTORS (CONTROL BOTH TAILS).

GAMMA = 0.500

N \P →	0.125	0.100	0.050	0.025	0.010	0.005	0.0005
2	2.505	2.702	3.246	3.717	4.264	4.635	5.699
3	1.912	2.072	2.512	2.893	3.335	3.636	4.497
4	1.726	1.876	2.288	2.645	3.060	3.341	4.148
5	1.629	1.774	2.174	2.520	2.921	3.194	3.976
6	1.568	1.710	2.102	2.442	2.836	3.104	3.872
7	1.524	1.665	2.052	2.388	2.777	3.042	3.801
8	1.492	1.631	2.015	2.348	2.734	2.997	3.749
9	1.466	1.605	1.986	2.317	2.700	2.961	3.708
10	1.446	1.583	1.963	2.292	2.673	2.933	3.677
11	1.428	1.565	1.944	2.271	2.651	2.910	3.650
12	1.414	1.550	1.927	2.253	2.632	2.890	3.628
13	1.401	1.537	1.913	2.239	2.616	2.874	3.610
14	1.390	1.526	1.901	2.226	2.603	2.859	3.594
15	1.380	1.516	1.890	2.214	2.590	2.847	3.580
16	1.372	1.507	1.881	2.204	2.580	2.835	3.567
17	1.364	1.499	1.872	2.195	2.570	2.825	3.556
18	1.357	1.492	1.864	2.187	2.561	2.816	3.546
19	1.350	1.485	1.857	2.179	2.553	2.808	3.537
20	1.345	1.479	1.851	2.173	2.546	2.801	3.529
21	1.339	1.473	1.845	2.166	2.540	2.794	3.522
22	1.334	1.468	1.839	2.161	2.534	2.788	3.515
23	1.330	1.464	1.834	2.155	2.528	2.782	3.508
24	1.325	1.459	1.830	2.150	2.523	2.776	3.502
25	1.321	1.455	1.825	2.146	2.518	2.772	3.497
26	1.317	1.451	1.821	2.142	2.514	2.767	3.492
27	1.314	1.448	1.817	2.138	2.509	2.763	3.487
28	1.311	1.444	1.814	2.134	2.505	2.758	3.483
29	1.307	1.441	1.810	2.130	2.502	2.755	3.479
30	1.305	1.438	1.807	2.127	2.498	2.751	3.475
31	1.302	1.435	1.804	2.124	2.495	2.748	3.471
32	1.299	1.432	1.801	2.121	2.492	2.744	3.467
33	1.297	1.430	1.799	2.118	2.489	2.741	3.464
34	1.294	1.427	1.796	2.115	2.486	2.738	3.461
35	1.292	1.425	1.793	2.113	2.483	2.736	3.458
36	1.290	1.423	1.791	2.110	2.481	2.733	3.455
37	1.288	1.421	1.789	2.108	2.478	2.730	3.452
38	1.286	1.419	1.787	2.106	2.476	2.728	3.450
39	1.284	1.417	1.785	2.103	2.474	2.726	3.447
40	1.282	1.415	1.783	2.101	2.471	2.723	3.445
41	1.280	1.413	1.781	2.099	2.469	2.721	3.443
42	1.278	1.411	1.779	2.097	2.467	2.719	3.440
43	1.277	1.410	1.777	2.096	2.466	2.717	3.438
44	1.275	1.408	1.776	2.094	2.464	2.715	3.436
45	1.274	1.406	1.774	2.092	2.462	2.714	3.434
46	1.272	1.405	1.772	2.091	2.460	2.712	3.432
47	1.271	1.404	1.771	2.089	2.459	2.710	3.431
48	1.269	1.402	1.769	2.087	2.457	2.709	3.429
49	1.268	1.401	1.768	2.086	2.455	2.707	3.427
50	1.267	1.399	1.767	2.085	2.454	2.705	3.426

TABLE 4.7.2.   TWO-SIDED TOLERANCE LIMIT FACTORS (CONTROL BOTH TAILS).

GAMMA = 0.500 (CONTINUED)

N \P →	0.125	0.100	0.050	0.025	0.010	0.005	0.0005
51	1.265	1.398	1.765	2.083	2.453	2.704	3.424
52	1.264	1.397	1.764	2.082	2.451	2.703	3.422
53	1.263	1.396	1.763	2.081	2.450	2.701	3.421
54	1.262	1.395	1.761	2.079	2.448	2.700	3.420
55	1.261	1.393	1.760	2.078	2.447	2.699	3.418
56	1.260	1.392	1.759	2.077	2.446	2.697	3.417
57	1.259	1.391	1.758	2.076	2.445	2.696	3.416
58	1.258	1.390	1.757	2.075	2.444	2.695	3.414
59	1.257	1.389	1.756	2.073	2.443	2.694	3.413
60	1.256	1.388	1.755	2.072	2.441	2.693	3.412
61	1.255	1.387	1.754	2.071	2.440	2.691	3.411
62	1.254	1.386	1.753	2.070	2.439	2.690	3.410
63	1.253	1.386	1.752	2.069	2.438	2.689	3.408
64	1.252	1.385	1.751	2.068	2.437	2.688	3.407
65	1.251	1.384	1.750	2.068	2.436	2.687	3.406
66	1.251	1.383	1.749	2.067	2.435	2.686	3.405
67	1.250	1.382	1.748	2.066	2.434	2.685	3.404
68	1.249	1.381	1.748	2.065	2.434	2.685	3.403
69	1.248	1.381	1.747	2.064	2.433	2.684	3.402
70	1.247	1.380	1.746	2.063	2.432	2.683	3.401
71	1.247	1.379	1.745	2.062	2.431	2.682	3.400
72	1.246	1.378	1.744	2.062	2.430	2.681	3.400
73	1.245	1.378	1.744	2.061	2.429	2.680	3.399
74	1.245	1.377	1.743	2.060	2.429	2.679	3.398
75	1.244	1.376	1.742	2.059	2.428	2.679	3.397
76	1.243	1.376	1.742	2.059	2.427	2.678	3.396
77	1.243	1.375	1.741	2.058	2.426	2.677	3.395
78	1.242	1.374	1.740	2.057	2.426	2.676	3.395
79	1.241	1.374	1.740	2.057	2.425	2.676	3.394
80	1.241	1.373	1.739	2.056	2.424	2.675	3.393
81	1.240	1.372	1.738	2.055	2.424	2.674	3.392
82	1.240	1.372	1.738	2.055	2.423	2.674	3.392
83	1.239	1.371	1.737	2.054	2.422	2.673	3.391
84	1.238	1.371	1.736	2.053	2.422	2.672	3.390
85	1.238	1.370	1.736	2.053	2.421	2.672	3.390
86	1.237	1.370	1.735	2.052	2.420	2.671	3.389
87	1.237	1.369	1.735	2.052	2.420	2.671	3.388
88	1.236	1.368	1.734	2.051	2.419	2.670	3.388
89	1.236	1.368	1.734	2.050	2.419	2.669	3.387
90	1.235	1.367	1.733	2.050	2.418	2.669	3.387
91	1.235	1.367	1.733	2.049	2.418	2.668	3.386
92	1.234	1.366	1.732	2.049	2.417	2.668	3.385
93	1.234	1.366	1.732	2.048	2.416	2.667	3.385
94	1.233	1.365	1.731	2.048	2.416	2.667	3.384
95	1.233	1.365	1.731	2.047	2.415	2.666	3.384
96	1.232	1.365	1.730	2.047	2.415	2.665	3.383
97	1.232	1.364	1.730	2.046	2.414	2.665	3.383
98	1.232	1.364	1.729	2.046	2.414	2.664	3.382
99	1.231	1.363	1.729	2.045	2.413	2.664	3.381

TABLE 4.7.3. TWO-SIDED TOLERANCE LIMIT FACTORS (CONTROL BOTH TAILS).

GAMMA = 0.500 (CONTINUED)

N \P →	0.125	0.100	0.050	0.025	0.010	0.005	0.0005
100	1.231	1.363	1.728	2.045	2.413	2.663	3.381
102	1.230	1.362	1.727	2.044	2.412	2.663	3.380
104	1.229	1.361	1.726	2.043	2.411	2.662	3.379
106	1.228	1.360	1.726	2.042	2.410	2.661	3.378
108	1.227	1.360	1.725	2.041	2.409	2.660	3.377
110	1.227	1.359	1.724	2.041	2.409	2.659	3.376
112	1.226	1.358	1.723	2.040	2.408	2.658	3.375
114	1.225	1.357	1.723	2.039	2.407	2.657	3.375
116	1.225	1.357	1.722	2.038	2.406	2.657	3.374
118	1.224	1.356	1.721	2.038	2.405	2.656	3.373
120	1.223	1.355	1.720	2.037	2.405	2.655	3.372
122	1.223	1.355	1.720	2.036	2.404	2.654	3.371
124	1.222	1.354	1.719	2.036	2.403	2.654	3.371
126	1.221	1.353	1.719	2.035	2.403	2.653	3.370
128	1.221	1.353	1.718	2.034	2.402	2.652	3.369
130	1.220	1.352	1.717	2.034	2.401	2.652	3.369
132	1.220	1.352	1.717	2.033	2.401	2.651	3.368
134	1.219	1.351	1.716	2.033	2.400	2.650	3.367
136	1.219	1.351	1.716	2.032	2.400	2.650	3.367
138	1.218	1.350	1.715	2.031	2.399	2.649	3.366
140	1.218	1.350	1.715	2.031	2.398	2.649	3.365
142	1.217	1.349	1.714	2.030	2.398	2.648	3.365
144	1.217	1.349	1.713	2.030	2.397	2.648	3.364
146	1.216	1.348	1.713	2.029	2.397	2.647	3.364
148	1.216	1.348	1.712	2.029	2.396	2.647	3.363
150	1.215	1.347	1.712	2.028	2.396	2.646	3.363
152	1.215	1.347	1.712	2.028	2.395	2.645	3.362
154	1.214	1.346	1.711	2.027	2.395	2.645	3.362
156	1.214	1.346	1.711	2.027	2.394	2.645	3.361
158	1.214	1.345	1.710	2.026	2.394	2.644	3.361
160	1.213	1.345	1.710	2.026	2.393	2.644	3.360
162	1.213	1.345	1.709	2.026	2.393	2.643	3.360
164	1.212	1.344	1.709	2.025	2.393	2.643	3.359
166	1.212	1.344	1.709	2.025	2.392	2.642	3.359
168	1.212	1.343	1.708	2.024	2.392	2.642	3.358
170	1.211	1.343	1.708	2.024	2.391	2.641	3.358
172	1.211	1.343	1.707	2.023	2.391	2.641	3.357
174	1.210	1.342	1.707	2.023	2.390	2.641	3.357
176	1.210	1.342	1.707	2.023	2.390	2.640	3.357
178	1.210	1.342	1.706	2.022	2.390	2.640	3.356
180	1.209	1.341	1.706	2.022	2.389	2.639	3.356
185	1.209	1.340	1.705	2.021	2.388	2.638	3.355
190	1.208	1.340	1.704	2.020	2.388	2.638	3.354
195	1.207	1.339	1.703	2.019	2.387	2.637	3.353
200	1.206	1.338	1.703	2.019	2.386	2.636	3.352
205	1.205	1.337	1.702	2.018	2.385	2.635	3.351
210	1.205	1.337	1.701	2.017	2.384	2.634	3.350
215	1.204	1.336	1.700	2.016	2.384	2.634	3.350
220	1.204	1.335	1.700	2.016	2.383	2.633	3.349

TABLE 4.7.4. *TWO-SIDED TOLERANCE LIMIT FACTORS (CONTROL BOTH TAILS).*

GAMMA = 0.500 (CONTINUED)

N \P →	0.125	0.100	0.050	0.025	0.010	0.005	0.0005
225	1.203	1.335	1.699	2.015	2.382	2.632	3.348
230	1.202	1.334	1.698	2.014	2.382	2.632	3.348
235	1.202	1.333	1.698	2.014	2.381	2.631	3.347
240	1.201	1.333	1.697	2.013	2.380	2.630	3.346
245	1.201	1.332	1.697	2.013	2.380	2.630	3.346
250	1.200	1.332	1.696	2.012	2.379	2.629	3.345
255	1.200	1.331	1.696	2.012	2.379	2.629	3.344
260	1.199	1.331	1.695	2.011	2.378	2.628	3.344
265	1.199	1.330	1.695	2.010	2.378	2.628	3.343
270	1.198	1.330	1.694	2.010	2.377	2.627	3.343
275	1.198	1.329	1.694	2.010	2.377	2.626	3.342
280	1.197	1.329	1.693	2.009	2.376	2.626	3.342
285	1.197	1.329	1.693	2.009	2.376	2.626	3.341
290	1.196	1.328	1.692	2.008	2.375	2.625	3.341
295	1.196	1.328	1.692	2.008	2.375	2.625	3.340
300	1.196	1.327	1.692	2.007	2.374	2.624	3.340
310	1.195	1.327	1.691	2.006	2.374	2.623	3.339
320	1.194	1.326	1.690	2.006	2.373	2.623	3.338
330	1.193	1.325	1.689	2.005	2.372	2.622	3.337
340	1.193	1.324	1.689	2.004	2.371	2.621	3.337
350	1.192	1.324	1.688	2.004	2.371	2.620	3.336
360	1.192	1.323	1.687	2.003	2.370	2.620	3.335
370	1.191	1.323	1.687	2.002	2.369	2.619	3.335
380	1.190	1.322	1.686	2.002	2.369	2.619	3.334
390	1.190	1.321	1.686	2.001	2.368	2.618	3.333
400	1.189	1.321	1.685	2.001	2.368	2.617	3.333
425	1.188	1.320	1.684	1.999	2.366	2.616	3.332
450	1.187	1.319	1.683	1.998	2.365	2.615	3.330
475	1.186	1.318	1.682	1.997	2.364	2.614	3.329
500	1.185	1.317	1.681	1.996	2.363	2.613	3.328
525	1.184	1.316	1.680	1.995	2.362	2.612	3.327
550	1.184	1.315	1.679	1.994	2.361	2.611	3.326
575	1.183	1.314	1.678	1.994	2.360	2.610	3.325
600	1.182	1.314	1.677	1.993	2.360	2.609	3.325
625	1.181	1.313	1.677	1.992	2.359	2.609	3.324
650	1.181	1.312	1.676	1.992	2.358	2.608	3.323
700	1.180	1.311	1.675	1.990	2.357	2.607	3.322
750	1.179	1.310	1.674	1.989	2.356	2.606	3.321
800	1.178	1.309	1.673	1.988	2.355	2.605	3.320
850	1.177	1.308	1.672	1.988	2.354	2.604	3.319
900	1.176	1.308	1.671	1.987	2.353	2.603	3.318
950	1.175	1.307	1.671	1.986	2.353	2.602	3.317
1000	1.175	1.306	1.670	1.985	2.352	2.602	3.317
1500	1.170	1.302	1.665	1.981	2.347	2.597	3.312
2000	1.168	1.299	1.662	1.978	2.344	2.594	3.309
3000	1.164	1.296	1.659	1.974	2.341	2.590	3.305
5000	1.161	1.292	1.656	1.971	2.338	2.587	3.302
10000	1.158	1.289	1.653	1.968	2.334	2.584	3.299
∞	1.150	1.282	1.645	1.960	2.326	2.576	3.291

TABLE 5

Two-sided sampling plan factors for a normal distribution
(control center) for values of P = 0.20, 0.10, 0.05, 0.025,
0.02, 0.01, 0.005; N = 2(1)25(5)50(10)100, 120, 150, 300,
500(100)1000, 1500, 2000, 3000, 5000, 10000, ∞; and
γ = 0.900.   *146*

TABLE 6

Two-sided sampling plan factors for a normal distribution
(control both tails) for values of P = 0.20, 0.10, 0.05,
0.025, 0.02, 0.01, 0.005; N = 2(1)25(5)50(10)100, 120, 150,
300, 500(100)1000, 1500, 2000, 3000, 5000, 10000, ∞; and
γ = 0.900.   *147*

TABLE 5. TWO-SIDED SAMPLING PLAN FACTORS (CONTROL CENTER).

GAMMA = 0.900

N \P →	0.20	0.10	0.05	0.025	0.02	0.01	0.005
2	6.987	10.253	13.090	15.586	16.331	18.500	20.486
3	3.039	4.258	5.311	6.244	6.523	7.340	8.092
4	2.295	3.188	3.957	4.637	4.841	5.438	5.988
5	1.976	2.742	3.400	3.981	4.156	4.666	5.136
6	1.806	2.494	3.092	3.620	3.779	4.243	4.669
7	1.721	2.334	2.894	3.389	3.538	3.972	4.372
8	1.666	2.227	2.755	3.227	3.369	3.783	4.164
9	1.626	2.158	2.652	3.106	3.242	3.641	4.009
10	1.595	2.112	2.576	3.012	3.144	3.532	3.888
11	1.570	2.075	2.520	2.938	3.066	3.444	3.792
12	1.550	2.045	2.479	2.879	3.004	3.371	3.712
13	1.533	2.020	2.446	2.833	2.953	3.312	3.646
14	1.519	1.999	2.419	2.796	2.912	3.261	3.589
15	1.506	1.981	2.395	2.767	2.880	3.219	3.541
16	1.496	1.965	2.374	2.742	2.853	3.184	3.499
17	1.486	1.950	2.356	2.720	2.830	3.155	3.463
18	1.478	1.938	2.340	2.701	2.810	3.130	3.433
19	1.470	1.927	2.325	2.683	2.791	3.109	3.406
20	1.463	1.916	2.312	2.667	2.775	3.090	3.383
21	1.457	1.907	2.300	2.653	2.760	3.073	3.364
22	1.451	1.899	2.290	2.640	2.746	3.057	3.346
23	1.446	1.891	2.280	2.628	2.733	3.043	3.330
24	1.441	1.884	2.270	2.617	2.721	3.029	3.315
25	1.437	1.877	2.262	2.606	2.711	3.017	3.301
30	1.419	1.851	2.227	2.565	2.667	2.967	3.245
35	1.406	1.831	2.202	2.534	2.634	2.929	3.203
40	1.396	1.816	2.182	2.510	2.609	2.901	3.171
45	1.387	1.804	2.166	2.491	2.589	2.878	3.146
50	1.381	1.794	2.154	2.476	2.573	2.859	3.124
60	1.370	1.778	2.133	2.451	2.547	2.829	3.091
70	1.362	1.766	2.118	2.433	2.528	2.807	3.066
80	1.356	1.757	2.106	2.418	2.513	2.790	3.047
90	1.351	1.750	2.097	2.407	2.500	2.776	3.031
100	1.347	1.744	2.089	2.397	2.490	2.764	3.018
120	1.341	1.734	2.076	2.382	2.474	2.746	2.997
150	1.334	1.723	2.062	2.365	2.457	2.726	2.975
300	1.317	1.699	2.030	2.326	2.416	2.678	2.922
500	1.309	1.686	2.013	2.306	2.394	2.654	2.895
600	1.306	1.682	2.008	2.300	2.388	2.647	2.887
700	1.304	1.679	2.005	2.295	2.383	2.641	2.880
800	1.303	1.677	2.002	2.292	2.379	2.637	2.875
900	1.301	1.675	1.999	2.289	2.376	2.633	2.871
1000	1.300	1.673	1.997	2.286	2.374	2.630	2.868
1500	1.297	1.668	1.990	2.278	2.365	2.620	2.856
2000	1.295	1.665	1.986	2.273	2.359	2.614	2.850
3000	1.292	1.661	1.981	2.267	2.353	2.607	2.842
5000	1.290	1.657	1.976	2.261	2.347	2.600	2.834
10000	1.287	1.654	1.971	2.255	2.341	2.593	2.826
∞	1.282	1.645	1.960	2.241	2.326	2.576	2.807

TABLE 6. TWO-SIDED SAMPLING PLAN FACTORS (CONTROL BOTH TAILS).

GAMMA = 0.900

N \P →	0.20	0.10	0.05	0.025	0.02	0.01	0.005
2	6.987	10.253	13.090	15.586	16.331	18.500	20.486
3	3.039	4.258	5.311	6.244	6.523	7.340	8.092
4	2.295	3.188	3.957	4.637	4.841	5.438	5.988
5	1.976	2.742	3.400	3.981	4.156	4.666	5.136
6	1.795	2.494	3.092	3.620	3.779	4.243	4.669
7	1.676	2.333	2.894	3.389	3.538	3.972	4.372
8	1.590	2.219	2.754	3.227	3.369	3.783	4.164
9	1.525	2.133	2.650	3.106	3.242	3.641	4.009
10	1.474	2.066	2.568	3.011	3.144	3.532	3.888
11	1.433	2.011	2.503	2.935	3.065	3.443	3.792
12	1.398	1.966	2.448	2.872	3.000	3.371	3.712
13	1.368	1.928	2.402	2.820	2.945	3.309	3.645
14	1.343	1.895	2.363	2.774	2.898	3.257	3.588
15	1.321	1.867	2.329	2.735	2.857	3.212	3.538
16	1.301	1.842	2.299	2.701	2.821	3.172	3.495
17	1.284	1.819	2.272	2.670	2.789	3.137	3.456
18	1.268	1.800	2.249	2.643	2.761	3.105	3.422
19	1.254	1.782	2.227	2.618	2.736	3.077	3.391
20	1.241	1.765	2.208	2.596	2.712	3.052	3.363
21	1.229	1.750	2.190	2.576	2.691	3.028	3.338
22	1.218	1.737	2.174	2.557	2.672	3.007	3.315
23	1.208	1.724	2.159	2.540	2.654	2.987	3.293
24	1.199	1.712	2.145	2.525	2.638	2.969	3.273
25	1.190	1.702	2.132	2.510	2.623	2.952	3.255
30	1.154	1.657	2.080	2.450	2.561	2.884	3.180
35	1.127	1.624	2.041	2.406	2.515	2.833	3.125
40	1.106	1.598	2.010	2.371	2.479	2.793	3.082
45	1.089	1.577	1.986	2.343	2.450	2.761	3.047
50	1.075	1.559	1.965	2.320	2.426	2.735	3.018
60	1.052	1.532	1.933	2.284	2.389	2.694	2.973
70	1.035	1.511	1.909	2.256	2.360	2.662	2.940
80	1.022	1.495	1.890	2.235	2.338	2.638	2.913
90	1.011	1.481	1.874	2.217	2.320	2.618	2.891
100	1.001	1.470	1.861	2.203	2.304	2.601	2.873
120	0.986	1.452	1.841	2.179	2.280	2.574	2.844
150	0.970	1.433	1.818	2.154	2.254	2.546	2.813
300	0.931	1.386	1.765	2.094	2.192	2.477	2.739
500	0.910	1.362	1.736	2.062	2.159	2.442	2.701
600	0.904	1.355	1.728	2.053	2.150	2.431	2.689
700	0.899	1.349	1.722	2.046	2.142	2.423	2.680
800	0.896	1.344	1.717	2.040	2.136	2.417	2.673
900	0.892	1.341	1.712	2.035	2.132	2.411	2.668
1000	0.890	1.338	1.709	2.031	2.127	2.407	2.663
1500	0.881	1.327	1.697	2.018	2.114	2.392	2.646
2000	0.875	1.321	1.690	2.010	2.105	2.383	2.637
3000	0.869	1.314	1.681	2.001	2.096	2.372	2.625
5000	0.863	1.306	1.673	1.991	2.086	2.362	2.614
10000	0.857	1.299	1.665	1.982	2.077	2.351	2.603
∞	0.842	1.282	1.645	1.960	2.054	2.326	2.576

TABLE 7

Confidence limits on the proportion in one tail of the normal distribution
for values of k = -3.0(0.20)6.0 and N = 2(1)18(3)30, 40(20)120, 240,
600, 1000, 1200.

TABLE 7.1.1. CONFIDENCE LIMITS ON THE PROPORTION IN ONE TAIL.

ETA = 0.500

K ↓	N = 2	N = 3	N = 4	N = 5	N = 6
-3.0	0.01883	0.00585	0.00369	0.00290	0.00250
-2.8	0.02576	0.00924	0.00617	0.00499	0.00438
-2.6	0.03469	0.01424	0.01002	0.00833	0.00744
-2.4	0.04600	0.02141	0.01582	0.01350	0.01225
-2.2	0.06008	0.03140	0.02429	0.02123	0.01955
-2.0	0.07734	0.04498	0.03629	0.03242	0.03025
-1.8	0.09819	0.06293	0.05275	0.04808	0.04542
-1.6	0.12304	0.08604	0.07466	0.06929	0.06619
-1.4	0.15228	0.11508	0.10297	0.09711	0.09368
-1.2	0.18631	0.15064	0.13845	0.13243	0.12885
-1.0	0.22552	0.19317	0.18165	0.17585	0.17237
-0.8	0.27021	0.24282	0.23273	0.22756	0.22443
-0.6	0.32056	0.29940	0.29138	0.28721	0.28466
-0.4	0.37639	0.36217	0.35666	0.35375	0.35197
-0.2	0.43683	0.42975	0.42696	0.42548	0.42457
0.0	0.50000	0.50000	0.50000	0.50000	0.50000
0.2	0.56317	0.57025	0.57304	0.57452	0.57543
0.4	0.62361	0.63783	0.64334	0.64625	0.64803
0.6	0.67944	0.70060	0.70862	0.71279	0.71534
0.8	0.72979	0.75718	0.76727	0.77244	0.77557
1.0	0.77448	0.80683	0.81835	0.82415	0.82763
1.2	0.81369	0.84936	0.86155	0.86757	0.87115
1.4	0.84772	0.88492	0.89703	0.90289	0.90632
1.6	0.87696	0.91396	0.92534	0.93071	0.93381
1.8	0.90181	0.93707	0.94725	0.95192	0.95458
2.0	0.92266	0.95502	0.96371	0.96758	0.96975
2.2	0.93992	0.96860	0.97571	0.97877	0.98045
2.4	0.95400	0.97859	0.98418	0.98650	0.98775
2.6	0.96531	0.98576	0.98998	0.99167	0.99256
2.8	0.97424	0.99076	0.99383	0.99501	0.99562
3.0	0.98117	0.99415	0.99631	0.99710	0.99750
3.2	0.98645	0.99639	0.99786	0.99837	0.99862
3.4	0.99042	0.99783	0.99879	0.99911	0.99926
3.6	0.99333	0.99872	0.99934	0.99953	0.99962
3.8	0.99543	0.99927	0.99965	0.99976	0.99981
4.0	0.99693	0.99959	0.99982	0.99988	0.99991
4.2	0.99797	0.99978	0.99991	0.99994	0.99996
4.4	0.99868	0.99988	0.99996	0.99997	0.99998
4.6	0.99915	0.99994	0.99998	0.99999	0.99999
4.8	0.99947	0.99997	0.99999	0.99999	1.00000
5.0	0.99967	0.99999	1.00000	1.00000	1.00000
5.2	0.99980	0.99999	1.00000	1.00000	1.00000
5.4	0.99988	1.00000	1.00000	1.00000	1.00000
5.6	0.99993	1.00000	1.00000	1.00000	1.00000
5.8	0.99996	1.00000	1.00000	1.00000	1.00000
6.0	0.99998	1.00000	1.00000	1.00000	1.00000

TABLE 7.1.2. CONFIDENCE LIMITS ON THE PROPORTION IN ONE TAIL.

ETA = 0.500 *(CONTINUED)*

K ↓	N = 7	N = 8	N = 9	N = 10	N = 11
-3.0	0.00226	0.00210	0.00199	0.00191	0.00184
-2.8	0.00401	0.00376	0.00359	0.00346	0.00335
-2.6	0.00689	0.00652	0.00626	0.00606	0.00590
-2.4	0.01147	0.01094	0.01056	0.01027	0.01004
-2.2	0.01849	0.01776	0.01723	0.01683	0.01651
-2.0	0.02887	0.02792	0.02722	0.02669	0.02627
-1.8	0.04371	0.04252	0.04165	0.04098	0.04045
-1.6	0.06418	0.06277	0.06172	0.06092	0.06029
-1.4	0.09144	0.08985	0.08868	0.08777	0.08705
-1.2	0.12650	0.12482	0.12358	0.12261	0.12185
-1.0	0.17006	0.16841	0.16718	0.16622	0.16546
-0.8	0.22234	0.22084	0.21972	0.21884	0.21814
-0.6	0.28295	0.28172	0.28079	0.28007	0.27949
-0.4	0.35076	0.34990	0.34924	0.34873	0.34832
-0.2	0.42395	0.42350	0.42316	0.42290	0.42268
0.0	0.50000	0.50000	0.50000	0.50000	0.50000
0.2	0.57605	0.57650	0.57684	0.57710	0.57732
0.4	0.64924	0.65010	0.65076	0.65127	0.65168
0.6	0.71705	0.71828	0.71921	0.71993	0.72051
0.8	0.77766	0.77916	0.78028	0.78116	0.78186
1.0	0.82994	0.83159	0.83282	0.83378	0.83454
1.2	0.87350	0.87518	0.87642	0.87739	0.87815
1.4	0.90856	0.91015	0.91132	0.91223	0.91295
1.6	0.93582	0.93723	0.93828	0.93908	0.93971
1.8	0.95629	0.95748	0.95835	0.95902	0.95955
2.0	0.97113	0.97208	0.97278	0.97331	0.97373
2.2	0.98151	0.98224	0.98277	0.98317	0.98349
2.4	0.98853	0.98906	0.98944	0.98973	0.98996
2.6	0.99311	0.99348	0.99374	0.99394	0.99410
2.8	0.99599	0.99624	0.99641	0.99654	0.99665
3.0	0.99774	0.99790	0.99801	0.99809	0.99816
3.2	0.99877	0.99887	0.99893	0.99898	0.99902
3.4	0.99935	0.99941	0.99945	0.99948	0.99950
3.6	0.99967	0.99970	0.99972	0.99974	0.99975
3.8	0.99984	0.99985	0.99987	0.99988	0.99988
4.0	0.99992	0.99993	0.99994	0.99994	0.99995
4.2	0.99996	0.99997	0.99997	0.99997	0.99998
4.4	0.99998	0.99999	0.99999	0.99999	0.99999
4.6	0.99999	0.99999	0.99999	1.00000	1.00000
4.8	1.00000	1.00000	1.00000	1.00000	1.00000
5.0	1.00000	1.00000	1.00000	1.00000	1.00000
5.2	1.00000	1.00000	1.00000	1.00000	1.00000
5.4	1.00000	1.00000	1.00000	1.00000	1.00000
5.6	1.00000	1.00000	1.00000	1.00000	1.00000
5.8	1.00000	1.00000	1.00000	1.00000	1.00000
6.0	1.00000	1.00000	1.00000	1.00000	1.00000

TABLE 7.1.3. CONFIDENCE LIMITS ON THE PROPORTION IN ONE TAIL.

ETA = 0.500 (CONTINUED)

K ↓	N = 12	N = 13	N = 14	N = 15	N = 16
-3.0	0.00179	0.00175	0.00172	0.00169	0.00166
-2.8	0.00327	0.00321	0.00315	0.00310	0.00307
-2.6	0.00578	0.00568	0.00559	0.00552	0.00546
-2.4	0.00986	0.00971	0.00958	0.00948	0.00939
-2.2	0.01626	0.01605	0.01588	0.01573	0.01560
-2.0	0.02593	0.02565	0.02542	0.02522	0.02505
-1.8	0.04002	0.03966	0.03936	0.03911	0.03889
-1.6	0.05977	0.05934	0.05898	0.05867	0.05841
-1.4	0.08646	0.08597	0.08556	0.08521	0.08491
-1.2	0.12122	0.12070	0.12026	0.11988	0.11956
-1.0	0.16483	0.16431	0.16387	0.16350	0.16317
-0.8	0.21757	0.21709	0.21669	0.21634	0.21604
-0.6	0.27902	0.27862	0.27828	0.27800	0.27775
-0.4	0.34798	0.34770	0.34746	0.34726	0.34708
-0.2	0.42251	0.42236	0.42224	0.42213	0.42204
0.0	0.50000	0.50000	0.50000	0.50000	0.50000
0.2	0.57749	0.57764	0.57776	0.57787	0.57796
0.4	0.65202	0.65230	0.65254	0.65274	0.65292
0.6	0.72098	0.72138	0.72172	0.72200	0.72225
0.8	0.78243	0.78291	0.78331	0.78366	0.78396
1.0	0.83517	0.83569	0.83613	0.83650	0.83683
1.2	0.87878	0.87930	0.87974	0.88012	0.88044
1.4	0.91354	0.91403	0.91444	0.91479	0.91509
1.6	0.94023	0.94066	0.94102	0.94133	0.94159
1.8	0.95998	0.96034	0.96064	0.96089	0.96111
2.0	0.97407	0.97435	0.97458	0.97478	0.97495
2.2	0.98374	0.98395	0.98412	0.98427	0.98440
2.4	0.99014	0.99029	0.99042	0.99052	0.99061
2.6	0.99422	0.99432	0.99441	0.99448	0.99454
2.8	0.99673	0.99679	0.99685	0.99690	0.99693
3.0	0.99821	0.99825	0.99828	0.99831	0.99834
3.2	0.99905	0.99908	0.99910	0.99912	0.99913
3.4	0.99952	0.99953	0.99954	0.99955	0.99956
3.6	0.99976	0.99977	0.99978	0.99978	0.99979
3.8	0.99989	0.99989	0.99989	0.99990	0.99990
4.0	0.99995	0.99995	0.99995	0.99995	0.99995
4.2	0.99998	0.99998	0.99998	0.99998	0.99998
4.4	0.99999	0.99999	0.99999	0.99999	0.99999
4.6	1.00000	1.00000	1.00000	1.00000	1.00000
4.8	1.00000	1.00000	1.00000	1.00000	1.00000
5.0	1.00000	1.00000	1.00000	1.00000	1.00000
5.2	1.00000	1.00000	1.00000	1.00000	1.00000
5.4	1.00000	1.00000	1.00000	1.00000	1.00000
5.6	1.00000	1.00000	1.00000	1.00000	1.00000
5.8	1.00000	1.00000	1.00000	1.00000	1.00000
6.0	1.00000	1.00000	1.00000	1.00000	1.00000

TABLE 7.1.4. CONFIDENCE LIMITS ON THE PROPORTION IN ONE TAIL.

ETA = 0.500 (CONTINUED)

K ↓	N = 17	N = 18	N = 21	N = 24	N = 27
-3.0	0.00164	0.00162	0.00158	0.00155	0.00152
-2.8	0.00303	0.00300	0.00293	0.00288	0.00284
-2.6	0.00541	0.00536	0.00525	0.00517	0.00511
-2.4	0.00931	0.00924	0.00908	0.00896	0.00887
-2.2	0.01549	0.01539	0.01516	0.01499	0.01486
-2.0	0.02490	0.02477	0.02446	0.02423	0.02405
-1.8	0.03870	0.03853	0.03813	0.03784	0.03761
-1.6	0.05818	0.05797	0.05749	0.05713	0.05686
-1.4	0.08465	0.08441	0.08386	0.08345	0.08313
-1.2	0.11927	0.11902	0.11842	0.11798	0.11764
-1.0	0.16289	0.16264	0.16204	0.16159	0.16125
-0.8	0.21578	0.21555	0.21499	0.21458	0.21427
-0.6	0.27753	0.27734	0.27688	0.27653	0.27627
-0.4	0.34692	0.34679	0.34646	0.34621	0.34603
-0.2	0.42196	0.42189	0.42172	0.42159	0.42149
0.0	0.50000	0.50000	0.50000	0.50000	0.50000
0.2	0.57804	0.57811	0.57828	0.57841	0.57851
0.4	0.65308	0.65321	0.65354	0.65379	0.65397
0.6	0.72247	0.72266	0.72312	0.72347	0.72373
0.8	0.78422	0.78445	0.78501	0.78542	0.78573
1.0	0.83711	0.83736	0.83796	0.83841	0.83875
1.2	0.88073	0.88098	0.88158	0.88202	0.88236
1.4	0.91535	0.91559	0.91614	0.91655	0.91687
1.6	0.94182	0.94203	0.94251	0.94287	0.94314
1.8	0.96130	0.96147	0.96187	0.96216	0.96239
2.0	0.97510	0.97523	0.97554	0.97577	0.97595
2.2	0.98451	0.98461	0.98484	0.98501	0.98514
2.4	0.99069	0.99076	0.99092	0.99104	0.99113
2.6	0.99459	0.99464	0.99475	0.99483	0.99489
2.8	0.99697	0.99700	0.99707	0.99712	0.99716
3.0	0.99836	0.99838	0.99842	0.99845	0.99848
3.2	0.99914	0.99915	0.99918	0.99920	0.99921
3.4	0.99957	0.99957	0.99959	0.99960	0.99961
3.6	0.99979	0.99979	0.99980	0.99981	0.99981
3.8	0.99990	0.99990	0.99991	0.99991	0.99991
4.0	0.99996	0.99996	0.99996	0.99996	0.99996
4.2	0.99998	0.99998	0.99998	0.99998	0.99998
4.4	0.99999	0.99999	0.99999	0.99999	0.99999
4.6	1.00000	1.00000	1.00000	1.00000	1.00000
4.8	1.00000	1.00000	1.00000	1.00000	1.00000
5.0	1.00000	1.00000	1.00000	1.00000	1.00000
5.2	1.00000	1.00000	1.00000	1.00000	1.00000
5.4	1.00000	1.00000	1.00000	1.00000	1.00000
5.6	1.00000	1.00000	1.00000	1.00000	1.00000
5.8	1.00000	1.00000	1.00000	1.00000	1.00000
6.0	1.00000	1.00000	1.00000	1.00000	1.00000

TABLE 7.1.5. CONFIDENCE LIMITS ON THE PROPORTION IN ONE TAIL.

ETA = 0.500 (CONTINUED)

K ↓	N = 30	N = 40	N = 60	N = 80	N = 100
-3.0	0.00150	0.00146	0.00142	0.00140	0.00139
-2.8	0.00281	0.00274	0.00268	0.00265	0.00263
-2.6	0.00506	0.00495	0.00485	0.00480	0.00477
-2.4	0.00879	0.00864	0.00849	0.00841	0.00837
-2.2	0.01476	0.01453	0.01432	0.01421	0.01415
-2.0	0.02391	0.02361	0.02332	0.02317	0.02309
-1.8	0.03744	0.03704	0.03666	0.03648	0.03637
-1.6	0.05664	0.05616	0.05570	0.05547	0.05533
-1.4	0.08288	0.08233	0.08180	0.08153	0.08138
-1.2	0.11738	0.11678	0.11620	0.11591	0.11574
-1.0	0.16098	0.16038	0.15980	0.15951	0.15934
-0.8	0.21402	0.21346	0.21292	0.21265	0.21249
-0.6	0.27606	0.27560	0.27514	0.27492	0.27478
-0.4	0.34588	0.34554	0.34522	0.34506	0.34496
-0.2	0.42142	0.42124	0.42107	0.42099	0.42094
0.0	0.50000	0.50000	0.50000	0.50000	0.50000
0.2	0.57858	0.57876	0.57893	0.57901	0.57906
0.4	0.65412	0.65446	0.65478	0.65494	0.65504
0.6	0.72394	0.72440	0.72486	0.72508	0.72522
0.8	0.78598	0.78654	0.78708	0.78735	0.78751
1.0	0.83902	0.83962	0.84020	0.84049	0.84066
1.2	0.88262	0.88322	0.88380	0.88409	0.88426
1.4	0.91712	0.91767	0.91820	0.91847	0.91862
1.6	0.94336	0.94384	0.94430	0.94453	0.94467
1.8	0.96256	0.96296	0.96334	0.96352	0.96363
2.0	0.97609	0.97639	0.97668	0.97683	0.97691
2.2	0.98524	0.98547	0.98568	0.98579	0.98585
2.4	0.99121	0.99136	0.99151	0.99159	0.99163
2.6	0.99494	0.99505	0.99515	0.99520	0.99523
2.8	0.99719	0.99726	0.99732	0.99735	0.99737
3.0	0.99850	0.99854	0.99858	0.99860	0.99861
3.2	0.99922	0.99925	0.99927	0.99928	0.99929
3.4	0.99961	0.99963	0.99964	0.99965	0.99965
3.6	0.99981	0.99982	0.99983	0.99983	0.99983
3.8	0.99991	0.99992	0.99992	0.99992	0.99992
4.0	0.99996	0.99996	0.99997	0.99997	0.99997
4.2	0.99998	0.99998	0.99999	0.99999	0.99999
4.4	0.99999	0.99999	0.99999	0.99999	0.99999
4.6	1.00000	1.00000	1.00000	1.00000	1.00000
4.8	1.00000	1.00000	1.00000	1.00000	1.00000
5.0	1.00000	1.00000	1.00000	1.00000	1.00000
5.2	1.00000	1.00000	1.00000	1.00000	1.00000
5.4	1.00000	1.00000	1.00000	1.00000	1.00000
5.6	1.00000	1.00000	1.00000	1.00000	1.00000
5.8	1.00000	1.00000	1.00000	1.00000	1.00000
6.0	1.00000	1.00000	1.00000	1.00000	1.00000

TABLE 7.1.6. *CONFIDENCE LIMITS ON THE PROPORTION IN ONE TAIL.*

ETA = 0.500 (*CONTINUED*)

K	N = 120	N = 240	N = 600	N = 1000	N = 1200
-3.0	0.00139	0.00137	0.00136	0.00135	0.00135
-2.8	0.00261	0.00258	0.00257	0.00256	0.00256
-2.6	0.00476	0.00471	0.00468	0.00467	0.00467
-2.4	0.00834	0.00827	0.00823	0.00821	0.00821
-2.2	0.01411	0.01400	0.01394	0.01393	0.01392
-2.0	0.02303	0.02289	0.02281	0.02278	0.02278
-1.8	0.03629	0.03611	0.03600	0.03597	0.03597
-1.6	0.05524	0.05502	0.05489	0.05485	0.05484
-1.4	0.08127	0.08101	0.08086	0.08082	0.08081
-1.2	0.11563	0.11535	0.11518	0.11514	0.11513
-1.0	0.15922	0.15894	0.15877	0.15872	0.15871
-0.8	0.21238	0.21212	0.21196	0.21192	0.21191
-0.6	0.27469	0.27447	0.27434	0.27431	0.27430
-0.4	0.34490	0.34474	0.34464	0.34462	0.34461
-0.2	0.42091	0.42082	0.42077	0.42076	0.42076
0.0	0.50000	0.50000	0.50000	0.50000	0.50000
0.2	0.57909	0.57918	0.57923	0.57924	0.57924
0.4	0.65510	0.65526	0.65536	0.65538	0.65539
0.6	0.72531	0.72553	0.72566	0.72569	0.72570
0.8	0.78762	0.78788	0.78804	0.78808	0.78809
1.0	0.84078	0.84106	0.84123	0.84128	0.84129
1.2	0.88437	0.88465	0.88482	0.88486	0.88487
1.4	0.91873	0.91899	0.91914	0.91918	0.91919
1.6	0.94476	0.94498	0.94511	0.94515	0.94516
1.8	0.96371	0.96389	0.96400	0.96403	0.96403
2.0	0.97697	0.97711	0.97719	0.97722	0.97722
2.2	0.98589	0.98600	0.98606	0.98607	0.98608
2.4	0.99166	0.99173	0.99177	0.99179	0.99179
2.6	0.99524	0.99529	0.99532	0.99533	0.99533
2.8	0.99739	0.99742	0.99743	0.99744	0.99744
3.0	0.99861	0.99863	0.99864	0.99865	0.99865
3.2	0.99929	0.99930	0.99931	0.99931	0.99931
3.4	0.99965	0.99966	0.99966	0.99966	0.99966
3.6	0.99983	0.99984	0.99984	0.99984	0.99984
3.8	0.99992	0.99993	0.99993	0.99993	0.99993
4.0	0.99997	0.99997	0.99997	0.99997	0.99997
4.2	0.99999	0.99999	0.99999	0.99999	0.99999
4.4	0.99999	0.99999	0.99999	0.99999	0.99999
4.6	1.00000	1.00000	1.00000	1.00000	1.00000
4.8	1.00000	1.00000	1.00000	1.00000	1.00000
5.0	1.00000	1.00000	1.00000	1.00000	1.00000
5.2	1.00000	1.00000	1.00000	1.00000	1.00000
5.4	1.00000	1.00000	1.00000	1.00000	1.00000
5.6	1.00000	1.00000	1.00000	1.00000	1.00000
5.8	1.00000	1.00000	1.00000	1.00000	1.00000
6.0	1.00000	1.00000	1.00000	1.00000	1.00000

TABLE 7.2.1. CONFIDENCE LIMITS ON THE PROPORTION IN ONE TAIL.

ETA = 0.750

K ↓	N = 2	N = 3	N = 4	N = 5	N = 6
-3.0	0.00020	0.00015	0.00017	0.00020	0.00022
-2.8	0.00045	0.00036	0.00040	0.00045	0.00051
-2.6	0.00097	0.00081	0.00089	0.00100	0.00110
-2.4	0.00200	0.00172	0.00188	0.00209	0.00228
-2.2	0.00393	0.00349	0.00379	0.00415	0.00450
-2.0	0.00734	0.00672	0.00726	0.00788	0.00846
-1.8	0.01306	0.01231	0.01323	0.01423	0.01515
-1.6	0.02215	0.02144	0.02294	0.02449	0.02589
-1.4	0.03580	0.03556	0.03790	0.04020	0.04223
-1.2	0.05526	0.05618	0.05971	0.06298	0.06580
-1.0	0.08160	0.08470	0.08982	0.09429	0.09805
-0.8	0.11555	0.12204	0.12919	0.13510	0.13994
-0.6	0.15725	0.16837	0.17800	0.18556	0.19158
-0.4	0.20602	0.22291	0.23543	0.24480	0.25211
-0.2	0.26009	0.28383	0.29962	0.31098	0.31963
0.0	0.31670	0.34848	0.36797	0.38146	0.39152
0.2	0.37277	0.41399	0.43758	0.45335	0.46486
0.4	0.42601	0.47791	0.50589	0.52400	0.53693
0.6	0.47531	0.53857	0.57099	0.59133	0.60555
0.8	0.52048	0.59508	0.63163	0.65391	0.66917
1.0	0.56175	0.64704	0.68713	0.71086	0.72679
1.2	0.59950	0.69439	0.73715	0.76173	0.77789
1.4	0.63410	0.73718	0.78163	0.80638	0.82230
1.6	0.66588	0.77558	0.82065	0.84491	0.86014
1.8	0.69515	0.80978	0.85443	0.87759	0.89177
2.0	0.72214	0.84002	0.88328	0.90484	0.91767
2.2	0.74705	0.86653	0.90758	0.92716	0.93847
2.4	0.77008	0.88958	0.92776	0.94512	0.95483
2.6	0.79135	0.90943	0.94428	0.95932	0.96744
2.8	0.81100	0.92638	0.95759	0.97033	0.97696
3.0	0.82915	0.94070	0.96816	0.97872	0.98400
3.2	0.84588	0.95269	0.97643	0.98499	0.98909
3.4	0.86130	0.96261	0.98279	0.98959	0.99271
3.6	0.87548	0.97073	0.98761	0.99291	0.99522
3.8	0.88850	0.97732	0.99121	0.99525	0.99693
4.0	0.90043	0.98260	0.99385	0.99687	0.99806
4.2	0.91133	0.98678	0.99577	0.99798	0.99880
4.4	0.92126	0.99006	0.99712	0.99872	0.99927
4.6	0.93029	0.99260	0.99808	0.99920	0.99957
4.8	0.93848	0.99455	0.99873	0.99951	0.99975
5.0	0.94587	0.99603	0.99918	0.99970	0.99986
5.2	0.95253	0.99713	0.99947	0.99982	0.99992
5.4	0.95851	0.99796	0.99967	0.99990	0.99996
5.6	0.96385	0.99856	0.99979	0.99994	0.99998
5.8	0.96861	0.99899	0.99987	0.99997	0.99999
6.0	0.97284	0.99930	0.99992	0.99998	0.99999

TABLE 7.2.2. CONFIDENCE LIMITS ON THE PROPORTION IN ONE TAIL.

ETA = 0.750 (*CONTINUED*)

K ↓	N = 7	N = 8	N = 9	N = 10	N = 11
-3.0	0.00025	0.00027	0.00029	0.00031	0.00033
-2.8	0.00056	0.00060	0.00065	0.00069	0.00072
-2.6	0.00120	0.00129	0.00137	0.00145	0.00152
-2.4	0.00246	0.00262	0.00277	0.00291	0.00303
-2.2	0.00481	0.00509	0.00535	0.00558	0.00579
-2.0	0.00897	0.00944	0.00986	0.01023	0.01058
-1.8	0.01597	0.01669	0.01734	0.01792	0.01845
-1.6	0.02712	0.02819	0.02915	0.03000	0.03077
-1.4	0.04398	0.04551	0.04686	0.04805	0.04912
-1.2	0.06820	0.07028	0.07210	0.07370	0.07512
-1.0	0.10122	0.10393	0.10627	0.10833	0.11015
-0.8	0.14395	0.14735	0.15027	0.15282	0.15506
-0.6	0.19652	0.20064	0.20416	0.20720	0.20987
-0.4	0.25800	0.26287	0.26699	0.27053	0.27362
-0.2	0.32650	0.33213	0.33684	0.34087	0.34436
0.0	0.39939	0.40576	0.41106	0.41555	0.41942
0.2	0.47372	0.48082	0.48667	0.49159	0.49582
0.4	0.54674	0.55451	0.56085	0.56615	0.57067
0.6	0.61618	0.62450	0.63124	0.63683	0.64156
0.8	0.68041	0.68910	0.69608	0.70182	0.70665
1.0	0.73834	0.74718	0.75420	0.75994	0.76474
1.2	0.78942	0.79814	0.80500	0.81056	0.81518
1.4	0.83348	0.84182	0.84832	0.85354	0.85785
1.6	0.87066	0.87840	0.88436	0.88912	0.89300
1.8	0.90137	0.90833	0.91364	0.91783	0.92122
2.0	0.92619	0.93227	0.93684	0.94042	0.94329
2.2	0.94581	0.95096	0.95479	0.95775	0.96011
2.4	0.96098	0.96523	0.96833	0.97071	0.97258
2.6	0.97246	0.97586	0.97830	0.98015	0.98159
2.8	0.98095	0.98359	0.98546	0.98686	0.98793
3.0	0.98708	0.98908	0.99048	0.99150	0.99228
3.2	0.99142	0.99290	0.99390	0.99463	0.99518
3.4	0.99442	0.99548	0.99619	0.99669	0.99706
3.6	0.99645	0.99718	0.99767	0.99801	0.99825
3.8	0.99778	0.99828	0.99861	0.99883	0.99899
4.0	0.99865	0.99898	0.99919	0.99933	0.99943
4.2	0.99919	0.99941	0.99954	0.99962	0.99968
4.4	0.99953	0.99966	0.99974	0.99979	0.99983
4.6	0.99973	0.99981	0.99986	0.99989	0.99991
4.8	0.99985	0.99990	0.99993	0.99994	0.99995
5.0	0.99992	0.99995	0.99996	0.99997	0.99998
5.2	0.99996	0.99997	0.99998	0.99999	0.99999
5.4	0.99998	0.99999	0.99999	0.99999	0.99999
5.6	0.99999	0.99999	1.00000	1.00000	1.00000
5.8	0.99999	1.00000	1.00000	1.00000	1.00000
6.0	1.00000	1.00000	1.00000	1.00000	1.00000

TABLE 7.2.3. CONFIDENCE LIMITS ON THE PROPORTION IN ONE TAIL.

ETA = 0.750 (CONTINUED)

K ↓	N = 12	N = 13	N = 14	N = 15	N = 16
-3.0	0.00035	0.00036	0.00038	0.00039	0.00041
-2.8	0.00076	0.00079	0.00082	0.00085	0.00088
-2.6	0.00158	0.00164	0.00170	0.00175	0.00180
-2.4	0.00315	0.00326	0.00336	0.00345	0.00354
-2.2	0.00599	0.00617	0.00634	0.00649	0.00664
-2.0	0.01089	0.01118	0.01144	0.01169	0.01192
-1.8	0.01892	0.01936	0.01976	0.02013	0.02048
-1.6	0.03146	0.03210	0.03268	0.03321	0.03371
-1.4	0.05009	0.05096	0.05176	0.05249	0.05317
-1.2	0.07640	0.07756	0.07861	0.07957	0.08046
-1.0	0.11178	0.11324	0.11457	0.11579	0.11690
-0.8	0.15705	0.15884	0.16046	0.16194	0.16329
-0.6	0.21224	0.21435	0.21626	0.21799	0.21957
-0.4	0.27635	0.27878	0.28096	0.28293	0.28473
-0.2	0.34743	0.35015	0.35259	0.35479	0.35679
0.0	0.42281	0.42580	0.42847	0.43087	0.43305
0.2	0.49949	0.50272	0.50559	0.50817	0.51049
0.4	0.57458	0.57801	0.58104	0.58375	0.58619
0.6	0.64564	0.64919	0.65232	0.65511	0.65761
0.8	0.71079	0.71438	0.71754	0.72033	0.72284
1.0	0.76882	0.77235	0.77544	0.77816	0.78060
1.2	0.81909	0.82245	0.82538	0.82796	0.83024
1.4	0.86147	0.86457	0.86726	0.86961	0.87170
1.6	0.89626	0.89902	0.90140	0.90349	0.90532
1.8	0.92405	0.92643	0.92848	0.93026	0.93182
2.0	0.94566	0.94766	0.94935	0.95082	0.95211
2.2	0.96203	0.96364	0.96501	0.96618	0.96720
2.4	0.97410	0.97535	0.97641	0.97732	0.97811
2.6	0.98275	0.98370	0.98450	0.98518	0.98576
2.8	0.98879	0.98949	0.99007	0.99056	0.99098
3.0	0.99289	0.99339	0.99380	0.99414	0.99444
3.2	0.99561	0.99595	0.99623	0.99646	0.99666
3.4	0.99735	0.99758	0.99776	0.99792	0.99805
3.6	0.99844	0.99859	0.99871	0.99881	0.99889
3.8	0.99911	0.99920	0.99928	0.99934	0.99938
4.0	0.99950	0.99956	0.99960	0.99964	0.99967
4.2	0.99973	0.99976	0.99979	0.99981	0.99983
4.4	0.99986	0.99988	0.99989	0.99990	0.99991
4.6	0.99993	0.99994	0.99994	0.99995	0.99996
4.8	0.99996	0.99997	0.99997	0.99998	0.99998
5.0	0.99998	0.99998	0.99999	0.99999	0.99999
5.2	0.99999	0.99999	0.99999	0.99999	1.00000
5.4	1.00000	1.00000	1.00000	1.00000	1.00000
5.6	1.00000	1.00000	1.00000	1.00000	1.00000
5.8	1.00000	1.00000	1.00000	1.00000	1.00000
6.0	1.00000	1.00000	1.00000	1.00000	1.00000

TABLE 7.2.4. CONFIDENCE LIMITS ON THE PROPORTION IN ONE TAIL.

ETA = 0.750 *(CONTINUED)*

K ↓	N = 17	N = 18	N = 21	N = 24	N = 27
-3.0	0.00042	0.00044	0.00047	0.00050	0.00053
-2.8	0.00090	0.00093	0.00100	0.00105	0.00110
-2.6	0.00185	0.00189	0.00202	0.00212	0.00221
-2.4	0.00362	0.00370	0.00391	0.00408	0.00424
-2.2	0.00677	0.00690	0.00724	0.00753	0.00779
-2.0	0.01213	0.01233	0.01287	0.01332	0.01371
-1.8	0.02080	0.02110	0.02189	0.02256	0.02314
-1.6	0.03417	0.03459	0.03572	0.03667	0.03748
-1.4	0.05380	0.05438	0.05592	0.05720	0.05829
-1.2	0.08128	0.08204	0.08403	0.08569	0.08709
-1.0	0.11793	0.11888	0.12137	0.12342	0.12516
-0.8	0.16453	0.16568	0.16866	0.17112	0.17319
-0.6	0.22102	0.22236	0.22583	0.22867	0.23105
-0.4	0.28638	0.28789	0.29180	0.29499	0.29765
-0.2	0.35861	0.36028	0.36459	0.36809	0.37100
0.0	0.43503	0.43684	0.44149	0.44525	0.44836
0.2	0.51260	0.51454	0.51946	0.52342	0.52669
0.4	0.58840	0.59041	0.59553	0.59962	0.60298
0.6	0.65987	0.66193	0.66713	0.67126	0.67465
0.8	0.72509	0.72714	0.73229	0.73636	0.73967
1.0	0.78278	0.78475	0.78971	0.79360	0.79676
1.2	0.83229	0.83414	0.83876	0.84236	0.84527
1.4	0.87356	0.87523	0.87939	0.88261	0.88520
1.6	0.90695	0.90842	0.91203	0.91482	0.91704
1.8	0.93320	0.93444	0.93748	0.93980	0.94164
2.0	0.95324	0.95425	0.95671	0.95858	0.96006
2.2	0.96810	0.96889	0.97082	0.97228	0.97341
2.4	0.97879	0.97940	0.98086	0.98195	0.98280
2.6	0.98627	0.98671	0.98778	0.98857	0.98918
2.8	0.99134	0.99166	0.99242	0.99297	0.99339
3.0	0.99469	0.99491	0.99542	0.99579	0.99608
3.2	0.99683	0.99697	0.99731	0.99756	0.99774
3.4	0.99816	0.99825	0.99847	0.99862	0.99874
3.6	0.99896	0.99902	0.99915	0.99924	0.99931
3.8	0.99943	0.99946	0.99954	0.99960	0.99964
4.0	0.99969	0.99971	0.99976	0.99979	0.99981
4.2	0.99984	0.99985	0.99988	0.99990	0.99991
4.4	0.99992	0.99993	0.99994	0.99995	0.99996
4.6	0.99996	0.99996	0.99997	0.99998	0.99998
4.8	0.99998	0.99998	0.99999	0.99999	0.99999
5.0	0.99999	0.99999	0.99999	1.00000	1.00000
5.2	1.00000	1.00000	1.00000	1.00000	1.00000
5.4	1.00000	1.00000	1.00000	1.00000	1.00000
5.6	1.00000	1.00000	1.00000	1.00000	1.00000
5.8	1.00000	1.00000	1.00000	1.00000	1.00000
6.0	1.00000	1.00000	1.00000	1.00000	1.00000

TABLE 7.2.5. CONFIDENCE LIMITS ON THE PROPORTION IN ONE TAIL.

ETA = 0.750 (CONTINUED)

K ↓	N = 30	N = 40	N = 60	N = 80	N = 100
-3.0	0.00055	0.00062	0.00071	0.00077	0.00082
-2.8	0.00115	0.00127	0.00144	0.00155	0.00163
-2.6	0.00229	0.00251	0.00280	0.00299	0.00313
-2.4	0.00438	0.00474	0.00522	0.00553	0.00576
-2.2	0.00801	0.00859	0.00935	0.00985	0.01020
-2.0	0.01405	0.01494	0.01610	0.01684	0.01737
-1.8	0.02364	0.02495	0.02662	0.02768	0.02843
-1.6	0.03818	0.04000	0.04231	0.04376	0.04479
-1.4	0.05923	0.06166	0.06470	0.06661	0.06795
-1.2	0.08830	0.09140	0.09526	0.09766	0.09934
-1.0	0.12666	0.13045	0.13515	0.13804	0.14006
-0.8	0.17496	0.17945	0.18495	0.18832	0.19066
-0.6	0.23308	0.23821	0.24445	0.24824	0.25087
-0.4	0.29993	0.30562	0.31248	0.31664	0.31950
-0.2	0.37347	0.37963	0.38702	0.39145	0.39450
0.0	0.45100	0.45753	0.46531	0.46994	0.47311
0.2	0.52945	0.53625	0.54427	0.54902	0.55226
0.4	0.60581	0.61275	0.62086	0.62564	0.62887
0.6	0.67748	0.68440	0.69242	0.69710	0.70025
0.8	0.74244	0.74916	0.75687	0.76134	0.76434
1.0	0.79938	0.80571	0.81290	0.81704	0.81980
1.2	0.84767	0.85344	0.85993	0.86364	0.86610
1.4	0.88733	0.89241	0.89807	0.90127	0.90338
1.6	0.91887	0.92318	0.92793	0.93060	0.93234
1.8	0.94315	0.94668	0.95053	0.95266	0.95405
2.0	0.96126	0.96404	0.96704	0.96869	0.96975
2.2	0.97433	0.97645	0.97870	0.97992	0.98071
2.4	0.98348	0.98503	0.98666	0.98753	0.98809
2.6	0.98967	0.99076	0.99190	0.99250	0.99288
2.8	0.99372	0.99447	0.99523	0.99563	0.99588
3.0	0.99630	0.99679	0.99728	0.99753	0.99769
3.2	0.99788	0.99819	0.99850	0.99865	0.99875
3.4	0.99882	0.99901	0.99920	0.99929	0.99935
3.6	0.99937	0.99948	0.99959	0.99964	0.99967
3.8	0.99967	0.99973	0.99979	0.99982	0.99984
4.0	0.99983	0.99987	0.99990	0.99991	0.99992
4.2	0.99992	0.99994	0.99995	0.99996	0.99996
4.4	0.99996	0.99997	0.99998	0.99998	0.99998
4.6	0.99998	0.99999	0.99999	0.99999	0.99999
4.8	0.99999	0.99999	1.00000	0.99999	0.99999
5.0	1.00000	1.00000	1.00000	1.00000	1.00000
5.2	1.00000	1.00000	1.00000	1.00000	1.00000
5.4	1.00000	1.00000	1.00000	1.00000	1.00000
5.6	1.00000	1.00000	1.00000	1.00000	1.00000
5.8	1.00000	1.00000	1.00000	1.00000	1.00000
6.0	1.00000	1.00000	1.00000	1.00000	1.00000

TABLE 7.2.6. CONFIDENCE LIMITS ON THE PROPORTION IN ONE TAIL.

ETA = 0.750 (CONTINUED)

K ↓	N = 120	N = 240	N = 600	N = 1000	N = 1200
-3.0	0.00085	0.00097	0.00110	0.00115	0.00116
-2.8	0.00169	0.00191	0.00212	0.00221	0.00224
-2.6	0.00323	0.00359	0.00395	0.00410	0.00414
-2.4	0.00594	0.00651	0.00708	0.00732	0.00739
-2.2	0.01047	0.01136	0.01223	0.01258	0.01269
-2.0	0.01777	0.01908	0.02034	0.02086	0.02101
-1.8	0.02900	0.03085	0.03261	0.03332	0.03354
-1.6	0.04557	0.04806	0.05041	0.05136	0.05165
-1.4	0.06896	0.07218	0.07520	0.07641	0.07678
-1.2	0.10060	0.10460	0.10831	0.10979	0.11024
-1.0	0.14158	0.14634	0.15073	0.15248	0.15301
-0.8	0.19241	0.19789	0.20290	0.20488	0.20548
-0.6	0.25283	0.25893	0.26446	0.26664	0.26729
-0.4	0.32163	0.32822	0.33416	0.33649	0.33719
-0.2	0.39675	0.40371	0.40993	0.41235	0.41308
0.0	0.47545	0.48264	0.48902	0.49149	0.49223
0.2	0.55464	0.56191	0.56832	0.57080	0.57154
0.4	0.63124	0.63845	0.64476	0.64719	0.64791
0.6	0.70256	0.70954	0.71560	0.71792	0.71861
0.8	0.76653	0.77310	0.77876	0.78091	0.78155
1.0	0.82181	0.82781	0.83293	0.83487	0.83545
1.2	0.86788	0.87317	0.87765	0.87934	0.87984
1.4	0.90490	0.90940	0.91318	0.91459	0.91501
1.6	0.93360	0.93728	0.94034	0.94148	0.94182
1.8	0.95505	0.95795	0.96033	0.96122	0.96147
2.0	0.97051	0.97271	0.97449	0.97515	0.97534
2.2	0.98127	0.98286	0.98414	0.98461	0.98475
2.4	0.98848	0.98959	0.99048	0.99080	0.99089
2.6	0.99314	0.99389	0.99447	0.99468	0.99475
2.8	0.99605	0.99653	0.99691	0.99704	0.99708
3.0	0.99780	0.99810	0.99833	0.99841	0.99843
3.2	0.99882	0.99899	0.99913	0.99917	0.99919
3.4	0.99938	0.99949	0.99956	0.99959	0.99959
3.6	0.99969	0.99975	0.99979	0.99980	0.99980
3.8	0.99985	0.99988	0.99990	0.99991	0.99991
4.0	0.99993	0.99994	0.99995	0.99996	0.99996
4.2	0.99997	0.99998	0.99998	0.99998	0.99998
4.4	0.99999	0.99999	0.99999	0.99999	0.99999
4.6	0.99999	1.00000	1.00000	1.00000	1.00000
4.8	1.00000	1.00000	1.00000	1.00000	1.00000
5.0	1.00000	1.00000	1.00000	1.00000	1.00000
5.2	1.00000	1.00000	1.00000	1.00000	1.00000
5.4	1.00000	1.00000	1.00000	1.00000	1.00000
5.6	1.00000	1.00000	1.00000	1.00000	1.00000
5.8	1.00000	1.00000	1.00000	1.00000	1.00000
6.0	1.00000	1.00000	1.00000	1.00000	1.00000

TABLE 7.3.1. *CONFIDENCE LIMITS ON THE PROPORTION IN ONE TAIL.*

ETA = 0.900

K ↓	N = 2	N = 3	N = 4	N = 5	N = 6
-3.0	0.00000	0.00000	0.00000	0.00001	0.00001
-2.8	0.00000	0.00001	0.00002	0.00003	0.00005
-2.6	0.00000	0.00002	0.00005	0.00009	0.00013
-2.4	0.00002	0.00007	0.00015	0.00025	0.00035
-2.2	0.00007	0.00022	0.00043	0.00065	0.00088
-2.0	0.00024	0.00063	0.00111	0.00160	0.00207
-1.8	0.00073	0.00165	0.00266	0.00363	0.00454
-1.6	0.00201	0.00395	0.00589	0.00768	0.00929
-1.4	0.00494	0.00866	0.01211	0.01514	0.01779
-1.2	0.01100	0.01745	0.02308	0.02783	0.03189
-1.0	0.02209	0.03235	0.04085	0.04780	0.05359
-0.8	0.04016	0.05532	0.06729	0.07683	0.08462
-0.6	0.06641	0.08750	0.10347	0.11588	0.12584
-0.4	0.10048	0.12868	0.14910	0.16460	0.17685
-0.2	0.14025	0.17702	0.20247	0.22130	0.23593
0.0	0.18242	0.22968	0.26083	0.28328	0.30042
0.2	0.22384	0.28367	0.32128	0.34764	0.36740
0.4	0.26268	0.33676	0.38141	0.41188	0.43431
0.6	0.29837	0.38764	0.43961	0.47420	0.49920
0.8	0.33105	0.43575	0.49494	0.53345	0.56078
1.0	0.36105	0.48091	0.54697	0.58896	0.61825
1.2	0.38874	0.52318	0.59548	0.64040	0.67114
1.4	0.41446	0.56268	0.64045	0.68760	0.71925
1.6	0.43848	0.59957	0.68190	0.73054	0.76249
1.8	0.46101	0.63399	0.71992	0.76928	0.80095
2.0	0.48225	0.66608	0.75462	0.80393	0.83477
2.2	0.50233	0.69598	0.78613	0.83466	0.86418
2.4	0.52138	0.72380	0.81458	0.86168	0.88948
2.6	0.53951	0.74965	0.84013	0.88522	0.91098
2.8	0.55679	0.77363	0.86294	0.90554	0.92904
3.0	0.57331	0.79583	0.88318	0.92293	0.94404
3.2	0.58912	0.81634	0.90102	0.93766	0.95634
3.4	0.60428	0.83524	0.91664	0.95001	0.96631
3.6	0.61883	0.85262	0.93024	0.96027	0.97428
3.8	0.63282	0.86855	0.94198	0.96871	0.98059
4.0	0.64628	0.88311	0.95206	0.97558	0.98552
4.2	0.65924	0.89637	0.96064	0.98112	0.98932
4.4	0.67174	0.90842	0.96790	0.98554	0.99221
4.6	0.68379	0.91934	0.97399	0.98902	0.99438
4.8	0.69543	0.92918	0.97907	0.99175	0.99600
5.0	0.70666	0.93803	0.98327	0.99386	0.99718
5.2	0.71752	0.94596	0.98672	0.99547	0.99804
5.4	0.72801	0.95303	0.98953	0.99669	0.99865
5.6	0.73815	0.95932	0.99180	0.99761	0.99908
5.8	0.74796	0.96489	0.99362	0.99829	0.99938
6.0	0.75745	0.96981	0.99508	0.99878	0.99959

TABLE 7.3.2. CONFIDENCE LIMITS ON THE PROPORTION IN ONE TAIL.

ETA = 0.900 (CONTINUED)

K ↓	N = 7	N = 8	N = 9	N = 10	N = 11
-3.0	0.00002	0.00003	0.00004	0.00005	0.00005
-2.8	0.00006	0.00008	0.00010	0.00012	0.00014
-2.6	0.00018	0.00022	0.00027	0.00032	0.00036
-2.4	0.00046	0.00056	0.00067	0.00077	0.00087
-2.2	0.00111	0.00133	0.00154	0.00175	0.00194
-2.0	0.00253	0.00296	0.00336	0.00374	0.00410
-1.8	0.00538	0.00616	0.00688	0.00754	0.00816
-1.6	0.01074	0.01206	0.01325	0.01434	0.01534
-1.4	0.02012	0.02219	0.02405	0.02572	0.02724
-1.2	0.03538	0.03844	0.04115	0.04356	0.04573
-1.0	0.05851	0.06275	0.06646	0.06974	0.07267
-0.8	0.09114	0.09670	0.10152	0.10575	0.10950
-0.6	0.13407	0.14102	0.14699	0.15219	0.15678
-0.4	0.18685	0.19520	0.20233	0.20850	0.21391
-0.2	0.24773	0.25749	0.26576	0.27287	0.27907
0.0	0.31406	0.32524	0.33462	0.34264	0.34960
0.2	0.38291	0.39550	0.40597	0.41486	0.42253
0.4	0.45167	0.46560	0.47710	0.48679	0.49509
0.6	0.51829	0.53345	0.54584	0.55620	0.56503
0.8	0.58136	0.59753	0.61062	0.62149	0.63069
1.0	0.63999	0.65686	0.67041	0.68156	0.69093
1.2	0.69363	0.71087	0.72457	0.73575	0.74508
1.4	0.74202	0.75926	0.77280	0.78376	0.79282
1.6	0.78509	0.80196	0.81505	0.82554	0.83414
1.8	0.82293	0.83909	0.85147	0.86127	0.86924
2.0	0.85574	0.87089	0.88234	0.89130	0.89851
2.2	0.88382	0.89774	0.90810	0.91611	0.92248
2.4	0.90752	0.92006	0.92925	0.93624	0.94174
2.6	0.92726	0.93834	0.94631	0.95229	0.95693
2.8	0.94348	0.95308	0.95986	0.96486	0.96869
3.0	0.95662	0.96479	0.97044	0.97453	0.97762
3.2	0.96712	0.97394	0.97855	0.98184	0.98427
3.4	0.97540	0.98099	0.98468	0.98726	0.98914
3.6	0.98182	0.98632	0.98922	0.99121	0.99263
3.8	0.98674	0.99030	0.99254	0.99403	0.99508
4.0	0.99046	0.99322	0.99491	0.99602	0.99678
4.2	0.99322	0.99533	0.99659	0.99739	0.99793
4.4	0.99525	0.99683	0.99775	0.99831	0.99869
4.6	0.99671	0.99788	0.99854	0.99893	0.99919
4.8	0.99776	0.99860	0.99906	0.99933	0.99950
5.0	0.99849	0.99909	0.99941	0.99959	0.99970
5.2	0.99900	0.99942	0.99964	0.99975	0.99983
5.4	0.99934	0.99964	0.99978	0.99985	0.99990
5.6	0.99957	0.99977	0.99987	0.99992	0.99994
5.8	0.99973	0.99986	0.99992	0.99995	0.99997
6.0	0.99983	0.99992	0.99995	0.99997	0.99998

TABLE 7.3.3. CONFIDENCE LIMITS ON THE PROPORTION IN ONE TAIL.

ETA = 0.900 (CONTINUED)

K ↓	N = 12	N = 13	N = 14	N = 15	N = 16
-3.0	0.00006	0.00007	0.00008	0.00009	0.00010
-2.8	0.00016	0.00018	0.00021	0.00022	0.00024
-2.6	0.00041	0.00045	0.00050	0.00054	0.00058
-2.4	0.00096	0.00105	0.00114	0.00123	0.00131
-2.2	0.00212	0.00230	0.00247	0.00263	0.00278
-2.0	0.00443	0.00475	0.00505	0.00534	0.00561
-1.8	0.00874	0.00928	0.00978	0.01025	0.01070
-1.6	0.01627	0.01712	0.01791	0.01865	0.01935
-1.4	0.02863	0.02991	0.03109	0.03218	0.03319
-1.2	0.04770	0.04950	0.05115	0.05267	0.05408
-1.0	0.07531	0.07770	0.07989	0.08189	0.08374
-0.8	0.11286	0.11589	0.11864	0.12116	0.12348
-0.6	0.16087	0.16454	0.16787	0.17090	0.17367
-0.4	0.21871	0.22300	0.22687	0.23038	0.23359
-0.2	0.28454	0.28942	0.29380	0.29777	0.30138
0.0	0.35571	0.36113	0.36598	0.37036	0.37434
0.2	0.42923	0.43515	0.44042	0.44517	0.44946
0.4	0.50231	0.50866	0.51430	0.51935	0.52390
0.6	0.57267	0.57935	0.58526	0.59053	0.59527
0.8	0.63859	0.64548	0.65154	0.65693	0.66176
1.0	0.69894	0.70587	0.71195	0.71733	0.72214
1.2	0.75300	0.75982	0.76577	0.77101	0.77567
1.4	0.80047	0.80701	0.81268	0.81766	0.82207
1.6	0.84134	0.84746	0.85273	0.85734	0.86139
1.8	0.87585	0.88143	0.88621	0.89036	0.89399
2.0	0.90444	0.90940	0.91362	0.91726	0.92043
2.2	0.92766	0.93197	0.93560	0.93871	0.94141
2.4	0.94617	0.94981	0.95286	0.95545	0.95768
2.6	0.96062	0.96363	0.96613	0.96823	0.97003
2.8	0.97170	0.97412	0.97612	0.97778	0.97919
3.0	0.98002	0.98192	0.98348	0.98476	0.98584
3.2	0.98614	0.98761	0.98879	0.98976	0.99056
3.4	0.99056	0.99166	0.99254	0.99325	0.99384
3.6	0.99369	0.99449	0.99513	0.99564	0.99606
3.8	0.99585	0.99643	0.99688	0.99724	0.99753
4.0	0.99733	0.99773	0.99805	0.99829	0.99848
4.2	0.99831	0.99859	0.99880	0.99896	0.99909
4.4	0.99895	0.99914	0.99928	0.99938	0.99946
4.6	0.99936	0.99948	0.99957	0.99964	0.99969
4.8	0.99962	0.99970	0.99975	0.99979	0.99983
5.0	0.99978	0.99982	0.99986	0.99989	0.99990
5.2	0.99987	0.99990	0.99992	0.99994	0.99995
5.4	0.99993	0.99995	0.99996	0.99997	0.99997
5.6	0.99996	0.99997	0.99998	0.99998	0.99999
5.8	0.99998	0.99998	0.99999	0.99999	0.99999
6.0	0.99999	0.99999	0.99999	1.00000	1.00000

TABLE 7.3.4. CONFIDENCE LIMITS ON THE PROPORTION IN ONE TAIL.

ETA = 0.900 (CONTINUED)

K ↓	N = 17	N = 18	N = 21	N = 24	N = 27
-3.0	0.00011	0.00011	0.00014	0.00016	0.00018
-2.8	0.00026	0.00028	0.00034	0.00039	0.00043
-2.6	0.00062	0.00066	0.00077	0.00087	0.00096
-2.4	0.00139	0.00146	0.00167	0.00187	0.00204
-2.2	0.00293	0.00307	0.00346	0.00380	0.00411
-2.0	0.00586	0.00610	0.00677	0.00735	0.00787
-1.8	0.01112	0.01151	0.01259	0.01352	0.01434
-1.6	0.02000	0.02061	0.02225	0.02366	0.02488
-1.4	0.03414	0.03503	0.03740	0.03941	0.04114
-1.2	0.05539	0.05661	0.05984	0.06256	0.06489
-1.0	0.08546	0.08705	0.09124	0.09474	0.09772
-0.8	0.12561	0.12760	0.13278	0.13708	0.14072
-0.6	0.17623	0.17859	0.18474	0.18981	0.19408
-0.4	0.23654	0.23926	0.24631	0.25209	0.25694
-0.2	0.30469	0.30773	0.31559	0.32199	0.32735
0.0	0.37797	0.38130	0.38987	0.39682	0.40260
0.2	0.45337	0.45696	0.46613	0.47352	0.47964
0.4	0.52804	0.53182	0.54145	0.54917	0.55554
0.6	0.59957	0.60348	0.61339	0.62129	0.62777
0.8	0.66612	0.67008	0.68007	0.68797	0.69442
1.0	0.72646	0.73037	0.74019	0.74790	0.75416
1.2	0.77985	0.78361	0.79301	0.80035	0.80626
1.4	0.82600	0.82953	0.83829	0.84507	0.85050
1.6	0.86499	0.86822	0.87616	0.88225	0.88709
1.8	0.89720	0.90007	0.90707	0.91238	0.91657
2.0	0.92322	0.92569	0.93169	0.93619	0.93972
2.2	0.94376	0.94584	0.95083	0.95454	0.95742
2.4	0.95962	0.96132	0.96536	0.96833	0.97060
2.6	0.97158	0.97293	0.97612	0.97843	0.98018
2.8	0.98040	0.98145	0.98389	0.98564	0.98694
3.0	0.98676	0.98755	0.98938	0.99066	0.99160
3.2	0.99124	0.99182	0.99315	0.99406	0.99473
3.4	0.99433	0.99474	0.99568	0.99631	0.99677
3.6	0.99640	0.99669	0.99734	0.99776	0.99807
3.8	0.99777	0.99796	0.99839	0.99868	0.99887
4.0	0.99864	0.99877	0.99905	0.99923	0.99936
4.2	0.99919	0.99928	0.99946	0.99957	0.99964
4.4	0.99953	0.99958	0.99969	0.99976	0.99981
4.6	0.99973	0.99977	0.99983	0.99987	0.99990
4.8	0.99985	0.99987	0.99991	0.99993	0.99995
5.0	0.99992	0.99993	0.99995	0.99997	0.99997
5.2	0.99996	0.99996	0.99998	0.99998	0.99999
5.4	0.99998	0.99998	0.99999	0.99999	0.99999
5.6	0.99999	0.99999	0.99999	1.00000	1.00000
5.8	0.99999	1.00000	1.00000	1.00000	1.00000
6.0	1.00000	1.00000	1.00000	1.00000	1.00000

TABLE 7.3.5. CONFIDENCE LIMITS ON THE PROPORTION IN ONE TAIL.

ETA = 0.900 (CONTINUED)

K ↓	N = 30	N = 40	N = 60	N = 80	N = 100
-3.0	0.00021	0.00027	0.00036	0.00043	0.00049
-2.8	0.00048	0.00060	0.00079	0.00093	0.00104
-2.6	0.00105	0.00129	0.00165	0.00190	0.00209
-2.4	0.00220	0.00264	0.00327	0.00372	0.00405
-2.2	0.00439	0.00515	0.00622	0.00695	0.00749
-2.0	0.00833	0.00958	0.01129	0.01243	0.01327
-1.8	0.01506	0.01700	0.01959	0.02130	0.02254
-1.6	0.02596	0.02880	0.03253	0.03495	0.03670
-1.4	0.04266	0.04662	0.05173	0.05501	0.05734
-1.2	0.06691	0.07216	0.07884	0.08305	0.08603
-1.0	0.10030	0.10693	0.11524	0.12043	0.12408
-0.8	0.14386	0.15186	0.16178	0.16792	0.17221
-0.6	0.19775	0.20704	0.21845	0.22545	0.23031
-0.4	0.26108	0.27152	0.28422	0.29195	0.29728
-0.2	0.33191	0.34332	0.35708	0.36539	0.37110
0.0	0.40750	0.41971	0.43430	0.44303	0.44901
0.2	0.48481	0.49762	0.51278	0.52179	0.52793
0.4	0.56089	0.57407	0.58952	0.59863	0.60480
0.6	0.63321	0.64649	0.66190	0.67090	0.67697
0.8	0.69981	0.71288	0.72788	0.73657	0.74238
1.0	0.75937	0.77190	0.78612	0.79426	0.79969
1.2	0.81115	0.82284	0.83592	0.84334	0.84824
1.4	0.85497	0.86554	0.87722	0.88375	0.88804
1.6	0.89105	0.90033	0.91042	0.91600	0.91962
1.8	0.91998	0.92787	0.93631	0.94091	0.94387
2.0	0.94256	0.94906	0.95590	0.95956	0.96190
2.2	0.95971	0.96491	0.97027	0.97309	0.97486
2.4	0.97241	0.97643	0.98049	0.98259	0.98389
2.6	0.98155	0.98457	0.98755	0.98905	0.98998
2.8	0.98796	0.99016	0.99227	0.99331	0.99395
3.0	0.99233	0.99388	0.99533	0.99603	0.99645
3.2	0.99523	0.99630	0.99726	0.99771	0.99798
3.4	0.99711	0.99782	0.99844	0.99872	0.99889
3.6	0.99829	0.99875	0.99913	0.99931	0.99940
3.8	0.99902	0.99930	0.99953	0.99963	0.99969
4.0	0.99945	0.99962	0.99976	0.99981	0.99984
4.2	0.99970	0.99980	0.99988	0.99991	0.99992
4.4	0.99984	0.99990	0.99994	0.99996	0.99996
4.6	0.99992	0.99995	0.99997	0.99998	0.99998
4.8	0.99996	0.99997	0.99999	0.99999	0.99999
5.0	0.99998	0.99999	0.99999	1.00000	1.00000
5.2	0.99999	0.99999	1.00000	1.00000	1.00000
5.4	1.00000	1.00000	1.00000	1.00000	1.00000
5.6	1.00000	1.00000	1.00000	1.00000	1.00000
5.8	1.00000	1.00000	1.00000	1.00000	1.00000
6.0	1.00000	1.00000	1.00000	1.00000	1.00000

TABLE 7.3.6. CONFIDENCE LIMITS ON THE PROPORTION IN ONE TAIL.

ETA = 0.900 (CONTINUED)

K ↓	N = 120	N = 240	N = 600	N = 1000	N = 1200
-3.0	0.00054	0.00071	0.00090	0.00099	0.00101
-2.8	0.00112	0.00144	0.00178	0.00193	0.00198
-2.6	0.00225	0.00279	0.00338	0.00363	0.00371
-2.4	0.00431	0.00522	0.00617	0.00658	0.00671
-2.2	0.00791	0.00936	0.01084	0.01147	0.01166
-2.0	0.01392	0.01611	0.01831	0.01924	0.01952
-1.8	0.02349	0.02665	0.02978	0.03108	0.03147
-1.6	0.03803	0.04239	0.04663	0.04837	0.04890
-1.4	0.05911	0.06486	0.07036	0.07260	0.07328
-1.2	0.08829	0.09553	0.10237	0.10512	0.10596
-1.0	0.12683	0.13558	0.14373	0.14700	0.14798
-0.8	0.17543	0.18558	0.19495	0.19867	0.19980
-0.6	0.23394	0.24532	0.25572	0.25983	0.26106
-0.4	0.30126	0.31364	0.32484	0.32924	0.33056
-0.2	0.37534	0.38846	0.40023	0.40482	0.40620
0.0	0.45343	0.46704	0.47914	0.48384	0.48524
0.2	0.53245	0.54627	0.55846	0.56317	0.56458
0.4	0.60933	0.62309	0.63513	0.63975	0.64113
0.6	0.68141	0.69480	0.70641	0.71084	0.71215
0.8	0.74662	0.75932	0.77022	0.77435	0.77558
1.0	0.80362	0.81532	0.82524	0.82898	0.83008
1.2	0.85177	0.86221	0.87095	0.87422	0.87518
1.4	0.89112	0.90011	0.90755	0.91031	0.91112
1.6	0.92221	0.92969	0.93579	0.93803	0.93869
1.8	0.94597	0.95198	0.95679	0.95854	0.95905
2.0	0.96354	0.96819	0.97184	0.97316	0.97354
2.2	0.97611	0.97957	0.98224	0.98319	0.98346
2.4	0.98480	0.98728	0.98916	0.98982	0.99001
2.6	0.99061	0.99233	0.99360	0.99404	0.99417
2.8	0.99438	0.99552	0.99635	0.99663	0.99671
3.0	0.99673	0.99747	0.99799	0.99816	0.99821
3.2	0.99816	0.99862	0.99893	0.99903	0.99906
3.4	0.99899	0.99927	0.99945	0.99951	0.99952
3.6	0.99947	0.99962	0.99973	0.99976	0.99977
3.8	0.99973	0.99981	0.99987	0.99988	0.99989
4.0	0.99986	0.99991	0.99994	0.99995	0.99995
4.2	0.99993	0.99996	0.99997	0.99998	0.99998
4.4	0.99997	0.99998	0.99999	0.99999	0.99999
4.6	0.99999	0.99999	1.00000	1.00000	1.00000
4.8	0.99999	1.00000	1.00000	1.00000	1.00000
5.0	1.00000	1.00000	1.00000	1.00000	1.00000
5.2	1.00000	1.00000	1.00000	1.00000	1.00000
5.4	1.00000	1.00000	1.00000	1.00000	1.00000
5.6	1.00000	1.00000	1.00000	1.00000	1.00000
5.8	1.00000	1.00000	1.00000	1.00000	1.00000
6.0	1.00000	1.00000	1.00000	1.00000	1.00000

TABLE 7.4.1. CONFIDENCE LIMITS ON THE PROPORTION IN ONE TAIL.

ETA = 0.950

K ↓	N = 2	N = 3	N = 4	N = 5	N = 6
-3.0	0.00000	0.00000	0.00000	0.00000	0.00000
-2.8	0.00000	0.00000	0.00000	0.00000	0.00001
-2.6	0.00000	0.00000	0.00001	0.00002	0.00003
-2.4	0.00000	0.00001	0.00003	0.00006	0.00010
-2.2	0.00000	0.00003	0.00009	0.00018	0.00029
-2.0	0.00002	0.00011	0.00029	0.00052	0.00079
-1.8	0.00008	0.00038	0.00085	0.00140	0.00199
-1.6	0.00030	0.00115	0.00226	0.00344	0.00461
-1.4	0.00106	0.00311	0.00543	0.00772	0.00987
-1.2	0.00317	0.00753	0.01188	0.01589	0.01951
-1.0	0.00820	0.01629	0.02364	0.03006	0.03565
-0.8	0.01831	0.03164	0.04292	0.05236	0.06034
-0.6	0.03544	0.05534	0.07131	0.08424	0.09493
-0.4	0.05997	0.08771	0.10903	0.12581	0.13941
-0.2	0.09006	0.12725	0.15462	0.17559	0.19228
0.0	0.12240	0.17114	0.20542	0.23099	0.25095
0.2	0.15403	0.21648	0.25856	0.28913	0.31255
0.4	0.18342	0.26118	0.31177	0.34762	0.37460
0.6	0.21024	0.30413	0.36360	0.40483	0.43534
0.8	0.23469	0.34491	0.41329	0.45976	0.49359
1.0	0.25712	0.38345	0.46048	0.51188	0.54869
1.2	0.27784	0.41982	0.50506	0.56090	0.60025
1.4	0.29713	0.45416	0.54703	0.60671	0.64806
1.6	0.31518	0.48661	0.58642	0.64929	0.69205
1.8	0.33219	0.51732	0.62332	0.68867	0.73223
2.0	0.34828	0.54640	0.65781	0.72489	0.76866
2.2	0.36356	0.57398	0.68997	0.75806	0.80144
2.4	0.37814	0.60013	0.71989	0.78826	0.83072
2.6	0.39207	0.62495	0.74766	0.81563	0.85668
2.8	0.40543	0.64851	0.77335	0.84028	0.87950
3.0	0.41828	0.67087	0.79706	0.86237	0.89942
3.2	0.43064	0.69210	0.81886	0.88204	0.91666
3.4	0.44258	0.71224	0.83886	0.89946	0.93146
3.6	0.45411	0.73134	0.85713	0.91478	0.94405
3.8	0.46527	0.74945	0.87376	0.92818	0.95468
4.0	0.47609	0.76661	0.88885	0.93983	0.96357
4.2	0.48658	0.78285	0.90249	0.94988	0.97095
4.4	0.49677	0.79821	0.91476	0.95849	0.97701
4.6	0.50668	0.81272	0.92577	0.96584	0.98196
4.8	0.51633	0.82641	0.93560	0.97205	0.98595
5.0	0.52572	0.83932	0.94434	0.97727	0.98915
5.2	0.53488	0.85147	0.95208	0.98164	0.99169
5.4	0.54381	0.86290	0.95890	0.98525	0.99368
5.6	0.55252	0.87363	0.96489	0.98823	0.99524
5.8	0.56103	0.88369	0.97013	0.99067	0.99644
6.0	0.56935	0.89311	0.97468	0.99265	0.99736

TABLE 7.4.2. CONFIDENCE LIMITS ON THE PROPORTION IN ONE TAIL.

ETA = 0.950 (*CONTINUED*)

K ↓	N = 7	N = 8	N = 9	N = 10	N = 11
-3.0	0.00000	0.00001	0.00001	0.00001	0.00002
-2.8	0.00001	0.00002	0.00003	0.00004	0.00005
-2.6	0.00005	0.00007	0.00009	0.00011	0.00014
-2.4	0.00014	0.00020	0.00026	0.00031	0.00038
-2.2	0.00041	0.00054	0.00067	0.00080	0.00094
-2.0	0.00107	0.00135	0.00163	0.00191	0.00218
-1.8	0.00257	0.00315	0.00371	0.00424	0.00476
-1.6	0.00574	0.00682	0.00783	0.00879	0.00969
-1.4	0.01187	0.01372	0.01543	0.01701	0.01848
-1.2	0.02277	0.02571	0.02837	0.03079	0.03301
-1.0	0.04055	0.04488	0.04874	0.05220	0.05534
-0.8	0.06719	0.07315	0.07839	0.08305	0.08723
-0.6	0.10394	0.11168	0.11840	0.12433	0.12959
-0.4	0.15071	0.16028	0.16853	0.17574	0.18211
-0.2	0.20594	0.21740	0.22718	0.23566	0.24310
0.0	0.26707	0.28044	0.29175	0.30148	0.30997
0.2	0.33120	0.34649	0.35932	0.37027	0.37977
0.4	0.39580	0.41299	0.42728	0.43938	0.44981
0.6	0.45898	0.47795	0.49357	0.50671	0.51795
0.8	0.51947	0.54001	0.55677	0.57076	0.58264
1.0	0.57648	0.59829	0.61592	0.63052	0.64284
1.2	0.62954	0.65226	0.67045	0.68537	0.69787
1.4	0.67839	0.70162	0.72002	0.73498	0.74741
1.6	0.72292	0.74625	0.76451	0.77921	0.79132
1.8	0.76315	0.78616	0.80395	0.81812	0.82967
2.0	0.79913	0.82146	0.83849	0.85187	0.86268
2.2	0.83104	0.85234	0.86834	0.88077	0.89067
2.4	0.85905	0.87906	0.89384	0.90516	0.91407
2.6	0.88341	0.90192	0.91535	0.92547	0.93334
2.8	0.90440	0.92125	0.93325	0.94215	0.94896
3.0	0.92229	0.93741	0.94796	0.95565	0.96145
3.2	0.93740	0.95077	0.95990	0.96643	0.97127
3.4	0.95003	0.96168	0.96946	0.97491	0.97888
3.6	0.96048	0.97049	0.97701	0.98149	0.98469
3.8	0.96903	0.97752	0.98291	0.98652	0.98906
4.0	0.97596	0.98306	0.98744	0.99031	0.99229
4.2	0.98151	0.98737	0.99088	0.99313	0.99464
4.4	0.98592	0.99069	0.99346	0.99519	0.99633
4.6	0.98938	0.99321	0.99537	0.99668	0.99752
4.8	0.99207	0.99510	0.99676	0.99774	0.99835
5.0	0.99413	0.99651	0.99776	0.99848	0.99892
5.2	0.99570	0.99754	0.99847	0.99899	0.99930
5.4	0.99688	0.99829	0.99897	0.99934	0.99955
5.6	0.99776	0.99882	0.99932	0.99958	0.99972
5.8	0.99841	0.99920	0.99955	0.99973	0.99983
6.0	0.99888	0.99946	0.99971	0.99983	0.99989

TABLE 7.4.3. CONFIDENCE LIMITS ON THE PROPORTION IN ONE TAIL.

ETA = 0.950 (CONTINUED)

K ↓	N = 12	N = 13	N = 14	N = 15	N = 16
-3.0	0.00002	0.00002	0.00003	0.00003	0.00004
-2.8	0.00006	0.00007	0.00008	0.00009	0.00011
-2.6	0.00017	0.00019	0.00022	0.00025	0.00028
-2.4	0.00044	0.00050	0.00056	0.00062	0.00068
-2.2	0.00107	0.00120	0.00133	0.00145	0.00158
-2.0	0.00245	0.00271	0.00296	0.00320	0.00343
-1.8	0.00525	0.00572	0.00617	0.00660	0.00701
-1.6	0.01054	0.01134	0.01210	0.01281	0.01349
-1.4	0.01984	0.02111	0.02230	0.02341	0.02446
-1.2	0.03505	0.03693	0.03868	0.04030	0.04182
-1.0	0.05820	0.06082	0.06322	0.06545	0.06752
-0.8	0.09100	0.09443	0.09757	0.10046	0.10312
-0.6	0.13432	0.13860	0.14249	0.14604	0.14932
-0.4	0.18778	0.19289	0.19751	0.20173	0.20559
-0.2	0.24970	0.25561	0.26094	0.26577	0.27019
0.0	0.31745	0.32412	0.33011	0.33553	0.34046
0.2	0.38809	0.39547	0.40207	0.40802	0.41342
0.4	0.45890	0.46693	0.47407	0.48048	0.48627
0.6	0.52770	0.53626	0.54384	0.55063	0.55673
0.8	0.59289	0.60184	0.60974	0.61677	0.62307
1.0	0.65339	0.66256	0.67060	0.67773	0.68410
1.2	0.70852	0.71770	0.72572	0.73280	0.73910
1.4	0.75791	0.76692	0.77474	0.78161	0.78768
1.6	0.80146	0.81011	0.81757	0.82407	0.82981
1.8	0.83927	0.84738	0.85434	0.86037	0.86565
2.0	0.87157	0.87903	0.88537	0.89084	0.89559
2.2	0.89875	0.90546	0.91112	0.91596	0.92014
2.4	0.92126	0.92717	0.93211	0.93630	0.93990
2.6	0.93961	0.94471	0.94893	0.95248	0.95551
2.8	0.95432	0.95864	0.96217	0.96512	0.96761
3.0	0.96594	0.96952	0.97242	0.97481	0.97681
3.2	0.97497	0.97788	0.98020	0.98210	0.98368
3.4	0.98187	0.98419	0.98602	0.98750	0.98871
3.6	0.98706	0.98887	0.99028	0.99141	0.99232
3.8	0.99090	0.99229	0.99336	0.99420	0.99487
4.0	0.99370	0.99474	0.99553	0.99615	0.99663
4.2	0.99570	0.99647	0.99704	0.99748	0.99783
4.4	0.99711	0.99767	0.99808	0.99839	0.99862
4.6	0.99809	0.99848	0.99877	0.99898	0.99914
4.8	0.99875	0.99903	0.99923	0.99937	0.99948
5.0	0.99920	0.99939	0.99952	0.99962	0.99969
5.2	0.99949	0.99962	0.99971	0.99977	0.99982
5.4	0.99969	0.99977	0.99983	0.99987	0.99989
5.6	0.99981	0.99986	0.99990	0.99992	0.99994
5.8	0.99988	0.99992	0.99994	0.99996	0.99997
6.0	0.99993	0.99995	0.99997	0.99998	0.99998

TABLE 7.4.4. CONFIDENCE LIMITS ON THE PROPORTION IN ONE TAIL.

ETA = 0.950 (CONTINUED)

K ↓	N = 17	N = 18	N = 21	N = 24	N = 27
-3.0	0.00004	0.00005	0.00006	0.00008	0.00009
-2.8	0.00012	0.00013	0.00017	0.00020	0.00024
-2.6	0.00030	0.00033	0.00041	0.00049	0.00056
-2.4	0.00074	0.00080	0.00097	0.00113	0.00128
-2.2	0.00170	0.00181	0.00214	0.00245	0.00273
-2.0	0.00365	0.00387	0.00447	0.00502	0.00551
-1.8	0.00740	0.00778	0.00881	0.00973	0.01056
-1.6	0.01413	0.01474	0.01641	0.01787	0.01916
-1.4	0.02544	0.02638	0.02889	0.03106	0.03295
-1.2	0.04324	0.04457	0.04814	0.05118	0.05381
-1.0	0.06945	0.07125	0.07602	0.08005	0.08351
-0.8	0.10559	0.10790	0.11396	0.11903	0.12335
-0.6	0.15234	0.15515	0.16250	0.16860	0.17377
-0.4	0.20915	0.21244	0.22101	0.22807	0.23402
-0.2	0.27425	0.27799	0.28768	0.29561	0.30226
0.0	0.34497	0.34912	0.35982	0.36853	0.37579
0.2	0.41834	0.42285	0.43444	0.44380	0.45158
0.4	0.49154	0.49636	0.50867	0.51856	0.52672
0.6	0.56227	0.56732	0.58014	0.59037	0.59878
0.8	0.62877	0.63395	0.64703	0.65740	0.66586
1.0	0.68984	0.69504	0.70809	0.71835	0.72668
1.2	0.74474	0.74984	0.76255	0.77247	0.78047
1.4	0.79311	0.79799	0.81008	0.81943	0.82691
1.6	0.83490	0.83946	0.85068	0.85926	0.86608
1.8	0.87032	0.87448	0.88463	0.89232	0.89836
2.0	0.89977	0.90348	0.91243	0.91914	0.92436
2.2	0.92380	0.92702	0.93473	0.94043	0.94482
2.4	0.94302	0.94576	0.95224	0.95696	0.96055
2.6	0.95811	0.96038	0.96569	0.96950	0.97237
2.8	0.96973	0.97157	0.97582	0.97882	0.98105
3.0	0.97850	0.97996	0.98328	0.98558	0.98727
3.2	0.98500	0.98612	0.98866	0.99038	0.99163
3.4	0.98972	0.99057	0.99246	0.99372	0.99461
3.6	0.99308	0.99370	0.99508	0.99598	0.99660
3.8	0.99542	0.99588	0.99685	0.99748	0.99791
4.0	0.99703	0.99735	0.99803	0.99845	0.99874
4.2	0.99810	0.99833	0.99879	0.99907	0.99926
4.4	0.99881	0.99896	0.99927	0.99945	0.99957
4.6	0.99927	0.99937	0.99957	0.99968	0.99976
4.8	0.99956	0.99962	0.99975	0.99982	0.99987
5.0	0.99974	0.99978	0.99986	0.99990	0.99993
5.2	0.99985	0.99987	0.99992	0.99995	0.99996
5.4	0.99991	0.99993	0.99996	0.99997	0.99998
5.6	0.99995	0.99996	0.99998	0.99999	0.99999
5.8	0.99997	0.99998	0.99999	0.99999	1.00000
6.0	0.99999	0.99999	0.99999	1.00000	1.00000

TABLE 7.4.5. CONFIDENCE LIMITS ON THE PROPORTION IN ONE TAIL.

ETA = 0.950 (CONTINUED)

K ↓	N = 30	N = 40	N = 60	N = 80	N = 100
-3.0	0.00011	0.00016	0.00024	0.00030	0.00036
-2.8	0.00027	0.00037	0.00054	0.00068	0.00078
-2.6	0.00064	0.00085	0.00118	0.00143	0.00163
-2.4	0.00142	0.00182	0.00244	0.00290	0.00325
-2.2	0.00299	0.00373	0.00481	0.00559	0.00618
-2.0	0.00597	0.00723	0.00904	0.01029	0.01123
-1.8	0.01130	0.01334	0.01617	0.01809	0.01951
-1.6	0.02031	0.02341	0.02761	0.03040	0.03243
-1.4	0.03463	0.03909	0.04499	0.04884	0.05162
-1.2	0.05612	0.06219	0.07005	0.07511	0.07872
-1.0	0.08653	0.09436	0.10434	0.11066	0.11513
-0.8	0.12710	0.13674	0.14884	0.15640	0.16172
-0.6	0.17822	0.18958	0.20367	0.21238	0.21846
-0.4	0.23913	0.25206	0.26790	0.27760	0.28433
-0.2	0.30795	0.32224	0.33956	0.35007	0.35732
0.0	0.38197	0.39740	0.41592	0.42705	0.43467
0.2	0.45816	0.47450	0.49389	0.50544	0.51331
0.4	0.53361	0.55056	0.57048	0.58223	0.59020
0.6	0.60583	0.62308	0.64311	0.65482	0.66270
0.8	0.67294	0.69010	0.70978	0.72116	0.72877
1.0	0.73361	0.75027	0.76913	0.77990	0.78706
1.2	0.78708	0.80283	0.82041	0.83033	0.83686
1.4	0.83305	0.84755	0.86347	0.87233	0.87811
1.6	0.87164	0.88461	0.89861	0.90628	0.91123
1.8	0.90325	0.91454	0.92648	0.93291	0.93702
2.0	0.92855	0.93809	0.94797	0.95319	0.95649
2.2	0.94831	0.95614	0.96407	0.96818	0.97074
2.4	0.96338	0.96963	0.97581	0.97893	0.98085
2.6	0.97460	0.97946	0.98412	0.98642	0.98781
2.8	0.98276	0.98642	0.98984	0.99148	0.99246
3.0	0.98855	0.99123	0.99366	0.99480	0.99546
3.2	0.99256	0.99447	0.99615	0.99691	0.99735
3.4	0.99527	0.99660	0.99772	0.99821	0.99849
3.6	0.99706	0.99796	0.99869	0.99900	0.99917
3.8	0.99821	0.99880	0.99926	0.99945	0.99955
4.0	0.99894	0.99932	0.99960	0.99971	0.99977
4.2	0.99938	0.99962	0.99979	0.99985	0.99988
4.4	0.99965	0.99979	0.99989	0.99992	0.99994
4.6	0.99981	0.99989	0.99994	0.99996	0.99997
4.8	0.99989	0.99994	0.99997	0.99998	0.99999
5.0	0.99994	0.99997	0.99999	0.99999	0.99999
5.2	0.99997	0.99999	0.99999	1.00000	1.00000
5.4	0.99999	0.99999	1.00000	1.00000	1.00000
5.6	0.99999	1.00000	1.00000	1.00000	1.00000
5.8	1.00000	1.00000	1.00000	1.00000	1.00000
6.0	1.00000	1.00000	1.00000	1.00000	1.00000

TABLE 7.4.6. CONFIDENCE LIMITS ON THE PROPORTION IN ONE TAIL.

ETA = 0.950 (CONTINUED)

K ↓	N = 120	N = 240	N = 600	N = 1000	N = 1200
-3.0	0.00040	0.00058	0.00080	0.00090	0.00093
-2.8	0.00087	0.00121	0.00160	0.00178	0.00184
-2.6	0.00180	0.00239	0.00307	0.00338	0.00348
-2.4	0.00353	0.00455	0.00567	0.00617	0.00633
-2.2	0.00665	0.00831	0.01007	0.01084	0.01108
-2.0	0.01197	0.01453	0.01718	0.01832	0.01867
-1.8	0.02062	0.02437	0.02818	0.02979	0.03028
-1.6	0.03401	0.03926	0.04447	0.04665	0.04731
-1.4	0.05375	0.06075	0.06757	0.07038	0.07124
-1.2	0.08147	0.09039	0.09892	0.10240	0.10346
-1.0	0.11852	0.12939	0.13964	0.14378	0.14503
-0.8	0.16572	0.17844	0.19028	0.19501	0.19644
-0.6	0.22302	0.23737	0.25056	0.25579	0.25737
-0.4	0.28935	0.30504	0.31931	0.32493	0.32662
-0.2	0.36270	0.37941	0.39445	0.40033	0.40210
0.0	0.44032	0.45772	0.47323	0.47926	0.48106
0.2	0.51911	0.53687	0.55255	0.55860	0.56041
0.4	0.59604	0.61380	0.62932	0.63527	0.63705
0.6	0.66846	0.68583	0.70084	0.70656	0.70826
0.8	0.73431	0.75088	0.76502	0.77037	0.77195
1.0	0.79224	0.80759	0.82053	0.82539	0.82682
1.2	0.84156	0.85536	0.86683	0.87109	0.87234
1.4	0.88225	0.89425	0.90406	0.90767	0.90873
1.6	0.91475	0.92484	0.93295	0.93589	0.93675
1.8	0.93991	0.94811	0.95456	0.95687	0.95755
2.0	0.95879	0.96521	0.97016	0.97191	0.97241
2.2	0.97251	0.97736	0.98102	0.98229	0.98265
2.4	0.98216	0.98571	0.98831	0.98920	0.98945
2.6	0.98875	0.99125	0.99303	0.99363	0.99380
2.8	0.99311	0.99481	0.99598	0.99636	0.99647
3.0	0.99590	0.99701	0.99776	0.99800	0.99806
3.2	0.99763	0.99833	0.99879	0.99893	0.99897
3.4	0.99867	0.99910	0.99937	0.99945	0.99947
3.6	0.99927	0.99953	0.99968	0.99973	0.99974
3.8	0.99962	0.99976	0.99984	0.99987	0.99988
4.0	0.99980	0.99988	0.99993	0.99994	0.99994
4.2	0.99990	0.99994	0.99997	0.99997	0.99997
4.4	0.99995	0.99997	0.99999	0.99999	0.99999
4.6	0.99998	0.99999	0.99999	1.00000	1.00000
4.8	0.99999	1.00000	1.00000	1.00000	1.00000
5.0	1.00000	1.00000	1.00000	1.00000	1.00000
5.2	1.00000	1.00000	1.00000	1.00000	1.00000
5.4	1.00000	1.00000	1.00000	1.00000	1.00000
5.6	1.00000	1.00000	1.00000	1.00000	1.00000
5.8	1.00000	1.00000	1.00000	1.00000	1.00000
6.0	1.00000	1.00000	1.00000	1.00000	1.00000

TABLE 7.5.1. CONFIDENCE LIMITS ON THE PROPORTION IN ONE TAIL.

ETA = 0.975

K ↓	N = 2	N = 3	N = 4	N = 5	N = 6
-3.0	0.00000	0.00000	0.00000	0.00000	0.00000
-2.8	0.00000	0.00000	0.00000	0.00000	0.00000
-2.6	0.00000	0.00000	0.00000	0.00000	0.00001
-2.4	0.00000	0.00000	0.00000	0.00001	0.00003
-2.2	0.00000	0.00000	0.00002	0.00005	0.00010
-2.0	0.00000	0.00002	0.00008	0.00018	0.00031
-1.8	0.00001	0.00009	0.00028	0.00057	0.00091
-1.6	0.00004	0.00034	0.00089	0.00160	0.00238
-1.4	0.00022	0.00114	0.00251	0.00407	0.00566
-1.2	0.00090	0.00332	0.00629	0.00934	0.01229
-1.0	0.00303	0.00837	0.01403	0.01939	0.02431
-0.8	0.00838	0.01846	0.02798	0.03647	0.04396
-0.6	0.01910	0.03565	0.05010	0.06241	0.07293
-0.4	0.03627	0.06079	0.08106	0.09772	0.11161
-0.2	0.05854	0.09275	0.11971	0.14120	0.15872
0.0	0.08289	0.12890	0.16355	0.19037	0.21181
0.2	0.10661	0.16649	0.20981	0.24247	0.26807
0.4	0.12847	0.20360	0.25636	0.29522	0.32515
0.6	0.14827	0.23930	0.30189	0.34710	0.38137
0.8	0.16626	0.27324	0.34575	0.39726	0.43572
1.0	0.18272	0.30542	0.38768	0.44523	0.48760
1.2	0.19792	0.33592	0.42761	0.49081	0.53669
1.4	0.21207	0.36487	0.46554	0.53393	0.58284
1.6	0.22533	0.39241	0.50154	0.57455	0.62597
1.8	0.23783	0.41865	0.53568	0.61272	0.66608
2.0	0.24967	0.44371	0.56804	0.64847	0.70318
2.2	0.26095	0.46768	0.59869	0.68186	0.73735
2.4	0.27172	0.49063	0.62770	0.71296	0.76865
2.6	0.28204	0.51264	0.65512	0.74183	0.79718
2.8	0.29196	0.53377	0.68103	0.76854	0.82304
3.0	0.30151	0.55406	0.70547	0.79318	0.84637
3.2	0.31073	0.57358	0.72850	0.81583	0.86730
3.4	0.31965	0.59235	0.75017	0.83656	0.88596
3.6	0.32829	0.61042	0.77053	0.85548	0.90251
3.8	0.33667	0.62781	0.78962	0.87267	0.91710
4.0	0.34482	0.64455	0.80749	0.88824	0.92988
4.2	0.35274	0.66068	0.82419	0.90226	0.94102
4.4	0.36046	0.67622	0.83977	0.91486	0.95066
4.6	0.36798	0.69118	0.85426	0.92612	0.95895
4.8	0.37532	0.70559	0.86772	0.93614	0.96604
5.0	0.38249	0.71946	0.88020	0.94502	0.97207
5.2	0.38950	0.73282	0.89173	0.95285	0.97716
5.4	0.39636	0.74568	0.90236	0.95974	0.98143
5.6	0.40307	0.75805	0.91215	0.96575	0.98499
5.8	0.40965	0.76995	0.92114	0.97099	0.98794
6.0	0.41609	0.78139	0.92937	0.97553	0.99036

TABLE 7.5.2. CONFIDENCE LIMITS ON THE PROPORTION IN ONE TAIL.

ETA = 0.975 (CONTINUED)

K ↓	N = 7	N = 8	N = 9	N = 10	N = 11
-3.0	0.00000	0.00000	0.00000	0.00000	0.00001
-2.8	0.00000	0.00001	0.00001	0.00001	0.00002
-2.6	0.00001	0.00002	0.00003	0.00004	0.00006
-2.4	0.00005	0.00007	0.00010	0.00014	0.00017
-2.2	0.00016	0.00023	0.00031	0.00039	0.00048
-2.0	0.00047	0.00065	0.00083	0.00102	0.00122
-1.8	0.00128	0.00168	0.00208	0.00248	0.00288
-1.6	0.00319	0.00400	0.00480	0.00558	0.00633
-1.4	0.00724	0.00876	0.01022	0.01160	0.01291
-1.2	0.01507	0.01767	0.02009	0.02234	0.02444
-1.0	0.02879	0.03286	0.03656	0.03995	0.04306
-0.8	0.05058	0.05646	0.06172	0.06647	0.07076
-0.6	0.08202	0.08996	0.09696	0.10319	0.10878
-0.4	0.12338	0.13351	0.14234	0.15012	0.15705
-0.2	0.17334	0.18574	0.19645	0.20580	0.21406
0.0	0.22941	0.24417	0.25677	0.26770	0.27728
0.2	0.28878	0.30595	0.32047	0.33294	0.34381
0.4	0.34902	0.36858	0.38498	0.39896	0.41106
0.6	0.40834	0.43022	0.44838	0.46375	0.47696
0.8	0.46562	0.48961	0.50935	0.52592	0.54007
1.0	0.52013	0.54596	0.56702	0.58456	0.59942
1.2	0.57148	0.59880	0.62086	0.63907	0.65440
1.4	0.61944	0.64785	0.67055	0.68913	0.70464
1.6	0.66391	0.69298	0.71596	0.73458	0.74998
1.8	0.70484	0.73415	0.75703	0.77538	0.79041
2.0	0.74228	0.77140	0.79384	0.81163	0.82605
2.2	0.77629	0.80483	0.82651	0.84347	0.85707
2.4	0.80697	0.83457	0.85522	0.87115	0.88377
2.6	0.83447	0.86082	0.88021	0.89495	0.90646
2.8	0.85893	0.88379	0.90174	0.91518	0.92553
3.0	0.88055	0.90371	0.92012	0.93218	0.94134
3.2	0.89951	0.92084	0.93564	0.94632	0.95430
3.4	0.91602	0.93543	0.94861	0.95794	0.96479
3.6	0.93029	0.94776	0.95935	0.96739	0.97318
3.8	0.94253	0.95808	0.96814	0.97497	0.97979
4.0	0.95294	0.96663	0.97526	0.98099	0.98495
4.2	0.96174	0.97366	0.98098	0.98571	0.98892
4.4	0.96911	0.97938	0.98551	0.98938	0.99194
4.6	0.97523	0.98400	0.98907	0.99219	0.99420
4.8	0.98029	0.98769	0.99184	0.99432	0.99588
5.0	0.98442	0.99061	0.99396	0.99591	0.99711
5.2	0.98778	0.99289	0.99558	0.99709	0.99799
5.4	0.99048	0.99467	0.99679	0.99795	0.99862
5.6	0.99265	0.99604	0.99770	0.99857	0.99907
5.8	0.99436	0.99708	0.99836	0.99902	0.99938
6.0	0.99570	0.99787	0.99885	0.99933	0.99959

TABLE 7.5.3. CONFIDENCE LIMITS ON THE PROPORTION IN ONE TAIL.

ETA = 0.975 (CONTINUED)

K ↓	N = 12	N = 13	N = 14	N = 15	N = 16
-3.0	0.00001	0.00001	0.00001	0.00001	0.00002
-2.8	0.00002	0.00003	0.00004	0.00004	0.00005
-2.6	0.00007	0.00009	0.00010	0.00012	0.00014
-2.4	0.00021	0.00025	0.00029	0.00033	0.00037
-2.2	0.00057	0.00066	0.00075	0.00084	0.00093
-2.0	0.00141	0.00161	0.00180	0.00199	0.00218
-1.8	0.00328	0.00366	0.00404	0.00440	0.00475
-1.6	0.00706	0.00776	0.00843	0.00907	0.00969
-1.4	0.01415	0.01532	0.01643	0.01749	0.01849
-1.2	0.02639	0.02822	0.02993	0.03155	0.03306
-1.0	0.04592	0.04857	0.05103	0.05333	0.05547
-0.8	0.07469	0.07828	0.08159	0.08466	0.08750
-0.6	0.11384	0.11844	0.12266	0.12653	0.13011
-0.4	0.16326	0.16889	0.17400	0.17868	0.18299
-0.2	0.22142	0.22805	0.23404	0.23951	0.24451
0.0	0.28577	0.29336	0.30020	0.30641	0.31207
0.2	0.35339	0.36191	0.36955	0.37645	0.38272
0.4	0.42165	0.43102	0.43939	0.44692	0.45374
0.6	0.48846	0.49858	0.50757	0.51563	0.52290
0.8	0.55231	0.56303	0.57251	0.58096	0.58856
1.0	0.61220	0.62333	0.63313	0.64182	0.64960
1.2	0.66749	0.67881	0.68873	0.69749	0.70529
1.4	0.71779	0.72909	0.73893	0.74757	0.75523
1.6	0.76294	0.77401	0.78357	0.79193	0.79930
1.8	0.80296	0.81358	0.82270	0.83062	0.83756
2.0	0.83796	0.84798	0.85651	0.86386	0.87026
2.2	0.86820	0.87746	0.88529	0.89198	0.89777
2.4	0.89398	0.90239	0.90944	0.91541	0.92054
2.6	0.91567	0.92318	0.92941	0.93464	0.93910
2.8	0.93370	0.94028	0.94568	0.95018	0.95398
3.0	0.94847	0.95414	0.95875	0.96254	0.96571
3.2	0.96042	0.96523	0.96908	0.97223	0.97483
3.4	0.96996	0.97397	0.97714	0.97969	0.98179
3.6	0.97748	0.98076	0.98332	0.98536	0.98701
3.8	0.98332	0.98596	0.98800	0.98960	0.99088
4.0	0.98779	0.98989	0.99148	0.99272	0.99369
4.2	0.99118	0.99282	0.99404	0.99497	0.99570
4.4	0.99370	0.99496	0.99589	0.99658	0.99711
4.6	0.99556	0.99651	0.99720	0.99771	0.99809
4.8	0.99691	0.99762	0.99812	0.99849	0.99876
5.0	0.99788	0.99840	0.99876	0.99902	0.99920
5.2	0.99856	0.99893	0.99919	0.99937	0.99950
5.4	0.99904	0.99930	0.99948	0.99960	0.99969
5.6	0.99936	0.99955	0.99967	0.99975	0.99981
5.8	0.99958	0.99971	0.99979	0.99985	0.99989
6.0	0.99973	0.99982	0.99987	0.99991	0.99993

TABLE 7.5.4. CONFIDENCE LIMITS ON THE PROPORTION IN ONE TAIL.

ETA = 0.975 (CONTINUED)

K ↓	N = 17	N = 18	N = 21	N = 24	N = 27
-3.0	0.00002	0.00002	0.00003	0.00004	0.00005
-2.8	0.00006	0.00006	0.00009	0.00011	0.00014
-2.6	0.00016	0.00018	0.00023	0.00029	0.00035
-2.4	0.00042	0.00046	0.00059	0.00071	0.00083
-2.2	0.00103	0.00112	0.00138	0.00164	0.00188
-2.0	0.00237	0.00255	0.00306	0.00355	0.00399
-1.8	0.00510	0.00543	0.00637	0.00722	0.00800
-1.6	0.01028	0.01085	0.01244	0.01385	0.01512
-1.4	0.01945	0.02036	0.02284	0.02502	0.02696
-1.2	0.03449	0.03585	0.03951	0.04267	0.04544
-1.0	0.05748	0.05937	0.06442	0.06873	0.07246
-0.8	0.09016	0.09264	0.09922	0.10476	0.10953
-0.6	0.13343	0.13653	0.14467	0.15147	0.15727
-0.4	0.18697	0.19066	0.20032	0.20832	0.21510
-0.2	0.24912	0.25338	0.26445	0.27357	0.28123
0.0	0.31726	0.32205	0.33444	0.34455	0.35301
0.2	0.38846	0.39373	0.40728	0.41828	0.42743
0.4	0.45995	0.46564	0.48019	0.49192	0.50162
0.6	0.52949	0.53552	0.55085	0.56311	0.57319
0.8	0.59543	0.60169	0.61751	0.63007	0.64033
1.0	0.65661	0.66296	0.67895	0.69154	0.70176
1.2	0.71229	0.71861	0.73442	0.74675	0.75670
1.4	0.76208	0.76824	0.78351	0.79532	0.80477
1.6	0.80585	0.81172	0.82615	0.83720	0.84596
1.8	0.84370	0.84917	0.86251	0.87260	0.88052
2.0	0.87589	0.88088	0.89293	0.90194	0.90894
2.2	0.90283	0.90728	0.91793	0.92579	0.93182
2.4	0.92499	0.92888	0.93810	0.94478	0.94986
2.6	0.94293	0.94627	0.95406	0.95963	0.96379
2.8	0.95722	0.96001	0.96647	0.97100	0.97434
3.0	0.96840	0.97070	0.97593	0.97953	0.98215
3.2	0.97701	0.97886	0.98301	0.98581	0.98781
3.4	0.98352	0.98498	0.98821	0.99034	0.99184
3.6	0.98837	0.98950	0.99196	0.99354	0.99464
3.8	0.99192	0.99278	0.99461	0.99576	0.99655
4.0	0.99447	0.99511	0.99645	0.99727	0.99782
4.2	0.99628	0.99674	0.99770	0.99828	0.99865
4.4	0.99753	0.99787	0.99854	0.99893	0.99918
4.6	0.99839	0.99862	0.99909	0.99935	0.99951
4.8	0.99897	0.99913	0.99944	0.99961	0.99971
5.0	0.99935	0.99946	0.99966	0.99977	0.99984
5.2	0.99959	0.99967	0.99980	0.99987	0.99991
5.4	0.99975	0.99980	0.99988	0.99993	0.99995
5.6	0.99985	0.99988	0.99993	0.99996	0.99997
5.8	0.99991	0.99993	0.99996	0.99998	0.99999
6.0	0.99995	0.99996	0.99998	0.99999	0.99999

TABLE 7.5.5. CONFIDENCE LIMITS ON THE PROPORTION IN ONE TAIL.

ETA = 0.975 (CONTINUED)

K ↓	N = 30	N = 40	N = 60	N = 80	N = 100
-3.0	0.00006	0.00010	0.00016	0.00022	0.00027
-2.8	0.00016	0.00024	0.00039	0.00051	0.00061
-2.6	0.00040	0.00058	0.00088	0.00111	0.00131
-2.4	0.00095	0.00131	0.00188	0.00232	0.00267
-2.2	0.00211	0.00278	0.00383	0.00460	0.00521
-2.0	0.00441	0.00561	0.00741	0.00869	0.00968
-1.8	0.00872	0.01073	0.01362	0.01564	0.01717
-1.6	0.01626	0.01943	0.02384	0.02684	0.02907
-1.4	0.02869	0.03337	0.03971	0.04394	0.04702
-1.2	0.04789	0.05442	0.06305	0.06868	0.07275
-1.0	0.07574	0.08435	0.09549	0.10264	0.10774
-0.8	0.11369	0.12446	0.13817	0.14683	0.15296
-0.6	0.16229	0.17519	0.19134	0.20141	0.20847
-0.4	0.22093	0.23579	0.25414	0.26545	0.27332
-0.2	0.28781	0.30440	0.32464	0.33698	0.34551
0.0	0.36023	0.37832	0.40012	0.41327	0.42231
0.2	0.43519	0.45450	0.47751	0.49124	0.50062
0.4	0.50981	0.53003	0.55383	0.56790	0.57743
0.6	0.58167	0.60242	0.62654	0.64064	0.65013
0.8	0.64892	0.66976	0.69366	0.70747	0.71670
1.0	0.71026	0.73071	0.75383	0.76702	0.77576
1.2	0.76491	0.78449	0.80628	0.81853	0.82658
1.4	0.81252	0.83079	0.85078	0.86184	0.86904
1.6	0.85309	0.86971	0.88753	0.89723	0.90347
1.8	0.88693	0.90164	0.91710	0.92535	0.93059
2.0	0.91454	0.92724	0.94026	0.94706	0.95133
2.2	0.93660	0.94726	0.95793	0.96337	0.96673
2.4	0.95384	0.96256	0.97105	0.97528	0.97785
2.6	0.96703	0.97398	0.98055	0.98373	0.98563
2.8	0.97690	0.98230	0.98724	0.98956	0.99092
3.0	0.98413	0.98821	0.99183	0.99347	0.99442
3.2	0.98931	0.99232	0.99489	0.99602	0.99666
3.4	0.99294	0.99511	0.99688	0.99764	0.99805
3.6	0.99543	0.99695	0.99815	0.99863	0.99890
3.8	0.99710	0.99814	0.99892	0.99923	0.99939
4.0	0.99820	0.99889	0.99939	0.99958	0.99967
4.2	0.99890	0.99935	0.99966	0.99977	0.99983
4.4	0.99935	0.99963	0.99982	0.99988	0.99991
4.6	0.99962	0.99980	0.99991	0.99994	0.99996
4.8	0.99978	0.99989	0.99995	0.99997	0.99998
5.0	0.99988	0.99994	0.99998	0.99999	0.99999
5.2	0.99993	0.99997	0.99999	0.99999	1.00000
5.4	0.99996	0.99998	0.99999	1.00000	1.00000
5.6	0.99998	0.99999	1.00000	1.00000	1.00000
5.8	0.99999	1.00000	1.00000	1.00000	1.00000
6.0	1.00000	1.00000	1.00000	1.00000	1.00000

TABLE 7.5.6. *CONFIDENCE LIMITS ON THE PROPORTION IN ONE TAIL.*

ETA = 0.975 (*CONTINUED*)

K ↓	N = 120	N = 240	N = 600	N = 1000	N = 1200
-3.0	0.00031	0.00049	0.00072	0.00083	0.00087
-2.8	0.00070	0.00103	0.00145	0.00166	0.00172
-2.6	0.00147	0.00209	0.00283	0.00317	0.00328
-2.4	0.00296	0.00404	0.00527	0.00584	0.00601
-2.2	0.00570	0.00748	0.00944	0.01032	0.01060
-2.0	0.01047	0.01326	0.01624	0.01755	0.01796
-1.8	0.01837	0.02252	0.02685	0.02871	0.02928
-1.6	0.03081	0.03669	0.04267	0.04519	0.04597
-1.4	0.04941	0.05735	0.06523	0.06850	0.06951
-1.2	0.07587	0.08608	0.09600	0.10008	0.10132
-1.0	0.11163	0.12419	0.13616	0.14102	0.14250
-0.8	0.15759	0.17239	0.18628	0.19186	0.19355
-0.6	0.21378	0.23058	0.24613	0.25232	0.25419
-0.4	0.27921	0.29767	0.31455	0.32121	0.32322
-0.2	0.35186	0.37162	0.38946	0.39645	0.39855
0.0	0.42900	0.44966	0.46811	0.47529	0.47744
0.2	0.50753	0.52871	0.54740	0.55462	0.55678
0.4	0.58443	0.60570	0.62426	0.63137	0.63349
0.6	0.65707	0.67796	0.69597	0.70282	0.70486
0.8	0.72341	0.74343	0.76046	0.76689	0.76879
1.0	0.78207	0.80074	0.81639	0.82223	0.82396
1.2	0.83236	0.84925	0.86318	0.86833	0.86984
1.4	0.87417	0.88897	0.90096	0.90534	0.90662
1.6	0.90788	0.92044	0.93041	0.93399	0.93504
1.8	0.93426	0.94456	0.95255	0.95538	0.95620
2.0	0.95429	0.96245	0.96863	0.97078	0.97140
2.2	0.96904	0.97529	0.97990	0.98147	0.98192
2.4	0.97959	0.98421	0.98752	0.98863	0.98894
2.6	0.98690	0.99021	0.99250	0.99325	0.99346
2.8	0.99182	0.99410	0.99563	0.99612	0.99626
3.0	0.99503	0.99656	0.99754	0.99784	0.99793
3.2	0.99706	0.99805	0.99866	0.99884	0.99889
3.4	0.99831	0.99893	0.99929	0.99940	0.99943
3.6	0.99906	0.99943	0.99964	0.99970	0.99971
3.8	0.99949	0.99970	0.99982	0.99985	0.99986
4.0	0.99973	0.99985	0.99991	0.99993	0.99994
4.2	0.99986	0.99993	0.99996	0.99997	0.99997
4.4	0.99993	0.99997	0.99998	0.99999	0.99999
4.6	0.99997	0.99998	0.99999	0.99999	0.99999
4.8	0.99998	0.99999	1.00000	1.00000	1.00000
5.0	0.99999	1.00000	1.00000	1.00000	1.00000
5.2	1.00000	1.00000	1.00000	1.00000	1.00000
5.4	1.00000	1.00000	1.00000	1.00000	1.00000
5.6	1.00000	1.00000	1.00000	1.00000	1.00000
5.8	1.00000	1.00000	1.00000	1.00000	1.00000
6.0	1.00000	1.00000	1.00000	1.00000	1.00000

TABLE 7.6.1. CONFIDENCE LIMITS ON THE PROPORTION IN ONE TAIL.

ETA = 0.990

K ↓	N = 2	N = 3	N = 4	N = 5	N = 6
-3.0	0.00000	0.00000	0.00000	0.00000	0.00000
-2.8	0.00000	0.00000	0.00000	0.00000	0.00000
-2.6	0.00000	0.00000	0.00000	0.00000	0.00000
-2.4	0.00000	0.00000	0.00000	0.00000	0.00001
-2.2	0.00000	0.00000	0.00000	0.00001	0.00003
-2.0	0.00000	0.00000	0.00001	0.00005	0.00010
-1.8	0.00000	0.00001	0.00007	0.00018	0.00034
-1.6	0.00000	0.00007	0.00027	0.00060	0.00103
-1.4	0.00003	0.00031	0.00094	0.00180	0.00281
-1.2	0.00017	0.00115	0.00279	0.00476	0.00687
-1.0	0.00080	0.00355	0.00721	0.01114	0.01504
-0.8	0.00298	0.00923	0.01626	0.02314	0.02958
-0.6	0.00852	0.02031	0.03205	0.04282	0.05247
-0.4	0.01890	0.03808	0.05573	0.07116	0.08453
-0.2	0.03356	0.06195	0.08659	0.10732	0.12484
0.0	0.04999	0.08962	0.12238	0.14908	0.17113
0.2	0.06594	0.11860	0.16054	0.19380	0.22071
0.4	0.08047	0.14725	0.19911	0.23938	0.27138
0.6	0.09354	0.17478	0.23698	0.28444	0.32162
0.8	0.10534	0.20098	0.27360	0.32827	0.37053
1.0	0.11611	0.22585	0.30878	0.37050	0.41763
1.2	0.12604	0.24949	0.34249	0.41097	0.46266
1.4	0.13527	0.27201	0.37475	0.44964	0.50550
1.6	0.14393	0.29352	0.40564	0.48651	0.54610
1.8	0.15209	0.31413	0.43522	0.52161	0.58446
2.0	0.15984	0.33391	0.46356	0.55499	0.62060
2.2	0.16721	0.35294	0.49072	0.58669	0.65454
2.4	0.17427	0.37128	0.51675	0.61676	0.68633
2.6	0.18103	0.38899	0.54172	0.64523	0.71602
2.8	0.18754	0.40612	0.56567	0.67216	0.74365
3.0	0.19382	0.42270	0.58864	0.69759	0.76930
3.2	0.19989	0.43877	0.61066	0.72157	0.79303
3.4	0.20576	0.45436	0.63178	0.74413	0.81491
3.6	0.21146	0.46950	0.65202	0.76533	0.83501
3.8	0.21700	0.48421	0.67141	0.78520	0.85341
4.0	0.22239	0.49852	0.68997	0.80379	0.87020
4.2	0.22764	0.51244	0.70774	0.82115	0.88546
4.4	0.23276	0.52600	0.72474	0.83732	0.89928
4.6	0.23775	0.53920	0.74098	0.85234	0.91173
4.8	0.24264	0.55206	0.75649	0.86628	0.92293
5.0	0.24742	0.56460	0.77130	0.87917	0.93294
5.2	0.25209	0.57682	0.78541	0.89107	0.94187
5.4	0.25668	0.58875	0.79886	0.90203	0.94979
5.6	0.26117	0.60038	0.81166	0.91208	0.95680
5.8	0.26558	0.61173	0.82383	0.92129	0.96296
6.0	0.26990	0.62280	0.83538	0.92971	0.96837

TABLE 7.6.2. *CONFIDENCE LIMITS ON THE PROPORTION IN ONE TAIL.*

ETA = 0.990 (*CONTINUED*)

K ↓	N = 7	N = 8	N = 9	N = 10	N = 11
-3.0	0.00000	0.00000	0.00000	0.00000	0.00000
-2.8	0.00000	0.00000	0.00000	0.00000	0.00001
-2.6	0.00000	0.00001	0.00001	0.00001	0.00002
-2.4	0.00001	0.00002	0.00003	0.00005	0.00007
-2.2	0.00005	0.00008	0.00012	0.00016	0.00021
-2.0	0.00017	0.00026	0.00036	0.00047	0.00059
-1.8	0.00053	0.00076	0.00101	0.00128	0.00155
-1.6	0.00152	0.00205	0.00261	0.00317	0.00374
-1.4	0.00389	0.00500	0.00611	0.00721	0.00828
-1.2	0.00899	0.01108	0.01309	0.01502	0.01685
-1.0	0.01877	0.02230	0.02561	0.02870	0.03160
-0.8	0.03551	0.04094	0.04592	0.05049	0.05470
-0.6	0.06109	0.06880	0.07574	0.08201	0.08771
-0.4	0.09619	0.10643	0.11550	0.12360	0.13089
-0.2	0.13981	0.15275	0.16406	0.17405	0.18296
0.0	0.18963	0.20540	0.21904	0.23097	0.24152
0.2	0.24294	0.26166	0.27767	0.29156	0.30374
0.4	0.29745	0.31915	0.33753	0.35334	0.36712
0.6	0.35152	0.37614	0.39682	0.41445	0.42971
0.8	0.40414	0.43153	0.45432	0.47362	0.49020
1.0	0.45470	0.48461	0.50928	0.53001	0.54770
1.2	0.50286	0.53498	0.56123	0.58312	0.60165
1.4	0.54845	0.58241	0.60991	0.63262	0.65172
1.6	0.59138	0.62677	0.65514	0.67837	0.69773
1.8	0.63160	0.66802	0.69688	0.72028	0.73960
2.0	0.66913	0.70614	0.73512	0.75836	0.77737
2.2	0.70399	0.74117	0.76991	0.79269	0.81112
2.4	0.73622	0.77317	0.80135	0.82339	0.84101
2.6	0.76591	0.80225	0.82955	0.85061	0.86724
2.8	0.79311	0.82851	0.85466	0.87455	0.89003
3.0	0.81793	0.85208	0.87688	0.89543	0.90967
3.2	0.84046	0.87311	0.89637	0.91348	0.92642
3.4	0.86082	0.89175	0.91336	0.92897	0.94057
3.6	0.87913	0.90818	0.92805	0.94213	0.95241
3.8	0.89551	0.92256	0.94065	0.95321	0.96222
4.0	0.91009	0.93506	0.95138	0.96248	0.97028
4.2	0.92299	0.94587	0.96045	0.97014	0.97682
4.4	0.93436	0.95514	0.96806	0.97643	0.98208
4.6	0.94431	0.96305	0.97438	0.98155	0.98628
4.8	0.95299	0.96975	0.97960	0.98567	0.98958
5.0	0.96051	0.97539	0.98387	0.98897	0.99217
5.2	0.96699	0.98010	0.98734	0.99157	0.99416
5.4	0.97254	0.98400	0.99014	0.99362	0.99569
5.6	0.97728	0.98722	0.99237	0.99521	0.99685
5.8	0.98129	0.98986	0.99415	0.99643	0.99772
6.0	0.98467	0.99200	0.99554	0.99736	0.99836

TABLE 7.6.3. *CONFIDENCE LIMITS ON THE PROPORTION IN ONE TAIL.*

ETA = 0.990 (*CONTINUED*)

K ↓	N = 12	N = 13	N = 14	N = 15	N = 16
-3.0	0.00000	0.00000	0.00000	0.00000	0.00001
-2.8	0.00001	0.00001	0.00001	0.00002	0.00002
-2.6	0.00003	0.00003	0.00004	0.00005	0.00006
-2.4	0.00009	0.00011	0.00013	0.00015	0.00018
-2.2	0.00026	0.00031	0.00037	0.00043	0.00049
-2.0	0.00072	0.00085	0.00098	0.00111	0.00125
-1.8	0.00183	0.00211	0.00239	0.00267	0.00295
-1.6	0.00430	0.00486	0.00540	0.00594	0.00646
-1.4	0.00932	0.01032	0.01129	0.01223	0.01313
-1.2	0.01861	0.02027	0.02186	0.02337	0.02480
-1.0	0.03432	0.03686	0.03925	0.04150	0.04363
-0.8	0.05859	0.06220	0.06555	0.06868	0.07161
-0.6	0.09292	0.09770	0.10211	0.10619	0.10999
-0.4	0.13748	0.14349	0.14899	0.15406	0.15874
-0.2	0.19097	0.19821	0.20480	0.21084	0.21639
0.0	0.25093	0.25939	0.26706	0.27403	0.28042
0.2	0.31454	0.32419	0.33289	0.34077	0.34795
0.4	0.37924	0.39003	0.39969	0.40841	0.41632
0.6	0.44307	0.45487	0.46540	0.47486	0.48342
0.8	0.50462	0.51731	0.52856	0.53862	0.54769
1.0	0.56299	0.57635	0.58815	0.59866	0.60808
1.2	0.61757	0.63140	0.64354	0.65430	0.66390
1.4	0.66800	0.68206	0.69434	0.70515	0.71475
1.6	0.71411	0.72815	0.74034	0.75101	0.76044
1.8	0.75582	0.76962	0.78151	0.79185	0.80094
2.0	0.79317	0.80652	0.81792	0.82777	0.83637
2.2	0.82630	0.83899	0.84975	0.85898	0.86698
2.4	0.85537	0.86726	0.87726	0.88576	0.89306
2.6	0.88063	0.89161	0.90074	0.90844	0.91501
2.8	0.90236	0.91234	0.92056	0.92743	0.93323
3.0	0.92085	0.92980	0.93709	0.94311	0.94816
3.2	0.93643	0.94434	0.95071	0.95591	0.96022
3.4	0.94942	0.95631	0.96179	0.96622	0.96984
3.6	0.96013	0.96606	0.97070	0.97441	0.97742
3.8	0.96887	0.97390	0.97778	0.98084	0.98329
4.0	0.97593	0.98014	0.98334	0.98583	0.98779
4.2	0.98157	0.98504	0.98764	0.98964	0.99119
4.4	0.98602	0.98885	0.99094	0.99251	0.99372
4.6	0.98951	0.99178	0.99343	0.99465	0.99558
4.8	0.99220	0.99401	0.99529	0.99623	0.99693
5.0	0.99426	0.99568	0.99666	0.99737	0.99789
5.2	0.99582	0.99691	0.99766	0.99819	0.99857
5.4	0.99698	0.99782	0.99838	0.99877	0.99905
5.6	0.99785	0.99848	0.99889	0.99917	0.99937
5.8	0.99848	0.99895	0.99925	0.99945	0.99959
6.0	0.99894	0.99928	0.99950	0.99964	0.99974

TABLE 7.6.4. CONFIDENCE LIMITS ON THE PROPORTION IN ONE TAIL.

ETA = 0.990 (CONTINUED)

K ↓	N = 17	N = 18	N = 21	N = 24	N = 27
-3.0	0.00001	0.00001	0.00001	0.00002	0.00002
-2.8	0.00002	0.00003	0.00004	0.00005	0.00007
-2.6	0.00007	0.00008	0.00012	0.00015	0.00019
-2.4	0.00021	0.00023	0.00032	0.00041	0.00050
-2.2	0.00055	0.00062	0.00081	0.00100	0.00120
-2.0	0.00139	0.00152	0.00193	0.00232	0.00270
-1.8	0.00323	0.00350	0.00428	0.00502	0.00571
-1.6	0.00697	0.00746	0.00887	0.01016	0.01134
-1.4	0.01400	0.01483	0.01716	0.01925	0.02113
-1.2	0.02617	0.02748	0.03106	0.03422	0.03702
-1.0	0.04563	0.04753	0.05268	0.05713	0.06103
-0.8	0.07435	0.07694	0.08385	0.08976	0.09488
-0.6	0.11353	0.11684	0.12563	0.13304	0.13941
-0.4	0.16308	0.16713	0.17777	0.18667	0.19424
-0.2	0.22151	0.22627	0.23870	0.24900	0.25771
0.0	0.28630	0.29173	0.30585	0.31744	0.32718
0.2	0.35454	0.36061	0.37627	0.38904	0.39969
0.4	0.42355	0.43019	0.44723	0.46101	0.47244
0.6	0.49120	0.49832	0.51649	0.53107	0.54309
0.8	0.55591	0.56340	0.58240	0.59752	0.60991
1.0	0.61659	0.62431	0.64378	0.65915	0.67165
1.2	0.67254	0.68035	0.69990	0.71520	0.72754
1.4	0.72335	0.73109	0.75033	0.76523	0.77715
1.6	0.76883	0.77635	0.79490	0.80912	0.82038
1.8	0.80898	0.81615	0.83369	0.84696	0.85737
2.0	0.84394	0.85065	0.86690	0.87903	0.88845
2.2	0.87397	0.88013	0.89489	0.90576	0.91409
2.4	0.89941	0.90496	0.91812	0.92765	0.93487
2.6	0.92067	0.92559	0.93708	0.94528	0.95140
2.8	0.93819	0.94246	0.95233	0.95924	0.96431
3.0	0.95243	0.95608	0.96439	0.97010	0.97422
3.2	0.96384	0.96691	0.97378	0.97840	0.98168
3.4	0.97286	0.97539	0.98097	0.98464	0.98720
3.6	0.97988	0.98194	0.98639	0.98925	0.99120
3.8	0.98528	0.98692	0.99041	0.99259	0.99406
4.0	0.98937	0.99065	0.99334	0.99498	0.99605
4.2	0.99242	0.99341	0.99544	0.99665	0.99742
4.4	0.99467	0.99542	0.99693	0.99780	0.99835
4.6	0.99630	0.99686	0.99796	0.99858	0.99896
4.8	0.99746	0.99788	0.99867	0.99910	0.99935
5.0	0.99828	0.99858	0.99914	0.99944	0.99961
5.2	0.99886	0.99907	0.99946	0.99965	0.99976
5.4	0.99925	0.99940	0.99966	0.99979	0.99986
5.6	0.99951	0.99961	0.99979	0.99988	0.99992
5.8	0.99969	0.99976	0.99987	0.99993	0.99995
6.0	0.99980	0.99985	0.99992	0.99996	0.99997

TABLE 7.6.5. CONFIDENCE LIMITS ON THE PROPORTION IN ONE TAIL.

ETA = 0.990 (CONTINUED)

K ↓	N = 30	N = 40	N = 60	N = 80	N = 100
-3.0	0.00003	0.00005	0.00010	0.00015	0.00019
-2.8	0.00009	0.00014	0.00026	0.00036	0.00045
-2.6	0.00023	0.00037	0.00061	0.00082	0.00100
-2.4	0.00059	0.00087	0.00137	0.00178	0.00212
-2.2	0.00138	0.00196	0.00291	0.00365	0.00425
-2.0	0.00306	0.00414	0.00583	0.00711	0.00811
-1.8	0.00636	0.00825	0.01109	0.01315	0.01474
-1.6	0.01243	0.01552	0.01999	0.02314	0.02551
-1.4	0.02285	0.02758	0.03420	0.03873	0.04208
-1.2	0.03953	0.04634	0.05558	0.06174	0.06624
-1.0	0.06449	0.07371	0.08590	0.09384	0.09957
-0.8	0.09937	0.11117	0.12643	0.13620	0.14316
-0.6	0.14495	0.15934	0.17759	0.18909	0.19720
-0.4	0.20080	0.21762	0.23862	0.25166	0.26078
-0.2	0.26520	0.28423	0.30765	0.32201	0.33197
0.0	0.33552	0.35650	0.38196	0.39740	0.40802
0.2	0.40876	0.43141	0.45851	0.47476	0.48586
0.4	0.48211	0.50607	0.53436	0.55112	0.56249
0.6	0.55322	0.57805	0.60699	0.62392	0.63532
0.8	0.62028	0.64550	0.67446	0.69119	0.70236
1.0	0.68206	0.70711	0.73542	0.75155	0.76222
1.2	0.73774	0.76206	0.78908	0.80423	0.81415
1.4	0.78694	0.80999	0.83512	0.84897	0.85795
1.6	0.82956	0.85090	0.87367	0.88599	0.89387
1.8	0.86578	0.88506	0.90517	0.91581	0.92253
2.0	0.89598	0.91298	0.93027	0.93921	0.94477
2.2	0.92069	0.93532	0.94980	0.95710	0.96156
2.4	0.94051	0.95282	0.96463	0.97041	0.97389
2.6	0.95613	0.96623	0.97561	0.98007	0.98269
2.8	0.96819	0.97628	0.98355	0.98689	0.98881
3.0	0.97732	0.98366	0.98914	0.99158	0.99295
3.2	0.98411	0.98897	0.99300	0.99472	0.99567
3.4	0.98907	0.99269	0.99558	0.99677	0.99740
3.6	0.99260	0.99526	0.99728	0.99807	0.99848
3.8	0.99509	0.99698	0.99836	0.99888	0.99914
4.0	0.99679	0.99812	0.99903	0.99936	0.99952
4.2	0.99795	0.99885	0.99944	0.99965	0.99974
4.4	0.99871	0.99931	0.99969	0.99981	0.99986
4.6	0.99920	0.99960	0.99983	0.99990	0.99993
4.8	0.99952	0.99977	0.99991	0.99995	0.99997
5.0	0.99971	0.99987	0.99995	0.99997	0.99998
5.2	0.99983	0.99993	0.99998	0.99999	0.99999
5.4	0.99990	0.99996	0.99999	0.99999	1.00000
5.6	0.99995	0.99998	0.99999	1.00000	1.00000
5.8	0.99997	0.99999	1.00000	1.00000	1.00000
6.0	0.99998	0.99999	1.00000	1.00000	1.00000

TABLE 7.6.6. *CONFIDENCE LIMITS ON THE PROPORTION IN ONE TAIL.*

ETA = 0.990 (*CONTINUED*)

K ↓	N = 120	N = 240	N = 600	N = 1000	N = 1200
-3.0	0.00023	0.00040	0.00063	0.00076	0.00080
-2.8	0.00053	0.00086	0.00130	0.00152	0.00159
-2.6	0.00116	0.00178	0.00256	0.00294	0.00307
-2.4	0.00240	0.00351	0.00484	0.00546	0.00566
-2.2	0.00475	0.00660	0.00876	0.00974	0.01006
-2.0	0.00893	0.01190	0.01521	0.01669	0.01716
-1.8	0.01601	0.02052	0.02537	0.02749	0.02815
-1.6	0.02739	0.03388	0.04064	0.04354	0.04444
-1.4	0.04471	0.05358	0.06257	0.06636	0.06753
-1.2	0.06972	0.08127	0.09269	0.09742	0.09887
-1.0	0.10396	0.11831	0.13218	0.13786	0.13959
-0.8	0.14845	0.16551	0.18170	0.18824	0.19023
-0.6	0.20332	0.22283	0.24103	0.24832	0.25052
-0.4	0.26763	0.28921	0.30905	0.31691	0.31928
-0.2	0.33941	0.36262	0.38368	0.39195	0.39443
0.0	0.41591	0.44032	0.46217	0.47068	0.47323
0.2	0.49406	0.51919	0.54142	0.55000	0.55256
0.4	0.57084	0.59621	0.61835	0.62683	0.62935
0.6	0.64365	0.66871	0.69027	0.69846	0.70089
0.8	0.71047	0.73463	0.75511	0.76280	0.76508
1.0	0.76992	0.79260	0.81150	0.81853	0.82060
1.2	0.82126	0.84194	0.85886	0.86507	0.86689
1.4	0.86432	0.88261	0.89727	0.90258	0.90412
1.6	0.89942	0.91509	0.92736	0.93173	0.93300
1.8	0.92721	0.94021	0.95013	0.95360	0.95461
2.0	0.94860	0.95904	0.96678	0.96943	0.97019
2.2	0.96460	0.97270	0.97853	0.98049	0.98104
2.4	0.97622	0.98232	0.98655	0.98794	0.98833
2.6	0.98443	0.98887	0.99183	0.99278	0.99304
2.8	0.99007	0.99319	0.99519	0.99582	0.99599
3.0	0.99383	0.99595	0.99726	0.99765	0.99776
3.2	0.99626	0.99766	0.99849	0.99873	0.99879
3.4	0.99780	0.99869	0.99919	0.99933	0.99937
3.6	0.99874	0.99929	0.99958	0.99966	0.99968
3.8	0.99929	0.99962	0.99979	0.99983	0.99985
4.0	0.99962	0.99981	0.99990	0.99992	0.99993
4.2	0.99980	0.99990	0.99995	0.99996	0.99997
4.4	0.99990	0.99995	0.99998	0.99998	0.99999
4.6	0.99995	0.99998	0.99999	0.99999	0.99999
4.8	0.99997	0.99999	1.00000	1.00000	1.00000
5.0	0.99999	1.00000	1.00000	1.00000	1.00000
5.2	0.99999	1.00000	1.00000	1.00000	1.00000
5.4	1.00000	1.00000	1.00000	1.00000	1.00000
5.6	1.00000	1.00000	1.00000	1.00000	1.00000
5.8	1.00000	1.00000	1.00000	1.00000	1.00000
6.0	1.00000	1.00000	1.00000	1.00000	1.00000

TABLE 7.7.1. CONFIDENCE LIMITS ON THE PROPORTION IN ONE TAIL.

ETA = 0.995

K ↓	N = 2	N = 3	N = 4	N = 5	N = 6
-3.0	0.00000	0.00000	0.00000	0.00000	0.00000
-2.8	0.00000	0.00000	0.00000	0.00000	0.00000
-2.6	0.00000	0.00000	0.00000	0.00000	0.00000
-2.4	0.00000	0.00000	0.00000	0.00000	0.00000
-2.2	0.00000	0.00000	0.00000	0.00000	0.00001
-2.0	0.00000	0.00000	0.00000	0.00002	0.00004
-1.8	0.00000	0.00000	0.00002	0.00008	0.00016
-1.6	0.00000	0.00002	0.00011	0.00029	0.00056
-1.4	0.00001	0.00012	0.00045	0.00099	0.00169
-1.2	0.00005	0.00052	0.00153	0.00291	0.00450
-1.0	0.00029	0.00187	0.00442	0.00744	0.01061
-0.8	0.00136	0.00553	0.01092	0.01661	0.02219
-0.6	0.00464	0.01341	0.02312	0.03256	0.04134
-0.4	0.01163	0.02699	0.04238	0.05650	0.06913
-0.2	0.02218	0.04601	0.06831	0.08789	0.10488
0.0	0.03427	0.06849	0.09889	0.12467	0.14650
0.2	0.04598	0.09216	0.13172	0.16436	0.19142
0.4	0.05658	0.11554	0.16499	0.20496	0.23752
0.6	0.06604	0.13800	0.19771	0.24523	0.28341
0.8	0.07456	0.15936	0.22940	0.28452	0.32827
1.0	0.08232	0.17964	0.25992	0.32250	0.37166
1.2	0.08946	0.19894	0.28925	0.35908	0.41338
1.4	0.09610	0.21735	0.31742	0.39420	0.45332
1.6	0.10232	0.23496	0.34449	0.42790	0.49146
1.8	0.10819	0.25187	0.37054	0.46019	0.52780
2.0	0.11376	0.26813	0.39561	0.49113	0.56235
2.2	0.11906	0.28382	0.41978	0.52074	0.59513
2.4	0.12414	0.29898	0.44308	0.54908	0.62619
2.6	0.12901	0.31366	0.46557	0.57619	0.65556
2.8	0.13369	0.32789	0.48728	0.60209	0.68327
3.0	0.13821	0.34171	0.50826	0.62683	0.70937
3.2	0.14258	0.35515	0.52852	0.65044	0.73389
3.4	0.14682	0.36823	0.54811	0.67295	0.75689
3.6	0.15093	0.38098	0.56705	0.69438	0.77840
3.8	0.15493	0.39341	0.58536	0.71478	0.79847
4.0	0.15882	0.40554	0.60305	0.73416	0.81716
4.2	0.16261	0.41739	0.62016	0.75255	0.83452
4.4	0.16630	0.42897	0.63670	0.76999	0.85059
4.6	0.16992	0.44030	0.65269	0.78650	0.86544
4.8	0.17345	0.45138	0.66813	0.80210	0.87911
5.0	0.17691	0.46223	0.68305	0.81683	0.89168
5.2	0.18029	0.47285	0.69745	0.83071	0.90318
5.4	0.18361	0.48327	0.71136	0.84378	0.91370
5.6	0.18687	0.49347	0.72478	0.85605	0.92327
5.8	0.19006	0.50347	0.73773	0.86757	0.93197
6.0	0.19320	0.51329	0.75021	0.87835	0.93984

TABLE 7.7.2. CONFIDENCE LIMITS ON THE PROPORTION IN ONE TAIL.

ETA = 0.995 (CONTINUED)

K ↓	N = 7	N = 8	N = 9	N = 10	N = 11
-3.0	0.00000	0.00000	0.00000	0.00000	0.00000
-2.8	0.00000	0.00000	0.00000	0.00000	0.00000
-2.6	0.00000	0.00000	0.00000	0.00001	0.00001
-2.4	0.00000	0.00001	0.00001	0.00002	0.00003
-2.2	0.00002	0.00004	0.00006	0.00008	0.00011
-2.0	0.00008	0.00013	0.00019	0.00027	0.00035
-1.8	0.00028	0.00043	0.00060	0.00079	0.00099
-1.6	0.00089	0.00127	0.00168	0.00211	0.00256
-1.4	0.00248	0.00334	0.00422	0.00512	0.00602
-1.2	0.00618	0.00790	0.00961	0.01128	0.01291
-1.0	0.01378	0.01686	0.01982	0.02264	0.02532
-0.8	0.02750	0.03248	0.03713	0.04146	0.04550
-0.6	0.04939	0.05673	0.06343	0.06956	0.07519
-0.4	0.08037	0.09041	0.09941	0.10753	0.11489
-0.2	0.11968	0.13265	0.14411	0.15433	0.16350
0.0	0.16513	0.18123	0.19528	0.20767	0.21869
0.2	0.21414	0.23351	0.25023	0.26483	0.27772
0.4	0.26449	0.28720	0.30661	0.32342	0.33814
0.6	0.31464	0.34067	0.36271	0.38165	0.39812
0.8	0.36367	0.39288	0.41741	0.43832	0.45639
1.0	0.41103	0.44321	0.47001	0.49269	0.51214
1.2	0.45644	0.49131	0.52010	0.54426	0.56485
1.4	0.49973	0.53696	0.56741	0.59278	0.61422
1.6	0.54084	0.58005	0.61182	0.63806	0.66006
1.8	0.57973	0.62053	0.65326	0.68003	0.70228
2.0	0.61641	0.65839	0.69170	0.71867	0.74088
2.2	0.65089	0.69364	0.72718	0.75402	0.77591
2.4	0.68320	0.72633	0.75972	0.78614	0.80745
2.6	0.71338	0.75649	0.78942	0.81514	0.83563
2.8	0.74148	0.78421	0.81637	0.84114	0.86063
3.0	0.76756	0.80957	0.84069	0.86430	0.88263
3.2	0.79167	0.83265	0.86249	0.88479	0.90183
3.4	0.81390	0.85357	0.88194	0.90278	0.91847
3.6	0.83431	0.87243	0.89918	0.91848	0.93277
3.8	0.85297	0.88935	0.91437	0.93208	0.94496
4.0	0.86999	0.90446	0.92766	0.94377	0.95527
4.2	0.88544	0.91788	0.93924	0.95374	0.96391
4.4	0.89941	0.92975	0.94924	0.96220	0.97110
4.6	0.91199	0.94017	0.95784	0.96932	0.97703
4.8	0.92328	0.94929	0.96518	0.97526	0.98188
5.0	0.93337	0.95723	0.97141	0.98018	0.98581
5.2	0.94234	0.96409	0.97666	0.98424	0.98898
5.4	0.95029	0.97000	0.98106	0.98754	0.99150
5.6	0.95731	0.97506	0.98472	0.99023	0.99350
5.8	0.96348	0.97937	0.98775	0.99238	0.99507
6.0	0.96888	0.98302	0.99023	0.99410	0.99628

TABLE 7.7.3. *CONFIDENCE LIMITS ON THE PROPORTION IN ONE TAIL.*

ETA = 0.995 *(CONTINUED)*

K ↓	N = 12	N = 13	N = 14	N = 15	N = 16
-3.0	0.00000	0.00000	0.00000	0.00000	0.00000
-2.8	0.00000	0.00000	0.00001	0.00001	0.00001
-2.6	0.00001	0.00002	0.00002	0.00003	0.00003
-2.4	0.00004	0.00006	0.00007	0.00009	0.00011
-2.2	0.00015	0.00018	0.00022	0.00027	0.00031
-2.0	0.00044	0.00053	0.00063	0.00074	0.00084
-1.8	0.00120	0.00142	0.00165	0.00187	0.00210
-1.6	0.00302	0.00347	0.00393	0.00439	0.00484
-1.4	0.00691	0.00778	0.00864	0.00947	0.01028
-1.2	0.01449	0.01600	0.01746	0.01886	0.02021
-1.0	0.02786	0.03027	0.03255	0.03472	0.03678
-0.8	0.04927	0.05279	0.05609	0.05919	0.06210
-0.6	0.08038	0.08517	0.08962	0.09375	0.09762
-0.4	0.12160	0.12775	0.13340	0.13863	0.14348
-0.2	0.17178	0.17931	0.18620	0.19252	0.19836
0.0	0.22857	0.23749	0.24559	0.25300	0.25980
0.2	0.28919	0.29948	0.30878	0.31724	0.32497
0.4	0.35116	0.36278	0.37322	0.38267	0.39127
0.6	0.41259	0.42543	0.43692	0.44726	0.45664
0.8	0.47217	0.48609	0.49848	0.50958	0.51961
1.0	0.52903	0.54384	0.55696	0.56866	0.57918
1.2	0.58261	0.59810	0.61174	0.62385	0.63468
1.4	0.63259	0.64851	0.66245	0.67476	0.68572
1.6	0.67877	0.69488	0.70889	0.72119	0.73208
1.8	0.72106	0.73710	0.75097	0.76306	0.77370
2.0	0.75946	0.77521	0.78871	0.80041	0.81065
2.2	0.79404	0.80928	0.82224	0.83339	0.84307
2.4	0.82492	0.83946	0.85172	0.86218	0.87119
2.6	0.85225	0.86595	0.87738	0.88705	0.89532
2.8	0.87625	0.88897	0.89950	0.90831	0.91578
3.0	0.89713	0.90881	0.91836	0.92628	0.93292
3.2	0.91514	0.92573	0.93428	0.94130	0.94713
3.4	0.93055	0.94002	0.94758	0.95372	0.95876
3.6	0.94360	0.95198	0.95858	0.96387	0.96817
3.8	0.95457	0.96189	0.96758	0.97208	0.97569
4.0	0.96370	0.97002	0.97486	0.97864	0.98163
4.2	0.97123	0.97662	0.98070	0.98382	0.98627
4.4	0.97738	0.98194	0.98532	0.98788	0.98985
4.6	0.98237	0.98617	0.98894	0.99101	0.99258
4.8	0.98637	0.98950	0.99175	0.99340	0.99464
5.0	0.98955	0.99211	0.99391	0.99521	0.99616
5.2	0.99206	0.99412	0.99555	0.99656	0.99729
5.4	0.99401	0.99566	0.99678	0.99755	0.99810
5.6	0.99553	0.99683	0.99769	0.99828	0.99869
5.8	0.99669	0.99770	0.99836	0.99880	0.99910
6.0	0.99757	0.99835	0.99885	0.99918	0.99939

TABLE 7.7.4. CONFIDENCE LIMITS ON THE PROPORTION IN ONE TAIL.

ETA = 0.995 (CONTINUED)

K ↓	N = 17	N = 18	N = 21	N = 24	N = 27
-3.0	0.00000	0.00000	0.00001	0.00001	0.00001
-2.8	0.00001	0.00001	0.00002	0.00003	0.00004
-2.6	0.00004	0.00005	0.00007	0.00010	0.00013
-2.4	0.00012	0.00014	0.00021	0.00027	0.00034
-2.2	0.00036	0.00040	0.00055	0.00071	0.00087
-2.0	0.00095	0.00106	0.00139	0.00172	0.00204
-1.8	0.00233	0.00256	0.00323	0.00388	0.00450
-1.6	0.00528	0.00572	0.00697	0.00815	0.00925
-1.4	0.01107	0.01184	0.01401	0.01599	0.01780
-1.2	0.02151	0.02275	0.02620	0.02927	0.03204
-1.0	0.03873	0.04060	0.04568	0.05014	0.05408
-0.8	0.06485	0.06744	0.07445	0.08048	0.08574
-0.6	0.10124	0.10463	0.11370	0.12140	0.12806
-0.4	0.14800	0.15221	0.16337	0.17274	0.18077
-0.2	0.20376	0.20879	0.22199	0.23297	0.24229
0.0	0.26607	0.27188	0.28703	0.29952	0.31005
0.2	0.33207	0.33862	0.35559	0.36947	0.38109
0.4	0.39913	0.40636	0.42498	0.44009	0.45265
0.6	0.46518	0.47301	0.49303	0.50915	0.52247
0.8	0.52871	0.53702	0.55814	0.57500	0.58883
1.0	0.58869	0.59733	0.61918	0.63647	0.65054
1.2	0.64443	0.65326	0.67542	0.69280	0.70683
1.4	0.69554	0.70439	0.72644	0.74355	0.75726
1.6	0.74179	0.75051	0.77203	0.78855	0.80167
1.8	0.78314	0.79157	0.81220	0.82784	0.84013
2.0	0.81967	0.82768	0.84709	0.86161	0.87289
2.2	0.85154	0.85903	0.87697	0.89019	0.90034
2.4	0.87903	0.88591	0.90220	0.91402	0.92297
2.6	0.90245	0.90866	0.92320	0.93357	0.94131
2.8	0.92217	0.92769	0.94044	0.94937	0.95593
3.0	0.93856	0.94339	0.95439	0.96194	0.96740
3.2	0.95203	0.95619	0.96552	0.97179	0.97624
3.4	0.96295	0.96648	0.97426	0.97938	0.98294
3.6	0.97170	0.97465	0.98104	0.98514	0.98794
3.8	0.97863	0.98105	0.98622	0.98945	0.99160
4.0	0.98404	0.98601	0.99011	0.99261	0.99424
4.2	0.98822	0.98979	0.99300	0.99490	0.99612
4.4	0.99140	0.99263	0.99511	0.99653	0.99742
4.6	0.99380	0.99475	0.99663	0.99768	0.99831
4.8	0.99558	0.99631	0.99771	0.99847	0.99891
5.0	0.99688	0.99743	0.99846	0.99900	0.99931
5.2	0.99783	0.99824	0.99898	0.99936	0.99957
5.4	0.99851	0.99880	0.99934	0.99960	0.99974
5.6	0.99898	0.99920	0.99957	0.99975	0.99984
5.8	0.99932	0.99947	0.99973	0.99985	0.99991
6.0	0.99955	0.99965	0.99983	0.99991	0.99994

TABLE 7.7.5. CONFIDENCE LIMITS ON THE PROPORTION IN ONE TAIL.

ETA = 0.995 (CONTINUED)

K ↓	N = 30	N = 40	N = 60	N = 80	N = 100
-3.0	0.00002	0.00004	0.00008	0.00012	0.00015
-2.8	0.00006	0.00010	0.00020	0.00029	0.00037
-2.6	0.00016	0.00026	0.00048	0.00067	0.00084
-2.4	0.00042	0.00066	0.00110	0.00148	0.00180
-2.2	0.00102	0.00153	0.00240	0.00311	0.00369
-2.0	0.00236	0.00334	0.00494	0.00618	0.00717
-1.8	0.00509	0.00685	0.00960	0.01165	0.01326
-1.6	0.01028	0.01325	0.01768	0.02086	0.02330
-1.4	0.01946	0.02413	0.03082	0.03547	0.03896
-1.2	0.03454	0.04140	0.05090	0.05733	0.06206
-1.0	0.05759	0.06706	0.07978	0.08817	0.09427
-0.8	0.09039	0.10271	0.11884	0.12926	0.13673
-0.6	0.13388	0.14909	0.16858	0.18096	0.18973
-0.4	0.18774	0.20572	0.22835	0.24249	0.25242
-0.2	0.25034	0.27087	0.29630	0.31197	0.32287
0.0	0.31908	0.34190	0.36974	0.38668	0.39836
0.2	0.39099	0.41581	0.44563	0.46355	0.47582
0.4	0.46330	0.48973	0.52106	0.53964	0.55226
0.6	0.53369	0.56129	0.59353	0.61241	0.62512
0.8	0.60042	0.62866	0.66113	0.67989	0.69241
1.0	0.66227	0.69054	0.72253	0.74073	0.75277
1.2	0.71845	0.74614	0.77690	0.79413	0.80540
1.4	0.76852	0.79504	0.82392	0.83980	0.85006
1.6	0.81235	0.83717	0.86363	0.87788	0.88697
1.8	0.85004	0.87276	0.89639	0.90884	0.91667
2.0	0.88190	0.90223	0.92281	0.93339	0.93993
2.2	0.90837	0.92614	0.94363	0.95237	0.95769
2.4	0.92996	0.94516	0.95965	0.96669	0.97088
2.6	0.94728	0.95999	0.97171	0.97721	0.98043
2.8	0.96093	0.97133	0.98057	0.98476	0.98716
3.0	0.97149	0.97982	0.98693	0.99004	0.99177
3.2	0.97953	0.98605	0.99139	0.99364	0.99485
3.4	0.98553	0.99053	0.99445	0.99603	0.99686
3.6	0.98994	0.99369	0.99650	0.99758	0.99813
3.8	0.99312	0.99588	0.99784	0.99856	0.99892
4.0	0.99537	0.99735	0.99869	0.99916	0.99939
4.2	0.99693	0.99833	0.99923	0.99952	0.99966
4.4	0.99800	0.99897	0.99955	0.99974	0.99982
4.6	0.99872	0.99938	0.99975	0.99986	0.99990
4.8	0.99919	0.99963	0.99986	0.99992	0.99995
5.0	0.99950	0.99978	0.99992	0.99996	0.99998
5.2	0.99970	0.99988	0.99996	0.99998	0.99999
5.4	0.99982	0.99993	0.99998	0.99999	0.99999
5.6	0.99989	0.99996	0.99999	1.00000	1.00000
5.8	0.99994	0.99998	0.99999	1.00000	1.00000
6.0	0.99996	0.99999	1.00000	1.00000	1.00000

TABLE 7.7.6. CONFIDENCE LIMITS ON THE PROPORTION IN ONE TAIL.

ETA = 0.995 (CONTINUED)

K ↓	N = 120	N = 240	N = 600	N = 1000	N = 1200
-3.0	0.00019	0.00035	0.00058	0.00071	0.00075
-2.8	0.00044	0.00076	0.00121	0.00144	0.00151
-2.6	0.00098	0.00159	0.00240	0.00280	0.00293
-2.4	0.00208	0.00318	0.00456	0.00522	0.00544
-2.2	0.00418	0.00606	0.00831	0.00937	0.00970
-2.0	0.00800	0.01105	0.01454	0.01613	0.01663
-1.8	0.01456	0.01924	0.02440	0.02668	0.02740
-1.6	0.02524	0.03206	0.03930	0.04244	0.04342
-1.4	0.04171	0.05113	0.06081	0.06494	0.06621
-1.2	0.06575	0.07811	0.09047	0.09564	0.09723
-1.0	0.09896	0.11442	0.12952	0.13574	0.13764
-0.8	0.14243	0.16093	0.17862	0.18580	0.18798
-0.6	0.19637	0.21763	0.23759	0.24561	0.24803
-0.4	0.25989	0.28351	0.30533	0.31399	0.31661
-0.2	0.33102	0.35654	0.37975	0.38889	0.39164
0.0	0.40705	0.43397	0.45813	0.46754	0.47036
0.2	0.48488	0.51271	0.53733	0.54684	0.54968
0.4	0.56153	0.58972	0.61431	0.62372	0.62652
0.6	0.63441	0.66235	0.68637	0.69547	0.69817
0.8	0.70151	0.72855	0.75143	0.76000	0.76253
1.0	0.76144	0.78695	0.80813	0.81598	0.81828
1.2	0.81346	0.83684	0.85587	0.86282	0.86486
1.4	0.85735	0.87815	0.89471	0.90066	0.90240
1.6	0.89336	0.91131	0.92524	0.93016	0.93159
1.8	0.92211	0.93711	0.94843	0.95236	0.95349
2.0	0.94443	0.95658	0.96547	0.96849	0.96935
2.2	0.96129	0.97082	0.97756	0.97979	0.98043
2.4	0.97369	0.98092	0.98585	0.98745	0.98789
2.6	0.98255	0.98787	0.99135	0.99245	0.99275
2.8	0.98871	0.99250	0.99487	0.99560	0.99580
3.0	0.99287	0.99549	0.99706	0.99752	0.99764
3.2	0.99562	0.99737	0.99836	0.99865	0.99872
3.4	0.99737	0.99851	0.99912	0.99928	0.99933
3.6	0.99846	0.99918	0.99954	0.99963	0.99966
3.8	0.99913	0.99956	0.99977	0.99982	0.99983
4.0	0.99951	0.99977	0.99989	0.99991	0.99992
4.2	0.99974	0.99988	0.99995	0.99996	0.99996
4.4	0.99986	0.99994	0.99998	0.99998	0.99998
4.6	0.99993	0.99997	0.99999	0.99999	0.99999
4.8	0.99996	0.99999	1.00000	1.00000	1.00000
5.0	0.99998	0.99999	1.00000	1.00000	1.00000
5.2	0.99999	1.00000	1.00000	1.00000	1.00000
5.4	1.00000	1.00000	1.00000	1.00000	1.00000
5.6	1.00000	1.00000	1.00000	1.00000	1.00000
5.8	1.00000	1.00000	1.00000	1.00000	1.00000
6.0	1.00000	1.00000	1.00000	1.00000	1.00000

PART B

TABLES FOR SCREENING PROCEDURES BASED
ON THE BIVARIATE DISTRIBUTION

TABLE 8

Screening proportion for normal conditioned on normal distribution for
$\rho = 0.35(0.05)1.00$.

TABLE 8.1. SCREENING PROPORTION FOR NORMAL CONDITIONED ON NORMAL.

DELTA = 0.950

γ \ ρ →	0.35	0.40	0.45	0.50	0.55	0.60	0.65
0.71	0.0058	0.0199	0.0474	0.0889	0.1427	0.2056	0.2744
0.72	0.0075	0.0242	0.0551	0.1002	0.1572	0.2226	0.2930
0.73	0.0096	0.0293	0.0640	0.1129	0.1731	0.2408	0.3125
0.74	0.0123	0.0355	0.0743	0.1271	0.1905	0.2603	0.3332
0.75	0.0158	0.0429	0.0860	0.1429	0.2093	0.2812	0.3550
0.76	0.0203	0.0517	0.0995	0.1604	0.2299	0.3035	0.3780
0.77	0.0259	0.0622	0.1150	0.1800	0.2522	0.3273	0.4022
0.78	0.0329	0.0748	0.1327	0.2016	0.2764	0.3527	0.4276
0.79	0.0419	0.0897	0.1529	0.2257	0.3026	0.3798	0.4543
0.80	0.0531	0.1073	0.1759	0.2523	0.3311	0.4086	0.4823
0.81	0.0671	0.1283	0.2021	0.2816	0.3618	0.4392	0.5117
0.82	0.0847	0.1530	0.2318	0.3141	0.3950	0.4716	0.5425
0.83	0.1065	0.1821	0.2654	0.3497	0.4307	0.5060	0.5746
0.84	0.1335	0.2162	0.3034	0.3889	0.4691	0.5423	0.6081
0.85	0.1669	0.2561	0.3462	0.4318	0.5103	0.5806	0.6429
0.86	0.2078	0.3024	0.3941	0.4786	0.5542	0.6209	0.6789
0.87	0.2576	0.3561	0.4476	0.5294	0.6010	0.6629	0.7162
0.88	0.3179	0.4177	0.5068	0.5841	0.6504	0.7067	0.7544
0.89	0.3898	0.4878	0.5719	0.6428	0.7022	0.7519	0.7933
0.90	0.4746	0.5666	0.6425	0.7049	0.7560	0.7981	0.8327
0.91	0.5725	0.6535	0.7180	0.7696	0.8111	0.8446	0.8718
0.92	0.6823	0.7469	0.7967	0.8356	0.8662	0.8905	0.9100
0.93	0.7997	0.8430	0.8754	0.9001	0.9192	0.9341	0.9459
0.94	0.9145	0.9339	0.9480	0.9586	0.9665	0.9727	0.9774

γ \ ρ →	0.70	0.75	0.80	0.85	0.90	0.95	1.00
0.71	0.3461	0.4184	0.4897	0.5588	0.6253	0.6894	0.7474
0.72	0.3653	0.4376	0.5081	0.5760	0.6408	0.7028	0.7579
0.73	0.3854	0.4574	0.5270	0.5934	0.6564	0.7162	0.7684
0.74	0.4064	0.4778	0.5463	0.6112	0.6723	0.7297	0.7789
0.75	0.4282	0.4989	0.5661	0.6292	0.6882	0.7432	0.7895
0.76	0.4509	0.5206	0.5863	0.6476	0.7043	0.7567	0.8000
0.77	0.4745	0.5430	0.6070	0.6661	0.7205	0.7703	0.8105
0.78	0.4991	0.5661	0.6281	0.6849	0.7368	0.7839	0.8211
0.79	0.5246	0.5898	0.6496	0.7040	0.7532	0.7975	0.8316
0.80	0.5511	0.6142	0.6715	0.7232	0.7696	0.8110	0.8421
0.81	0.5785	0.6392	0.6938	0.7427	0.7862	0.8246	0.8526
0.82	0.6069	0.6648	0.7165	0.7623	0.8028	0.8381	0.8632
0.83	0.6362	0.6910	0.7395	0.7821	0.8194	0.8516	0.8737
0.84	0.6664	0.7178	0.7628	0.8020	0.8360	0.8651	0.8842
0.85	0.6975	0.7450	0.7863	0.8220	0.8526	0.8784	0.8947
0.86	0.7293	0.7727	0.8100	0.8419	0.8691	0.8917	0.9053
0.87	0.7618	0.8007	0.8338	0.8618	0.8855	0.9048	0.9158
0.88	0.7947	0.8288	0.8575	0.8816	0.9016	0.9177	0.9263
0.89	0.8280	0.8569	0.8811	0.9011	0.9176	0.9305	0.9368
0.90	0.8612	0.8848	0.9043	0.9202	0.9331	0.9430	0.9474
0.91	0.8940	0.9121	0.9268	0.9388	0.9482	0.9552	0.9579
0.92	0.9256	0.9382	0.9484	0.9565	0.9627	0.9671	0.9684
0.93	0.9552	0.9626	0.9684	0.9730	0.9763	0.9785	0.9789
0.94	0.9811	0.9840	0.9861	0.9878	0.9889	0.9894	0.9895

TABLE 8.2. *SCREENING PROPORTION FOR NORMAL CONDITIONED ON NORMAL.*

DELTA = 0.990

γ ρ →	0.35	0.40	0.45	0.50	0.55	0.60	0.65
0.75	0.0000	0.0003	0.0021	0.0082	0.0225	0.0491	0.0903
0.76	0.0000	0.0004	0.0027	0.0098	0.0261	0.0553	0.0995
0.77	0.0000	0.0006	0.0034	0.0118	0.0303	0.0623	0.1097
0.78	0.0001	0.0008	0.0043	0.0142	0.0351	0.0702	0.1208
0.79	0.0001	0.0011	0.0054	0.0172	0.0408	0.0792	0.1332
0.80	0.0002	0.0015	0.0068	0.0207	0.0473	0.0893	0.1468
0.81	0.0002	0.0020	0.0087	0.0250	0.0550	0.1007	0.1618
0.82	0.0004	0.0027	0.0110	0.0301	0.0639	0.1137	0.1784
0.83	0.0006	0.0038	0.0140	0.0365	0.0743	0.1283	0.1969
0.84	0.0009	0.0052	0.0179	0.0442	0.0866	0.1450	0.2173
0.85	0.0013	0.0071	0.0229	0.0536	0.1009	0.1641	0.2399
0.86	0.0020	0.0098	0.0294	0.0651	0.1178	0.1857	0.2651
0.87	0.0031	0.0136	0.0378	0.0793	0.1378	0.2105	0.2931
0.88	0.0048	0.0189	0.0488	0.0967	0.1614	0.2388	0.3244
0.89	0.0076	0.0265	0.0631	0.1184	0.1894	0.2714	0.3593
0.90	0.0120	0.0373	0.0821	0.1454	0.2227	0.3088	0.3984
0.91	0.0191	0.0528	0.1073	0.1791	0.2626	0.3520	0.4423
0.92	0.0307	0.0754	0.1408	0.2215	0.3105	0.4020	0.4916
0.93	0.0500	0.1085	0.1861	0.2750	0.3681	0.4600	0.5470
0.94	0.0825	0.1575	0.2474	0.3431	0.4379	0.5275	0.6095
0.95	0.1377	0.2309	0.3311	0.4300	0.5225	0.6060	0.6797
0.96	0.2332	0.3415	0.4459	0.5411	0.6249	0.6971	0.7583
0.97	0.3983	0.5081	0.6026	0.6819	0.7474	0.8010	0.8446
0.98	0.6732	0.7490	0.8071	0.8518	0.8863	0.9130	0.9337

γ ρ →	0.70	0.75	0.80	0.85	0.90	0.95	1.00
0.75	0.1471	0.2186	0.3029	0.3978	0.5014	0.6141	0.7576
0.76	0.1592	0.2331	0.3190	0.4145	0.5177	0.6286	0.7677
0.77	0.1723	0.2485	0.3359	0.4319	0.5343	0.6433	0.7778
0.78	0.1864	0.2649	0.3536	0.4499	0.5514	0.6582	0.7879
0.79	0.2017	0.2824	0.3723	0.4685	0.5689	0.6733	0.7980
0.80	0.2183	0.3010	0.3918	0.4879	0.5868	0.6886	0.8081
0.81	0.2362	0.3208	0.4124	0.5079	0.6052	0.7042	0.8182
0.82	0.2557	0.3420	0.4340	0.5287	0.6241	0.7199	0.8283
0.83	0.2768	0.3646	0.4567	0.5504	0.6434	0.7358	0.8384
0.84	0.2997	0.3887	0.4806	0.5728	0.6633	0.7519	0.8485
0.85	0.3247	0.4144	0.5058	0.5961	0.6836	0.7682	0.8586
0.86	0.3518	0.4420	0.5323	0.6203	0.7044	0.7847	0.8687
0.87	0.3814	0.4715	0.5602	0.6454	0.7258	0.8014	0.8788
0.88	0.4137	0.5030	0.5896	0.6715	0.7476	0.8183	0.8889
0.89	0.4490	0.5369	0.6206	0.6985	0.7700	0.8354	0.8990
0.90	0.4876	0.5732	0.6533	0.7266	0.7929	0.8526	0.9091
0.91	0.5298	0.6121	0.6877	0.7558	0.8163	0.8699	0.9192
0.92	0.5761	0.6539	0.7240	0.7860	0.8402	0.8874	0.9293
0.93	0.6270	0.6988	0.7622	0.8172	0.8645	0.9049	0.9394
0.94	0.6827	0.7469	0.8022	0.8494	0.8891	0.9224	0.9495
0.95	0.7436	0.7981	0.8441	0.8824	0.9139	0.9397	0.9596
0.96	0.8097	0.8522	0.8872	0.9157	0.9385	0.9567	0.9697
0.97	0.8798	0.9081	0.9306	0.9484	0.9624	0.9730	0.9798
0.98	0.9497	0.9620	0.9715	0.9787	0.9841	0.9880	0.9899

TABLE 8.3. SCREENING PROPORTION FOR NORMAL CONDITIONED ON NORMAL.

DELTA = 0.999

γ ρ →	0.35	0.40	0.45	0.50	0.55	0.60	0.65
0.76	0.0000	0.0000	0.0000	0.0001	0.0007	0.0034	0.0115
0.77	0.0000	0.0000	0.0000	0.0001	0.0009	0.0040	0.0132
0.78	0.0000	0.0000	0.0000	0.0002	0.0011	0.0048	0.0151
0.79	0.0000	0.0000	0.0000	0.0002	0.0014	0.0057	0.0174
0.80	0.0000	0.0000	0.0000	0.0003	0.0017	0.0068	0.0200
0.81	0.0000	0.0000	0.0000	0.0004	0.0021	0.0081	0.0231
0.82	0.0000	0.0000	0.0000	0.0005	0.0026	0.0097	0.0267
0.83	0.0000	0.0000	0.0001	0.0006	0.0033	0.0116	0.0309
0.84	0.0000	0.0000	0.0001	0.0009	0.0042	0.0140	0.0358
0.85	0.0000	0.0000	0.0001	0.0011	0.0053	0.0168	0.0417
0.86	0.0000	0.0000	0.0002	0.0016	0.0067	0.0204	0.0486
0.87	0.0000	0.0000	0.0003	0.0021	0.0086	0.0248	0.0569
0.88	0.0000	0.0000	0.0005	0.0029	0.0110	0.0303	0.0668
0.89	0.0000	0.0001	0.0007	0.0040	0.0143	0.0373	0.0787
0.90	0.0000	0.0001	0.0011	0.0057	0.0187	0.0462	0.0933
0.91	0.0000	0.0002	0.0018	0.0081	0.0246	0.0575	0.1111
0.92	0.0000	0.0004	0.0029	0.0117	0.0329	0.0724	0.1333
0.93	0.0001	0.0008	0.0048	0.0172	0.0446	0.0920	0.1612
0.94	0.0002	0.0016	0.0081	0.0259	0.0615	0.1185	0.1968
0.95	0.0004	0.0034	0.0143	0.0402	0.0867	0.1552	0.2433
0.96	0.0013	0.0076	0.0265	0.0649	0.1258	0.2077	0.3056
0.97	0.0043	0.0190	0.0529	0.1103	0.1899	0.2863	0.3922
0.98	0.0182	0.0549	0.1177	0.2033	0.3043	0.4121	0.5192
0.99	0.1104	0.2060	0.3167	0.4305	0.5389	0.6369	0.7221

γ ρ →	0.70	0.75	0.80	0.85	0.90	0.95	1.00
0.76	0.0302	0.0662	0.1258	0.2146	0.3365	0.4970	0.7608
0.77	0.0338	0.0724	0.1352	0.2270	0.3510	0.5119	0.7708
0.78	0.0378	0.0792	0.1453	0.2401	0.3662	0.5271	0.7808
0.79	0.0423	0.0868	0.1562	0.2540	0.3820	0.5427	0.7908
0.80	0.0474	0.0951	0.1679	0.2688	0.3985	0.5587	0.8008
0.81	0.0532	0.1043	0.1807	0.2844	0.4157	0.5752	0.8108
0.82	0.0598	0.1145	0.1945	0.3012	0.4337	0.5920	0.8208
0.83	0.0672	0.1258	0.2095	0.3189	0.4526	0.6094	0.8308
0.84	0.0758	0.1383	0.2258	0.3379	0.4723	0.6272	0.8408
0.85	0.0855	0.1523	0.2436	0.3582	0.4930	0.6456	0.8509
0.86	0.0967	0.1680	0.2631	0.3800	0.5148	0.6645	0.8609
0.87	0.1096	0.1856	0.2844	0.4033	0.5377	0.6839	0.8709
0.88	0.1246	0.2054	0.3079	0.4284	0.5618	0.7040	0.8809
0.89	0.1421	0.2279	0.3338	0.4554	0.5872	0.7247	0.8909
0.90	0.1625	0.2534	0.3625	0.4847	0.6141	0.7461	0.9009
0.91	0.1868	0.2826	0.3944	0.5164	0.6426	0.7682	0.9109
0.92	0.2157	0.3164	0.4302	0.5510	0.6728	0.7911	0.9209
0.93	0.2505	0.3556	0.4705	0.5890	0.7051	0.8148	0.9309
0.94	0.2930	0.4017	0.5163	0.6308	0.7397	0.8395	0.9409
0.95	0.3459	0.4566	0.5689	0.6771	0.7768	0.8651	0.9510
0.96	0.4130	0.5231	0.6300	0.7291	0.8169	0.8918	0.9610
0.97	0.5006	0.6055	0.7023	0.7879	0.8604	0.9196	0.9710
0.98	0.6201	0.7110	0.7897	0.8553	0.9079	0.9485	0.9810
0.99	0.7935	0.8518	0.8979	0.9332	0.9593	0.9779	0.9910

TABLE 9

Screening factors for normal conditioned on t-distribution for ρ = 0.70 (0.05)1.00 for degrees of freedom = 2(2)30, 40, 60(30)150, ∞ .

TABLE 9.1	δ = 0.900	γ = 0.74(0.01)0.87	*200*
TABLE 9.2	δ = 0.950	γ = 0.80(0.01)0.93	*207*
TABLE 9.3	δ = 0.990	γ = 0.85(0.01)0.98	*214*
TABLE 9.4	δ = 0.995	γ = 0.85(0.01)0.98	*221*

TABLE 9.1.1. SCREENING FACTORS FOR NORMAL CONDITIONED ON t-DISTRIBUTION.

2 DEGREES OF FREEDOM   DELTA = 0.90

ρ → γ ↓	0.70	0.75	0.80	0.85	0.90	0.95	1.00
0.74	0.324	0.455	0.576	0.691	0.801	0.910	1.019
0.75	0.390	0.518	0.637	0.750	0.859	0.967	1.076
0.76	0.459	0.584	0.701	0.813	0.922	1.029	1.137
0.77	0.532	0.654	0.770	0.881	0.989	1.096	1.204
0.78	0.610	0.730	0.844	0.954	1.062	1.169	1.277
0.79	0.693	0.811	0.925	1.035	1.143	1.250	1.358
0.80	0.784	0.901	1.014	1.125	1.233	1.341	1.450
0.81	0.884	1.001	1.115	1.225	1.335	1.444	1.554
0.82	0.997	1.115	1.229	1.341	1.452	1.563	1.675
0.83	1.128	1.246	1.362	1.477	1.590	1.704	1.818
0.84	1.282	1.404	1.522	1.640	1.757	1.874	1.992
0.85	1.473	1.599	1.722	1.844	1.966	2.089	2.213
0.86	1.721	1.854	1.984	2.114	2.244	2.375	2.507
0.87	2.071	2.214	2.357	2.499	2.642	2.785	2.931

4 DEGREES OF FREEDOM   DELTA = 0.90

ρ → γ ↓	0.70	0.75	0.80	0.85	0.90	0.95	1.00
0.74	0.314	0.444	0.565	0.678	0.785	0.887	0.986
0.75	0.380	0.506	0.624	0.735	0.840	0.940	1.037
0.76	0.448	0.571	0.686	0.794	0.898	0.996	1.091
0.77	0.519	0.639	0.752	0.858	0.959	1.056	1.150
0.78	0.595	0.712	0.822	0.926	1.025	1.120	1.212
0.79	0.675	0.789	0.897	0.999	1.097	1.190	1.281
0.80	0.761	0.873	0.979	1.079	1.175	1.267	1.356
0.81	0.855	0.964	1.068	1.166	1.261	1.351	1.439
0.82	0.959	1.066	1.167	1.264	1.357	1.446	1.533
0.83	1.075	1.180	1.280	1.375	1.467	1.555	1.640
0.84	1.208	1.312	1.410	1.504	1.595	1.682	1.765
0.85	1.366	1.468	1.565	1.659	1.748	1.834	1.917
0.86	1.560	1.661	1.758	1.851	1.940	2.025	2.107
0.87	1.814	1.915	2.012	2.105	2.194	2.280	2.362

6 DEGREES OF FREEDOM   DELTA = 0.90

ρ → γ ↓	0.70	0.75	0.80	0.85	0.90	0.95	1.00
0.74	0.311	0.441	0.561	0.674	0.780	0.881	0.973
0.75	0.376	0.502	0.620	0.730	0.834	0.932	1.022
0.76	0.444	0.567	0.681	0.789	0.891	0.986	1.074
0.77	0.515	0.635	0.746	0.851	0.950	1.043	1.129
0.78	0.590	0.706	0.815	0.918	1.014	1.105	1.188
0.79	0.669	0.783	0.889	0.988	1.082	1.171	1.251
0.80	0.754	0.865	0.968	1.065	1.157	1.242	1.321
0.81	0.846	0.953	1.054	1.148	1.238	1.321	1.397
0.82	0.947	1.051	1.149	1.241	1.327	1.408	1.481
0.83	1.059	1.160	1.255	1.344	1.428	1.506	1.577
0.84	1.186	1.284	1.376	1.463	1.544	1.619	1.687
0.85	1.333	1.428	1.517	1.601	1.680	1.753	1.818
0.86	1.511	1.603	1.690	1.771	1.846	1.916	1.978
0.87	1.738	1.827	1.911	1.989	2.061	2.128	2.186

TABLE 9.1.2. *SCREENING FACTORS FOR NORMAL CONDITIONED ON* t-*DISTRIBUTION.*

8 DEGREES OF FREEDOM   DELTA = 0.90

$\rho \rightarrow$ $\gamma \downarrow$	0.70	0.75	0.80	0.85	0.90	0.95	1.00
0.74	0.309	0.439	0.559	0.672	0.778	0.877	0.965
0.75	0.374	0.501	0.618	0.728	0.831	0.928	1.013
0.76	0.442	0.565	0.679	0.787	0.887	0.981	1.063
0.77	0.513	0.632	0.744	0.848	0.946	1.037	1.116
0.78	0.588	0.704	0.812	0.914	1.009	1.097	1.174
0.79	0.667	0.779	0.885	0.983	1.076	1.161	1.235
0.80	0.751	0.860	0.963	1.058	1.148	1.230	1.301
0.81	0.842	0.948	1.047	1.140	1.226	1.305	1.374
0.82	0.941	1.044	1.140	1.229	1.313	1.389	1.454
0.83	1.051	1.150	1.243	1.329	1.409	1.482	1.543
0.84	1.175	1.270	1.359	1.442	1.519	1.588	1.646
0.85	1.317	1.409	1.494	1.574	1.647	1.712	1.766
0.86	1.487	1.575	1.657	1.732	1.801	1.862	1.912
0.87	1.701	1.785	1.862	1.933	1.997	2.053	2.099

10 DEGREES OF FREEDOM   DELTA = 0.90

$\rho \rightarrow$ $\gamma \downarrow$	0.70	0.75	0.80	0.85	0.90	0.95	1.00
0.74	0.308	0.438	0.558	0.671	0.777	0.876	0.959
0.75	0.373	0.500	0.617	0.727	0.830	0.926	1.006
0.76	0.441	0.564	0.678	0.785	0.885	0.978	1.056
0.77	0.512	0.631	0.742	0.846	0.944	1.033	1.108
0.78	0.586	0.702	0.810	0.911	1.006	1.092	1.164
0.79	0.665	0.777	0.882	0.980	1.072	1.155	1.224
0.80	0.749	0.858	0.960	1.054	1.143	1.223	1.288
0.81	0.840	0.945	1.043	1.135	1.220	1.296	1.358
0.82	0.938	1.040	1.134	1.223	1.304	1.377	1.436
0.83	1.047	1.144	1.235	1.320	1.398	1.467	1.522
0.84	1.168	1.262	1.349	1.430	1.504	1.569	1.620
0.85	1.308	1.398	1.481	1.557	1.627	1.687	1.734
0.86	1.474	1.559	1.637	1.709	1.774	1.829	1.872
0.87	1.680	1.760	1.833	1.900	1.959	2.009	2.046

12 DEGREES OF FREEDOM   DELTA = 0.90

$\rho \rightarrow$ $\gamma \downarrow$	0.70	0.75	0.80	0.85	0.90	0.95	1.00
0.74	0.308	0.437	0.558	0.671	0.776	0.874	0.955
0.75	0.373	0.499	0.616	0.726	0.829	0.924	1.001
0.76	0.440	0.563	0.677	0.784	0.884	0.976	1.050
0.77	0.511	0.630	0.741	0.845	0.942	1.031	1.102
0.78	0.585	0.701	0.809	0.910	1.004	1.089	1.157
0.79	0.664	0.776	0.881	0.978	1.069	1.151	1.216
0.80	0.748	0.857	0.958	1.052	1.139	1.218	1.279
0.81	0.838	0.943	1.041	1.131	1.215	1.290	1.348
0.82	0.936	1.037	1.131	1.218	1.298	1.369	1.423
0.83	1.044	1.141	1.231	1.314	1.390	1.457	1.507
0.84	1.164	1.257	1.343	1.422	1.494	1.556	1.602
0.85	1.302	1.390	1.472	1.546	1.613	1.671	1.712
0.86	1.465	1.548	1.624	1.694	1.756	1.808	1.844
0.87	1.666	1.744	1.815	1.878	1.934	1.980	2.011

TABLE 9.1.3. SCREENING FACTORS FOR NORMAL CONDITIONED ON t-DISTRIBUTION.

14 DEGREES OF FREEDOM   DELTA = 0.90

ρ → γ ↓	0.70	0.75	0.80	0.85	0.90	0.95	1.00
0.74	0.307	0.437	0.557	0.670	0.776	0.874	0.951
0.75	0.372	0.498	0.616	0.725	0.828	0.923	0.998
0.76	0.440	0.562	0.677	0.783	0.883	0.975	1.046
0.77	0.510	0.630	0.741	0.844	0.941	1.029	1.097
0.78	0.585	0.700	0.808	0.909	1.002	1.087	1.152
0.79	0.663	0.775	0.880	0.977	1.067	1.148	1.210
0.80	0.747	0.855	0.956	1.050	1.137	1.214	1.272
0.81	0.837	0.941	1.039	1.129	1.212	1.285	1.339
0.82	0.934	1.035	1.128	1.215	1.294	1.363	1.413
0.83	1.041	1.138	1.227	1.310	1.385	1.449	1.496
0.84	1.161	1.253	1.338	1.416	1.487	1.547	1.589
0.85	1.298	1.385	1.465	1.539	1.604	1.659	1.696
0.86	1.458	1.540	1.615	1.683	1.743	1.792	1.824
0.87	1.656	1.732	1.801	1.863	1.916	1.959	1.985

16 DEGREES OF FREEDOM   DELTA = 0.90

ρ → γ ↓	0.70	0.75	0.80	0.85	0.90	0.95	1.00
0.74	0.307	0.436	0.557	0.670	0.775	0.873	0.949
0.75	0.372	0.498	0.615	0.725	0.828	0.922	0.994
0.76	0.439	0.562	0.676	0.783	0.883	0.973	1.043
0.77	0.510	0.629	0.740	0.844	0.940	1.028	1.094
0.78	0.584	0.700	0.808	0.908	1.001	1.085	1.147
0.79	0.663	0.775	0.879	0.976	1.066	1.146	1.205
0.80	0.746	0.855	0.955	1.049	1.135	1.211	1.266
0.81	0.836	0.940	1.037	1.127	1.210	1.282	1.333
0.82	0.933	1.034	1.127	1.213	1.291	1.359	1.406
0.83	1.040	1.136	1.225	1.307	1.381	1.444	1.487
0.84	1.159	1.250	1.335	1.412	1.481	1.540	1.578
0.85	1.294	1.381	1.461	1.533	1.597	1.650	1.683
0.86	1.453	1.534	1.609	1.675	1.733	1.780	1.809
0.87	1.649	1.724	1.791	1.851	1.903	1.943	1.966

18 DEGREES OF FREEDOM   DELTA = 0.90

ρ → γ ↓	0.70	0.75	0.80	0.85	0.90	0.95	1.00
0.74	0.307	0.436	0.557	0.669	0.775	0.872	0.946
0.75	0.371	0.498	0.615	0.725	0.827	0.921	0.992
0.76	0.439	0.562	0.676	0.783	0.882	0.973	1.040
0.77	0.510	0.629	0.740	0.843	0.940	1.027	1.090
0.78	0.584	0.699	0.807	0.907	1.000	1.084	1.144
0.79	0.662	0.774	0.878	0.975	1.065	1.144	1.201
0.80	0.746	0.854	0.954	1.048	1.134	1.209	1.262
0.81	0.835	0.940	1.036	1.126	1.208	1.279	1.328
0.82	0.932	1.032	1.125	1.211	1.288	1.355	1.400
0.83	1.039	1.134	1.223	1.304	1.377	1.440	1.480
0.84	1.157	1.248	1.332	1.409	1.477	1.534	1.570
0.85	1.292	1.378	1.457	1.528	1.591	1.643	1.674
0.86	1.450	1.530	1.603	1.669	1.726	1.771	1.797
0.87	1.643	1.717	1.784	1.843	1.893	1.931	1.951

TABLE 9.1.4. *SCREENING FACTORS FOR NORMAL CONDITIONED ON* t-*DISTRIBUTION.*

20 *DEGREES OF FREEDOM*   *DELTA* = 0.90

ρ → γ ↓	0.70	0.75	0.80	0.85	0.90	0.95	1.00
0.74	0.306	0.436	0.556	0.669	0.775	0.872	0.944
0.75	0.371	0.497	0.615	0.725	0.827	0.921	0.990
0.76	0.439	0.561	0.676	0.782	0.882	0.972	1.038
0.77	0.509	0.629	0.739	0.843	0.939	1.026	1.088
0.78	0.584	0.699	0.807	0.907	1.000	1.082	1.141
0.79	0.662	0.774	0.878	0.975	1.064	1.143	1.197
0.80	0.745	0.853	0.954	1.047	1.132	1.207	1.258
0.81	0.835	0.939	1.035	1.125	1.206	1.277	1.324
0.82	0.932	1.032	1.124	1.209	1.287	1.353	1.395
0.83	1.038	1.133	1.221	1.302	1.375	1.436	1.474
0.84	1.156	1.247	1.330	1.406	1.474	1.530	1.563
0.85	1.290	1.375	1.454	1.525	1.587	1.637	1.666
0.86	1.447	1.527	1.599	1.664	1.720	1.764	1.787
0.87	1.638	1.712	1.778	1.836	1.884	1.921	1.939

22 *DEGREES OF FREEDOM*   *DELTA* = 0.90

ρ → γ ↓	0.70	0.75	0.80	0.85	0.90	0.95	1.00
0.74	0.306	0.436	0.556	0.669	0.775	0.872	0.943
0.75	0.371	0.497	0.615	0.724	0.827	0.920	0.988
0.76	0.439	0.561	0.675	0.782	0.881	0.971	1.036
0.77	0.509	0.628	0.739	0.843	0.939	1.025	1.086
0.78	0.583	0.699	0.806	0.907	0.999	1.082	1.138
0.79	0.662	0.774	0.878	0.974	1.063	1.142	1.195
0.80	0.745	0.853	0.953	1.046	1.131	1.206	1.255
0.81	0.834	0.938	1.035	1.124	1.205	1.275	1.320
0.82	0.931	1.031	1.123	1.208	1.285	1.350	1.391
0.83	1.037	1.132	1.220	1.301	1.373	1.433	1.470
0.84	1.155	1.245	1.328	1.404	1.471	1.526	1.558
0.85	1.288	1.373	1.451	1.522	1.583	1.632	1.659
0.86	1.444	1.524	1.596	1.660	1.715	1.758	1.779
0.87	1.635	1.707	1.773	1.830	1.878	1.913	1.929

24 *DEGREES OF FREEDOM*   *DELTA* = 0.90

ρ → γ ↓	0.70	0.75	0.80	0.85	0.90	0.95	1.00
0.74	0.306	0.436	0.556	0.669	0.775	0.871	0.941
0.75	0.371	0.497	0.614	0.724	0.827	0.920	0.987
0.76	0.438	0.561	0.675	0.782	0.881	0.971	1.034
0.77	0.509	0.628	0.739	0.842	0.938	1.024	1.084
0.78	0.583	0.699	0.806	0.906	0.999	1.081	1.136
0.79	0.661	0.773	0.877	0.974	1.063	1.141	1.192
0.80	0.745	0.853	0.953	1.046	1.131	1.205	1.252
0.81	0.834	0.938	1.034	1.123	1.204	1.274	1.317
0.82	0.931	1.030	1.122	1.207	1.284	1.348	1.388
0.83	1.036	1.131	1.219	1.299	1.371	1.431	1.466
0.84	1.154	1.244	1.327	1.402	1.469	1.523	1.553
0.85	1.287	1.372	1.449	1.519	1.580	1.628	1.654
0.86	1.442	1.521	1.593	1.657	1.711	1.753	1.773
0.87	1.632	1.704	1.768	1.825	1.872	1.906	1.921

TABLE 9.1.5. *SCREENING FACTORS FOR NORMAL CONDITIONED ON t-DISTRIBUTION.*

26 DEGREES OF FREEDOM   DELTA = 0.90

ρ → γ ↓	0.70	0.75	0.80	0.85	0.90	0.95	1.00
0.74	0.306	0.435	0.556	0.669	0.774	0.871	0.940
0.75	0.371	0.497	0.614	0.724	0.827	0.920	0.985
0.76	0.438	0.561	0.675	0.782	0.881	0.971	1.032
0.77	0.509	0.628	0.739	0.842	0.938	1.024	1.082
0.78	0.583	0.698	0.806	0.906	0.998	1.080	1.134
0.79	0.661	0.773	0.877	0.973	1.062	1.140	1.190
0.80	0.744	0.852	0.953	1.045	1.130	1.204	1.250
0.81	0.834	0.938	1.034	1.123	1.203	1.272	1.314
0.82	0.930	1.030	1.122	1.206	1.283	1.347	1.385
0.83	1.036	1.131	1.218	1.298	1.370	1.429	1.462
0.84	1.153	1.243	1.326	1.401	1.467	1.521	1.549
0.85	1.286	1.370	1.448	1.517	1.578	1.625	1.649
0.86	1.441	1.519	1.591	1.654	1.708	1.748	1.767
0.87	1.629	1.701	1.765	1.821	1.867	1.900	1.914

28 DEGREES OF FREEDOM   DELTA = 0.90

ρ → γ ↓	0.70	0.75	0.80	0.85	0.90	0.95	1.00
0.74	0.306	0.435	0.556	0.669	0.774	0.871	0.939
0.75	0.371	0.497	0.614	0.724	0.826	0.920	0.984
0.76	0.438	0.561	0.675	0.782	0.881	0.970	1.031
0.77	0.509	0.628	0.739	0.842	0.938	1.024	1.081
0.78	0.583	0.698	0.806	0.906	0.998	1.080	1.133
0.79	0.661	0.773	0.877	0.973	1.062	1.139	1.188
0.80	0.744	0.852	0.952	1.045	1.130	1.203	1.248
0.81	0.833	0.937	1.033	1.122	1.202	1.271	1.312
0.82	0.930	1.029	1.121	1.206	1.282	1.345	1.382
0.83	1.035	1.130	1.218	1.297	1.368	1.427	1.459
0.84	1.152	1.242	1.325	1.400	1.465	1.518	1.546
0.85	1.285	1.369	1.446	1.516	1.576	1.622	1.645
0.86	1.439	1.518	1.588	1.651	1.705	1.745	1.762
0.87	1.627	1.698	1.762	1.817	1.863	1.895	1.908

30 DEGREES OF FREEDOM   DELTA = 0.90

ρ → γ ↓	0.70	0.75	0.80	0.85	0.90	0.95	1.00
0.74	0.306	0.435	0.556	0.669	0.774	0.871	0.938
0.75	0.371	0.497	0.614	0.724	0.826	0.919	0.983
0.76	0.438	0.561	0.675	0.781	0.881	0.970	1.030
0.77	0.509	0.628	0.739	0.842	0.938	1.023	1.079
0.78	0.583	0.698	0.806	0.905	0.998	1.079	1.131
0.79	0.661	0.773	0.876	0.973	1.061	1.139	1.187
0.80	0.744	0.852	0.952	1.045	1.129	1.202	1.246
0.81	0.833	0.937	1.033	1.122	1.202	1.270	1.310
0.82	0.929	1.029	1.121	1.205	1.281	1.344	1.380
0.83	1.035	1.130	1.217	1.297	1.367	1.426	1.457
0.84	1.152	1.241	1.324	1.398	1.464	1.516	1.543
0.85	1.284	1.368	1.445	1.514	1.574	1.620	1.641
0.86	1.438	1.516	1.587	1.649	1.702	1.742	1.758
0.87	1.625	1.696	1.759	1.814	1.860	1.891	1.903

TABLE 9.1.6. *SCREENING FACTORS FOR NORMAL CONDITIONED ON t-DISTRIBUTION.*

40 *DEGREES OF FREEDOM*   *DELTA* = 0.90

ρ → γ ↓	0.70	0.75	0.80	0.85	0.90	0.95	1.00
0.74	0.305	0.435	0.555	0.668	0.774	0.870	0.935
0.75	0.370	0.496	0.614	0.723	0.826	0.919	0.979
0.76	0.438	0.560	0.674	0.781	0.880	0.969	1.026
0.77	0.508	0.627	0.738	0.841	0.937	1.022	1.075
0.78	0.582	0.698	0.805	0.905	0.997	1.078	1.126
0.79	0.660	0.772	0.876	0.972	1.060	1.137	1.181
0.80	0.743	0.851	0.951	1.043	1.127	1.199	1.240
0.81	0.832	0.936	1.032	1.120	1.200	1.267	1.303
0.82	0.928	1.028	1.119	1.203	1.278	1.340	1.372
0.83	1.033	1.128	1.215	1.294	1.364	1.420	1.447
0.84	1.150	1.239	1.321	1.395	1.459	1.510	1.532
0.85	1.281	1.365	1.441	1.509	1.567	1.611	1.629
0.86	1.433	1.511	1.580	1.642	1.693	1.730	1.744
0.87	1.618	1.688	1.750	1.804	1.847	1.876	1.885

60 *DEGREES OF FREEDOM*   *DELTA* = 0.90

ρ → γ ↓	0.70	0.75	0.80	0.85	0.90	0.95	1.00
0.74	0.305	0.435	0.555	0.668	0.774	0.870	0.931
0.75	0.370	0.496	0.613	0.723	0.825	0.918	0.975
0.76	0.437	0.560	0.674	0.781	0.880	0.968	1.022
0.77	0.508	0.627	0.738	0.841	0.936	1.021	1.070
0.78	0.582	0.697	0.804	0.904	0.996	1.076	1.121
0.79	0.660	0.772	0.875	0.971	1.059	1.134	1.175
0.80	0.743	0.850	0.950	1.042	1.126	1.197	1.233
0.81	0.832	0.935	1.031	1.118	1.197	1.263	1.296
0.82	0.927	1.026	1.117	1.201	1.275	1.336	1.364
0.83	1.032	1.126	1.212	1.291	1.360	1.415	1.438
0.84	1.148	1.236	1.318	1.391	1.454	1.503	1.522
0.85	1.278	1.361	1.436	1.503	1.560	1.602	1.617
0.86	1.429	1.506	1.574	1.635	1.684	1.719	1.729
0.87	1.612	1.680	1.741	1.793	1.835	1.861	1.868

90 *DEGREES OF FREEDOM*   *DELTA* = 0.90

ρ → γ ↓	0.70	0.75	0.80	0.85	0.90	0.95	1.00
0.74	0.305	0.434	0.555	0.668	0.773	0.869	0.929
0.75	0.370	0.496	0.613	0.723	0.825	0.917	0.973
0.76	0.437	0.560	0.674	0.780	0.879	0.967	1.019
0.77	0.507	0.627	0.737	0.840	0.936	1.020	1.067
0.78	0.581	0.697	0.804	0.904	0.995	1.075	1.118
0.79	0.660	0.771	0.875	0.970	1.058	1.133	1.172
0.80	0.742	0.850	0.950	1.041	1.125	1.195	1.229
0.81	0.831	0.934	1.030	1.117	1.196	1.261	1.291
0.82	0.927	1.025	1.116	1.199	1.273	1.333	1.358
0.83	1.031	1.125	1.211	1.289	1.357	1.411	1.432
0.84	1.146	1.235	1.315	1.388	1.451	1.498	1.515
0.85	1.276	1.359	1.433	1.500	1.556	1.596	1.609
0.86	1.426	1.502	1.570	1.630	1.678	1.711	1.720
0.87	1.607	1.675	1.735	1.786	1.827	1.851	1.856

TABLE 9.1.7. SCREENING FACTORS FOR NORMAL CONDITIONED ON t-DISTRIBUTION.

120 *DEGREES OF FREEDOM*   *DELTA* = 0.90

ρ → γ ↓	0.70	0.75	0.80	0.85	0.90	0.95	1.00
0.74	0.305	0.434	0.555	0.668	0.773	0.869	0.927
0.75	0.369	0.496	0.613	0.723	0.825	0.917	0.971
0.76	0.437	0.560	0.674	0.780	0.879	0.967	1.017
0.77	0.507	0.626	0.737	0.840	0.936	1.019	1.065
0.78	0.581	0.697	0.804	0.903	0.995	1.074	1.116
0.79	0.659	0.771	0.874	0.970	1.058	1.132	1.170
0.80	0.742	0.850	0.949	1.041	1.124	1.194	1.227
0.81	0.831	0.934	1.029	1.117	1.195	1.260	1.289
0.82	0.926	1.025	1.116	1.199	1.272	1.331	1.356
0.83	1.030	1.124	1.210	1.288	1.356	1.409	1.429
0.84	1.146	1.234	1.314	1.387	1.449	1.496	1.511
0.85	1.275	1.357	1.432	1.498	1.554	1.593	1.605
0.86	1.425	1.500	1.568	1.627	1.675	1.707	1.715
0.87	1.605	1.673	1.732	1.783	1.822	1.846	1.851

150 *DEGREES OF FREEDOM*   *DELTA* = 0.90

ρ → γ ↓	0.70	0.75	0.80	0.85	0.90	0.95	1.00
0.74	0.305	0.434	0.555	0.668	0.773	0.869	0.927
0.75	0.369	0.495	0.613	0.723	0.825	0.917	0.971
0.76	0.437	0.559	0.674	0.780	0.879	0.967	1.016
0.77	0.507	0.626	0.737	0.840	0.935	1.019	1.064
0.78	0.581	0.697	0.804	0.903	0.995	1.074	1.115
0.79	0.659	0.771	0.874	0.970	1.057	1.132	1.169
0.80	0.742	0.850	0.949	1.041	1.124	1.193	1.226
0.81	0.831	0.934	1.029	1.117	1.195	1.259	1.287
0.82	0.926	1.025	1.115	1.198	1.272	1.330	1.354
0.83	1.030	1.124	1.210	1.287	1.355	1.408	1.427
0.84	1.145	1.233	1.314	1.386	1.448	1.494	1.509
0.85	1.274	1.357	1.431	1.497	1.552	1.592	1.603
0.86	1.424	1.499	1.567	1.626	1.673	1.705	1.712
0.87	1.604	1.671	1.731	1.781	1.820	1.843	1.847

∞ *DEGREES OF FREEDOM*   *DELTA* = 0.90

ρ → γ ↓	0.70	0.75	0.80	0.85	0.90	0.95	1.00
0.74	0.304	0.434	0.554	0.667	0.773	0.868	0.924
0.75	0.369	0.495	0.613	0.722	0.825	0.916	0.967
0.76	0.436	0.559	0.673	0.780	0.879	0.966	1.013
0.77	0.507	0.626	0.737	0.840	0.935	1.018	1.061
0.78	0.581	0.696	0.803	0.903	0.994	1.073	1.111
0.79	0.659	0.770	0.874	0.969	1.056	1.130	1.164
0.80	0.742	0.849	0.948	1.040	1.122	1.191	1.221
0.81	0.830	0.933	1.028	1.115	1.193	1.256	1.282
0.82	0.925	1.024	1.114	1.197	1.269	1.327	1.348
0.83	1.029	1.122	1.208	1.285	1.352	1.404	1.420
0.84	1.144	1.231	1.311	1.383	1.444	1.489	1.501
0.85	1.272	1.354	1.428	1.493	1.547	1.584	1.593
0.86	1.420	1.495	1.562	1.620	1.666	1.696	1.701
0.87	1.598	1.665	1.723	1.773	1.810	1.831	1.834

TABLE 9.2.1. *SCREENING FACTORS FOR NORMAL CONDITIONED ON* t-*DISTRIBUTION.*

2 DEGREES OF FREEDOM   DELTA = 0.95

ρ → γ ↓	0.70	0.75	0.80	0.85	0.90	0.95	1.00
0.80	0.140	0.304	0.453	0.591	0.722	0.850	0.975
0.81	0.212	0.370	0.514	0.649	0.778	0.903	1.027
0.82	0.287	0.438	0.578	0.710	0.837	0.961	1.083
0.83	0.364	0.510	0.647	0.776	0.902	1.024	1.145
0.84	0.445	0.587	0.721	0.848	0.972	1.094	1.214
0.85	0.532	0.670	0.801	0.927	1.050	1.171	1.291
0.86	0.627	0.762	0.891	1.016	1.138	1.259	1.378
0.87	0.731	0.864	0.992	1.116	1.239	1.360	1.479
0.88	0.850	0.981	1.109	1.233	1.356	1.478	1.599
0.89	0.988	1.119	1.247	1.373	1.498	1.621	1.745
0.90	1.156	1.288	1.418	1.547	1.674	1.802	1.928
0.91	1.370	1.507	1.641	1.774	1.907	2.040	2.173
0.92	1.668	1.812	1.955	2.098	2.240	2.383	2.526
0.93	2.142	2.303	2.463	2.624	2.785	2.946	3.109

4 DEGREES OF FREEDOM   DELTA = 0.95

ρ → γ ↓	0.70	0.75	0.80	0.85	0.90	0.95	1.00
0.80	0.134	0.297	0.448	0.590	0.726	0.859	0.992
0.81	0.205	0.363	0.510	0.648	0.781	0.911	1.042
0.82	0.279	0.432	0.574	0.709	0.840	0.967	1.095
0.83	0.356	0.504	0.643	0.775	0.902	1.027	1.152
0.84	0.438	0.581	0.716	0.845	0.970	1.092	1.214
0.85	0.524	0.663	0.795	0.921	1.043	1.163	1.283
0.86	0.618	0.753	0.881	1.005	1.124	1.242	1.359
0.87	0.720	0.852	0.977	1.098	1.215	1.331	1.446
0.88	0.834	0.962	1.085	1.203	1.319	1.432	1.545
0.89	0.965	1.089	1.209	1.325	1.439	1.551	1.662
0.90	1.117	1.239	1.357	1.472	1.584	1.694	1.803
0.91	1.304	1.424	1.541	1.654	1.765	1.874	1.982
0.92	1.548	1.667	1.783	1.896	2.006	2.115	2.223
0.93	1.903	2.023	2.140	2.254	2.366	2.476	2.586

6 DEGREES OF FREEDOM   DELTA = 0.95

ρ → γ ↓	0.70	0.75	0.80	0.85	0.90	0.95	1.00
0.80	0.132	0.295	0.447	0.590	0.729	0.865	1.001
0.81	0.203	0.361	0.508	0.649	0.784	0.917	1.049
0.82	0.276	0.430	0.573	0.710	0.842	0.972	1.101
0.83	0.354	0.502	0.642	0.775	0.904	1.030	1.156
0.84	0.435	0.579	0.715	0.845	0.971	1.094	1.216
0.85	0.522	0.662	0.794	0.920	1.043	1.163	1.282
0.86	0.615	0.751	0.879	1.003	1.122	1.239	1.354
0.87	0.717	0.849	0.973	1.093	1.209	1.323	1.435
0.88	0.830	0.957	1.079	1.195	1.308	1.418	1.526
0.89	0.958	1.081	1.198	1.312	1.421	1.528	1.632
0.90	1.106	1.225	1.339	1.449	1.555	1.658	1.759
0.91	1.284	1.399	1.510	1.616	1.719	1.819	1.915
0.92	1.512	1.623	1.730	1.833	1.932	2.028	2.120
0.93	1.831	1.939	2.042	2.141	2.237	2.329	2.417

TABLE 9.2.2. *SCREENING FACTORS FOR NORMAL CONDITIONED ON* t-*DISTRIBUTION*.

8 *DEGREES OF FREEDOM*   *DELTA* = 0.95

ρ → γ ↓	0.70	0.75	0.80	0.85	0.90	0.95	1.00
0.80	0.131	0.294	0.446	0.591	0.731	0.868	1.005
0.81	0.202	0.360	0.508	0.649	0.786	0.920	1.053
0.82	0.275	0.429	0.573	0.711	0.844	0.974	1.104
0.83	0.352	0.501	0.642	0.776	0.906	1.033	1.158
0.84	0.434	0.578	0.715	0.846	0.972	1.095	1.217
0.85	0.521	0.661	0.794	0.921	1.043	1.163	1.281
0.86	0.614	0.750	0.879	1.002	1.121	1.237	1.351
0.87	0.716	0.847	0.972	1.092	1.207	1.319	1.428
0.88	0.828	0.955	1.076	1.191	1.303	1.411	1.516
0.89	0.955	1.077	1.193	1.305	1.413	1.516	1.616
0.90	1.101	1.218	1.330	1.438	1.541	1.640	1.735
0.91	1.275	1.388	1.495	1.598	1.696	1.790	1.880
0.92	1.494	1.602	1.704	1.802	1.895	1.983	2.066
0.93	1.796	1.898	1.995	2.087	2.174	2.255	2.331

10 *DEGREES OF FREEDOM*   *DELTA* = 0.95

ρ → γ ↓	0.70	0.75	0.80	0.85	0.90	0.95	1.00
0.80	0.131	0.293	0.446	0.591	0.732	0.870	1.008
0.81	0.201	0.359	0.508	0.650	0.787	0.922	1.056
0.82	0.274	0.428	0.573	0.711	0.845	0.976	1.106
0.83	0.351	0.501	0.642	0.777	0.907	1.034	1.160
0.84	0.433	0.578	0.715	0.846	0.973	1.097	1.217
0.85	0.520	0.660	0.793	0.921	1.044	1.164	1.280
0.86	0.613	0.749	0.878	1.002	1.121	1.237	1.348
0.87	0.715	0.846	0.971	1.091	1.206	1.317	1.424
0.88	0.827	0.954	1.074	1.189	1.300	1.407	1.508
0.89	0.953	1.075	1.191	1.301	1.408	1.509	1.605
0.90	1.097	1.214	1.325	1.431	1.533	1.629	1.719
0.91	1.269	1.380	1.486	1.587	1.683	1.773	1.856
0.92	1.484	1.589	1.689	1.783	1.873	1.956	2.032
0.93	1.776	1.874	1.967	2.054	2.136	2.211	2.277

12 *DEGREES OF FREEDOM*   *DELTA* = 0.95

ρ → γ ↓	0.70	0.75	0.80	0.85	0.90	0.95	1.00
0.80	0.130	0.293	0.445	0.591	0.733	0.872	1.010
0.81	0.200	0.359	0.507	0.650	0.788	0.923	1.057
0.82	0.274	0.428	0.573	0.712	0.846	0.978	1.107
0.83	0.351	0.500	0.642	0.777	0.908	1.036	1.160
0.84	0.433	0.578	0.715	0.846	0.974	1.097	1.217
0.85	0.519	0.660	0.793	0.921	1.044	1.164	1.279
0.86	0.613	0.749	0.878	1.002	1.121	1.236	1.346
0.87	0.714	0.846	0.971	1.090	1.205	1.316	1.420
0.88	0.826	0.953	1.073	1.188	1.299	1.404	1.503
0.89	0.952	1.073	1.189	1.299	1.404	1.505	1.597
0.90	1.095	1.212	1.322	1.427	1.527	1.621	1.707
0.91	1.266	1.376	1.480	1.580	1.674	1.761	1.840
0.92	1.477	1.581	1.679	1.771	1.858	1.938	2.008
0.93	1.763	1.858	1.948	2.033	2.111	2.181	2.241

TABLE 9.2.3. *SCREENING FACTORS FOR NORMAL CONDITIONED ON t-DISTRIBUTION.*

14 DEGREES OF FREEDOM   DELTA = 0.95

ρ → γ ↓	0.70	0.75	0.80	0.85	0.90	0.95	1.00
0.80	0.130	0.292	0.445	0.591	0.733	0.873	1.011
0.81	0.200	0.358	0.507	0.650	0.789	0.925	1.058
0.82	0.273	0.427	0.573	0.712	0.847	0.979	1.108
0.83	0.351	0.500	0.642	0.777	0.908	1.037	1.160
0.84	0.432	0.577	0.715	0.847	0.974	1.098	1.217
0.85	0.519	0.660	0.793	0.921	1.045	1.164	1.278
0.86	0.613	0.749	0.878	1.002	1.121	1.236	1.344
0.87	0.714	0.846	0.970	1.090	1.205	1.315	1.417
0.88	0.826	0.952	1.073	1.187	1.297	1.402	1.499
0.89	0.951	1.072	1.187	1.297	1.402	1.501	1.591
0.90	1.094	1.210	1.320	1.424	1.523	1.616	1.699
0.91	1.263	1.373	1.476	1.575	1.667	1.753	1.828
0.92	1.472	1.575	1.672	1.763	1.848	1.925	1.990
0.93	1.753	1.847	1.935	2.017	2.093	2.160	2.214

16 DEGREES OF FREEDOM   DELTA = 0.95

ρ → γ ↓	0.70	0.75	0.80	0.85	0.90	0.95	1.00
0.80	0.130	0.292	0.445	0.591	0.734	0.874	1.012
0.81	0.200	0.358	0.507	0.650	0.789	0.926	1.058
0.82	0.273	0.427	0.573	0.712	0.847	0.980	1.108
0.83	0.350	0.500	0.642	0.777	0.909	1.037	1.160
0.84	0.432	0.577	0.715	0.847	0.975	1.099	1.216
0.85	0.519	0.660	0.793	0.921	1.045	1.164	1.277
0.86	0.612	0.749	0.878	1.002	1.121	1.236	1.343
0.87	0.714	0.845	0.970	1.090	1.204	1.314	1.415
0.88	0.825	0.952	1.072	1.187	1.297	1.401	1.495
0.89	0.950	1.071	1.187	1.296	1.400	1.499	1.586
0.90	1.093	1.208	1.318	1.422	1.520	1.612	1.692
0.91	1.261	1.370	1.473	1.571	1.662	1.746	1.818
0.92	1.469	1.570	1.666	1.756	1.840	1.915	1.976
0.93	1.746	1.839	1.926	2.006	2.079	2.143	2.193

18 DEGREES OF FREEDOM   DELTA = 0.95

ρ → γ ↓	0.70	0.75	0.80	0.85	0.90	0.95	1.00
0.80	0.130	0.292	0.445	0.592	0.734	0.875	1.012
0.81	0.200	0.358	0.507	0.650	0.790	0.926	1.059
0.82	0.273	0.427	0.573	0.712	0.848	0.980	1.108
0.83	0.350	0.500	0.642	0.778	0.909	1.038	1.160
0.84	0.432	0.577	0.715	0.847	0.975	1.099	1.216
0.85	0.519	0.660	0.793	0.921	1.045	1.165	1.276
0.86	0.612	0.749	0.878	1.002	1.121	1.236	1.341
0.87	0.714	0.845	0.970	1.089	1.204	1.313	1.413
0.88	0.825	0.952	1.072	1.186	1.296	1.400	1.492
0.89	0.950	1.071	1.186	1.295	1.399	1.497	1.582
0.90	1.092	1.207	1.317	1.420	1.518	1.608	1.686
0.91	1.260	1.368	1.471	1.568	1.659	1.741	1.810
0.92	1.466	1.567	1.662	1.751	1.833	1.907	1.965
0.93	1.740	1.832	1.918	1.997	2.069	2.131	2.176

TABLE 9.2.4. SCREENING FACTORS FOR NORMAL CONDITIONED ON t-DISTRIBUTION.

20 DEGREES OF FREEDOM   DELTA = 0.95

ρ → γ ↓	0.70	0.75	0.80	0.85	0.90	0.95	1.00
0.80	0.130	0.292	0.445	0.592	0.735	0.875	1.012
0.81	0.200	0.358	0.507	0.651	0.790	0.927	1.059
0.82	0.273	0.427	0.573	0.712	0.848	0.981	1.108
0.83	0.350	0.500	0.642	0.778	0.910	1.038	1.160
0.84	0.432	0.577	0.715	0.847	0.975	1.099	1.215
0.85	0.518	0.660	0.793	0.922	1.045	1.165	1.275
0.86	0.612	0.748	0.878	1.002	1.121	1.236	1.340
0.87	0.713	0.845	0.970	1.089	1.204	1.313	1.411
0.88	0.825	0.951	1.071	1.186	1.295	1.399	1.489
0.89	0.949	1.070	1.185	1.294	1.398	1.495	1.578
0.90	1.091	1.207	1.315	1.419	1.516	1.606	1.681
0.91	1.258	1.367	1.469	1.565	1.656	1.737	1.803
0.92	1.464	1.564	1.659	1.747	1.828	1.900	1.956
0.93	1.736	1.827	1.912	1.990	2.060	2.120	2.163

22 DEGREES OF FREEDOM   DELTA = 0.95

ρ → γ ↓	0.70	0.75	0.80	0.85	0.90	0.95	1.00
0.80	0.129	0.292	0.445	0.592	0.735	0.876	1.013
0.81	0.199	0.358	0.507	0.651	0.790	0.927	1.059
0.82	0.273	0.427	0.573	0.713	0.848	0.982	1.108
0.83	0.350	0.500	0.642	0.778	0.910	1.039	1.160
0.84	0.431	0.577	0.715	0.847	0.975	1.100	1.215
0.85	0.518	0.660	0.793	0.922	1.045	1.165	1.274
0.86	0.612	0.748	0.878	1.002	1.121	1.236	1.338
0.87	0.713	0.845	0.970	1.089	1.204	1.313	1.409
0.88	0.825	0.951	1.071	1.186	1.295	1.398	1.487
0.89	0.949	1.070	1.185	1.294	1.397	1.494	1.575
0.90	1.091	1.206	1.315	1.418	1.514	1.604	1.677
0.91	1.257	1.366	1.468	1.564	1.653	1.734	1.797
0.92	1.462	1.562	1.656	1.744	1.824	1.895	1.948
0.93	1.733	1.823	1.907	1.984	2.053	2.112	2.152

24 DEGREES OF FREEDOM   DELTA = 0.95

ρ → γ ↓	0.70	0.75	0.80	0.85	0.90	0.95	1.00
0.80	0.129	0.291	0.445	0.592	0.735	0.876	1.013
0.81	0.199	0.357	0.507	0.651	0.790	0.928	1.059
0.82	0.273	0.427	0.573	0.713	0.849	0.982	1.108
0.83	0.350	0.500	0.642	0.778	0.910	1.039	1.159
0.84	0.431	0.577	0.715	0.847	0.976	1.100	1.214
0.85	0.518	0.659	0.793	0.922	1.046	1.165	1.273
0.86	0.612	0.748	0.878	1.002	1.121	1.236	1.337
0.87	0.713	0.845	0.970	1.089	1.204	1.312	1.407
0.88	0.825	0.951	1.071	1.185	1.295	1.397	1.485
0.89	0.949	1.070	1.184	1.293	1.396	1.493	1.572
0.90	1.091	1.205	1.314	1.417	1.513	1.602	1.673
0.91	1.257	1.365	1.466	1.562	1.651	1.731	1.793
0.92	1.460	1.560	1.654	1.741	1.821	1.891	1.941
0.93	1.730	1.819	1.903	1.979	2.048	2.105	2.142

TABLE 9.2.5.   SCREENING FACTORS FOR NORMAL CONDITIONED ON   t-DISTRIBUTION.

26 DEGREES OF FREEDOM   DELTA = 0.95

ρ → γ ↓	0.70	0.75	0.80	0.85	0.90	0.95	1.00
0.80	0.129	0.291	0.445	0.592	0.735	0.877	1.013
0.81	0.199	0.357	0.507	0.651	0.791	0.928	1.059
0.82	0.272	0.427	0.573	0.713	0.849	0.982	1.107
0.83	0.350	0.500	0.642	0.778	0.910	1.039	1.159
0.84	0.431	0.577	0.715	0.848	0.976	1.100	1.214
0.85	0.518	0.659	0.793	0.922	1.046	1.165	1.273
0.86	0.612	0.748	0.878	1.002	1.121	1.236	1.336
0.87	0.713	0.845	0.970	1.089	1.204	1.312	1.406
0.88	0.824	0.951	1.071	1.185	1.294	1.397	1.483
0.89	0.949	1.070	1.184	1.293	1.396	1.492	1.570
0.90	1.090	1.205	1.313	1.416	1.512	1.600	1.670
0.91	1.256	1.364	1.465	1.561	1.649	1.729	1.788
0.92	1.459	1.559	1.652	1.739	1.818	1.887	1.936
0.93	1.727	1.817	1.899	1.975	2.043	2.099	2.134

28 DEGREES OF FREEDOM   DELTA = 0.95

ρ → γ ↓	0.70	0.75	0.80	0.85	0.90	0.95	1.00
0.80	0.129	0.291	0.445	0.592	0.735	0.877	1.013
0.81	0.199	0.357	0.507	0.651	0.791	0.928	1.059
0.82	0.272	0.426	0.573	0.713	0.849	0.983	1.107
0.83	0.349	0.499	0.642	0.778	0.911	1.040	1.158
0.84	0.431	0.577	0.715	0.848	0.976	1.100	1.213
0.85	0.518	0.659	0.793	0.922	1.046	1.165	1.272
0.86	0.612	0.748	0.878	1.002	1.121	1.236	1.335
0.87	0.713	0.845	0.970	1.089	1.203	1.312	1.405
0.88	0.824	0.951	1.071	1.185	1.294	1.396	1.481
0.89	0.949	1.069	1.184	1.292	1.395	1.491	1.568
0.90	1.090	1.204	1.313	1.415	1.511	1.599	1.667
0.91	1.255	1.363	1.464	1.559	1.648	1.726	1.785
0.92	1.458	1.557	1.650	1.737	1.816	1.884	1.931
0.93	1.725	1.814	1.896	1.972	2.039	2.093	2.127

30 DEGREES OF FREEDOM   DELTA = 0.95

ρ → γ ↓	0.70	0.75	0.80	0.85	0.90	0.95	1.00
0.80	0.129	0.291	0.445	0.592	0.735	0.877	1.012
0.81	0.199	0.357	0.507	0.651	0.791	0.929	1.058
0.82	0.272	0.426	0.573	0.713	0.849	0.983	1.107
0.83	0.349	0.499	0.642	0.778	0.911	1.040	1.158
0.84	0.431	0.577	0.715	0.848	0.976	1.101	1.213
0.85	0.518	0.659	0.793	0.922	1.046	1.165	1.271
0.86	0.611	0.748	0.878	1.002	1.121	1.235	1.334
0.87	0.713	0.845	0.970	1.089	1.203	1.312	1.404
0.88	0.824	0.951	1.071	1.185	1.294	1.396	1.480
0.89	0.948	1.069	1.184	1.292	1.395	1.490	1.566
0.90	1.090	1.204	1.312	1.415	1.511	1.598	1.664
0.91	1.255	1.362	1.464	1.558	1.646	1.725	1.781
0.92	1.457	1.556	1.649	1.735	1.813	1.881	1.926
0.93	1.723	1.812	1.894	1.969	2.035	2.089	2.121

TABLE 9.2.6. SCREENING FACTORS FOR NORMAL CONDITIONED ON t-DISTRIBUTION.

40 DEGREES OF FREEDOM   DELTA = 0.95

ρ → γ ↓	0.70	0.75	0.80	0.85	0.90	0.95	1.00
0.80	0.129	0.291	0.444	0.592	0.736	0.878	1.012
0.81	0.199	0.357	0.507	0.651	0.791	0.930	1.057
0.82	0.272	0.426	0.573	0.713	0.850	0.984	1.106
0.83	0.349	0.499	0.642	0.779	0.911	1.041	1.156
0.84	0.431	0.577	0.715	0.848	0.977	1.101	1.210
0.85	0.518	0.659	0.793	0.922	1.046	1.166	1.268
0.86	0.611	0.748	0.878	1.002	1.121	1.235	1.331
0.87	0.713	0.844	0.970	1.089	1.203	1.311	1.399
0.88	0.824	0.950	1.070	1.184	1.293	1.394	1.474
0.89	0.948	1.069	1.183	1.291	1.393	1.487	1.558
0.90	1.089	1.203	1.311	1.413	1.508	1.594	1.655
0.91	1.253	1.360	1.461	1.555	1.642	1.718	1.769
0.92	1.454	1.552	1.644	1.729	1.806	1.871	1.910
0.93	1.717	1.804	1.885	1.958	2.022	2.073	2.099

60 DEGREES OF FREEDOM   DELTA = 0.95

ρ → γ ↓	0.70	0.75	0.80	0.85	0.90	0.95	1.00
0.80	0.129	0.291	0.444	0.592	0.737	0.879	1.010
0.81	0.199	0.357	0.507	0.651	0.792	0.931	1.056
0.82	0.272	0.426	0.572	0.713	0.850	0.985	1.103
0.83	0.349	0.499	0.642	0.779	0.912	1.042	1.154
0.84	0.430	0.577	0.715	0.848	0.977	1.102	1.207
0.85	0.518	0.659	0.794	0.922	1.047	1.166	1.264
0.86	0.611	0.748	0.878	1.002	1.122	1.235	1.326
0.87	0.712	0.844	0.969	1.089	1.203	1.310	1.393
0.88	0.824	0.950	1.070	1.184	1.292	1.393	1.467
0.89	0.947	1.068	1.182	1.290	1.392	1.485	1.550
0.90	1.088	1.202	1.309	1.411	1.505	1.589	1.644
0.91	1.251	1.358	1.458	1.551	1.637	1.711	1.756
0.92	1.450	1.548	1.639	1.723	1.798	1.861	1.893
0.93	1.710	1.797	1.876	1.948	2.009	2.057	2.077

90 DEGREES OF FREEDOM   DELTA = 0.95

ρ → γ ↓	0.70	0.75	0.80	0.85	0.90	0.95	1.00
0.80	0.129	0.291	0.444	0.592	0.737	0.880	1.008
0.81	0.198	0.357	0.507	0.651	0.792	0.932	1.054
0.82	0.272	0.426	0.572	0.713	0.851	0.985	1.101
0.83	0.349	0.499	0.642	0.779	0.912	1.042	1.151
0.84	0.430	0.576	0.715	0.848	0.977	1.102	1.204
0.85	0.517	0.659	0.794	0.923	1.047	1.166	1.261
0.86	0.611	0.748	0.878	1.002	1.122	1.235	1.322
0.87	0.712	0.844	0.969	1.089	1.203	1.310	1.388
0.88	0.823	0.950	1.070	1.184	1.292	1.392	1.461
0.89	0.947	1.068	1.181	1.289	1.391	1.483	1.543
0.90	1.087	1.201	1.308	1.409	1.503	1.586	1.636
0.91	1.250	1.357	1.456	1.549	1.634	1.707	1.746
0.92	1.448	1.546	1.636	1.719	1.793	1.854	1.881
0.93	1.706	1.792	1.870	1.941	2.001	2.046	2.062

TABLE 9.2.7.   *SCREENING FACTORS FOR NORMAL CONDITIONED ON   t-DISTRIBUTION.*

120 DEGREES OF FREEDOM   DELTA = 0.95

ρ → γ ↓	0.70	0.75	0.80	0.85	0.90	0.95	1.00
0.80	0.129	0.290	0.444	0.592	0.737	0.881	1.007
0.81	0.198	0.357	0.507	0.651	0.793	0.932	1.052
0.82	0.271	0.426	0.572	0.714	0.851	0.986	1.100
0.83	0.349	0.499	0.642	0.779	0.912	1.042	1.149
0.84	0.430	0.576	0.715	0.849	0.978	1.103	1.202
0.85	0.517	0.659	0.794	0.923	1.047	1.167	1.259
0.86	0.611	0.748	0.878	1.002	1.122	1.235	1.320
0.87	0.712	0.844	0.969	1.089	1.203	1.310	1.386
0.88	0.823	0.950	1.070	1.184	1.292	1.391	1.458
0.89	0.947	1.067	1.181	1.289	1.390	1.482	1.539
0.90	1.087	1.201	1.308	1.409	1.502	1.585	1.632
0.91	1.250	1.356	1.455	1.548	1.633	1.705	1.741
0.92	1.447	1.544	1.634	1.717	1.791	1.850	1.875
0.93	1.704	1.789	1.867	1.937	1.997	2.040	2.054

150 DEGREES OF FREEDOM   DELTA = 0.95

ρ → γ ↓	0.70	0.75	0.80	0.85	0.90	0.95	1.00
0.80	0.129	0.290	0.444	0.592	0.737	0.881	1.006
0.81	0.198	0.356	0.507	0.652	0.793	0.932	1.051
0.82	0.271	0.426	0.572	0.714	0.851	0.986	1.099
0.83	0.349	0.499	0.642	0.779	0.913	1.043	1.148
0.84	0.430	0.576	0.715	0.849	0.978	1.103	1.201
0.85	0.517	0.659	0.794	0.923	1.047	1.167	1.257
0.86	0.611	0.748	0.878	1.002	1.122	1.235	1.318
0.87	0.712	0.844	0.969	1.089	1.203	1.310	1.384
0.88	0.823	0.950	1.069	1.183	1.292	1.391	1.456
0.89	0.947	1.067	1.181	1.289	1.390	1.482	1.537
0.90	1.087	1.200	1.307	1.408	1.502	1.584	1.630
0.91	1.249	1.355	1.455	1.547	1.632	1.703	1.738
0.92	1.447	1.543	1.633	1.716	1.789	1.848	1.872
0.93	1.702	1.788	1.865	1.935	1.994	2.037	2.050

∞ DEGREES OF FREEDOM   DELTA = 0.95

ρ → γ ↓	0.70	0.75	0.80	0.85	0.90	0.95	1.00
0.80	0.128	0.290	0.444	0.592	0.738	0.882	1.003
0.81	0.198	0.356	0.507	0.652	0.793	0.933	1.048
0.82	0.271	0.426	0.572	0.714	0.852	0.987	1.095
0.83	0.348	0.499	0.642	0.779	0.913	1.043	1.144
0.84	0.430	0.576	0.715	0.849	0.978	1.103	1.196
0.85	0.517	0.659	0.794	0.923	1.048	1.167	1.252
0.86	0.611	0.748	0.878	1.002	1.122	1.235	1.312
0.87	0.712	0.844	0.969	1.089	1.203	1.309	1.377
0.88	0.823	0.949	1.069	1.183	1.291	1.390	1.449
0.89	0.946	1.067	1.180	1.288	1.389	1.479	1.529
0.90	1.086	1.199	1.306	1.407	1.500	1.580	1.620
0.91	1.248	1.354	1.453	1.545	1.628	1.698	1.727
0.92	1.444	1.540	1.629	1.711	1.783	1.839	1.858
0.93	1.697	1.782	1.858	1.926	1.984	2.023	2.032

TABLE 9.3.1.   SCREENING FACTORS FOR NORMAL CONDITIONED ON t-DISTRIBUTION.

2 DEGREES OF FREEDOM   DELTA = 0.99

ρ → γ ↓	0.70	0.75	0.80	0.85	0.90	0.95	1.00
0.85	-0.715	-0.276	0.016	0.246	0.442	0.617	0.774
0.86	-0.551	-0.178	0.088	0.304	0.491	0.659	0.812
0.87	-0.407	-0.084	0.160	0.363	0.543	0.705	0.854
0.88	-0.276	0.009	0.233	0.426	0.597	0.755	0.900
0.89	-0.154	0.101	0.309	0.491	0.656	0.809	0.951
0.90	-0.036	0.194	0.389	0.562	0.721	0.869	1.008
0.91	0.082	0.292	0.474	0.640	0.793	0.937	1.073
0.92	0.203	0.396	0.568	0.726	0.875	1.015	1.149
0.93	0.331	0.510	0.674	0.826	0.970	1.108	1.240
0.94	0.474	0.642	0.798	0.945	1.086	1.222	1.352
0.95	0.642	0.801	0.951	1.095	1.234	1.368	1.499
0.96	0.857	1.010	1.157	1.299	1.438	1.574	1.707
0.97	1.172	1.324	1.473	1.618	1.762	1.903	2.043
0.98	1.797	1.964	2.129	2.294	2.457	2.619	2.783

4 DEGREES OF FREEDOM   DELTA = 0.99

ρ → γ ↓	0.70	0.75	0.80	0.85	0.90	0.95	1.00
0.85	-0.553	-0.243	0.015	0.245	0.457	0.659	0.856
0.86	-0.449	-0.161	0.085	0.305	0.510	0.706	0.897
0.87	-0.347	-0.077	0.156	0.367	0.566	0.756	0.941
0.88	-0.244	0.008	0.230	0.433	0.625	0.809	0.989
0.89	-0.140	0.097	0.308	0.503	0.688	0.867	1.042
0.90	-0.033	0.189	0.391	0.579	0.757	0.931	1.100
0.91	0.078	0.288	0.480	0.661	0.834	1.001	1.166
0.92	0.198	0.396	0.579	0.752	0.919	1.082	1.242
0.93	0.328	0.515	0.689	0.856	1.017	1.175	1.330
0.94	0.474	0.651	0.818	0.978	1.134	1.287	1.437
0.95	0.647	0.814	0.973	1.127	1.278	1.427	1.573
0.96	0.863	1.022	1.174	1.323	1.470	1.614	1.757
0.97	1.167	1.319	1.466	1.611	1.754	1.896	2.037
0.98	1.713	1.862	2.009	2.154	2.299	2.443	2.586

6 DEGREES OF FREEDOM   DELTA = 0.99

ρ → γ ↓	0.70	0.75	0.80	0.85	0.90	0.95	1.00
0.85	-0.515	-0.234	0.015	0.244	0.463	0.679	0.899
0.86	-0.424	-0.156	0.083	0.305	0.518	0.727	0.941
0.87	-0.331	-0.075	0.154	0.369	0.575	0.779	0.986
0.88	-0.235	0.008	0.229	0.436	0.637	0.835	1.035
0.89	-0.136	0.095	0.308	0.508	0.702	0.894	1.089
0.90	-0.033	0.188	0.392	0.586	0.774	0.960	1.148
0.91	0.077	0.287	0.483	0.670	0.852	1.032	1.214
0.92	0.196	0.396	0.583	0.763	0.939	1.114	1.289
0.93	0.326	0.516	0.696	0.869	1.039	1.207	1.377
0.94	0.474	0.655	0.827	0.993	1.156	1.319	1.482
0.95	0.649	0.819	0.983	1.143	1.300	1.456	1.612
0.96	0.867	1.028	1.184	1.336	1.487	1.636	1.786
0.97	1.168	1.320	1.468	1.613	1.757	1.900	2.043
0.98	1.687	1.831	1.972	2.110	2.248	2.384	2.520

TABLE 9.3.2. SCREENING FACTORS FOR NORMAL CONDITIONED ON t-DISTRIBUTION.

8 DEGREES OF FREEDOM   DELTA = 0.99

$\rho \rightarrow$ $\gamma \downarrow$	0.70	0.75	0.80	0.85	0.90	0.95	1.00
0.85	-0.498	-0.229	0.015	0.244	0.466	0.690	0.926
0.86	-0.412	-0.153	0.083	0.305	0.522	0.740	0.969
0.87	-0.323	-0.074	0.154	0.370	0.581	0.793	1.015
0.88	-0.231	0.008	0.228	0.438	0.643	0.849	1.065
0.89	-0.134	0.095	0.308	0.511	0.710	0.910	1.119
0.90	-0.032	0.187	0.392	0.589	0.783	0.977	1.178
0.91	0.077	0.287	0.484	0.675	0.862	1.050	1.244
0.92	0.195	0.396	0.586	0.770	0.951	1.133	1.319
0.93	0.326	0.517	0.700	0.877	1.052	1.227	1.406
0.94	0.475	0.657	0.832	1.002	1.170	1.338	1.509
0.95	0.650	0.823	0.989	1.152	1.313	1.474	1.637
0.96	0.869	1.032	1.190	1.344	1.497	1.650	1.804
0.97	1.169	1.321	1.470	1.615	1.760	1.903	2.047
0.98	1.675	1.816	1.953	2.089	2.222	2.354	2.486

10 DEGREES OF FREEDOM   DELTA = 0.99

$\rho \rightarrow$ $\gamma \downarrow$	0.70	0.75	0.80	0.85	0.90	0.95	1.00
0.85	-0.489	-0.226	0.015	0.244	0.468	0.697	0.945
0.86	-0.405	-0.152	0.082	0.306	0.525	0.748	0.989
0.87	-0.318	-0.074	0.153	0.370	0.584	0.802	1.035
0.88	-0.228	0.008	0.228	0.439	0.648	0.859	1.085
0.89	-0.133	0.094	0.307	0.513	0.715	0.921	1.140
0.90	-0.032	0.186	0.393	0.592	0.789	0.989	1.199
0.91	0.076	0.286	0.485	0.678	0.869	1.063	1.266
0.92	0.194	0.396	0.588	0.774	0.959	1.146	1.341
0.93	0.325	0.518	0.702	0.882	1.060	1.240	1.427
0.94	0.475	0.658	0.835	1.007	1.178	1.351	1.529
0.95	0.651	0.825	0.993	1.158	1.321	1.486	1.654
0.96	0.870	1.034	1.193	1.349	1.504	1.660	1.817
0.97	1.169	1.322	1.471	1.617	1.762	1.906	2.051
0.98	1.668	1.807	1.943	2.076	2.207	2.336	2.465

12 DEGREES OF FREEDOM   DELTA = 0.99

$\rho \rightarrow$ $\gamma \downarrow$	0.70	0.75	0.80	0.85	0.90	0.95	1.00
0.85	-0.483	-0.225	0.015	0.244	0.470	0.702	0.960
0.86	-0.401	-0.151	0.082	0.306	0.527	0.754	1.004
0.87	-0.315	-0.073	0.153	0.371	0.587	0.808	1.050
0.88	-0.226	0.008	0.228	0.440	0.650	0.866	1.101
0.89	-0.132	0.094	0.307	0.514	0.719	0.929	1.155
0.90	-0.032	0.186	0.393	0.594	0.793	0.997	1.215
0.91	0.076	0.286	0.486	0.680	0.874	1.071	1.282
0.92	0.194	0.396	0.589	0.777	0.964	1.155	1.356
0.93	0.325	0.518	0.704	0.885	1.066	1.250	1.442
0.94	0.475	0.659	0.837	1.011	1.185	1.360	1.543
0.95	0.651	0.826	0.996	1.162	1.328	1.495	1.667
0.96	0.871	1.036	1.196	1.353	1.509	1.667	1.827
0.97	1.170	1.323	1.472	1.619	1.764	1.908	2.054
0.98	1.663	1.801	1.936	2.068	2.197	2.324	2.450

TABLE 9.3.3. SCREENING FACTORS FOR NORMAL CONDITIONED ON t-DISTRIBUTION.

14 DEGREES OF FREEDOM DELTA = 0.99

ρ → γ ↓	0.70	0.75	0.80	0.85	0.90	0.95	1.00
0.85	-0.478	-0.223	0.015	0.244	0.471	0.706	0.971
0.86	-0.398	-0.150	0.082	0.306	0.528	0.758	1.015
0.87	-0.313	-0.073	0.153	0.371	0.588	0.813	1.062
0.88	-0.225	0.008	0.228	0.441	0.653	0.872	1.113
0.89	-0.132	0.094	0.307	0.515	0.721	0.935	1.168
0.90	-0.032	0.186	0.393	0.595	0.796	1.003	1.228
0.91	0.076	0.286	0.486	0.682	0.877	1.078	1.294
0.92	0.194	0.396	0.589	0.779	0.968	1.162	1.369
0.93	0.325	0.519	0.705	0.888	1.070	1.257	1.454
0.94	0.475	0.660	0.839	1.014	1.189	1.368	1.555
0.95	0.652	0.828	0.998	1.165	1.332	1.501	1.677
0.96	0.872	1.037	1.198	1.356	1.513	1.672	1.835
0.97	1.171	1.324	1.473	1.620	1.765	1.910	2.057
0.98	1.660	1.797	1.931	2.062	2.190	2.315	2.439

16 DEGREES OF FREEDOM DELTA = 0.99

ρ → γ ↓	0.70	0.75	0.80	0.85	0.90	0.95	1.00
0.85	-0.475	-0.222	0.015	0.244	0.472	0.709	0.980
0.86	-0.395	-0.149	0.082	0.306	0.529	0.762	1.025
0.87	-0.312	-0.073	0.153	0.371	0.590	0.817	1.072
0.88	-0.224	0.008	0.227	0.441	0.654	0.876	1.123
0.89	-0.131	0.094	0.307	0.515	0.723	0.939	1.178
0.90	-0.032	0.186	0.393	0.596	0.798	1.008	1.238
0.91	0.076	0.286	0.487	0.683	0.880	1.083	1.304
0.92	0.193	0.396	0.590	0.780	0.971	1.167	1.379
0.93	0.325	0.519	0.706	0.890	1.074	1.262	1.464
0.94	0.475	0.661	0.840	1.016	1.193	1.373	1.564
0.95	0.652	0.828	0.999	1.168	1.336	1.507	1.685
0.96	0.872	1.038	1.200	1.358	1.516	1.676	1.841
0.97	1.171	1.324	1.474	1.621	1.766	1.912	2.059
0.98	1.658	1.794	1.927	2.057	2.184	2.309	2.430

18 DEGREES OF FREEDOM DELTA = 0.99

ρ → γ ↓	0.70	0.75	0.80	0.85	0.90	0.95	1.00
0.85	-0.473	-0.222	0.015	0.244	0.472	0.712	0.988
0.86	-0.394	-0.149	0.082	0.306	0.530	0.764	1.033
0.87	-0.311	-0.073	0.152	0.372	0.591	0.820	1.080
0.88	-0.223	0.008	0.227	0.441	0.656	0.879	1.131
0.89	-0.131	0.093	0.307	0.516	0.725	0.943	1.186
0.90	-0.032	0.186	0.393	0.596	0.800	1.012	1.246
0.91	0.076	0.286	0.487	0.684	0.882	1.087	1.313
0.92	0.193	0.396	0.591	0.782	0.974	1.172	1.387
0.93	0.325	0.519	0.707	0.892	1.077	1.267	1.472
0.94	0.475	0.661	0.841	1.018	1.196	1.378	1.571
0.95	0.652	0.829	1.001	1.170	1.339	1.511	1.692
0.96	0.873	1.039	1.201	1.360	1.519	1.680	1.845
0.97	1.171	1.325	1.474	1.622	1.767	1.913	2.060
0.98	1.656	1.792	1.925	2.054	2.180	2.303	2.423

TABLE 9.3.4. SCREENING FACTORS FOR NORMAL CONDITIONED ON t-DISTRIBUTION.

20 DEGREES OF FREEDOM   DELTA = 0.99

ρ → γ ↓	0.70	0.75	0.80	0.85	0.90	0.95	1.00
0.85	-0.471	-0.221	0.015	0.244	0.473	0.714	0.994
0.86	-0.392	-0.149	0.082	0.306	0.531	0.766	1.039
0.87	-0.310	-0.073	0.152	0.372	0.592	0.822	1.087
0.88	-0.223	0.008	0.227	0.442	0.657	0.882	1.138
0.89	-0.131	0.093	0.307	0.516	0.726	0.946	1.193
0.90	-0.032	0.185	0.393	0.597	0.802	1.015	1.253
0.91	0.076	0.286	0.487	0.685	0.884	1.091	1.320
0.92	0.193	0.396.	0.591	0.783	0.976	1.175	1.394
0.93	0.325	0.519	0.708	0.893	1.079	1.271	1.479
0.94	0.475	0.661	0.842	1.020	1.198	1.381	1.578
0.95	0.652	0.830	1.002	1.171	1.341	1.514	1.697
0.96	0.873	1.040	1.202	1.362	1.521	1.682	1.849
0.97	1.171	1.325	1.475	1.622	1.768	1.914	2.061
0.98	1.655	1.790	1.922	2.051	2.177	2.299	2.417

22 DEGREES OF FREEDOM   DELTA = 0.99

ρ → γ ↓	0.70	0.75	0.80	0.85	0.90	0.95	1.00
0.85	-0.469	-0.221	0.015	0.244	0.473	0.715	1.000
0.86	-0.391	-0.149	0.082	0.306	0.531	0.768	1.045
0.87	-0.309	-0.073	0.152	0.372	0.592	0.824	1.093
0.88	-0.222	0.008	0.227	0.442	0.658	0.884	1.144
0.89	-0.130	0.093	0.307	0.517	0.727	0.948	1.199
0.90	-0.032	0.185	0.393	0.598	0.803	1.018	1.259
0.91	0.075	0.285	0.487	0.686	0.886	1.094	1.326
0.92	0.193	0.396	0.591	0.784	0.977	1.178	1.400
0.93	0.325	0.520	0.708	0.894	1.081	1.274	1.484
0.94	0.475	0.662	0.843	1.021	1.200	1.385	1.583
0.95	0.653	0.830	1.003	1.173	1.343	1.517	1.702
0.96	0.873	1.040	1.203	1.363	1.523	1.685	1.853
0.97	1.172	1.325	1.475	1.623	1.769	1.915	2.062
0.98	1.653	1.789	1.920	2.049	2.174	2.295	2.412

24 DEGREES OF FREEDOM   DELTA = 0.99

ρ → γ ↓	0.70	0.75	0.80	0.85	0.90	0.95	1.00
0.85	-0.468	-0.220	0.015	0.243	0.474	0.717	1.005
0.86	-0.390	-0.148	0.082	0.306	0.532	0.770	1.050
0.87	-0.308	-0.073	0.152	0.372	0.593	0.826	1.098
0.88	-0.222	0.008	0.227	0.442	0.658	0.886	1.149
0.89	-0.130	0.093	0.307	0.517	0.728	0.950	1.204
0.90	-0.032	0.185	0.393	0.598	0.804	1.020	1.265
0.91	0.075	0.285	0.488	0.686	0.887	1.096	1.331
0.92	0.193	0.396	0.592	0.784	0.979	1.181	1.405
0.93	0.324	0.520	0.709	0.895	1.082	1.277	1.489
0.94	0.475	0.662	0.843	1.022	1.202	1.387	1.588
0.95	0.653	0.830	1.003	1.174	1.345	1.520	1.706
0.96	0.874	1.041	1.203	1.364	1.524	1.687	1.856
0.97	1.172	1.326	1.476	1.623	1.770	1.916	2.063
0.98	1.652	1.788	1.919	2.047	2.172	2.292	2.407

TABLE 9.3.5. *SCREENING FACTORS FOR NORMAL CONDITIONED ON* t-*DISTRIBUTION.*

26 *DEGREES OF FREEDOM* DELTA = 0.99

ρ →	0.70	0.75	0.80	0.85	0.90	0.95	1.00
γ ↓							
0.85	-0.467	-0.220	0.015	0.243	0.474	0.718	1.009
0.86	-0.389	-0.148	0.082	0.306	0.532	0.771	1.054
0.87	-0.308	-0.072	0.152	0.372	0.594	0.827	1.102
0.88	-0.222	0.008	0.227	0.442	0.659	0.888	1.153
0.89	-0.130	0.093	0.307	0.517	0.729	0.952	1.209
0.90	-0.032	0.185	0.394	0.598	0.805	1.022	1.269
0.91	0.075	0.285	0.488	0.687	0.888	1.098	1.335
0.92	0.193	0.396	0.592	0.785	0.980	1.183	1.410
0.93	0.324	0.520	0.709	0.895	1.084	1.279	1.494
0.94	0.475	0.662	0.844	1.023	1.203	1.390	1.592
0.95	0.653	0.831	1.004	1.175	1.346	1.522	1.709
0.96	0.874	1.041	1.204	1.365	1.525	1.689	1.858
0.97	1.172	1.326	1.476	1.624	1.770	1.917	2.064
0.98	1.652	1.787	1.918	2.045	2.170	2.290	2.403

28 *DEGREES OF FREEDOM* DELTA = 0.99

ρ →	0.70	0.75	0.80	0.85	0.90	0.95	1.00
γ ↓							
0.85	-0.466	-0.220	0.015	0.243	0.474	0.719	1.013
0.86	-0.389	-0.148	0.082	0.306	0.532	0.772	1.058
0.87	-0.307	-0.072	0.152	0.372	0.594	0.829	1.106
0.88	-0.221	0.008	0.227	0.442	0.660	0.889	1.157
0.89	-0.130	0.093	0.307	0.517	0.730	0.954	1.213
0.90	-0.031	0.185	0.394	0.599	0.806	1.023	1.273
0.91	0.075	0.285	0.488	0.687	0.889	1.100	1.340
0.92	0.193	0.396	0.592	0.785	0.981	1.185	1.413
0.93	0.324	0.520	0.709	0.896	1.085	1.281	1.497
0.94	0.475	0.662	0.844	1.024	1.205	1.392	1.595
0.95	0.653	0.831	1.004	1.175	1.347	1.524	1.712
0.96	0.874	1.041	1.205	1.366	1.527	1.690	1.861
0.97	1.172	1.326	1.476	1.624	1.771	1.917	2.064
0.98	1.651	1.786	1.917	2.044	2.168	2.288	2.400

30 *DEGREES OF FREEDOM* DELTA = 0.99

ρ →	0.70	0.75	0.80	0.85	0.90	0.95	1.00
γ ↓							
0.85	-0.465	-0.219	0.015	0.243	0.474	0.720	1.016
0.86	-0.388	-0.148	0.081	0.306	0.533	0.773	1.061
0.87	-0.307	-0.072	0.152	0.372	0.594	0.830	1.110
0.88	-0.221	0.008	0.227	0.442	0.660	0.890	1.161
0.89	-0.130	0.093	0.307	0.518	0.730	0.955	1.216
0.90	-0.031	0.185	0.394	0.599	0.806	1.025	1.277
0.91	0.075	0.285	0.488	0.688	0.890	1.102	1.343
0.92	0.193	0.396	0.592	0.786	0.982	1.187	1.417
0.93	0.324	0.520	0.709	0.897	1.086	1.283	1.501
0.94	0.475	0.662	0.844	1.024	1.206	1.393	1.598
0.95	0.653	0.831	1.005	1.176	1.349	1.526	1.715
0.96	0.874	1.042	1.205	1.366	1.528	1.692	1.863
0.97	1.172	1.326	1.477	1.624	1.771	1.918	2.065
0.98	1.650	1.785	1.916	2.043	2.166	2.286	2.397

TABLE 9.3.6. *SCREENING FACTORS FOR NORMAL CONDITIONED ON* t-*DISTRIBUTION.*

40 *DEGREES OF FREEDOM*   *DELTA* = 0.99

$\rho \rightarrow$ $\gamma \downarrow$	0.70	0.75	0.80	0.85	0.90	0.95	1.00
0.85	-0.463	-0.219	0.015	0.243	0.475	0.723	1.029
0.86	-0.386	-0.147	0.081	0.306	0.534	0.777	1.074
0.87	-0.306	-0.072	0.152	0.372	0.596	0.834	1.123
0.88	-0.220	0.008	0.227	0.443	0.662	0.895	1.174
0.89	-0.129	0.093	0.307	0.518	0.732	0.960	1.230
0.90	-0.031	0.185	0.394	0.600	0.809	1.030	1.290
0.91	0.075	0.285	0.488	0.689	0.892	1.107	1.356
0.92	0.193	0.396	0.593	0.788	0.985	1.193	1.430
0.93	0.324	0.520	0.710	0.899	1.089	1.289	1.513
0.94	0.475	0.663	0.846	1.027	1.209	1.400	1.610
0.95	0.653	0.832	1.006	1.179	1.352	1.532	1.725
0.96	0.875	1.043	1.207	1.369	1.531	1.697	1.869
0.97	1.173	1.327	1.478	1.626	1.773	1.920	2.066
0.98	1.648	1.782	1.912	2.039	2.161	2.279	2.385

60 *DEGREES OF FREEDOM*   *DELTA* = 0.99

$\rho \rightarrow$ $\gamma \downarrow$	0.70	0.75	0.80	0.85	0.90	0.95	1.00
0.85	-0.460	-0.218	0.015	0.243	0.476	0.726	1.043
0.86	-0.384	-0.147	0.081	0.306	0.535	0.780	1.089
0.87	-0.304	-0.072	0.152	0.372	0.597	0.838	1.137
0.88	-0.220	0.008	0.227	0.443	0.663	0.899	1.189
0.89	-0.129	0.093	0.307	0.519	0.735	0.965	1.245
0.90	-0.031	0.185	0.394	0.601	0.811	1.036	1.305
0.91	0.075	0.285	0.489	0.690	0.895	1.113	1.371
0.92	0.192	0.396	0.593	0.789	0.988	1.199	1.444
0.93	0.324	0.520	0.711	0.901	1.093	1.296	1.527
0.94	0.475	0.663	0.847	1.029	1.214	1.407	1.622
0.95	0.654	0.833	1.008	1.181	1.357	1.538	1.735
0.96	0.875	1.044	1.208	1.371	1.535	1.702	1.876
0.97	1.173	1.328	1.478	1.627	1.775	1.922	2.066
0.98	1.646	1.780	1.909	2.035	2.156	2.272	2.371

90 *DEGREES OF FREEDOM*   *DELTA* = 0.99

$\rho \rightarrow$ $\gamma \downarrow$	0.70	0.75	0.80	0.85	0.90	0.95	1.00
0.85	-0.458	-0.217	0.015	0.243	0.477	0.728	1.053
0.86	-0.383	-0.147	0.081	0.306	0.536	0.783	1.099
0.87	-0.304	-0.072	0.152	0.373	0.598	0.841	1.148
0.88	-0.219	0.008	0.227	0.443	0.665	0.902	1.200
0.89	-0.129	0.093	0.307	0.519	0.736	0.968	1.256
0.90	-0.031	0.185	0.394	0.601	0.813	1.040	1.316
0.91	0.075	0.285	0.489	0.691	0.897	1.117	1.382
0.92	0.192	0.396	0.594	0.790	0.991	1.204	1.455
0.93	0.324	0.521	0.712	0.902	1.096	1.300	1.537
0.94	0.475	0.664	0.848	1.031	1.216	1.411	1.631
0.95	0.654	0.834	1.009	1.183	1.359	1.543	1.742
0.96	0.876	1.045	1.210	1.373	1.537	1.706	1.880
0.97	1.173	1.328	1.479	1.628	1.776	1.924	2.065
0.98	1.645	1.778	1.907	2.032	2.153	2.267	2.358

TABLE 9.3.7. SCREENING FACTORS FOR NORMAL CONDITIONED ON t-DISTRIBUTION.

<u>120 *DEGREES OF FREEDOM*</u>   *DELTA* = 0.99

ρ → γ ↓	0.70	0.75	0.80	0.85	0.90	0.95	1.00
0.85	-0.457	-0.217	0.015	0.243	0.477	0.729	1.059
0.86	-0.382	-0.146	0.081	0.306	0.536	0.784	1.105
0.87	-0.303	-0.072	0.152	0.373	0.599	0.842	1.154
0.88	-0.219	0.008	0.227	0.444	0.665	0.904	1.206
0.89	-0.129	0.093	0.307	0.520	0.737	0.970	1.262
0.90	-0.031	0.185	0.394	0.602	0.814	1.042	1.322
0.91	0.075	0.285	0.489	0.692	0.898	1.120	1.388
0.92	0.192	0.396	0.594	0.791	0.992	1.206	1.460
0.93	0.324	0.521	0.712	0.903	1.097	1.303	1.542
0.94	0.475	0.664	0.848	1.031	1.218	1.414	1.635
0.95	0.654	0.834	1.010	1.184	1.361	1.545	1.745
0.96	0.876	1.045	1.210	1.374	1.538	1.708	1.882
0.97	1.174	1.328	1.480	1.628	1.777	1.925	2.064
0.98	1.645	1.778	1.907	2.032	2.153	2.268	2.358

<u>150 *DEGREES OF FREEDOM*</u>   *DELTA* = 0.99

ρ → γ ↓	0.70	0.75	0.80	0.85	0.90	0.95	1.00
0.85	-0.457	-0.217	0.015	0.243	0.477	0.730	1.063
0.86	-0.382	-0.146	0.081	0.306	0.536	0.785	1.109
0.87	-0.303	-0.072	0.152	0.373	0.599	0.843	1.158
0.88	-0.219	0.008	0.227	0.444	0.666	0.905	1.210
0.89	-0.129	0.093	0.307	0.520	0.737	0.971	1.266
0.90	-0.031	0.185	0.394	0.602	0.814	1.043	1.326
0.91	0.075	0.285	0.489	0.692	0.899	1.121	1.391
0.92	0.192	0.396	0.594	0.791	0.993	1.207	1.463
0.93	0.324	0.521	0.713	0.903	1.098	1.304	1.545
0.94	0.475	0.664	0.849	1.032	1.219	1.415	1.637
0.95	0.654	0.834	1.010	1.185	1.362	1.547	1.747
0.96	0.876	1.045	1.210	1.374	1.539	1.709	1.882
0.97	1.174	1.329	1.480	1.629	1.777	1.925	2.062
0.98	1.644	1.777	1.906	2.031	2.151	2.264	2.351

<u>∞ *DEGREES OF FREEDOM*</u>   *DELTA* = 0.99

ρ → γ ↓	0.70	0.75	0.80	0.85	0.90	0.95	1.00
0.85	-0.455	-0.216	0.015	0.243	0.478	0.733	1.074
0.86	-0.380	-0.146	0.081	0.306	0.537	0.788	1.120
0.87	-0.302	-0.072	0.151	0.373	0.600	0.847	1.169
0.88	-0.218	0.008	0.226	0.444	0.667	0.909	1.221
0.89	-0.128	0.093	0.307	0.520	0.739	0.976	1.276
0.90	-0.031	0.184	0.394	0.603	0.817	1.048	1.335
0.91	0.075	0.285	0.489	0.693	0.901	1.126	1.400
0.92	0.192	0.396	0.595	0.793	0.995	1.213	1.471
0.93	0.324	0.521	0.713	0.905	1.101	1.310	1.550
0.94	0.475	0.665	0.850	1.034	1.222	1.421	1.640
0.95	0.654	0.835	1.011	1.187	1.365	1.552	1.746
0.96	0.877	1.046	1.212	1.376	1.543	1.714	1.876
0.97	1.174	1.329	1.481	1.630	1.779	1.927	2.050
0.98	1.644	1.777	1.905	2.030	2.150	2.263	2.346

TABLE 9.4.1. *SCREENING FACTORS FOR NORMAL CONDITIONED ON* t*-DISTRIBUTION.*

2 DEGREES OF FREEDOM   DELTA = 0.995

ρ → γ ↓	0.70	0.75	0.80	0.85	0.90	0.95	1.00
0.85	-4.583	-0.876	-0.301	0.039	0.293	0.502	0.678
0.86	-2.038	-0.679	-0.205	0.102	0.341	0.540	0.711
0.87	-1.364	-0.517	-0.115	0.166	0.390	0.580	0.746
0.88	-0.987	-0.376	-0.028	0.230	0.441	0.623	0.783
0.89	-0.724	-0.249	0.058	0.295	0.495	0.669	0.825
0.90	-0.516	-0.130	0.143	0.363	0.552	0.720	0.871
0.91	-0.340	-0.015	0.230	0.435	0.614	0.776	0.923
0.92	-0.180	0.099	0.322	0.513	0.683	0.839	0.982
0.93	-0.028	0.216	0.419	0.599	0.761	0.912	1.051
0.94	0.125	0.341	0.528	0.697	0.853	0.998	1.135
0.95	0.287	0.480	0.654	0.814	0.964	1.105	1.240
0.96	0.471	0.647	0.810	0.963	1.108	1.246	1.379
0.97	0.705	0.869	1.023	1.171	1.313	1.451	1.584
0.98	1.065	1.223	1.375	1.523	1.668	1.810	1.950

4 DEGREES OF FREEDOM   DELTA = 0.995

ρ → γ ↓	0.70	0.75	0.80	0.85	0.90	0.95	1.00
0.85	-1.119	-0.622	-0.260	0.037	0.299	0.541	0.769
0.86	-0.964	-0.519	-0.183	0.099	0.351	0.584	0.805
0.87	-0.820	-0.419	-0.105	0.163	0.404	0.630	0.843
0.88	-0.684	-0.319	-0.026	0.229	0.460	0.678	0.884
0.89	-0.552	-0.220	0.055	0.297	0.520	0.729	0.929
0.90	-0.423	-0.118	0.139	0.370	0.583	0.785	0.979
0.91	-0.294	-0.014	0.228	0.447	0.652	0.847	1.034
0.92	-0.163	0.095	0.323	0.532	0.728	0.916	1.096
0.93	-0.026	0.212	0.426	0.625	0.813	0.994	1.169
0.94	0.120	0.341	0.542	0.731	0.911	1.085	1.254
0.95	0.283	0.487	0.676	0.856	1.028	1.195	1.359
0.96	0.474	0.662	0.840	1.011	1.176	1.337	1.495
0.97	0.716	0.891	1.059	1.221	1.379	1.534	1.687
0.98	1.078	1.241	1.400	1.555	1.708	1.858	2.008

6 DEGREES OF FREEDOM   DELTA = 0.995

ρ → γ ↓	0.70	0.75	0.80	0.85	0.90	0.95	1.00
0.85	-0.959	-0.570	-0.248	0.036	0.301	0.559	0.818
0.86	-0.847	-0.483	-0.176	0.098	0.355	0.604	0.855
0.87	-0.736	-0.394	-0.102	0.162	0.410	0.652	0.895
0.88	-0.625	-0.304	-0.025	0.228	0.469	0.703	0.939
0.89	-0.513	-0.211	0.054	0.298	0.531	0.758	0.985
0.90	-0.399	-0.115	0.138	0.373	0.597	0.817	1.036
0.91	-0.281	-0.014	0.227	0.452	0.669	0.881	1.093
0.92	-0.157	0.094	0.323	0.539	0.748	0.953	1.157
0.93	-0.025	0.211	0.429	0.636	0.836	1.033	1.230
0.94	0.119	0.341	0.547	0.745	0.937	1.127	1.316
0.95	0.282	0.489	0.685	0.874	1.058	1.240	1.421
0.96	0.475	0.669	0.853	1.032	1.208	1.382	1.555
0.97	0.721	0.901	1.075	1.244	1.411	1.576	1.741
0.98	1.085	1.251	1.413	1.572	1.729	1.885	2.042

TABLE 9.4.2. *SCREENING FACTORS FOR NORMAL CONDITIONED ON t-DISTRIBUTION.*

8 *DEGREES OF FREEDOM*   *DELTA* = 0.995

ρ → γ ↓	0.70	0.75	0.80	0.85	0.90	0.95	1.00
0.85	-0.899	-0.548	-0.243	0.036	0.303	0.569	0.850
0.86	-0.800	-0.466	-0.173	0.097	0.357	0.616	0.888
0.87	-0.701	-0.383	-0.100	0.161	0.414	0.666	0.929
0.88	-0.600	-0.297	-0.025	0.228	0.473	0.718	0.974
0.89	-0.496	-0.207	0.054	0.299	0.537	0.775	1.022
0.90	-0.388	-0.113	0.137	0.374	0.604	0.835	1.074
0.91	-0.275	-0.014	0.227	0.455	0.678	0.901	1.132
0.92	-0.155	0.093	0.324	0.544	0.759	0.975	1.196
0.93	-0.025	0.210	0.430	0.642	0.849	1.057	1.270
0.94	0.118	0.341	0.551	0.753	0.953	1.153	1.357
0.95	0.281	0.491	0.691	0.884	1.075	1.266	1.461
0.96	0.476	0.672	0.861	1.045	1.227	1.409	1.594
0.97	0.724	0.907	1.084	1.258	1.429	1.601	1.775
0.98	1.089	1.257	1.421	1.582	1.742	1.903	2.064

10 *DEGREES OF FREEDOM*   *DELTA* = 0.995

ρ → γ ↓	0.70	0.75	0.80	0.85	0.90	0.95	1.00
0.85	-0.867	-0.535	-0.240	0.036	0.303	0.576	0.873
0.86	-0.776	-0.457	-0.171	0.097	0.358	0.624	0.912
0.87	-0.682	-0.376	-0.099	0.161	0.416	0.675	0.954
0.88	-0.586	-0.292	-0.025	0.228	0.476	0.728	0.999
0.89	-0.486	-0.205	0.054	0.299	0.540	0.786	1.048
0.90	-0.382	-0.112	0.137	0.375	0.609	0.848	1.101
0.91	-0.271	-0.014	0.226	0.457	0.684	0.915	1.159
0.92	-0.153	0.093	0.324	0.546	0.766	0.990	1.224
0.93	-0.025	0.210	0.431	0.646	0.858	1.073	1.299
0.94	0.118	0.341	0.553	0.759	0.963	1.170	1.385
0.95	0.281	0.492	0.694	0.891	1.087	1.285	1.490
0.96	0.476	0.674	0.866	1.053	1.239	1.427	1.621
0.97	0.726	0.911	1.090	1.267	1.442	1.619	1.800
0.98	1.092	1.261	1.426	1.589	1.752	1.915	2.080

12 *DEGREES OF FREEDOM*   *DELTA* = 0.995

ρ → γ ↓	0.70	0.75	0.80	0.85	0.90	0.95	1.00
0.85	-0.848	-0.527	-0.238	0.036	0.304	0.580	0.890
0.86	-0.760	-0.451	-0.170	0.097	0.359	0.629	0.930
0.87	-0.670	-0.372	-0.099	0.161	0.417	0.681	0.973
0.88	-0.577	-0.290	-0.025	0.228	0.478	0.735	1.018
0.89	-0.480	-0.203	0.053	0.299	0.543	0.794	1.067
0.90	-0.377	-0.112	0.137	0.376	0.613	0.856	1.121
0.91	-0.269	-0.014	0.226	0.458	0.688	0.925	1.180
0.92	-0.152	0.093	0.324	0.548	0.771	1.000	1.246
0.93	-0.025	0.209	0.432	0.648	0.864	1.085	1.320
0.94	0.117	0.341	0.554	0.762	0.970	1.183	1.407
0.95	0.281	0.492	0.696	0.896	1.095	1.298	1.511
0.96	0.476	0.676	0.869	1.059	1.248	1.441	1.642
0.97	0.727	0.913	1.094	1.273	1.451	1.632	1.818
0.98	1.094	1.264	1.430	1.595	1.759	1.924	2.092

TABLE 9.4.3.   SCREENING FACTORS FOR NORMAL CONDITIONED ON t-DISTRIBUTION.

14 DEGREES OF FREEDOM   DELTA = 0.995

ρ →	0.70	0.75	0.80	0.85	0.90	0.95	1.00
γ ↓							
0.85	-0.834	-0.522	-0.236	0.036	0.304	0.584	0.904
0.86	-0.749	-0.447	-0.169	0.097	0.360	0.633	0.944
0.87	-0.662	-0.369	-0.098	0.161	0.418	0.685	0.987
0.88	-0.571	-0.288	-0.025	0.228	0.479	0.741	1.034
0.89	-0.475	-0.202	0.053	0.300	0.545	0.800	1.083
0.90	-0.375	-0.111	0.137	0.376	0.615	0.863	1.137
0.91	-0.267	-0.014	0.226	0.459	0.691	0.932	1.196
0.92	-0.152	0.092	0.324	0.549	0.775	1.008	1.263
0.93	-0.025	0.209	0.432	0.650	0.869	1.094	1.338
0.94	0.117	0.341	0.555	0.765	0.975	1.192	1.424
0.95	0.280	0.493	0.698	0.899	1.101	1.308	1.528
0.96	0.477	0.677	0.871	1.063	1.255	1.451	1.658
0.97	0.728	0.915	1.098	1.278	1.458	1.642	1.833
0.98	1.095	1.266	1.433	1.599	1.764	1.931	2.101

16 DEGREES OF FREEDOM   DELTA = 0.995

ρ →	0.70	0.75	0.80	0.85	0.90	0.95	1.00
γ ↓							
0.85	-0.825	-0.518	-0.235	0.036	0.305	0.587	0.915
0.86	-0.742	-0.444	-0.168	0.097	0.360	0.636	0.956
0.87	-0.656	-0.367	-0.098	0.161	0.419	0.689	1.000
0.88	-0.566	-0.286	-0.025	0.228	0.481	0.745	1.046
0.89	-0.472	-0.201	0.053	0.300	0.546	0.804	1.096
0.90	-0.373	-0.111	0.136	0.376	0.617	0.868	1.150
0.91	-0.266	-0.014	0.226	0.460	0.694	0.938	1.210
0.92	-0.151	0.092	0.324	0.551	0.778	1.015	1.276
0.93	-0.025	0.209	0.433	0.652	0.872	1.101	1.352
0.94	0.117	0.341	0.556	0.767	0.979	1.200	1.438
0.95	0.280	0.493	0.699	0.902	1.106	1.316	1.542
0.96	0.477	0.678	0.873	1.066	1.260	1.460	1.671
0.97	0.729	0.917	1.100	1.281	1.463	1.650	1.844
0.98	1.096	1.267	1.435	1.602	1.768	1.936	2.109

18 DEGREES OF FREEDOM   DELTA = 0.995

ρ →	0.70	0.75	0.80	0.85	0.90	0.95	1.00
γ ↓							
0.85	-0.817	-0.515	-0.234	0.036	0.305	0.589	0.925
0.86	-0.736	-0.442	-0.168	0.097	0.361	0.639	0.966
0.87	-0.651	-0.365	-0.098	0.160	0.419	0.692	1.010
0.88	-0.563	-0.285	-0.025	0.228	0.481	0.748	1.057
0.89	-0.470	-0.201	0.053	0.300	0.548	0.808	1.107
0.90	-0.371	-0.110	0.136	0.377	0.619	0.872	1.161
0.91	-0.265	-0.013	0.226	0.460	0.696	0.943	1.221
0.92	-0.151	0.092	0.324	0.551	0.780	1.020	1.288
0.93	-0.025	0.209	0.433	0.653	0.875	1.106	1.363
0.94	0.117	0.341	0.556	0.769	0.983	1.206	1.450
0.95	0.280	0.494	0.700	0.904	1.109	1.322	1.554
0.96	0.477	0.679	0.875	1.069	1.264	1.466	1.682
0.97	0.729	0.918	1.102	1.284	1.468	1.656	1.854
0.98	1.097	1.269	1.437	1.604	1.771	1.941	2.115

TABLE 9.4.4. *SCREENING FACTORS FOR NORMAL CONDITIONED ON* t-*DISTRIBUTION.*

20 DEGREES OF FREEDOM   DELTA = 0.995

ρ → γ ↓	0.70	0.75	0.80	0.85	0.90	0.95	1.00
0.85	-0.812	-0.512	-0.233	0.036	0.305	0.590	0.933
0.86	-0.731	-0.440	-0.167	0.096	0.361	0.641	0.975
0.87	-0.647	-0.364	-0.098	0.160	0.420	0.694	1.019
0.88	-0.560	-0.284	-0.025	0.228	0.482	0.751	1.066
0.89	-0.468	-0.200	0.053	0.300	0.549	0.811	1.116
0.90	-0.370	-0.110	0.136	0.377	0.620	0.876	1.171
0.91	-0.264	-0.013	0.226	0.460	0.697	0.946	1.231
0.92	-0.150	0.092	0.324	0.552	0.782	1.024	1.298
0.93	-0.025	0.209	0.433	0.654	0.877	1.111	1.373
0.94	0.117	0.341	0.557	0.770	0.985	1.211	1.460
0.95	0.280	0.494	0.701	0.905	1.112	1.328	1.563
0.96	0.477	0.679	0.876	1.071	1.268	1.472	1.692
0.97	0.730	0.919	1.103	1.287	1.471	1.661	1.862
0.98	1.098	1.270	1.439	1.606	1.774	1.944	2.121

22 DEGREES OF FREEDOM   DELTA = 0.995

ρ → γ ↓	0.70	0.75	0.80	0.85	0.90	0.95	1.00
0.85	-0.807	-0.510	-0.233	0.035	0.305	0.592	0.940
0.86	-0.727	-0.438	-0.167	0.096	0.361	0.643	0.982
0.87	-0.645	-0.363	-0.098	0.160	0.420	0.696	1.026
0.88	-0.558	-0.284	-0.025	0.228	0.483	0.753	1.073
0.89	-0.466	-0.200	0.053	0.300	0.549	0.813	1.124
0.90	-0.369	-0.110	0.136	0.377	0.621	0.879	1.179
0.91	-0.264	-0.013	0.226	0.461	0.698	0.950	1.239
0.92	-0.150	0.092	0.324	0.553	0.784	1.028	1.306
0.93	-0.025	0.209	0.433	0.655	0.879	1.115	1.382
0.94	0.117	0.341	0.557	0.771	0.988	1.215	1.469
0.95	0.280	0.494	0.701	0.907	1.115	1.332	1.572
0.96	0.477	0.680	0.877	1.073	1.271	1.476	1.700
0.97	0.730	0.920	1.105	1.289	1.474	1.666	1.870
0.98	1.098	1.271	1.440	1.608	1.776	1.948	2.125

24 DEGREES OF FREEDOM   DELTA = 0.995

ρ → γ ↓	0.70	0.75	0.80	0.85	0.90	0.95	1.00
0.85	-0.803	-0.509	-0.232	0.035	0.305	0.593	0.946
0.86	-0.724	-0.437	-0.167	0.096	0.361	0.644	0.988
0.87	-0.642	-0.362	-0.097	0.160	0.421	0.698	1.033
0.88	-0.556	-0.283	-0.024	0.228	0.483	0.755	1.080
0.89	-0.465	-0.199	0.053	0.300	0.550	0.816	1.131
0.90	-0.368	-0.110	0.136	0.377	0.622	0.881	1.186
0.91	-0.263	-0.013	0.226	0.461	0.700	0.952	1.247
0.92	-0.150	0.092	0.324	0.553	0.785	1.031	1.314
0.93	-0.025	0.209	0.433	0.655	0.881	1.118	1.389
0.94	0.117	0.341	0.557	0.772	0.990	1.218	1.476
0.95	0.280	0.494	0.702	0.908	1.117	1.336	1.579
0.96	0.477	0.680	0.878	1.074	1.273	1.480	1.707
0.97	0.730	0.920	1.106	1.290	1.477	1.669	1.876
0.98	1.099	1.271	1.441	1.609	1.778	1.950	2.129

TABLE 9.4.5. *SCREENING FACTORS FOR NORMAL CONDITIONED ON* t *-DISTRIBUTION.*

26 *DEGREES OF FREEDOM* DELTA = 0.995

ρ →	0.70	0.75	0.80	0.85	0.90	0.95	1.00
γ ↓							
0.85	-0.800	-0.507	-0.232	0.035	0.305	0.594	0.952
0.86	-0.722	-0.436	-0.166	0.096	0.362	0.645	0.994
0.87	-0.640	-0.361	-0.097	0.160	0.421	0.699	1.038
0.88	-0.554	-0.283	-0.024	0.228	0.484	0.756	1.086
0.89	-0.464	-0.199	0.053	0.300	0.551	0.817	1.137
0.90	-0.367	-0.110	0.136	0.377	0.622	0.883	1.192
0.91	-0.263	-0.013	0.226	0.461	0.700	0.955	1.253
0.92	-0.150	0.092	0.324	0.553	0.786	1.033	1.320
0.93	-0.025	0.209	0.434	0.656	0.882	1.121	1.396
0.94	0.117	0.341	0.558	0.773	0.991	1.221	1.483
0.95	0.280	0.494	0.702	0.909	1.119	1.339	1.586
0.96	0.477	0.680	0.878	1.075	1.275	1.484	1.713
0.97	0.731	0.921	1.107	1.292	1.479	1.673	1.881
0.98	1.099	1.272	1.442	1.610	1.780	1.953	2.133

28 *DEGREES OF FREEDOM* DELTA = 0.995

ρ →	0.70	0.75	0.80	0.85	0.90	0.95	1.00
γ ↓							
0.85	-0.797	-0.506	-0.232	0.035	0.306	0.595	0.957
0.86	-0.720	-0.435	-0.166	0.096	0.362	0.646	0.999
0.87	-0.638	-0.361	-0.097	0.160	0.421	0.700	1.044
0.88	-0.553	-0.282	-0.024	0.228	0.484	0.758	1.091
0.89	-0.463	-0.199	0.053	0.300	0.551	0.819	1.143
0.90	-0.366	-0.110	0.136	0.378	0.623	0.885	1.198
0.91	-0.263	-0.013	0.226	0.462	0.701	0.957	1.259
0.92	-0.150	0.092	0.324	0.554	0.787	1.035	1.326
0.93	-0.025	0.208	0.434	0.656	0.883	1.124	1.402
0.94	0.116	0.341	0.558	0.773	0.993	1.224	1.489
0.95	0.280	0.494	0.703	0.910	1.121	1.342	1.592
0.96	0.477	0.681	0.879	1.076	1.277	1.487	1.718
0.97	0.731	0.921	1.108	1.293	1.481	1.676	1.886
0.98	1.099	1.273	1.443	1.612	1.782	1.955	2.136

30 *DEGREES OF FREEDOM* DELTA = 0.995

ρ →	0.70	0.75	0.80	0.85	0.90	0.95	1.00
γ ↓							
0.85	-0.795	-0.505	-0.231	0.035	0.306	0.596	0.961
0.86	-0.718	-0.434	-0.166	0.096	0.362	0.647	1.003
0.87	-0.637	-0.360	-0.097	0.160	0.421	0.702	1.048
0.88	-0.552	-0.282	-0.024	0.228	0.484	0.759	1.096
0.89	-0.462	-0.199	0.053	0.300	0.551	0.820	1.148
0.90	-0.366	-0.110	0.136	0.378	0.624	0.887	1.203
0.91	-0.262	-0.013	0.226	0.462	0.702	0.958	1.264
0.92	-0.149	0.092	0.324	0.554	0.788	1.037	1.331
0.93	-0.025	0.208	0.434	0.657	0.884	1.126	1.407
0.94	0.116	0.341	0.558	0.774	0.994	1.227	1.494
0.95	0.280	0.494	0.703	0.911	1.122	1.345	1.597
0.96	0.477	0.681	0.880	1.077	1.279	1.489	1.723
0.97	0.731	0.922	1.108	1.294	1.483	1.678	1.890
0.98	1.100	1.273	1.443	1.613	1.783	1.957	2.139

TABLE 9.4.6. SCREENING FACTORS FOR NORMAL CONDITIONED ON t-DISTRIBUTION.

40 DEGREES OF FREEDOM DELTA = 0.995

ρ → γ ↓	0.70	0.75	0.80	0.85	0.90	0.95	1.00
0.85	-0.787	-0.502	-0.231	0.035	0.306	0.599	0.978
0.86	-0.711	-0.432	-0.165	0.096	0.362	0.651	1.021
0.87	-0.632	-0.358	-0.097	0.160	0.422	0.705	1.066
0.88	-0.548	-0.281	-0.024	0.228	0.485	0.763	1.115
0.89	-0.459	-0.198	0.053	0.300	0.553	0.825	1.166
0.90	-0.364	-0.109	0.136	0.378	0.626	0.892	1.222
0.91	-0.261	-0.013	0.226	0.462	0.704	0.965	1.284
0.92	-0.149	0.092	0.324	0.555	0.791	1.044	1.351
0.93	-0.024	0.208	0.434	0.658	0.888	1.134	1.427
0.94	0.116	0.341	0.559	0.776	0.998	1.235	1.514
0.95	0.280	0.495	0.704	0.913	1.127	1.354	1.616
0.96	0.477	0.682	0.881	1.081	1.284	1.499	1.742
0.97	0.732	0.923	1.111	1.298	1.489	1.687	1.906
0.98	1.101	1.275	1.446	1.616	1.788	1.964	2.149

60 DEGREES OF FREEDOM DELTA = 0.995

ρ → γ ↓	0.70	0.75	0.80	0.85	0.90	0.95	1.00
0.85	-0.779	-0.498	-0.230	0.035	0.306	0.602	0.998
0.86	-0.705	-0.429	-0.165	0.096	0.363	0.654	1.041
0.87	-0.627	-0.356	-0.097	0.160	0.423	0.709	1.087
0.88	-0.544	-0.279	-0.024	0.228	0.487	0.768	1.136
0.89	-0.456	-0.197	0.053	0.300	0.554	0.831	1.188
0.90	-0.362	-0.109	0.136	0.378	0.627	0.898	1.245
0.91	-0.260	-0.013	0.226	0.463	0.707	0.971	1.306
0.92	-0.148	0.091	0.324	0.556	0.794	1.052	1.374
0.93	-0.024	0.208	0.434	0.660	0.892	1.142	1.450
0.94	0.116	0.341	0.560	0.778	1.003	1.244	1.537
0.95	0.280	0.495	0.706	0.916	1.132	1.364	1.639
0.96	0.478	0.682	0.883	1.084	1.290	1.509	1.763
0.97	0.733	0.925	1.113	1.302	1.495	1.697	1.924
0.98	1.102	1.276	1.448	1.619	1.793	1.971	2.159

90 DEGREES OF FREEDOM DELTA = 0.995

ρ → γ ↓	0.70	0.75	0.80	0.85	0.90	0.95	1.00
0.85	-0.774	-0.496	-0.229	0.035	0.306	0.604	1.013
0.86	-0.701	-0.427	-0.164	0.096	0.363	0.656	1.057
0.87	-0.623	-0.355	-0.096	0.160	0.423	0.712	1.104
0.88	-0.542	-0.278	-0.024	0.228	0.487	0.771	1.153
0.89	-0.454	-0.197	0.053	0.300	0.555	0.834	1.206
0.90	-0.361	-0.109	0.136	0.379	0.629	0.902	1.263
0.91	-0.259	-0.013	0.226	0.463	0.709	0.976	1.324
0.92	-0.148	0.091	0.324	0.557	0.796	1.057	1.392
0.93	-0.024	0.208	0.435	0.661	0.894	1.147	1.468
0.94	0.116	0.341	0.560	0.780	1.006	1.250	1.555
0.95	0.280	0.495	0.707	0.918	1.136	1.370	1.656
0.96	0.478	0.683	0.885	1.087	1.294	1.516	1.779
0.97	0.733	0.926	1.115	1.305	1.499	1.704	1.937
0.98	1.103	1.278	1.450	1.622	1.796	1.976	2.166

TABLE 9.4.7. SCREENING FACTORS FOR NORMAL CONDITIONED ON t-DISTRIBUTION.

120 DEGREES OF FREEDOM   DELTA = 0.995

ρ → γ ↓	0.70	0.75	0.80	0.85	0.90	0.95	1.00
0.85	-0.772	-0.495	-0.229	0.035	0.306	0.605	1.022
0.86	-0.699	-0.427	-0.164	0.096	.0.364	0.657	1.066
0.87	-0.622	-0.354	-0.096	0.160	0.424	0.713	1.113
0.88	-0.540	-0.278	-0.024	0.228	0.488	0.773	1.163
0.89	-0.454	-0.196	0.053	0.300	0.556	0.836	1.216
0.90	-0.360	-0.109	0.136	0.379	0.629	0.904	1.273
0.91	-0.259	-0.013	0.225	0.464	0.709	0.978	1.335
0.92	-0.148	0.091	0.324	0.557	0.797	1.060	1.403
0.93	-0.024	0.208	0.435	0.661	0.896	1.150	1.478
0.94	0.116	0.341	0.560	0.780	1.007	1.254	1.565
0.95	0.280	0.495	0.707	0.919	1.138	1.374	1.665
0.96	0.478	0.683	0.885	1.088	1.296	1.520	1.787
0.97	0.733	0.926	1.116	1.306	1.501	1.708	1.944
0.98	1.103	1.278	1.451	1.623	1.798	1.979	2.170

150 DEGREES OF FREEDOM   DELTA = 0.995

ρ → γ ↓	0.70	0.75	0.80	0.85	0.90	0.95	1.00
0.85	-0.770	-0.494	-0.228	0.035	0.307	0.605	1.028
0.86	-0.697	-0.426	-0.164	0.096	0.364	0.658	1.072
0.87	-0.621	-0.354	-0.096	0.160	0.424	0.714	1.119
0.88	-0.540	-0.278	-0.024	0.228	0.488	0.774	1.169
0.89	-0.453	-0.196	0.053	0.300	0.556	0.837	1.222
0.90	-0.360	-0.109	0.135	0.379	0.630	0.905	1.279
0.91	-0.259	-0.013	0.225	0.464	0.710	0.980	1.341
0.92	-0.148	0.091	0.324	0.557	0.798	1.061	1.409
0.93	-0.024	0.208	0.435	0.662	0.896	1.152	1.485
0.94	0.116	0.341	0.560	0.781	1.008	1.256	1.571
0.95	0.280	0.496	0.707	0.920	1.139	1.376	1.671
0.96	0.478	0.684	0.886	1.089	1.298	1.522	1.793
0.97	0.733	0.926	1.116	1.307	1.503	1.710	1.948
0.98	1.103	1.279	1.451	1.624	1.799	1.981	2.171

∞ DEGREES OF FREEDOM   DELTA = 0.995

ρ → γ ↓	0.70	0.75	0.80	0.85	0.90	0.95	1.00
0.85	-0.764	-0.492	-0.228	0.035	0.307	0.608	1.055
0.86	-0.693	-0.424	-0.164	0.096	0.364	0.661	1.100
0.87	-0.617	-0.353	-0.096	0.160	0.425	0.718	1.147
0.88	-0.537	-0.277	-0.024	0.228	0.489	0.778	1.197
0.89	-0.451	-0.196	0.053	0.301	0.558	0.842	1.251
0.90	-0.358	-0.108	0.135	0.379	0.632	0.911	1.308
0.91	-0.258	-0.013	0.225	0.464	0.712	0.986	1.369
0.92	-0.148	0.091	0.324	0.558	0.801	1.068	1.437
0.93	-0.024	0.208	0.435	0.663	0.899	1.160	1.512
0.94	0.116	0.341	0.561	0.782	1.012	1.264	1.596
0.95	0.279	0.496	0.708	0.922	1.143	1.385	1.693
0.96	0.478	0.684	0.887	1.092	1.303	1.531	1.810
0.97	0.734	0.928	1.119	1.311	1.508	1.719	1.958
0.98	1.104	1.280	1.453	1.627	1.804	1.988	2.168

TABLE 10

Confidence limits on the correlation coefficient from a bivariate normal distribution for r = -0.95(0.05)0.95 and α = 0.005, 0.01, 0.025, 0.05, 0.1, 0.25, 0.75, 0.90, 0.95, 0.975, 0.99, 0.995.

TABLE 10.1. CONFIDENCE LIMITS ON THE CORRELATION COEFFICIENT.

SAMPLE SIZE = 3

α →	.005 ↓+	.01 ↓+	.025 ↓+	.05 ↓+	.10 ↓+	.25 ↓+	
R ↓							
-0.95	-0.9995	-0.9990	-0.9975	-0.9947	-0.9887	-0.9636	0.95
-0.90	-0.9990	-0.9980	-0.9950	-0.9897	-0.9779	-0.9308	0.90
-0.85	-0.9986	-0.9971	-0.9926	-0.9848	-0.9676	-0.9007	0.85
-0.80	-0.9981	-0.9962	-0.9902	-0.9799	-0.9575	-0.8724	0.80
-0.75	-0.9976	-0.9952	-0.9879	-0.9752	-0.9478	-0.8456	0.75
-0.70	-0.9972	-0.9943	-0.9856	-0.9705	-0.9382	-0.8201	0.70
-0.65	-0.9967	-0.9934	-0.9833	-0.9659	-0.9287	-0.7954	0.65
-0.60	-0.9963	-0.9925	-0.9810	-0.9612	-0.9193	-0.7715	0.60
-0.55	-0.9958	-0.9916	-0.9787	-0.9566	-0.9100	-0.7482	0.55
-0.50	-0.9953	-0.9906	-0.9763	-0.9519	-0.9006	-0.7254	0.50
-0.45	-0.9949	-0.9897	-0.9740	-0.9472	-0.8913	-0.7028	0.45
-0.40	-0.9944	-0.9887	-0.9716	-0.9424	-0.8818	-0.6805	0.40
-0.35	-0.9939	-0.9877	-0.9691	-0.9376	-0.8723	-0.6584	0.35
-0.30	-0.9934	-0.9867	-0.9666	-0.9326	-0.8627	-0.6363	0.30
-0.25	-0.9929	-0.9857	-0.9641	-0.9276	-0.8529	-0.6142	0.25
-0.20	-0.9923	-0.9846	-0.9615	-0.9224	-0.8429	-0.5919	0.20
-0.15	-0.9918	-0.9835	-0.9587	-0.9171	-0.8326	-0.5695	0.15
-0.10	-0.9912	-0.9824	-0.9559	-0.9116	-0.8221	-0.5467	0.10
-0.05	-0.9906	-0.9812	-0.9530	-0.9059	-0.8112	-0.5236	0.05
0.00	-0.9900	-0.9800	-0.9500	-0.9000	-0.8000	-0.5000	0.00
0.05	-0.9894	-0.9787	-0.9468	-0.8938	-0.7883	-0.4758	-0.05
0.10	-0.9887	-0.9773	-0.9435	-0.8873	-0.7761	-0.4509	-0.10
0.15	-0.9879	-0.9759	-0.9399	-0.8805	-0.7633	-0.4252	-0.15
0.20	-0.9872	-0.9744	-0.9362	-0.8732	-0.7498	-0.3985	-0.20
0.25	-0.9863	-0.9727	-0.9322	-0.8655	-0.7354	-0.3707	-0.25
0.30	-0.9854	-0.9709	-0.9278	-0.8571	-0.7201	-0.3414	-0.30
0.35	-0.9845	-0.9690	-0.9231	-0.8481	-0.7037	-0.3106	-0.35
0.40	-0.9834	-0.9669	-0.9180	-0.8383	-0.6858	-0.2778	-0.40
0.45	-0.9822	-0.9646	-0.9123	-0.8275	-0.6664	-0.2427	-0.45
0.50	-0.9809	-0.9619	-0.9060	-0.8155	-0.6449	-0.2048	-0.50
0.55	-0.9794	-0.9590	-0.8988	-0.8020	-0.6208	-0.1635	-0.55
0.60	-0.9776	-0.9555	-0.8905	-0.7864	-0.5935	-0.1180	-0.60
0.65	-0.9756	-0.9514	-0.8807	-0.7682	-0.5619	-0.0670	-0.65
0.70	-0.9731	-0.9465	-0.8689	-0.7464	-0.5246	-0.0090	-0.70
0.75	-0.9699	-0.9403	-0.8542	-0.7194	-0.4791	0.0586	-0.75
0.80	-0.9657	-0.9321	-0.8348	-0.6842	-0.4211	0.1397	-0.80
0.85	-0.9597	-0.9204	-0.8074	-0.6352	-0.3427	0.2411	-0.85
0.90	-0.9499	-0.9012	-0.7634	-0.5583	-0.2250	0.3766	-0.90
0.95	-0.9283	-0.8597	-0.6704	-0.4031	-0.0070	0.5802	-0.95

							↑ R
α →	.995 ↑-	.99 ↑-	.975 ↑-	.95 ↑-	.90 ↑-	.75 ↑-	

TABLE 10.2. *CONFIDENCE LIMITS ON THE CORRELATION COEFFICIENT.*

SAMPLE SIZE = 4

α →	.005 ↓+	.01 ↓+	.025 ↓+	.05 ↓+	.10 ↓+	.25 ↓+	
R ↓							
-0.95	-0.9985	-0.9975	-0.9953	-0.9921	-0.9863	-0.9677	0.95
-0.90	-0.9969	-0.9951	-0.9905	-0.9842	-0.9726	-0.9363	0.90
-0.85	-0.9954	-0.9925	-0.9857	-0.9762	-0.9590	-0.9058	0.85
-0.80	-0.9938	-0.9899	-0.9808	-0.9681	-0.9452	-0.8759	0.80
-0.75	-0.9921	-0.9873	-0.9758	-0.9599	-0.9314	-0.8465	0.75
-0.70	-0.9904	-0.9846	-0.9707	-0.9516	-0.9175	-0.8175	0.70
-0.65	-0.9887	-0.9818	-0.9655	-0.9431	-0.9034	-0.7887	0.65
-0.60	-0.9869	-0.9790	-0.9602	-0.9344	-0.8891	-0.7602	0.60
-0.55	-0.9851	-0.9760	-0.9547	-0.9255	-0.8746	-0.7317	0.55
-0.50	-0.9832	-0.9730	-0.9490	-0.9164	-0.8598	-0.7033	0.50
-0.45	-0.9812	-0.9699	-0.9432	-0.9070	-0.8448	-0.6749	0.45
-0.40	-0.9791	-0.9666	-0.9371	-0.8974	-0.8293	-0.6463	0.40
-0.35	-0.9770	-0.9632	-0.9309	-0.8874	-0.8135	-0.6175	0.35
-0.30	-0.9748	-0.9597	-0.9244	-0.8771	-0.7973	-0.5885	0.30
-0.25	-0.9724	-0.9560	-0.9176	-0.8664	-0.7805	-0.5591	0.25
-0.20	-0.9700	-0.9521	-0.9105	-0.8553	-0.7632	-0.5293	0.20
-0.15	-0.9674	-0.9480	-0.9031	-0.8437	-0.7453	-0.4989	0.15
-0.10	-0.9647	-0.9437	-0.8953	-0.8316	-0.7267	-0.4680	0.10
-0.05	-0.9618	-0.9392	-0.8871	-0.8188	-0.7073	-0.4364	0.05
0.00	-0.9587	-0.9343	-0.8783	-0.8054	-0.6870	-0.4040	0.00
0.05	-0.9555	-0.9292	-0.8691	-0.7912	-0.6658	-0.3706	-0.05
0.10	-0.9519	-0.9237	-0.8592	-0.7761	-0.6434	-0.3362	-0.10
0.15	-0.9481	-0.9177	-0.8486	-0.7601	-0.6198	-0.3007	-0.15
0.20	-0.9440	-0.9113	-0.8372	-0.7428	-0.5948	-0.2637	-0.20
0.25	-0.9395	-0.9043	-0.8248	-0.7243	-0.5680	-0.2252	-0.25
0.30	-0.9346	-0.8966	-0.8114	-0.7042	-0.5394	-0.1850	-0.30
0.35	-0.9291	-0.8881	-0.7966	-0.6824	-0.5085	-0.1426	-0.35
0.40	-0.9230	-0.8786	-0.7802	-0.6583	-0.4750	-0.0980	-0.40
0.45	-0.9161	-0.8680	-0.7618	-0.6317	-0.4384	-0.0506	-0.45
0.50	-0.9082	-0.8559	-0.7411	-0.6019	-0.3981	0.0000	-0.50
0.55	-0.8990	-0.8419	-0.7174	-0.5682	-0.3532	0.0543	-0.55
0.60	-0.8882	-0.8254	-0.6899	-0.5294	-0.3026	0.1132	-0.60
0.65	-0.8752	-0.8057	-0.6572	-0.4842	-0.2447	0.1774	-0.65
0.70	-0.8590	-0.7813	-0.6175	-0.4301	-0.1775	0.2484	-0.70
0.75	-0.8382	-0.7502	-0.5676	-0.3639	-0.0977	0.3276	-0.75
0.80	-0.8098	-0.7084	-0.5023	-0.2794	0.0000	0.4176	-0.80
0.85	-0.7682	-0.6480	-0.4109	-0.1659	0.1243	0.5218	-0.85
0.90	-0.6983	-0.5490	-0.2691	0.0000	0.2922	0.6457	-0.90
0.95	-0.5431	-0.3406	0.0000	0.2820	0.5421	0.7989	-0.95

↑ *R*

α →	.995 ↑-	.99 ↑-	.975 ↑-	.95 ↑-	.90 ↑-	.75 ↑-

TABLE 10.3. CONFIDENCE LIMITS ON THE CORRELATION COEFFICIENT.

SAMPLE SIZE = 5

α →	.005 ↓+	.01 ↓+	.025 ↓+	.05 ↓+	.10 ↓+	.25 ↓+	
R ↓							
-0.95	-0.9972	-0.9960	-0.9932	-0.9898	-0.9840	-0.9677	0.95
-0.90	-0.9944	-0.9918	-0.9863	-0.9794	-0.9679	-0.9357	0.90
-0.85	-0.9914	-0.9875	-0.9793	-0.9688	-0.9516	-0.9041	0.85
-0.80	-0.9884	-0.9832	-0.9720	-0.9580	-0.9351	-0.8727	0.80
-0.75	-0.9852	-0.9786	-0.9646	-0.9470	-0.9183	-0.8415	0.75
-0.70	-0.9820	-0.9739	-0.9569	-0.9357	-0.9012	-0.8103	0.70
-0.65	-0.9786	-0.9691	-0.9490	-0.9241	-0.8838	-0.7792	0.65
-0.60	-0.9751	-0.9641	-0.9409	-0.9122	-0.8661	-0.7480	0.60
-0.55	-0.9715	-0.9589	-0.9324	-0.8999	-0.8480	-0.7167	0.55
-0.50	-0.9677	-0.9535	-0.9237	-0.8872	-0.8295	-0.6853	0.50
-0.45	-0.9637	-0.9479	-0.9146	-0.8741	-0.8105	-0.6536	0.45
-0.40	-0.9596	-0.9420	-0.9052	-0.8606	-0.7909	-0.6216	0.40
-0.35	-0.9553	-0.9358	-0.8954	-0.8466	-0.7708	-0.5893	0.35
-0.30	-0.9507	-0.9293	-0.8851	-0.8320	-0.7501	-0.5566	0.30
-0.25	-0.9459	-0.9226	-0.8744	-0.8168	-0.7287	-0.5234	0.25
-0.20	-0.9409	-0.9154	-0.8631	-0.8009	-0.7065	-0.4896	0.20
-0.15	-0.9355	-0.9078	-0.8512	-0.7843	-0.6835	-0.4552	0.15
-0.10	-0.9298	-0.8998	-0.8387	-0.7669	-0.6595	-0.4201	0.10
-0.05	-0.9237	-0.8913	-0.8255	-0.7486	-0.6345	-0.3841	0.05
0.00	-0.9172	-0.8822	-0.8114	-0.7293	-0.6084	-0.3473	0.00
0.05	-0.9102	-0.8724	-0.7964	-0.7088	-0.5810	-0.3094	-0.05
0.10	-0.9027	-0.8620	-0.7804	-0.6871	-0.5522	-0.2705	-0.10
0.15	-0.8945	-0.8506	-0.7632	-0.6640	-0.5218	-0.2302	-0.15
0.20	-0.8856	-0.8383	-0.7447	-0.6391	-0.4896	-0.1886	-0.20
0.25	-0.8758	-0.8249	-0.7245	-0.6125	-0.4554	-0.1453	-0.25
0.30	-0.8651	-0.8101	-0.7026	-0.5836	-0.4189	-0.1003	-0.30
0.35	-0.8531	-0.7938	-0.6785	-0.5523	-0.3797	-0.0533	-0.35
0.40	-0.8397	-0.7755	-0.6519	-0.5180	-0.3374	-0.0040	-0.40
0.45	-0.8245	-0.7550	-0.6222	-0.4802	-0.2916	0.0478	-0.45
0.50	-0.8070	-0.7316	-0.5888	-0.4382	-0.2415	0.1026	-0.50
0.55	-0.7868	-0.7046	-0.5508	-0.3911	-0.1864	0.1608	-0.55
0.60	-0.7629	-0.6729	-0.5069	-0.3376	-0.1253	0.2229	-0.60
0.65	-0.7341	-0.6352	-0.4555	-0.2761	-0.0567	0.2895	-0.65
0.70	-0.6985	-0.5890	-0.3939	-0.2041	0.0212	0.3615	-0.70
0.75	-0.6528	-0.5307	-0.3184	-0.1182	0.1110	0.4399	-0.75
0.80	-0.5916	-0.4540	-0.2224	-0.0128	0.2166	0.5262	-0.80
0.85	-0.5036	-0.3467	-0.0945	0.1211	0.3438	0.6221	-0.85
0.90	-0.3624	-0.1819	0.0883	0.3005	0.5021	0.7302	-0.90
0.95	-0.0827	0.1195	0.3831	0.5609	0.7088	0.8543	-0.95

↑ R

α →	.995 ↑-	.99 ↑-	.975 ↑-	.95 ↑-	.90 ↑-	.75 ↑-

TABLE 10.4. CONFIDENCE LIMITS ON THE CORRELATION COEFFICIENT.

SAMPLE SIZE = 6

α →	.005 ↓+	.01 ↓+	.025 ↓+	.05 ↓+	.10 ↓+	.25 ↓+	
R ↓							
-0.95	-0.9959	-0.9944	-0.9914	-0.9878	-0.9821	-0.9671	0.95
-0.90	-0.9917	-0.9887	-0.9826	-0.9754	-0.9640	-0.9343	0.90
-0.85	-0.9874	-0.9828	-0.9736	-0.9627	-0.9456	-0.9016	0.85
-0.80	-0.9828	-0.9767	-0.9643	-0.9497	-0.9269	-0.8690	0.80
-0.75	-0.9782	-0.9703	-0.9547	-0.9363	-0.9078	-0.8364	0.75
-0.70	-0.9733	-0.9638	-0.9448	-0.9226	-0.8883	-0.8037	0.70
-0.65	-0.9683	-0.9570	-0.9346	-0.9085	-0.8685	-0.7709	0.65
-0.60	-0.9630	-0.9499	-0.9240	-0.8940	-0.8482	-0.7379	0.60
-0.55	-0.9576	-0.9426	-0.9131	-0.8790	-0.8274	-0.7047	0.55
-0.50	-0.9518	-0.9349	-0.9017	-0.8635	-0.8060	-0.6712	0.50
-0.45	-0.9459	-0.9269	-0.8899	-0.8475	-0.7841	-0.6374	0.45
-0.40	-0.9396	-0.9185	-0.8775	-0.8309	-0.7616	-0.6033	0.40
-0.35	-0.9330	-0.9098	-0.8647	-0.8136	-0.7384	-0.5687	0.35
-0.30	-0.9260	-0.9005	-0.8512	-0.7957	-0.7144	-0.5336	0.30
-0.25	-0.9187	-0.8908	-0.8371	-0.7770	-0.6897	-0.4979	0.25
-0.20	-0.9109	-0.8806	-0.8224	-0.7575	-0.6640	-0.4617	0.20
-0.15	-0.9027	-0.8697	-0.8068	-0.7371	-0.6374	-0.4248	0.15
-0.10	-0.8939	-0.8582	-0.7903	-0.7157	-0.6098	-0.3871	0.10
-0.05	-0.8846	-0.8460	-0.7729	-0.6932	-0.5810	-0.3485	0.05
0.00	-0.8745	-0.8329	-0.7545	-0.6694	-0.5509	-0.3091	0.00
0.05	-0.8637	-0.8189	-0.7348	-0.6444	-0.5194	-0.2686	-0.05
0.10	-0.8521	-0.8038	-0.7139	-0.6178	-0.4863	-0.2270	-0.10
0.15	-0.8395	-0.7875	-0.6913	-0.5895	-0.4515	-0.1841	-0.15
0.20	-0.8257	-0.7698	-0.6671	-0.5593	-0.4148	-0.1399	-0.20
0.25	-0.8106	-0.7505	-0.6409	-0.5269	-0.3759	-0.0941	-0.25
0.30	-0.7940	-0.7294	-0.6124	-0.4920	-0.3346	-0.0466	-0.30
0.35	-0.7755	-0.7060	-0.5813	-0.4543	-0.2905	0.0027	-0.35
0.40	-0.7549	-0.6800	-0.5470	-0.4133	-0.2433	0.0542	-0.40
0.45	-0.7315	-0.6509	-0.5090	-0.3684	-0.1924	0.1079	-0.45
0.50	-0.7049	-0.6178	-0.4666	-0.3190	-0.1374	0.1643	-0.50
0.55	-0.6741	-0.5800	-0.4188	-0.2641	-0.0774	0.2237	-0.55
0.60	-0.6380	-0.5361	-0.3642	-0.2025	-0.0117	0.2865	-0.60
0.65	-0.5948	-0.4842	-0.3011	-0.1328	0.0610	0.3531	-0.65
0.70	-0.5420	-0.4217	-0.2270	-0.0528	0.1420	0.4241	-0.70
0.75	-0.4755	-0.3444	-0.1381	0.0404	0.2333	0.5003	-0.75
0.80	-0.3885	-0.2456	-0.0287	0.1512	0.3377	0.5825	-0.80
0.85	-0.2680	-0.1133	0.1106	0.2862	0.4590	0.6718	-0.85
0.90	-0.0868	0.0767	0.2966	0.4563	0.6028	0.7698	-0.90
0.95	0.2290	0.3820	0.5640	0.6808	0.7781	0.8783	-0.95

↑ R

α →	.995 ↑-	.99 ↑-	.975 ↑-	.95 ↑-	.90 ↑-	.75 ↑-

TABLE 10.5. *CONFIDENCE LIMITS ON THE CORRELATION COEFFICIENT.*

SAMPLE SIZE = 7

α →	.005 ↓+	.01 ↓+	.025 ↓+	.05 ↓+	.10 ↓+	.25 ↓+	
R ↓							
-0.95	-0.9947	-0.9930	-0.9898	-0.9862	-0.9805	-0.9664	0.95
-0.90	-0.9892	-0.9858	-0.9794	-0.9720	-0.9608	-0.9329	0.90
-0.85	-0.9835	-0.9784	-0.9687	-0.9575	-0.9406	-0.8993	0.85
-0.80	-0.9776	-0.9707	-0.9576	-0.9426	-0.9201	-0.8656	0.80
-0.75	-0.9714	-0.9628	-0.9461	-0.9273	-0.8992	-0.8319	0.75
-0.70	-0.9651	-0.9545	-0.9343	-0.9116	-0.8778	-0.7980	0.70
-0.65	-0.9584	-0.9459	-0.9221	-0.8955	-0.8560	-0.7639	0.65
-0.60	-0.9515	-0.9370	-0.9095	-0.8788	-0.8337	-0.7296	0.60
-0.55	-0.9443	-0.9277	-0.8964	-0.8616	-0.8108	-0.6950	0.55
-0.50	-0.9367	-0.9180	-0.8828	-0.8438	-0.7873	-0.6600	0.50
-0.45	-0.9288	-0.9079	-0.8686	-0.8254	-0.7631	-0.6247	0.45
-0.40	-0.9205	-0.8973	-0.8538	-0.8063	-0.7383	-0.5889	0.40
-0.35	-0.9118	-0.8861	-0.8384	-0.7866	-0.7127	-0.5527	0.35
-0.30	-0.9026	-0.8744	-0.8223	-0.7660	-0.6864	-0.5160	0.30
-0.25	-0.8929	-0.8621	-0.8055	-0.7445	-0.6591	-0.4787	0.25
-0.20	-0.8826	-0.8491	-0.7877	-0.7222	-0.6309	-0.4407	0.20
-0.15	-0.8716	-0.8354	-0.7691	-0.6988	-0.6016	-0.4020	0.15
-0.10	-0.8600	-0.8208	-0.7495	-0.6743	-0.5713	-0.3626	0.10
-0.05	-0.8476	-0.8053	-0.7287	-0.6486	-0.5397	-0.3223	0.05
0.00	-0.8343	-0.7887	-0.7067	-0.6215	-0.5067	-0.2811	0.00
0.05	-0.8201	-0.7710	-0.6834	-0.5929	-0.4723	-0.2389	-0.05
0.10	-0.8047	-0.7520	-0.6584	-0.5627	-0.4363	-0.1956	-0.10
0.15	-0.7880	-0.7315	-0.6318	-0.5307	-0.3985	-0.1511	-0.15
0.20	-0.7699	-0.7093	-0.6032	-0.4966	-0.3587	-0.1052	-0.20
0.25	-0.7500	-0.6851	-0.5723	-0.4602	-0.3168	-0.0579	-0.25
0.30	-0.7282	-0.6587	-0.5389	-0.4211	-0.2724	-0.0091	-0.30
0.35	-0.7041	-0.6296	-0.5026	-0.3791	-0.2252	0.0416	-0.35
0.40	-0.6771	-0.5974	-0.4628	-0.3336	-0.1750	0.0941	-0.40
0.45	-0.6468	-0.5615	-0.4190	-0.2842	-0.1212	0.1489	-0.45
0.50	-0.6125	-0.5211	-0.3705	-0.2302	-0.0635	0.2059	-0.50
0.55	-0.5730	-0.4752	-0.3162	-0.1708	-0.0011	0.2656	-0.55
0.60	-0.5271	-0.4223	-0.2550	-0.1049	0.0666	0.3283	-0.60
0.65	-0.4729	-0.3607	-0.1851	-0.0311	0.1406	0.3943	-0.65
0.70	-0.4075	-0.2877	-0.1043	0.0522	0.2220	0.4640	-0.70
0.75	-0.3267	-0.1991	-0.0094	0.1474	0.3123	0.5380	-0.75
0.80	-0.2237	-0.0890	0.1044	0.2580	0.4136	0.6169	-0.80
0.85	-0.0863	0.0532	0.2443	0.3886	0.5283	0.7015	-0.85
0.90	0.1089	0.2461	0.4222	0.5466	0.6603	0.7927	-0.90
0.95	0.4156	0.5285	0.6595	0.7435	0.8148	0.8917	-0.95

↑ R

α →	.995 ↑-	.99 ↑-	.975 ↑-	.95 ↑-	.90 ↑-	.75 ↑-

TABLE 10.6. *CONFIDENCE LIMITS ON THE CORRELATION COEFFICIENT.*

SAMPLE SIZE = 8

α →	.005 ↓+	.01 ↓+	.025 ↓+	.05 ↓+	.10 ↓+	.25 ↓+	
R ↓							
-0.95	-0.9935	-0.9918	-0.9884	-0.9847	-0.9792	-0.9658	0.95
-0.90	-0.9868	-0.9832	-0.9766	-0.9691	-0.9580	-0.9315	0.90
-0.85	-0.9799	-0.9744	-0.9643	-0.9531	-0.9365	-0.8972	0.85
-0.80	-0.9727	-0.9653	-0.9517	-0.9366	-0.9145	-0.8627	0.80
-0.75	-0.9652	-0.9559	-0.9387	-0.9197	-0.8920	-0.8280	0.75
-0.70	-0.9574	-0.9461	-0.9252	-0.9023	-0.8691	-0.7931	0.70
-0.65	-0.9493	-0.9359	-0.9113	-0.8844	-0.8457	-0.7580	0.65
-0.60	-0.9408	-0.9253	-0.8968	-0.8659	-0.8217	-0.7226	0.60
-0.55	-0.9320	-0.9143	-0.8819	-0.8469	-0.7971	-0.6869	0.55
-0.50	-0.9227	-0.9028	-0.8663	-0.8272	-0.7718	-0.6508	0.50
-0.45	-0.9130	-0.8907	-0.8502	-0.8069	-0.7459	-0.6143	0.45
-0.40	-0.9029	-0.8782	-0.8334	-0.7858	-0.7192	-0.5774	0.40
-0.35	-0.8922	-0.8650	-0.8158	-0.7639	-0.6918	-0.5400	0.35
-0.30	-0.8809	-0.8511	-0.7975	-0.7411	-0.6635	-0.5020	0.30
-0.25	-0.8691	-0.8365	-0.7783	-0.7175	-0.6343	-0.4634	0.25
-0.20	-0.8565	-0.8211	-0.7582	-0.6928	-0.6041	-0.4242	0.20
-0.15	-0.8432	-0.8048	-0.7370	-0.6671	-0.5728	-0.3842	0.15
-0.10	-0.8290	-0.7876	-0.7147	-0.6401	-0.5403	-0.3435	0.10
-0.05	-0.8139	-0.7693	-0.6912	-0.6119	-0.5067	-0.3020	0.05
0.00	-0.7977	-0.7498	-0.6664	-0.5822	-0.4716	-0.2596	0.00
0.05	-0.7803	-0.7289	-0.6400	-0.5510	-0.4350	-0.2161	-0.05
0.10	-0.7616	-0.7066	-0.6120	-0.5180	-0.3969	-0.1717	-0.10
0.15	-0.7415	-0.6826	-0.5821	-0.4832	-0.3569	-0.1260	-0.15
0.20	-0.7195	-0.6566	-0.5501	-0.4462	-0.3150	-0.0791	-0.20
0.25	-0.6956	-0.6285	-0.5157	-0.4068	-0.2709	-0.0308	-0.25
0.30	-0.6694	-0.5978	-0.4786	-0.3648	-0.2244	0.0190	-0.30
0.35	-0.6405	-0.5642	-0.4384	-0.3197	-0.1752	0.0704	-0.35
0.40	-0.6084	-0.5272	-0.3946	-0.2712	-0.1231	0.1236	-0.40
0.45	-0.5725	-0.4861	-0.3468	-0.2189	-0.0676	0.1788	-0.45
0.50	-0.5320	-0.4402	-0.2940	-0.1620	-0.0083	0.2361	-0.50
0.55	-0.4858	-0.3884	-0.2355	-0.0998	0.0553	0.2958	-0.55
0.60	-0.4326	-0.3294	-0.1702	-0.0315	0.1238	0.3581	-0.60
0.65	-0.3704	-0.2613	-0.0964	0.0441	0.1980	0.4234	-0.65
0.70	-0.2965	-0.1818	-0.0123	0.1285	0.2788	0.4919	-0.70
0.75	-0.2067	-0.0871	0.0850	0.2235	0.3674	0.5640	-0.75
0.80	-0.0948	0.0282	0.1991	0.3319	0.4653	0.6403	-0.80
0.85	0.0496	0.1724	0.3358	0.4572	0.5744	0.7213	-0.85
0.90	0.2455	0.3598	0.5036	0.6044	0.6973	0.8077	-0.90
0.95	0.5307	0.6168	0.7166	0.7813	0.8375	0.9002	-0.95

↑ R

α →	.995 ↑-	.99 ↑-	.975 ↑-	.95 ↑-	.90 ↑-	.75 ↑-

TABLE 10.7. CONFIDENCE LIMITS ON THE CORRELATION COEFFICIENT.

SAMPLE SIZE = 9

α →	.005 ↓+	.01 ↓+	.025 ↓+	.05 ↓+	.10 ↓+	.25 ↓+	
R ↓							
-0.95	-0.9924	-0.9906	-0.9872	-0.9835	-0.9780	-0.9652	0.95
-0.90	-0.9846	-0.9809	-0.9740	-0.9666	-0.9557	-0.9303	0.90
-0.85	-0.9765	-0.9708	-0.9605	-0.9492	-0.9329	-0.8953	0.85
-0.80	-0.9681	-0.9604	-0.9465	-0.9314	-0.9096	-0.8601	0.80
-0.75	-0.9594	-0.9497	-0.9320	-0.9131	-0.8859	-0.8246	0.75
-0.70	-0.9503	-0.9385	-0.9171	-0.8942	-0.8617	-0.7890	0.70
-0.65	-0.9408	-0.9269	-0.9017	-0.8749	-0.8369	-0.7530	0.65
-0.60	-0.9309	-0.9148	-0.8857	-0.8549	-0.8115	-0.7168	0.60
-0.55	-0.9206	-0.9022	-0.8692	-0.8343	-0.7855	-0.6801	0.55
-0.50	-0.9099	-0.8891	-0.8520	-0.8130	-0.7588	-0.6431	0.50
-0.45	-0.8986	-0.8754	-0.8341	-0.7910	-0.7314	-0.6057	0.45
-0.40	-0.8867	-0.8610	-0.8155	-0.7682	-0.7032	-0.5678	0.40
-0.35	-0.8743	-0.8460	-0.7961	-0.7445	-0.6743	-0.5294	0.35
-0.30	-0.8612	-0.8302	-0.7759	-0.7200	-0.6444	-0.4905	0.30
-0.25	-0.8474	-0.8136	-0.7547	-0.6945	-0.6136	-0.4509	0.25
-0.20	-0.8327	-0.7962	-0.7326	-0.6680	-0.5818	-0.4107	0.20
-0.15	-0.8173	-0.7777	-0.7093	-0.6403	-0.5489	-0.3698	0.15
-0.10	-0.8008	-0.7582	-0.6849	-0.6114	-0.5148	-0.3281	0.10
-0.05	-0.7833	-0.7375	-0.6591	-0.5811	-0.4795	-0.2857	0.05
0.00	-0.7646	-0.7155	-0.6319	-0.5494	-0.4428	-0.2423	0.00
0.05	-0.7446	-0.6920	-0.6031	-0.5160	-0.4046	-0.1980	-0.05
0.10	-0.7230	-0.6669	-0.5726	-0.4809	-0.3648	-0.1526	-0.10
0.15	-0.6998	-0.6399	-0.5400	-0.4439	-0.3232	-0.1062	-0.15
0.20	-0.6747	-0.6109	-0.5053	-0.4047	-0.2797	-0.0585	-0.20
0.25	-0.6473	-0.5795	-0.4682	-0.3631	-0.2340	-0.0095	-0.25
0.30	-0.6174	-0.5454	-0.4283	-0.3189	-0.1861	0.0408	-0.30
0.35	-0.5846	-0.5082	-0.3852	-0.2716	-0.1355	0.0928	-0.35
0.40	-0.5483	-0.4674	-0.3385	-0.2210	-0.0821	0.1463	-0.40
0.45	-0.5079	-0.4224	-0.2877	-0.1666	-0.0255	0.2017	-0.45
0.50	-0.4627	-0.3723	-0.2321	-0.1079	0.0346	0.2591	-0.50
0.55	-0.4114	-0.3163	-0.1708	-0.0441	0.0988	0.3186	-0.55
0.60	-0.3528	-0.2531	-0.1028	0.0255	0.1675	0.3806	-0.60
0.65	-0.2850	-0.1808	-0.0269	0.1018	0.2414	0.4451	-0.65
0.70	-0.2054	-0.0974	0.0588	0.1862	0.3213	0.5125	-0.70
0.75	-0.1103	0.0005	0.1564	0.2803	0.4080	0.5832	-0.75
0.80	0.0060	0.1173	0.2693	0.3860	0.5028	0.6574	-0.80
0.85	0.1521	0.2600	0.4016	0.5061	0.6072	0.7356	-0.85
0.90	0.3430	0.4395	0.5599	0.6445	0.7231	0.8184	-0.90
0.95	0.6061	0.6744	0.7540	0.8065	0.8529	0.9063	-0.95
							↑ R
α →	.995 ↑-	.99 ↑-	.975 ↑-	.95 ↑-	.90 ↑-	.75 ↑-	

TABLE 10.8. *CONFIDENCE LIMITS ON THE CORRELATION COEFFICIENT.*

SAMPLE SIZE = 10

α →	.005 ↓+	.01 ↓+	.025 ↓+	.05 ↓+	.10 ↓+	.25 ↓+	
R ↓							
-0.95	-0.9915	-0.9895	-0.9861	-0.9824	-0.9770	-0.9647	0.95
-0.90	-0.9826	-0.9787	-0.9718	-0.9643	-0.9536	-0.9292	0.90
-0.85	-0.9734	-0.9676	-0.9570	-0.9458	-0.9298	-0.8936	0.85
-0.80	-0.9639	-0.9560	-0.9418	-0.9268	-0.9054	-0.8578	0.80
-0.75	-0.9540	-0.9440	-0.9262	-0.9072	-0.8806	-0.8217	0.75
-0.70	-0.9437	-0.9316	-0.9100	-0.8872	-0.8552	-0.7853	0.70
-0.65	-0.9330	-0.9187	-0.8932	-0.8665	-0.8292	-0.7487	0.65
-0.60	-0.9219	-0.9053	-0.8759	-0.8452	-0.8027	-0.7117	0.60
-0.55	-0.9102	-0.8913	-0.8579	-0.8232	-0.7755	-0.6743	0.55
-0.50	-0.8981	-0.8767	-0.8393	-0.8006	-0.7476	-0.6366	0.50
-0.45	-0.8853	-0.8615	-0.8199	-0.7772	-0.7190	-0.5984	0.45
-0.40	-0.8720	-0.8456	-0.7998	-0.7529	-0.6895	-0.5597	0.40
-0.35	-0.8579	-0.8290	-0.7788	-0.7278	-0.6593	-0.5205	0.35
-0.30	-0.8432	-0.8115	-0.7570	-0.7018	-0.6282	-0.4808	0.30
-0.25	-0.8276	-0.7932	-0.7341	-0.6747	-0.5961	-0.4405	0.25
-0.20	-0.8112	-0.7739	-0.7102	-0.6466	-0.5629	-0.3995	0.20
-0.15	-0.7938	-0.7536	-0.6852	-0.6173	-0.5287	-0.3578	0.15
-0.10	-0.7753	-0.7320	-0.6589	-0.5867	-0.4933	-0.3154	0.10
-0.05	-0.7557	-0.7093	-0.6312	-0.5548	-0.4567	-0.2722	0.05
0.00	-0.7348	-0.6851	-0.6021	-0.5214	-0.4187	-0.2281	0.00
0.05	-0.7124	-0.6594	-0.5713	-0.4864	-0.3792	-0.1831	-0.05
0.10	-0.6884	-0.6319	-0.5387	-0.4495	-0.3381	-0.1371	-0.10
0.15	-0.6626	-0.6025	-0.5040	-0.4108	-0.2952	-0.0900	-0.15
0.20	-0.6347	-0.5709	-0.4672	-0.3699	-0.2505	-0.0418	-0.20
0.25	-0.6045	-0.5368	-0.4279	-0.3266	-0.2037	0.0077	-0.25
0.30	-0.5716	-0.5000	-0.3857	-0.2807	-0.1546	0.0585	-0.30
0.35	-0.5355	-0.4599	-0.3404	-0.2318	-0.1031	0.1107	-0.35
0.40	-0.4958	-0.4162	-0.2916	-0.1797	-0.0488	0.1645	-0.40
0.45	-0.4519	-0.3681	-0.2386	-0.1239	0.0085	0.2200	-0.45
0.50	-0.4029	-0.3150	-0.1809	-0.0639	0.0691	0.2774	-0.50
0.55	-0.3478	-0.2560	-0.1178	0.0009	0.1335	0.3367	-0.55
0.60	-0.2853	-0.1897	-0.0482	0.0711	0.2022	0.3982	-0.60
0.65	-0.2136	-0.1148	0.0289	0.1476	0.2755	0.4620	-0.65
0.70	-0.1303	-0.0291	0.1151	0.2315	0.3543	0.5285	-0.70
0.75	-0.0321	0.0702	0.2123	0.3241	0.4393	0.5979	-0.75
0.80	0.0859	0.1868	0.3231	0.4271	0.5314	0.6704	-0.80
0.85	0.2310	0.3265	0.4510	0.5427	0.6319	0.7464	-0.85
0.90	0.4151	0.4978	0.6010	0.6738	0.7421	0.8264	-0.90
0.95	0.6584	0.7144	0.7803	0.8244	0.8640	0.9107	-0.95
							↑ R
α →	.995 ↑-	.99 ↑-	.975 ↑-	.95 ↑-	.90 ↑-	.75 ↑-	

TABLE 10.9. *CONFIDENCE LIMITS ON THE CORRELATION COEFFICIENT.*

SAMPLE SIZE = 11

α →	.005 ↓+	.01 ↓+	.025 ↓+	.05 ↓+	.10 ↓+	.25 ↓+	
R ↓							
-0.95	-0.9905	-0.9886	-0.9851	-0.9814	-0.9761	-0.9642	0.95
-0.90	-0.9807	-0.9768	-0.9698	-0.9623	-0.9518	-0.9283	0.90
-0.85	-0.9706	-0.9646	-0.9540	-0.9428	-0.9270	-0.8921	0.85
-0.80	-0.9600	-0.9520	-0.9377	-0.9227	-0.9017	-0.8557	0.80
-0.75	-0.9491	-0.9389	-0.9209	-0.9021	-0.8759	-0.8191	0.75
-0.70	-0.9377	-0.9253	-0.9035	-0.8809	-0.8495	-0.7821	0.70
-0.65	-0.9259	-0.9112	-0.8856	-0.8591	-0.8226	-0.7449	0.65
-0.60	-0.9135	-0.8966	-0.8671	-0.8366	-0.7950	-0.7073	0.60
-0.55	-0.9007	-0.8814	-0.8479	-0.8135	-0.7668	-0.6693	0.55
-0.50	-0.8872	-0.8655	-0.8280	-0.7896	-0.7378	-0.6309	0.50
-0.45	-0.8732	-0.8490	-0.8073	-0.7650	-0.7081	-0.5920	0.45
-0.40	-0.8584	-0.8317	-0.7858	-0.7395	-0.6776	-0.5527	0.40
-0.35	-0.8430	-0.8136	-0.7635	-0.7131	-0.6463	-0.5129	0.35
-0.30	-0.8267	-0.7947	-0.7402	-0.6858	-0.6141	-0.4725	0.30
-0.25	-0.8096	-0.7748	-0.7159	-0.6574	-0.5809	-0.4315	0.25
-0.20	-0.7915	-0.7539	-0.6905	-0.6280	-0.5467	-0.3899	0.20
-0.15	-0.7724	-0.7319	-0.6639	-0.5973	-0.5114	-0.3476	0.15
-0.10	-0.7522	-0.7087	-0.6360	-0.5654	-0.4749	-0.3045	0.10
-0.05	-0.7307	-0.6841	-0.6068	-0.5321	-0.4372	-0.2607	0.05
0.00	-0.7079	-0.6581	-0.5760	-0.4973	-0.3981	-0.2161	0.00
0.05	-0.6835	-0.6304	-0.5435	-0.4608	-0.3575	-0.1705	-0.05
0.10	-0.6574	-0.6009	-0.5092	-0.4226	-0.3154	-0.1240	-0.10
0.15	-0.6293	-0.5695	-0.4728	-0.3824	-0.2715	-0.0765	-0.15
0.20	-0.5991	-0.5358	-0.4342	-0.3402	-0.2259	-0.0278	-0.20
0.25	-0.5664	-0.4995	-0.3932	-0.2955	-0.1782	0.0220	-0.25
0.30	-0.5310	-0.4604	-0.3493	-0.2483	-0.1283	0.0731	-0.30
0.35	-0.4923	-0.4180	-0.3022	-0.1983	-0.0761	0.1256	-0.35
0.40	-0.4498	-0.3719	-0.2517	-0.1450	-0.0212	0.1795	-0.40
0.45	-0.4030	-0.3215	-0.1971	-0.0882	0.0365	0.2350	-0.45
0.50	-0.3511	-0.2661	-0.1380	-0.0274	0.0975	0.2922	-0.50
0.55	-0.2931	-0.2048	-0.0736	0.0380	0.1619	0.3513	-0.55
0.60	-0.2277	-0.1365	-0.0031	0.1084	0.2303	0.4124	-0.60
0.65	-0.1533	-0.0598	0.0746	0.1848	0.3031	0.4757	-0.65
0.70	-0.0677	0.0271	0.1608	0.2680	0.3809	0.5413	-0.70
0.75	0.0321	0.1267	0.2570	0.3591	0.4642	0.6096	-0.75
0.80	0.1503	0.2422	0.3656	0.4596	0.5540	0.6807	-0.80
0.85	0.2931	0.3785	0.4894	0.5712	0.6511	0.7549	-0.85
0.90	0.4700	0.5421	0.6323	0.6962	0.7568	0.8326	-0.90
0.95	0.6965	0.7437	0.7998	0.8378	0.8724	0.9142	-0.95
							↑ R
α →	.995 ↑-	.99 ↑-	.975 ↑-	.95 ↑-	.90 ↑-	.75 ↑-	

TABLE 10.10. *CONFIDENCE LIMITS ON THE CORRELATION COEFFICIENT.*

SAMPLE SIZE = 12

α →	.005 ↓+	.01 ↓+	.025 ↓+	.05 ↓+	.10 ↓+	.25 ↓+	
R ↓							
-0.95	-0.9897	-0.9877	-0.9842	-0.9805	-0.9753	-0.9638	0.95
-0.90	-0.9790	-0.9750	-0.9679	-0.9605	-0.9502	-0.9274	0.90
-0.85	-0.9679	-0.9618	-0.9512	-0.9400	-0.9246	-0.8908	0.85
-0.80	-0.9564	-0.9482	-0.9339	-0.9190	-0.8984	-0.8539	0.80
-0.75	-0.9445	-0.9342	-0.9161	-0.8974	-0.8717	-0.8167	0.75
-0.70	-0.9321	-0.9196	-0.8977	-0.8753	-0.8445	-0.7793	0.70
-0.65	-0.9193	-0.9044	-0.8788	-0.8525	-0.8167	-0.7415	0.65
-0.60	-0.9058	-0.8887	-0.8591	-0.8290	-0.7882	-0.7034	0.60
-0.55	-0.8919	-0.8724	-0.8388	-0.8048	-0.7591	-0.6648	0.55
-0.50	-0.8773	-0.8553	-0.8178	-0.7799	-0.7292	-0.6259	0.50
-0.45	-0.8620	-0.8376	-0.7960	-0.7542	-0.6986	-0.5865	0.45
-0.40	-0.8460	-0.8191	-0.7733	-0.7276	-0.6672	-0.5466	0.40
-0.35	-0.8292	-0.7997	-0.7497	-0.7001	-0.6349	-0.5062	0.35
-0.30	-0.8116	-0.7794	-0.7252	-0.6717	-0.6018	-0.4653	0.30
-0.25	-0.7931	-0.7582	-0.6996	-0.6422	-0.5676	-0.4238	0.25
-0.20	-0.7736	-0.7358	-0.6729	-0.6115	-0.5325	-0.3816	0.20
-0.15	-0.7530	-0.7124	-0.6450	-0.5797	-0.4963	-0.3388	0.15
-0.10	-0.7312	-0.6876	-0.6158	-0.5466	-0.4589	-0.2952	0.10
-0.05	-0.7081	-0.6615	-0.5851	-0.5121	-0.4202	-0.2509	0.05
0.00	-0.6835	-0.6339	-0.5529	-0.4762	-0.3802	-0.2058	0.00
0.05	-0.6574	-0.6045	-0.5190	-0.4385	-0.3388	-0.1598	-0.05
0.10	-0.6294	-0.5734	-0.4833	-0.3992	-0.2958	-0.1129	-0.10
0.15	-0.5994	-0.5401	-0.4455	-0.3579	-0.2512	-0.0650	-0.15
0.20	-0.5672	-0.5046	-0.4055	-0.3145	-0.2048	-0.0160	-0.20
0.25	-0.5325	-0.4665	-0.3629	-0.2688	-0.1564	0.0341	-0.25
0.30	-0.4948	-0.4256	-0.3176	-0.2205	-0.1059	0.0854	-0.30
0.35	-0.4539	-0.3813	-0.2692	-0.1695	-0.0531	0.1381	-0.35
0.40	-0.4092	-0.3333	-0.2174	-0.1154	0.0022	0.1921	-0.40
0.45	-0.3601	-0.2811	-0.1616	-0.0579	0.0602	0.2476	-0.45
0.50	-0.3059	-0.2239	-0.1014	0.0035	0.1213	0.3047	-0.50
0.55	-0.2457	-0.1609	-0.0362	0.0691	0.1857	0.3635	-0.55
0.60	-0.1781	-0.0912	0.0349	0.1396	0.2538	0.4242	-0.60
0.65	-0.1019	-0.0134	0.1128	0.2157	0.3259	0.4870	-0.65
0.70	-0.0149	0.0741	0.1986	0.2981	0.4027	0.5519	-0.70
0.75	0.0856	0.1734	0.2937	0.3877	0.4845	0.6192	-0.75
0.80	0.2031	0.2874	0.4001	0.4859	0.5722	0.6891	-0.80
0.85	0.3430	0.4200	0.5200	0.5940	0.6665	0.7618	-0.85
0.90	0.5130	0.5768	0.6568	0.7140	0.7684	0.8377	-0.90
0.95	0.7253	0.7660	0.8148	0.8482	0.8791	0.9169	-0.95
							↑ R
α →	.995 ↑-	.99 ↑-	.975 ↑-	.95 ↑-	.90 ↑-	.75 ↑-	

TABLE 10.11. *CONFIDENCE LIMITS ON THE CORRELATION COEFFICIENT.*

SAMPLE SIZE = 13

α →	.005 ↓+	.01 ↓+	.025 ↓+	.05 ↓+	.10 ↓+	.25 ↓+	
R ↓							
-0.95	-0.9889	-0.9869	-0.9834	-0.9797	-0.9746	-0.9634	0.95
-0.90	-0.9774	-0.9733	-0.9662	-0.9589	-0.9487	-0.9266	0.90
-0.85	-0.9655	-0.9593	-0.9486	-0.9376	-0.9223	-0.8895	0.85
-0.80	-0.9531	-0.9448	-0.9305	-0.9157	-0.8955	-0.8522	0.80
-0.75	-0.9403	-0.9298	-0.9118	-0.8933	-0.8680	-0.8146	0.75
-0.70	-0.9270	-0.9143	-0.8925	-0.8702	-0.8400	-0.7767	0.70
-0.65	-0.9131	-0.8982	-0.8725	-0.8465	-0.8114	-0.7385	0.65
-0.60	-0.8987	-0.8815	-0.8519	-0.8221	-0.7821	-0.6999	0.60
-0.55	-0.8837	-0.8641	-0.8306	-0.7970	-0.7522	-0.6609	0.55
-0.50	-0.8680	-0.8460	-0.8086	-0.7712	-0.7215	-0.6214	0.50
-0.45	-0.8517	-0.8271	-0.7857	-0.7445	-0.6901	-0.5816	0.45
-0.40	-0.8346	-0.8075	-0.7620	-0.7169	-0.6579	-0.5412	0.40
-0.35	-0.8166	-0.7870	-0.7373	-0.6885	-0.6248	-0.5003	0.35
-0.30	-0.7978	-0.7655	-0.7117	-0.6590	-0.5908	-0.4589	0.30
-0.25	-0.7780	-0.7430	-0.6850	-0.6285	-0.5559	-0.4169	0.25
-0.20	-0.7572	-0.7195	-0.6571	-0.5969	-0.5200	-0.3743	0.20
-0.15	-0.7352	-0.6947	-0.6281	-0.5641	-0.4829	-0.3311	0.15
-0.10	-0.7120	-0.6686	-0.5977	-0.5300	-0.4447	-0.2871	0.10
-0.05	-0.6875	-0.6411	-0.5658	-0.4945	-0.4053	-0.2424	0.05
0.00	-0.6614	-0.6120	-0.5324	-0.4575	-0.3646	-0.1968	0.00
0.05	-0.6336	-0.5813	-0.4973	-0.4189	-0.3224	-0.1505	-0.05
0.10	-0.6041	-0.5486	-0.4603	-0.3785	-0.2787	-0.1032	-0.10
0.15	-0.5724	-0.5139	-0.4213	-0.3363	-0.2335	-0.0550	-0.15
0.20	-0.5385	-0.4768	-0.3801	-0.2920	-0.1864	-0.0058	-0.20
0.25	-0.5020	-0.4372	-0.3363	-0.2454	-0.1375	0.0446	-0.25
0.30	-0.4625	-0.3947	-0.2899	-0.1963	-0.0865	0.0960	-0.30
0.35	-0.4198	-0.3489	-0.2404	-0.1446	-0.0333	0.1488	-0.35
0.40	-0.3732	-0.2994	-0.1875	-0.0898	0.0223	0.2028	-0.40
0.45	-0.3223	-0.2456	-0.1309	-0.0318	0.0805	0.2583	-0.45
0.50	-0.2662	-0.1871	-0.0699	0.0299	0.1416	0.3152	-0.50
0.55	-0.2042	-0.1229	-0.0040	0.0957	0.2059	0.3738	-0.55
0.60	-0.1351	-0.0522	0.0674	0.1662	0.2736	0.4342	-0.60
0.65	-0.0576	0.0262	0.1452	0.2418	0.3451	0.4965	-0.65
0.70	0.0302	0.1138	0.2304	0.3233	0.4209	0.5607	-0.70
0.75	0.1307	0.2125	0.3243	0.4115	0.5015	0.6272	-0.75
0.80	0.2471	0.3248	0.4285	0.5076	0.5874	0.6961	-0.80
0.85	0.3839	0.4540	0.5451	0.6126	0.6793	0.7675	-0.85
0.90	0.5475	0.6047	0.6766	0.7283	0.7780	0.8418	-0.90
0.95	0.7478	0.7835	0.8266	0.8566	0.8845	0.9192	-0.95

↑ R

α →	.995 ↑-	.99 ↑-	.975 ↑-	.95 ↑-	.90 ↑-	.75 ↑-

TABLE 10.12. CONFIDENCE LIMITS ON THE CORRELATION COEFFICIENT.

SAMPLE SIZE = 14

α →	.005 ↓+	.01 ↓+	.025 ↓+	.05 ↓+	.10 ↓+	.25 ↓+	
R ↓							
-0.95	-0.9881	-0.9861	-0.9826	-0.9790	-0.9739	-0.9630	0.95
-0.90	-0.9759	-0.9718	-0.9647	-0.9574	-0.9474	-0.9259	0.90
-0.85	-0.9632	-0.9570	-0.9463	-0.9353	-0.9203	-0.8884	0.85
-0.80	-0.9500	-0.9417	-0.9273	-0.9127	-0.8928	-0.8507	0.80
-0.75	-0.9363	-0.9258	-0.9078	-0.8894	-0.8646	-0.8127	0.75
-0.70	-0.9222	-0.9094	-0.8876	-0.8656	-0.8359	-0.7744	0.70
-0.65	-0.9074	-0.8924	-0.8668	-0.8411	-0.8066	-0.7358	0.65
-0.60	-0.8921	-0.8748	-0.8453	-0.8159	-0.7766	-0.6967	0.60
-0.55	-0.8762	-0.8564	-0.8231	-0.7900	-0.7460	-0.6573	0.55
-0.50	-0.8595	-0.8374	-0.8002	-0.7633	-0.7146	-0.6175	0.50
-0.45	-0.8421	-0.8176	-0.7764	-0.7357	-0.6825	-0.5772	0.45
-0.40	-0.8240	-0.7969	-0.7517	-0.7073	-0.6495	-0.5364	0.40
-0.35	-0.8050	-0.7753	-0.7261	-0.6780	-0.6157	-0.4951	0.35
-0.30	-0.7850	-0.7528	-0.6994	-0.6477	-0.5811	-0.4533	0.30
-0.25	-0.7641	-0.7292	-0.6717	-0.6163	-0.5454	-0.4109	0.25
-0.20	-0.7421	-0.7045	-0.6429	-0.5838	-0.5088	-0.3679	0.20
-0.15	-0.7189	-0.6785	-0.6128	-0.5501	-0.4710	-0.3242	0.15
-0.10	-0.6944	-0.6513	-0.5813	-0.5151	-0.4322	-0.2799	0.10
-0.05	-0.6686	-0.6226	-0.5484	-0.4787	-0.3921	-0.2348	0.05
0.00	-0.6411	-0.5923	-0.5140	-0.4409	-0.3507	-0.1890	0.00
0.05	-0.6120	-0.5602	-0.4778	-0.4014	-0.3079	-0.1423	-0.05
0.10	-0.5810	-0.5263	-0.4398	-0.3602	-0.2637	-0.0948	-0.10
0.15	-0.5479	-0.4903	-0.3997	-0.3172	-0.2178	-0.0463	-0.15
0.20	-0.5125	-0.4519	-0.3575	-0.2721	-0.1703	0.0032	-0.20
0.25	-0.4744	-0.4109	-0.3128	-0.2248	-0.1209	0.0536	-0.25
0.30	-0.4334	-0.3671	-0.2653	-0.1751	-0.0695	0.1053	-0.30
0.35	-0.3891	-0.3200	-0.2150	-0.1227	-0.0160	0.1580	-0.35
0.40	-0.3410	-0.2693	-0.1613	-0.0675	0.0398	0.2121	-0.40
0.45	-0.2885	-0.2144	-0.1039	-0.0091	0.0981	0.2675	-0.45
0.50	-0.2311	-0.1547	-0.0424	0.0529	0.1592	0.3243	-0.50
0.55	-0.1677	-0.0896	0.0239	0.1188	0.2233	0.3827	-0.55
0.60	-0.0975	-0.0183	0.0954	0.1890	0.2906	0.4428	-0.60
0.65	-0.0191	0.0605	0.1730	0.2641	0.3616	0.5046	-0.65
0.70	0.0691	0.1480	0.2576	0.3448	0.4365	0.5683	-0.70
0.75	0.1692	0.2459	0.3503	0.4317	0.5159	0.6340	-0.75
0.80	0.2842	0.3564	0.4525	0.5259	0.6002	0.7020	-0.80
0.85	0.4179	0.4823	0.5660	0.6282	0.6899	0.7724	-0.85
0.90	0.5758	0.6276	0.6929	0.7402	0.7859	0.8453	-0.90
0.95	0.7658	0.7976	0.8363	0.8634	0.8889	0.9211	-0.95

↑ R

α →	.995 ↑-	.99 ↑-	.975 ↑-	.95 ↑-	.90 ↑-	.75 ↑-

TABLE 10.13. *CONFIDENCE LIMITS ON THE CORRELATION COEFFICIENT.*

SAMPLE SIZE = 15

α →	.005 ↓+	.01 ↓+	.025 ↓+	.05 ↓+	.10 ↓+	.25 ↓+	
R ↓							
-0.95	-0.9874	-0.9854	-0.9819	-0.9783	-0.9733	-0.9627	0.95
-0.90	-0.9745	-0.9703	-0.9633	-0.9560	-0.9462	-0.9252	0.90
-0.85	-0.9610	-0.9548	-0.9441	-0.9333	-0.9185	-0.8874	0.85
-0.80	-0.9471	-0.9387	-0.9244	-0.9099	-0.8903	-0.8494	0.80
-0.75	-0.9327	-0.9221	-0.9041	-0.8860	-0.8616	-0.8110	0.75
-0.70	-0.9177	-0.9049	-0.8832	-0.8614	-0.8322	-0.7723	0.70
-0.65	-0.9021	-0.8871	-0.8616	-0.8361	-0.8023	-0.7333	0.65
-0.60	-0.8860	-0.8686	-0.8393	-0.8102	-0.7717	-0.6939	0.60
-0.55	-0.8691	-0.8494	-0.8163	-0.7835	-0.7404	-0.6541	0.55
-0.50	-0.8516	-0.8295	-0.7925	-0.7560	-0.7083	-0.6139	0.50
-0.45	-0.8333	-0.8087	-0.7678	-0.7277	-0.6756	-0.5732	0.45
-0.40	-0.8142	-0.7871	-0.7423	-0.6986	-0.6420	-0.5321	0.40
-0.35	-0.7942	-0.7646	-0.7158	-0.6685	-0.6076	-0.4904	0.35
-0.30	-0.7732	-0.7411	-0.6883	-0.6373	-0.5722	-0.4482	0.30
-0.25	-0.7513	-0.7165	-0.6597	-0.6052	-0.5360	-0.4055	0.25
-0.20	-0.7282	-0.6908	-0.6299	-0.5719	-0.4987	-0.3621	0.20
-0.15	-0.7039	-0.6638	-0.5989	-0.5374	-0.4604	-0.3181	0.15
-0.10	-0.6783	-0.6354	-0.5665	-0.5017	-0.4209	-0.2735	0.10
-0.05	-0.6512	-0.6056	-0.5327	-0.4645	-0.3802	-0.2281	0.05
0.00	-0.6226	-0.5742	-0.4973	-0.4259	-0.3383	-0.1820	0.00
0.05	-0.5922	-0.5411	-0.4602	-0.3857	-0.2950	-0.1350	-0.05
0.10	-0.5600	-0.5060	-0.4213	-0.3438	-0.2502	-0.0873	-0.10
0.15	-0.5256	-0.4688	-0.3803	-0.3001	-0.2039	-0.0386	-0.15
0.20	-0.4888	-0.4293	-0.3372	-0.2543	-0.1559	0.0110	-0.20
0.25	-0.4495	-0.3873	-0.2916	-0.2064	-0.1062	0.0617	-0.25
0.30	-0.4071	-0.3423	-0.2434	-0.1562	-0.0545	0.1134	-0.30
0.35	-0.3615	-0.2941	-0.1923	-0.1033	-0.0008	0.1662	-0.35
0.40	-0.3121	-0.2424	-0.1380	-0.0477	0.0552	0.2203	-0.40
0.45	-0.2584	-0.1865	-0.0800	0.0110	0.1136	0.2756	-0.45
0.50	-0.1997	-0.1260	-0.0181	0.0731	0.1746	0.3323	-0.50
0.55	-0.1354	-0.0603	0.0484	0.1389	0.2385	0.3905	-0.55
0.60	-0.0643	0.0115	0.1199	0.2089	0.3055	0.4502	-0.60
0.65	0.0146	0.0904	0.1972	0.2835	0.3759	0.5116	-0.65
0.70	0.1029	0.1776	0.2811	0.3634	0.4500	0.5748	-0.70
0.75	0.2025	0.2746	0.3726	0.4491	0.5283	0.6399	-0.75
0.80	0.3160	0.3833	0.4730	0.5415	0.6111	0.7071	-0.80
0.85	0.4467	0.5062	0.5837	0.6415	0.6990	0.7765	-0.85
0.90	0.5994	0.6467	0.7066	0.7502	0.7927	0.8483	-0.90
0.95	0.7806	0.8092	0.8443	0.8692	0.8927	0.9228	-0.95

↑ *R*

α →	.995 ↑-	.99 ↑-	.975 ↑-	.95 ↑-	.90 ↑-	.75 ↑-

TABLE 10.14. *CONFIDENCE LIMITS ON THE CORRELATION COEFFICIENT.*

SAMPLE SIZE = 16

α →	.005 ↓+	.01 ↓+	.025 ↓+	.05 ↓+	.10 ↓+	.25 ↓+	
R ↓							
-0.95	-0.9868	-0.9847	-0.9812	-0.9776	-0.9728	-0.9624	0.95
-0.90	-0.9731	-0.9690	-0.9619	-0.9548	-0.9451	-0.9246	0.90
-0.85	-0.9590	-0.9528	-0.9421	-0.9314	-0.9168	-0.8865	0.85
-0.80	-0.9444	-0.9360	-0.9217	-0.9074	-0.8881	-0.8481	0.80
-0.75	-0.9292	-0.9186	-0.9007	-0.8827	-0.8588	-0.8094	0.75
-0.70	-0.9135	-0.9007	-0.8791	-0.8575	-0.8288	-0.7704	0.70
-0.65	-0.8972	-0.8821	-0.8567	-0.8316	-0.7983	-0.7310	0.65
-0.60	-0.8802	-0.8628	-0.8337	-0.8050	-0.7671	-0.6913	0.60
-0.55	-0.8626	-0.8428	-0.8099	-0.7776	-0.7352	-0.6512	0.55
-0.50	-0.8442	-0.8221	-0.7854	-0.7494	-0.7026	-0.6106	0.50
-0.45	-0.8250	-0.8005	-0.7599	-0.7204	-0.6693	-0.5696	0.45
-0.40	-0.8050	-0.7781	-0.7336	-0.6906	-0.6351	-0.5281	0.40
-0.35	-0.7841	-0.7547	-0.7063	-0.6597	-0.6001	-0.4862	0.35
-0.30	-0.7623	-0.7303	-0.6780	-0.6279	-0.5642	-0.4436	0.30
-0.25	-0.7393	-0.7048	-0.6486	-0.5951	-0.5274	-0.4006	0.25
-0.20	-0.7153	-0.6781	-0.6180	-0.5611	-0.4896	-0.3569	0.20
-0.15	-0.6900	-0.6502	-0.5862	-0.5259	-0.4507	-0.3126	0.15
-0.10	-0.6633	-0.6209	-0.5530	-0.4895	-0.4107	-0.2677	0.10
-0.05	-0.6352	-0.5901	-0.5184	-0.4516	-0.3695	-0.2221	0.05
0.00	-0.6055	-0.5577	-0.4821	-0.4124	-0.3271	-0.1757	0.00
0.05	-0.5740	-0.5236	-0.4442	-0.3715	-0.2833	-0.1285	-0.05
0.10	-0.5406	-0.4875	-0.4045	-0.3290	-0.2381	-0.0806	-0.10
0.15	-0.5051	-0.4493	-0.3628	-0.2846	-0.1914	-0.0317	-0.15
0.20	-0.4672	-0.4088	-0.3189	-0.2384	-0.1431	0.0181	-0.20
0.25	-0.4267	-0.3658	-0.2726	-0.1899	-0.0930	0.0688	-0.25
0.30	-0.3832	-0.3199	-0.2237	-0.1392	-0.0411	0.1206	-0.30
0.35	-0.3365	-0.2708	-0.1720	-0.0860	0.0128	0.1735	-0.35
0.40	-0.2860	-0.2182	-0.1172	-0.0301	0.0689	0.2275	-0.40
0.45	-0.2312	-0.1616	-0.0588	0.0288	0.1273	0.2827	-0.45
0.50	-0.1716	-0.1004	0.0035	0.0909	0.1883	0.3393	-0.50
0.55	-0.1064	-0.0342	0.0700	0.1567	0.2519	0.3973	-0.55
0.60	-0.0348	0.0379	0.1415	0.2264	0.3185	0.4567	-0.60
0.65	0.0445	0.1168	0.2185	0.3006	0.3884	0.5178	-0.65
0.70	0.1327	0.2036	0.3017	0.3796	0.4618	0.5805	-0.70
0.75	0.2316	0.2996	0.3920	0.4642	0.5390	0.6451	-0.75
0.80	0.3435	0.4066	0.4906	0.5550	0.6206	0.7115	-0.80
0.85	0.4714	0.5267	0.5989	0.6529	0.7069	0.7801	-0.85
0.90	0.6193	0.6629	0.7183	0.7588	0.7985	0.8509	-0.90
0.95	0.7929	0.8189	0.8511	0.8740	0.8959	0.9242	-0.95
							↑ R
α →	.995 ↑-	.99 ↑-	.975 ↑-	.95 ↑-	.90 ↑-	.75 ↑-	

TABLE 10.15. *CONFIDENCE LIMITS ON THE CORRELATION COEFFICIENT.*

SAMPLE SIZE = 17

α →	.005 ↓+	.01 ↓+	.025 ↓+	.05 ↓+	.10 ↓+	.25 ↓+	
R ↓							
-0.95	-0.9862	-0.9841	-0.9806	-0.9771	-0.9723	-0.9621	0.95
-0.90	-0.9719	-0.9677	-0.9607	-0.9536	-0.9440	-0.9240	0.90
-0.85	-0.9571	-0.9508	-0.9402	-0.9296	-0.9153	-0.8856	0.85
-0.80	-0.9418	-0.9334	-0.9192	-0.9050	-0.8860	-0.8470	0.80
-0.75	-0.9260	-0.9154	-0.8975	-0.8798	-0.8562	-0.8080	0.75
-0.70	-0.9096	-0.8967	-0.8752	-0.8539	-0.8257	-0.7686	0.70
-0.65	-0.8925	-0.8774	-0.8522	-0.8274	-0.7947	-0.7290	0.65
-0.60	-0.8748	-0.8575	-0.8285	-0.8001	-0.7629	-0.6889	0.60
-0.55	-0.8564	-0.8367	-0.8041	-0.7721	-0.7305	-0.6485	0.55
-0.50	-0.8373	-0.8152	-0.7788	-0.7433	-0.6974	-0.6076	0.50
-0.45	-0.8173	-0.7929	-0.7527	-0.7137	-0.6635	-0.5663	0.45
-0.40	-0.7965	-0.7696	-0.7256	-0.6832	-0.6288	-0.5245	0.40
-0.35	-0.7748	-0.7455	-0.6976	-0.6518	-0.5933	-0.4823	0.35
-0.30	-0.7520	-0.7202	-0.6686	-0.6193	-0.5569	-0.4395	0.30
-0.25	-0.7282	-0.6939	-0.6384	-0.5858	-0.5196	-0.3961	0.25
-0.20	-0.7033	-0.6664	-0.6071	-0.5512	-0.4813	-0.3522	0.20
-0.15	-0.6771	-0.6376	-0.5745	-0.5154	-0.4419	-0.3077	0.15
-0.10	-0.6495	-0.6075	-0.5406	-0.4783	-0.4014	-0.2625	0.10
-0.05	-0.6204	-0.5758	-0.5052	-0.4399	-0.3598	-0.2166	0.05
0.00	-0.5897	-0.5426	-0.4683	-0.4000	-0.3170	-0.1700	0.00
0.05	-0.5573	-0.5075	-0.4297	-0.3586	-0.2728	-0.1227	-0.05
0.10	-0.5229	-0.4706	-0.3892	-0.3155	-0.2272	-0.0745	-0.10
0.15	-0.4863	-0.4315	-0.3468	-0.2707	-0.1802	-0.0255	-0.15
0.20	-0.4474	-0.3901	-0.3023	-0.2239	-0.1315	0.0244	-0.20
0.25	-0.4059	-0.3462	-0.2554	-0.1751	-0.0812	0.0752	-0.25
0.30	-0.3614	-0.2995	-0.2059	-0.1240	-0.0291	0.1270	-0.30
0.35	-0.3137	-0.2496	-0.1537	-0.0705	0.0250	0.1799	-0.35
0.40	-0.2622	-0.1963	-0.0984	-0.0143	0.0812	0.2339	-0.40
0.45	-0.2066	-0.1390	-0.0397	0.0447	0.1396	0.2891	-0.45
0.50	-0.1463	-0.0774	0.0228	0.1069	0.2004	0.3456	-0.50
0.55	-0.0804	-0.0108	0.0894	0.1725	0.2638	0.4034	-0.55
0.60	-0.0084	0.0615	0.1607	0.2420	0.3301	0.4626	-0.60
0.65	0.0710	0.1402	0.2373	0.3156	0.3995	0.5233	-0.65
0.70	0.1590	0.2265	0.3198	0.3940	0.4722	0.5856	-0.70
0.75	0.2572	0.3216	0.4091	0.4775	0.5485	0.6496	-0.75
0.80	0.3675	0.4269	0.5061	0.5669	0.6289	0.7155	-0.80
0.85	0.4928	0.5445	0.6120	0.6628	0.7138	0.7833	-0.85
0.90	0.6364	0.6768	0.7284	0.7663	0.8035	0.8532	-0.90
0.95	0.8033	0.8272	0.8569	0.8782	0.8987	0.9254	-0.95

↑ R

α →	.995 ↑-	.99 ↑-	.975 ↑-	.95 ↑-	.90 ↑-	.75 ↑-

TABLE 10.16. *CONFIDENCE LIMITS ON THE CORRELATION COEFFICIENT.*

SAMPLE SIZE = 18

α →	.005 ↓+	.01 ↓+	.025 ↓+	.05 ↓+	.10 ↓+	.25 ↓+	
R ↓							
-0.95	-0.9856	-0.9835	-0.9800	-0.9765	-0.9718	-0.9619	0.95
-0.90	-0.9707	-0.9666	-0.9596	-0.9525	-0.9431	-0.9235	0.90
-0.85	-0.9553	-0.9491	-0.9385	-0.9280	-0.9139	-0.8848	0.85
-0.80	-0.9394	-0.9310	-0.9169	-0.9028	-0.8841	-0.8459	0.80
-0.75	-0.9229	-0.9123	-0.8946	-0.8770	-0.8538	-0.8066	0.75
-0.70	-0.9059	-0.8931	-0.8717	-0.8506	-0.8228	-0.7670	0.70
-0.65	-0.8882	-0.8731	-0.8481	-0.8235	-0.7913	-0.7271	0.65
-0.60	-0.8698	-0.8524	-0.8237	-0.7957	-0.7591	-0.6867	0.60
-0.55	-0.8507	-0.8310	-0.7986	-0.7671	-0.7262	-0.6460	0.55
-0.50	-0.8308	-0.8088	-0.7727	-0.7377	-0.6926	-0.6049	0.50
-0.45	-0.8101	-0.7858	-0.7459	-0.7075	-0.6582	-0.5633	0.45
-0.40	-0.7885	-0.7618	-0.7182	-0.6764	-0.6231	-0.5213	0.40
-0.35	-0.7660	-0.7369	-0.6895	-0.6444	-0.5871	-0.4787	0.35
-0.30	-0.7425	-0.7109	-0.6598	-0.6113	-0.5502	-0.4357	0.30
-0.25	-0.7179	-0.6838	-0.6290	-0.5773	-0.5124	-0.3921	0.25
-0.20	-0.6921	-0.6556	-0.5970	-0.5421	-0.4736	-0.3479	0.20
-0.15	-0.6650	-0.6260	-0.5638	-0.5057	-0.4339	-0.3031	0.15
-0.10	-0.6366	-0.5950	-0.5292	-0.4681	-0.3930	-0.2577	0.10
-0.05	-0.6066	-0.5626	-0.4931	-0.4291	-0.3509	-0.2116	0.05
0.00	-0.5751	-0.5285	-0.4555	-0.3887	-0.3077	-0.1649	0.00
0.05	-0.5417	-0.4927	-0.4163	-0.3468	-0.2632	-0.1173	-0.05
0.10	-0.5064	-0.4549	-0.3752	-0.3033	-0.2173	-0.0690	-0.10
0.15	-0.4690	-0.4151	-0.3322	-0.2580	-0.1699	-0.0199	-0.15
0.20	-0.4291	-0.3729	-0.2871	-0.2108	-0.1210	0.0301	-0.20
0.25	-0.3867	-0.3283	-0.2396	-0.1615	-0.0705	0.0810	-0.25
0.30	-0.3414	-0.2808	-0.1897	-0.1101	-0.0182	0.1329	-0.30
0.35	-0.2928	-0.2303	-0.1371	-0.0564	0.0360	0.1858	-0.35
0.40	-0.2406	-0.1764	-0.0814	0.0000	0.0922	0.2397	-0.40
0.45	-0.1842	-0.1186	-0.0224	0.0591	0.1506	0.2949	-0.45
0.50	-0.1232	-0.0566	0.0402	0.1213	0.2113	0.3512	-0.50
0.55	-0.0569	0.0103	0.1068	0.1868	0.2745	0.4088	-0.55
0.60	0.0154	0.0826	0.1779	0.2559	0.3405	0.4677	-0.60
0.65	0.0949	0.1612	0.2541	0.3291	0.4093	0.5281	-0.65
0.70	0.1825	0.2470	0.3360	0.4067	0.4814	0.5901	-0.70
0.75	0.2798	0.3410	0.4242	0.4893	0.5570	0.6536	-0.75
0.80	0.3887	0.4448	0.5197	0.5773	0.6363	0.7189	-0.80
0.85	0.5115	0.5600	0.6236	0.6716	0.7198	0.7861	-0.85
0.90	0.6513	0.6889	0.7372	0.7728	0.8080	0.8552	-0.90
0.95	0.8122	0.8343	0.8620	0.8819	0.9012	0.9265	-0.95

↑ R

α →	.995 ↑-	.99 ↑-	.975 ↑-	.95 ↑-	.90 ↑-	.75 ↑-

TABLE 10.17. *CONFIDENCE LIMITS ON THE CORRELATION COEFFICIENT.*

SAMPLE SIZE = 19

α →	.005 ↓+	.01 ↓+	.025 ↓+	.05 ↓+	.10 ↓+	.25 ↓+	
R ↓							
-0.95	-0.9850	-0.9830	-0.9795	-0.9760	-0.9714	-0.9616	0.95
-0.90	-0.9696	-0.9655	-0.9585	-0.9515	-0.9422	-0.9230	0.90
-0.85	-0.9536	-0.9474	-0.9369	-0.9264	-0.9125	-0.8841	0.85
-0.80	-0.9371	-0.9287	-0.9147	-0.9007	-0.8823	-0.8449	0.80
-0.75	-0.9201	-0.9095	-0.8918	-0.8744	-0.8515	-0.8054	0.75
-0.70	-0.9024	-0.8896	-0.8683	-0.8475	-0.8202	-0.7655	0.70
-0.65	-0.8840	-0.8690	-0.8442	-0.8199	-0.7882	-0.7253	0.65
-0.60	-0.8650	-0.8477	-0.8192	-0.7915	-0.7555	-0.6847	0.60
-0.55	-0.8453	-0.8257	-0.7935	-0.7624	-0.7222	-0.6437	0.55
-0.50	-0.8247	-0.8028	-0.7670	-0.7325	-0.6881	-0.6023	0.50
-0.45	-0.8033	-0.7791	-0.7396	-0.7017	-0.6533	-0.5605	0.45
-0.40	-0.7810	-0.7545	-0.7113	-0.6701	-0.6177	-0.5182	0.40
-0.35	-0.7578	-0.7288	-0.6820	-0.6375	-0.5813	-0.4754	0.35
-0.30	-0.7336	-0.7022	-0.6517	-0.6040	-0.5440	-0.4322	0.30
-0.25	-0.7082	-0.6744	-0.6203	-0.5694	-0.5058	-0.3883	0.25
-0.20	-0.6816	-0.6454	-0.5877	-0.5337	-0.4666	-0.3440	0.20
-0.15	-0.6538	-0.6151	-0.5538	-0.4968	-0.4264	-0.2990	0.15
-0.10	-0.6246	-0.5835	-0.5186	-0.4587	-0.3852	-0.2534	0.10
-0.05	-0.5938	-0.5503	-0.4819	-0.4192	-0.3428	-0.2071	0.05
0.00	-0.5614	-0.5155	-0.4438	-0.3783	-0.2992	-0.1602	0.00
0.05	-0.5273	-0.4790	-0.4039	-0.3360	-0.2544	-0.1125	-0.05
0.10	-0.4911	-0.4405	-0.3623	-0.2920	-0.2082	-0.0640	-0.10
0.15	-0.4529	-0.3999	-0.3188	-0.2463	-0.1606	-0.0148	-0.15
0.20	-0.4123	-0.3571	-0.2731	-0.1987	-0.1114	0.0353	-0.20
0.25	-0.3690	-0.3118	-0.2252	-0.1492	-0.0607	0.0863	-0.25
0.30	-0.3229	-0.2637	-0.1749	-0.0975	-0.0082	0.1382	-0.30
0.35	-0.2736	-0.2126	-0.1219	-0.0435	0.0460	0.1911	-0.35
0.40	-0.2207	-0.1582	-0.0659	0.0130	0.1023	0.2450	-0.40
0.45	-0.1637	-0.1000	-0.0068	0.0721	0.1606	0.3001	-0.45
0.50	-0.1023	-0.0376	0.0559	0.1343	0.2212	0.3563	-0.50
0.55	-0.0356	0.0294	0.1225	0.1996	0.2842	0.4137	-0.55
0.60	0.0370	0.1018	0.1935	0.2685	0.3498	0.4724	-0.60
0.65	0.1163	0.1801	0.2693	0.3412	0.4182	0.5325	-0.65
0.70	0.2036	0.2653	0.3504	0.4181	0.4897	0.5941	-0.70
0.75	0.3001	0.3584	0.4377	0.4998	0.5645	0.6573	-0.75
0.80	0.4075	0.4607	0.5319	0.5866	0.6429	0.7220	-0.80
0.85	0.5280	0.5738	0.6339	0.6793	0.7252	0.7886	-0.85
0.90	0.6642	0.6995	0.7449	0.7785	0.8119	0.8570	-0.90
0.95	0.8199	0.8405	0.8664	0.8851	0.9033	0.9274	-0.95

↑ R

α →	.995 ↑-	.99 ↑-	.975 ↑-	.95 ↑-	.90 ↑-	.75 ↑-

TABLE 10.18.   CONFIDENCE LIMITS ON THE CORRELATION COEFFICIENT.

SAMPLE SIZE = 20

α →	.005 ↓+	.01 ↓+	.025 ↓+	.05 ↓+	.10 ↓+	.25 ↓+	
R ↓							
-0.95	-0.9845	-0.9825	-0.9790	-0.9756	-0.9709	-0.9614	0.95
-0.90	-0.9686	-0.9644	-0.9575	-0.9506	-0.9414	-0.9225	0.90
-0.85	-0.9521	-0.9458	-0.9353	-0.9250	-0.9113	-0.8834	0.85
-0.80	-0.9350	-0.9266	-0.9126	-0.8988	-0.8807	-0.8440	0.80
-0.75	-0.9174	-0.9068	-0.8892	-0.8720	-0.8495	-0.8042	0.75
-0.70	-0.8991	-0.8863	-0.8652	-0.8446	-0.8177	-0.7641	0.70
-0.65	-0.8801	-0.8652	-0.8405	-0.8165	-0.7852	-0.7237	0.65
-0.60	-0.8605	-0.8433	-0.8150	-0.7876	-0.7522	-0.6828	0.60
-0.55	-0.8401	-0.8207	-0.7888	-0.7580	-0.7184	-0.6416	0.55
-0.50	-0.8190	-0.7972	-0.7617	-0.7276	-0.6840	-0.6000	0.50
-0.45	-0.7969	-0.7728	-0.7337	-0.6964	-0.6487	-0.5579	0.45
-0.40	-0.7740	-0.7476	-0.7049	-0.6642	-0.6127	-0.5154	0.40
-0.35	-0.7501	-0.7213	-0.6750	-0.6312	-0.5759	-0.4724	0.35
-0.30	-0.7251	-0.6940	-0.6441	-0.5971	-0.5383	-0.4289	0.30
-0.25	-0.6991	-0.6656	-0.6121	-0.5620	-0.4997	-0.3849	0.25
-0.20	-0.6718	-0.6360	-0.5789	-0.5259	-0.4601	-0.3403	0.20
-0.15	-0.6433	-0.6050	-0.5445	-0.4885	-0.4196	-0.2951	0.15
-0.10	-0.6133	-0.5727	-0.5088	-0.4499	-0.3780	-0.2494	0.10
-0.05	-0.5818	-0.5389	-0.4716	-0.4100	-0.3353	-0.2029	0.05
0.00	-0.5487	-0.5034	-0.4329	-0.3687	-0.2914	-0.1558	0.00
0.05	-0.5138	-0.4662	-0.3925	-0.3260	-0.2462	-0.1080	-0.05
0.10	-0.4769	-0.4271	-0.3504	-0.2816	-0.1998	-0.0595	-0.10
0.15	-0.4379	-0.3859	-0.3063	-0.2355	-0.1519	-0.0101	-0.15
0.20	-0.3966	-0.3425	-0.2603	-0.1877	-0.1026	0.0400	-0.20
0.25	-0.3527	-0.2966	-0.2120	-0.1378	-0.0517	0.0911	-0.25
0.30	-0.3059	-0.2479	-0.1613	-0.0859	0.0008	0.1430	-0.30
0.35	-0.2559	-0.1963	-0.1079	-0.0318	0.0552	0.1959	-0.35
0.40	-0.2024	-0.1415	-0.0517	0.0248	0.1114	0.2498	-0.40
0.45	-0.1449	-0.0829	0.0076	0.0840	0.1697	0.3048	-0.45
0.50	-0.0830	-0.0203	0.0703	0.1461	0.2302	0.3609	-0.50
0.55	-0.0161	0.0468	0.1369	0.2113	0.2930	0.4181	-0.55
0.60	0.0566	0.1191	0.2076	0.2799	0.3583	0.4767	-0.60
0.65	0.1358	0.1972	0.2829	0.3521	0.4263	0.5365	-0.65
0.70	0.2227	0.2818	0.3635	0.4284	0.4972	0.5978	-0.70
0.75	0.3183	0.3740	0.4498	0.5092	0.5713	0.6605	-0.75
0.80	0.4243	0.4749	0.5427	0.5950	0.6488	0.7248	-0.80
0.85	0.5426	0.5860	0.6430	0.6863	0.7301	0.7908	-0.85
0.90	0.6757	0.7089	0.7518	0.7837	0.8154	0.8586	-0.90
0.95	0.8267	0.8460	0.8703	0.8879	0.9053	0.9283	-0.95

↑ R

α →	.995 ↑-	.99 ↑-	.975 ↑-	.95 ↑-	.90 ↑-	.75 ↑-

TABLE 10.19. CONFIDENCE LIMITS ON THE CORRELATION COEFFICIENT.

SAMPLE SIZE = 21

α →	.005 ↓+	.01 ↓+	.025 ↓+	.05 ↓+	.10 ↓+	.25 ↓+	
R ↓							
-0.95	-0.9840	-0.9820	-0.9785	-0.9751	-0.9706	-0.9612	0.95
-0.90	-0.9676	-0.9634	-0.9565	-0.9497	-0.9406	-0.9221	0.90
-0.85	-0.9505	-0.9443	-0.9339	-0.9236	-0.9101	-0.8827	0.85
-0.80	-0.9330	-0.9246	-0.9107	-0.8970	-0.8791	-0.8431	0.80
-0.75	-0.9148	-0.9042	-0.8868	-0.8698	-0.8475	-0.8031	0.75
-0.70	-0.8960	-0.8832	-0.8623	-0.8419	-0.8153	-0.7628	0.70
-0.65	-0.8765	-0.8615	-0.8370	-0.8133	-0.7825	-0.7221	0.65
-0.60	-0.8563	-0.8391	-0.8111	-0.7840	-0.7491	-0.6811	0.60
-0.55	-0.8353	-0.8159	-0.7843	-0.7539	-0.7149	-0.6396	0.55
-0.50	-0.8135	-0.7919	-0.7567	-0.7230	-0.6801	-0.5978	0.50
-0.45	-0.7909	-0.7670	-0.7282	-0.6913	-0.6445	-0.5555	0.45
-0.40	-0.7673	-0.7411	-0.6988	-0.6587	-0.6081	-0.5128	0.40
-0.35	-0.7428	-0.7143	-0.6684	-0.6252	-0.5709	-0.4696	0.35
-0.30	-0.7172	-0.6864	-0.6370	-0.5907	-0.5329	-0.4259	0.30
-0.25	-0.6905	-0.6573	-0.6045	-0.5552	-0.4940	-0.3817	0.25
-0.20	-0.6626	-0.6271	-0.5708	-0.5186	-0.4541	-0.3369	0.20
-0.15	-0.6334	-0.5955	-0.5359	-0.4808	-0.4132	-0.2916	0.15
-0.10	-0.6028	-0.5626	-0.4996	-0.4418	-0.3713	-0.2456	0.10
-0.05	-0.5706	-0.5282	-0.4619	-0.4015	-0.3283	-0.1991	0.05
0.00	-0.5368	-0.4921	-0.4227	-0.3598	-0.2841	-0.1518	0.00
0.05	-0.5012	-0.4543	-0.3819	-0.3167	-0.2387	-0.1039	-0.05
0.10	-0.4637	-0.4146	-0.3393	-0.2720	-0.1921	-0.0552	-0.10
0.15	-0.4240	-0.3728	-0.2949	-0.2256	-0.1440	-0.0058	-0.15
0.20	-0.3820	-0.3289	-0.2484	-0.1774	-0.0945	0.0444	-0.20
0.25	-0.3374	-0.2824	-0.1997	-0.1274	-0.0435	0.0955	-0.25
0.30	-0.2900	-0.2333	-0.1487	-0.0753	0.0092	0.1474	-0.30
0.35	-0.2395	-0.1813	-0.0951	-0.0210	0.0636	0.2003	-0.35
0.40	-0.1855	-0.1261	-0.0387	0.0357	0.1199	0.2542	-0.40
0.45	-0.1276	-0.0672	0.0207	0.0950	0.1781	0.3091	-0.45
0.50	-0.0654	-0.0044	0.0835	0.1570	0.2384	0.3651	-0.50
0.55	0.0018	0.0628	0.1500	0.2220	0.3010	0.4222	-0.55
0.60	0.0745	0.1350	0.2204	0.2902	0.3660	0.4805	-0.60
0.65	0.1536	0.2128	0.2954	0.3621	0.4336	0.5402	-0.65
0.70	0.2400	0.2969	0.3753	0.4378	0.5040	0.6011	-0.70
0.75	0.3348	0.3882	0.4608	0.5178	0.5775	0.6635	-0.75
0.80	0.4395	0.4878	0.5525	0.6026	0.6542	0.7274	-0.80
0.85	0.5558	0.5969	0.6512	0.6925	0.7345	0.7929	-0.85
0.90	0.6859	0.7173	0.7579	0.7883	0.8186	0.8601	-0.90
0.95	0.8327	0.8508	0.8737	0.8905	0.9070	0.9291	-0.95

↑ R

α →	.995 ↑-	.99 ↑-	.975 ↑-	.95 ↑-	.90 ↑-	.75 ↑-

TABLE 10.20. CONFIDENCE LIMITS ON THE CORRELATION COEFFICIENT.

SAMPLE SIZE = 22

α →	.005 ↓+	.01 ↓+	.025 ↓+	.05 ↓+	.10 ↓+	.25 ↓+	
R ↓							
-0.95	-0.9836	-0.9815	-0.9781	-0.9747	-0.9702	-0.9610	0.95
-0.90	-0.9666	-0.9625	-0.9556	-0.9488	-0.9399	-0.9217	0.90
-0.85	-0.9491	-0.9429	-0.9325	-0.9224	-0.9090	-0.8821	0.85
-0.80	-0.9310	-0.9227	-0.9088	-0.8953	-0.8776	-0.8423	0.80
-0.75	-0.9123	-0.9018	-0.8845	-0.8677	-0.8457	-0.8021	0.75
-0.70	-0.8930	-0.8803	-0.8595	-0.8393	-0.8131	-0.7616	0.70
-0.65	-0.8730	-0.8581	-0.8338	-0.8103	-0.7800	-0.7207	0.65
-0.60	-0.8522	-0.8352	-0.8073	-0.7806	-0.7461	-0.6794	0.60
-0.55	-0.8307	-0.8114	-0.7801	-0.7500	-0.7117	-0.6378	0.55
-0.50	-0.8084	-0.7869	-0.7520	-0.7187	-0.6765	-0.5958	0.50
-0.45	-0.7852	-0.7614	-0.7230	-0.6866	-0.6405	-0.5533	0.45
-0.40	-0.7611	-0.7350	-0.6931	-0.6536	-0.6038	-0.5104	0.40
-0.35	-0.7359	-0.7076	-0.6623	-0.6197	-0.5663	-0.4670	0.35
-0.30	-0.7098	-0.6792	-0.6304	-0.5847	-0.5279	-0.4231	0.30
-0.25	-0.6825	-0.6496	-0.5974	-0.5488	-0.4887	-0.3787	0.25
-0.20	-0.6539	-0.6187	-0.5632	-0.5118	-0.4485	-0.3338	0.20
-0.15	-0.6241	-0.5866	-0.5278	-0.4736	-0.4073	-0.2883	0.15
-0.10	-0.5928	-0.5531	-0.4910	-0.4342	-0.3651	-0.2422	0.10
-0.05	-0.5600	-0.5181	-0.4529	-0.3936	-0.3218	-0.1955	0.05
0.00	-0.5256	-0.4815	-0.4132	-0.3515	-0.2774	-0.1481	0.00
0.05	-0.4894	-0.4432	-0.3720	-0.3081	-0.2318	-0.1001	-0.05
0.10	-0.4513	-0.4029	-0.3290	-0.2630	-0.1849	-0.0513	-0.10
0.15	-0.4110	-0.3607	-0.2842	-0.2164	-0.1367	-0.0018	-0.15
0.20	-0.3684	-0.3162	-0.2373	-0.1680	-0.0870	0.0485	-0.20
0.25	-0.3232	-0.2693	-0.1883	-0.1177	-0.0358	0.0996	-0.25
0.30	-0.2753	-0.2198	-0.1370	-0.0654	0.0169	0.1515	-0.30
0.35	-0.2243	-0.1674	-0.0832	-0.0110	0.0714	0.2044	-0.35
0.40	-0.1698	-0.1118	-0.0267	0.0458	0.1276	0.2582	-0.40
0.45	-0.1116	-0.0527	0.0329	0.1050	0.1858	0.3131	-0.45
0.50	-0.0490	0.0102	0.0956	0.1669	0.2460	0.3690	-0.50
0.55	0.0183	0.0775	0.1620	0.2318	0.3083	0.4259	-0.55
0.60	0.0910	0.1495	0.2322	0.2998	0.3731	0.4841	-0.60
0.65	0.1699	0.2270	0.3068	0.3712	0.4403	0.5435	-0.65
0.70	0.2558	0.3106	0.3861	0.4464	0.5102	0.6042	-0.70
0.75	0.3498	0.4010	0.4708	0.5256	0.5831	0.6662	-0.75
0.80	0.4532	0.4994	0.5614	0.6094	0.6591	0.7297	-0.80
0.85	0.5676	0.6068	0.6587	0.6982	0.7384	0.7947	-0.85
0.90	0.6950	0.7248	0.7635	0.7924	0.8215	0.8614	-0.90
0.95	0.8381	0.8551	0.8768	0.8928	0.9086	0.9298	-0.95

↑ R

α →	.995 ↑-	.99 ↑-	.975 ↑-	.95 ↑-	.90 ↑-	.75 ↑-

TABLE 10.21. *CONFIDENCE LIMITS ON THE CORRELATION COEFFICIENT.*

SAMPLE SIZE = 23

α →	.005 ↓+	.01 ↓+	.025 ↓+	.05 ↓+	.10 ↓+	.25 ↓+	
R ↓							
-0.95	-0.9831	-0.9811	-0.9777	-0.9743	-0.9698	-0.9608	0.95
-0.90	-0.9657	-0.9616	-0.9547	-0.9480	-0.9392	-0.9213	0.90
-0.85	-0.9477	-0.9415	-0.9312	-0.9212	-0.9080	-0.8816	0.85
-0.80	-0.9292	-0.9208	-0.9071	-0.8937	-0.8763	-0.8415	0.80
-0.75	-0.9100	-0.8995	-0.8823	-0.8656	-0.8440	-0.8011	0.75
-0.70	-0.8902	-0.8776	-0.8569	-0.8369	-0.8111	-0.7604	0.70
-0.65	-0.8696	-0.8549	-0.8307	-0.8075	-0.7776	-0.7193	0.65
-0.60	-0.8484	-0.8314	-0.8038	-0.7773	-0.7434	-0.6779	0.60
-0.55	-0.8264	-0.8072	-0.7761	-0.7464	-0.7086	-0.6361	0.55
-0.50	-0.8035	-0.7821	-0.7476	-0.7147	-0.6731	-0.5938	0.50
-0.45	-0.7798	-0.7561	-0.7181	-0.6822	-0.6368	-0.5512	0.45
-0.40	-0.7551	-0.7292	-0.6878	-0.6488	-0.5998	-0.5081	0.40
-0.35	-0.7294	-0.7013	-0.6565	-0.6144	-0.5619	-0.4645	0.35
-0.30	-0.7027	-0.6723	-0.6241	-0.5791	-0.5233	-0.4205	0.30
-0.25	-0.6748	-0.6422	-0.5907	-0.5428	-0.4837	-0.3759	0.25
-0.20	-0.6457	-0.6109	-0.5560	-0.5054	-0.4432	-0.3309	0.20
-0.15	-0.6153	-0.5782	-0.5202	-0.4669	-0.4018	-0.2852	0.15
-0.10	-0.5835	-0.5442	-0.4830	-0.4271	-0.3593	-0.2390	0.10
-0.05	-0.5501	-0.5087	-0.4444	-0.3861	-0.3158	-0.1921	0.05
0.00	-0.5151	-0.4716	-0.4044	-0.3438	-0.2711	-0.1447	0.00
0.05	-0.4783	-0.4327	-0.3627	-0.3000	-0.2253	-0.0965	-0.05
0.10	-0.4396	-0.3920	-0.3194	-0.2547	-0.1782	-0.0477	-0.10
0.15	-0.3988	-0.3493	-0.2742	-0.2078	-0.1298	0.0019	-0.15
0.20	-0.3556	-0.3043	-0.2270	-0.1591	-0.0800	0.0522	-0.20
0.25	-0.3100	-0.2570	-0.1778	-0.1087	-0.0288	0.1034	-0.25
0.30	-0.2616	-0.2071	-0.1262	-0.0562	0.0241	0.1553	-0.30
0.35	-0.2101	-0.1544	-0.0722	-0.0017	0.0785	0.2082	-0.35
0.40	-0.1553	-0.0986	-0.0155	0.0551	0.1348	0.2620	-0.40
0.45	-0.0967	-0.0393	0.0441	0.1143	0.1929	0.3167	-0.45
0.50	-0.0339	0.0238	0.1069	0.1761	0.2529	0.3725	-0.50
0.55	0.0335	0.0910	0.1731	0.2408	0.3151	0.4294	-0.55
0.60	0.1062	0.1630	0.2431	0.3085	0.3796	0.4874	-0.60
0.65	0.1848	0.2401	0.3173	0.3795	0.4465	0.5465	-0.65
0.70	0.2703	0.3231	0.3960	0.4542	0.5160	0.6070	-0.70
0.75	0.3635	0.4128	0.4799	0.5328	0.5882	0.6687	-0.75
0.80	0.4656	0.5100	0.5695	0.6157	0.6635	0.7318	-0.80
0.85	0.5783	0.6158	0.6654	0.7033	0.7421	0.7964	-0.85
0.90	0.7033	0.7316	0.7685	0.7962	0.8241	0.8626	-0.90
0.95	0.8428	0.8590	0.8796	0.8949	0.9100	0.9304	-0.95

↑ R

α →	.995 ↑-	.99 ↑-	.975 ↑-	.95 ↑-	.90 ↑-	.75 ↑-

TABLE 10.22. CONFIDENCE LIMITS ON THE CORRELATION COEFFICIENT.

SAMPLE SIZE = 24

α →	.005 ↓+	.01 ↓+	.025 ↓+	.05 ↓+	.10 ↓+	.25 ↓+	
R ↓							
-0.95	-0.9827	-0.9806	-0.9772	-0.9739	-0.9695	-0.9606	0.95
-0.90	-0.9648	-0.9607	-0.9539	-0.9473	-0.9385	-0.9210	0.90
-0.85	-0.9464	-0.9402	-0.9300	-0.9200	-0.9070	-0.8810	0.85
-0.80	-0.9274	-0.9191	-0.9055	-0.8922	-0.8750	-0.8408	0.80
-0.75	-0.9078	-0.8974	-0.8803	-0.8637	-0.8423	-0.8002	0.75
-0.70	-0.8875	-0.8749	-0.8544	-0.8346	-0.8091	-0.7593	0.70
-0.65	-0.8665	-0.8518	-0.8278	-0.8048	-0.7753	-0.7181	0.65
-0.60	-0.8447	-0.8279	-0.8005	-0.7743	-0.7408	-0.6764	0.60
-0.55	-0.8222	-0.8032	-0.7723	-0.7430	-0.7057	-0.6344	0.55
-0.50	-0.7989	-0.7776	-0.7434	-0.7109	-0.6698	-0.5920	0.50
-0.45	-0.7746	-0.7512	-0.7135	-0.6780	-0.6333	-0.5492	0.45
-0.40	-0.7494	-0.7238	-0.6828	-0.6442	-0.5959	-0.5060	0.40
-0.35	-0.7232	-0.6954	-0.6510	-0.6095	-0.5578	-0.4622	0.35
-0.30	-0.6960	-0.6659	-0.6182	-0.5739	-0.5189	-0.4180	0.30
-0.25	-0.6676	-0.6353	-0.5843	-0.5372	-0.4790	-0.3733	0.25
-0.20	-0.6379	-0.6035	-0.5493	-0.4994	-0.4383	-0.3281	0.20
-0.15	-0.6070	-0.5703	-0.5130	-0.4606	-0.3966	-0.2823	0.15
-0.10	-0.5746	-0.5358	-0.4755	-0.4205	-0.3539	-0.2360	0.10
-0.05	-0.5407	-0.4998	-0.4365	-0.3792	-0.3101	-0.1890	0.05
0.00	-0.5052	-0.4622	-0.3961	-0.3365	-0.2653	-0.1415	0.00
0.05	-0.4679	-0.4229	-0.3541	-0.2925	-0.2192	-0.0932	-0.05
0.10	-0.4286	-0.3818	-0.3104	-0.2469	-0.1720	-0.0443	-0.10
0.15	-0.3873	-0.3386	-0.2648	-0.1998	-0.1234	0.0053	-0.15
0.20	-0.3437	-0.2933	-0.2174	-0.1509	-0.0735	0.0557	-0.20
0.25	-0.2975	-0.2456	-0.1679	-0.1003	-0.0222	0.1069	-0.25
0.30	-0.2487	-0.1953	-0.1161	-0.0477	0.0307	0.1589	-0.30
0.35	-0.1968	-0.1423	-0.0619	0.0069	0.0852	0.2117	-0.35
0.40	-0.1417	-0.0862	-0.0051	0.0637	0.1414	0.2655	-0.40
0.45	-0.0828	-0.0268	0.0545	0.1229	0.1994	0.3201	-0.45
0.50	-0.0199	0.0364	0.1172	0.1847	0.2594	0.3758	-0.50
0.55	0.0476	0.1036	0.1833	0.2492	0.3214	0.4326	-0.55
0.60	0.1202	0.1753	0.2531	0.3166	0.3856	0.4904	-0.60
0.65	0.1986	0.2522	0.3269	0.3873	0.4522	0.5494	-0.65
0.70	0.2836	0.3347	0.4051	0.4614	0.5212	0.6095	-0.70
0.75	0.3760	0.4235	0.4883	0.5393	0.5930	0.6710	-0.75
0.80	0.4770	0.5196	0.5769	0.6214	0.6676	0.7338	-0.80
0.85	0.5881	0.6240	0.6716	0.7080	0.7454	0.7980	-0.85
0.90	0.7107	0.7378	0.7731	0.7996	0.8265	0.8637	-0.90
0.95	0.8471	0.8625	0.8822	0.8968	0.9113	0.9310	-0.95
							↑ R
α →	.995 ↑-	.99 ↑-	.975 ↑-	.95 ↑-	.90 ↑-	.75 ↑-	

TABLE 10.23. CONFIDENCE LIMITS ON THE CORRELATION COEFFICIENT.

SAMPLE SIZE = 25

α →	.005 ↓+	.01 ↓+	.025 ↓+	.05 ↓+	.10 ↓+	.25 ↓+	
R ↓							
-0.95	-0.9823	-0.9802	-0.9769	-0.9736	-0.9692	-0.9604	0.95
-0.90	-0.9640	-0.9599	-0.9531	-0.9465	-0.9379	-0.9206	0.90
-0.85	-0.9452	-0.9390	-0.9288	-0.9190	-0.9061	-0.8805	0.85
-0.80	-0.9257	-0.9175	-0.9039	-0.8908	-0.8737	-0.8401	0.80
-0.75	-0.9057	-0.8953	-0.8783	-0.8620	-0.8408	-0.7993	0.75
-0.70	-0.8849	-0.8724	-0.8520	-0.8325	-0.8073	-0.7583	0.70
-0.65	-0.8635	-0.8488	-0.8251	-0.8023	-0.7732	-0.7169	0.65
-0.60	-0.8413	-0.8245	-0.7973	-0.7714	-0.7384	-0.6751	0.60
-0.55	-0.8183	-0.7993	-0.7688	-0.7398	-0.7030	-0.6329	0.55
-0.50	-0.7944	-0.7733	-0.7394	-0.7073	-0.6668	-0.5903	0.50
-0.45	-0.7697	-0.7464	-0.7091	-0.6741	-0.6300	-0.5474	0.45
-0.40	-0.7441	-0.7186	-0.6780	-0.6399	-0.5924	-0.5040	0.40
-0.35	-0.7174	-0.6897	-0.6458	-0.6049	-0.5540	-0.4601	0.35
-0.30	-0.6896	-0.6598	-0.6126	-0.5689	-0.5147	-0.4157	0.30
-0.25	-0.6607	-0.6287	-0.5784	-0.5319	-0.4746	-0.3709	0.25
-0.20	-0.6306	-0.5964	-0.5429	-0.4938	-0.4337	-0.3255	0.20
-0.15	-0.5991	-0.5629	-0.5063	-0.4546	-0.3917	-0.2796	0.15
-0.10	-0.5662	-0.5279	-0.4684	-0.4142	-0.3488	-0.2332	0.10
-0.05	-0.5318	-0.4914	-0.4290	-0.3726	-0.3048	-0.1861	0.05
0.00	-0.4958	-0.4534	-0.3882	-0.3297	-0.2598	-0.1384	0.00
0.05	-0.4580	-0.4137	-0.3459	-0.2854	-0.2136	-0.0901	-0.05
0.10	-0.4183	-0.3721	-0.3019	-0.2396	-0.1661	-0.0411	-0.10
0.15	-0.3765	-0.3285	-0.2561	-0.1922	-0.1175	0.0085	-0.15
0.20	-0.3324	-0.2828	-0.2084	-0.1432	-0.0674	0.0590	-0.20
0.25	-0.2859	-0.2348	-0.1586	-0.0924	-0.0160	0.1101	-0.25
0.30	-0.2366	-0.1843	-0.1067	-0.0397	0.0369	0.1621	-0.30
0.35	-0.1844	-0.1310	-0.0523	0.0149	0.0914	0.2150	-0.35
0.40	-0.1289	-0.0747	0.0046	0.0718	0.1476	0.2687	-0.40
0.45	-0.0699	-0.0151	0.0642	0.1310	0.2056	0.3233	-0.45
0.50	-0.0068	0.0481	0.1269	0.1926	0.2654	0.3789	-0.50
0.55	0.0608	0.1153	0.1929	0.2569	0.3272	0.4355	-0.55
0.60	0.1333	0.1868	0.2624	0.3241	0.3912	0.4932	-0.60
0.65	0.2114	0.2633	0.3358	0.3944	0.4574	0.5520	-0.65
0.70	0.2959	0.3453	0.4136	0.4681	0.5261	0.6119	-0.70
0.75	0.3876	0.4334	0.4960	0.5454	0.5973	0.6731	-0.75
0.80	0.4875	0.5285	0.5837	0.6267	0.6714	0.7356	-0.80
0.85	0.5970	0.6315	0.6772	0.7124	0.7484	0.7995	-0.85
0.90	0.7175	0.7434	0.7772	0.8028	0.8287	0.8648	-0.90
0.95	0.8511	0.8657	0.8845	0.8985	0.9125	0.9316	-0.95

↑ R

α →	.995 ↑-	.99 ↑-	.975 ↑-	.95 ↑-	.90 ↑-	.75 ↑-

TABLE 10.24. CONFIDENCE LIMITS ON THE CORRELATION COEFFICIENT.

SAMPLE SIZE = 26

α →	.005 ↓+	.01 ↓+	.025 ↓+	.05 ↓+	.10 ↓+	.25 ↓+	
R ↓							
-0.95	-0.9819	-0.9798	-0.9765	-0.9732	-0.9689	-0.9603	0.95
-0.90	-0.9632	-0.9591	-0.9524	-0.9459	-0.9373	-0.9203	0.90
-0.85	-0.9440	-0.9378	-0.9277	-0.9179	-0.9052	-0.8800	0.85
-0.80	-0.9241	-0.9159	-0.9024	-0.8894	-0.8726	-0.8394	0.80
-0.75	-0.9036	-0.8933	-0.8765	-0.8602	-0.8394	-0.7985	0.75
-0.70	-0.8825	-0.8700	-0.8498	-0.8304	-0.8055	-0.7573	0.70
-0.65	-0.8606	-0.8460	-0.8224	-0.7999	-0.7711	-0.7157	0.65
-0.60	-0.8379	-0.8213	-0.7943	-0.7687	-0.7361	-0.6738	0.60
-0.55	-0.8145	-0.7957	-0.7654	-0.7367	-0.7004	-0.6315	0.55
-0.50	-0.7902	-0.7693	-0.7356	-0.7039	-0.6640	-0.5887	0.50
-0.45	-0.7651	-0.7419	-0.7050	-0.6703	-0.6268	-0.5456	0.45
-0.40	-0.7389	-0.7136	-0.6734	-0.6358	-0.5890	-0.5021	0.40
-0.35	-0.7118	-0.6844	-0.6409	-0.6005	-0.5503	-0.4580	0.35
-0.30	-0.6836	-0.6540	-0.6074	-0.5642	-0.5108	-0.4136	0.30
-0.25	-0.6542	-0.6225	-0.5727	-0.5268	-0.4705	-0.3686	0.25
-0.20	-0.6236	-0.5898	-0.5369	-0.4885	-0.4293	-0.3231	0.20
-0.15	-0.5916	-0.5558	-0.4999	-0.4490	-0.3871	-0.2771	0.15
-0.10	-0.5583	-0.5204	-0.4616	-0.4083	-0.3440	-0.2305	0.10
-0.05	-0.5234	-0.4835	-0.4220	-0.3665	-0.2998	-0.1834	0.05
0.00	-0.4869	-0.4451	-0.3809	-0.3233	-0.2546	-0.1356	0.00
0.05	-0.4487	-0.4049	-0.3382	-0.2787	-0.2082	-0.0872	-0.05
0.10	-0.4085	-0.3630	-0.2939	-0.2327	-0.1607	-0.0382	-0.10
0.15	-0.3663	-0.3191	-0.2478	-0.1852	-0.1119	0.0116	-0.15
0.20	-0.3218	-0.2730	-0.1999	-0.1360	-0.0617	0.0620	-0.20
0.25	-0.2749	-0.2247	-0.1499	-0.0851	-0.0102	0.1132	-0.25
0.30	-0.2253	-0.1739	-0.0978	-0.0323	0.0427	0.1652	-0.30
0.35	-0.1727	-0.1204	-0.0433	0.0225	0.0973	0.2180	-0.35
0.40	-0.1170	-0.0639	0.0136	0.0793	0.1534	0.2717	-0.40
0.45	-0.0577	-0.0042	0.0733	0.1385	0.2113	0.3263	-0.45
0.50	0.0055	0.0590	0.1359	0.2000	0.2710	0.3818	-0.50
0.55	0.0730	0.1261	0.2018	0.2641	0.3327	0.4383	-0.55
0.60	0.1454	0.1975	0.2710	0.3311	0.3964	0.4958	-0.60
0.65	0.2233	0.2737	0.3441	0.4010	0.4623	0.5544	-0.65
0.70	0.3073	0.3552	0.4214	0.4742	0.5306	0.6142	-0.70
0.75	0.3983	0.4426	0.5032	0.5510	0.6014	0.6751	-0.75
0.80	0.4972	0.5367	0.5900	0.6316	0.6749	0.7373	-0.80
0.85	0.6052	0.6383	0.6824	0.7164	0.7513	0.8008	-0.85
0.90	0.7237	0.7486	0.7811	0.8057	0.8307	0.8657	-0.90
0.95	0.8546	0.8686	0.8866	0.9001	0.9136	0.9321	-0.95

↑ R

α →	.995 ↑-	.99 ↑-	.975 ↑-	.95 ↑-	.90 ↑-	.75 ↑-	

TABLE 10.25. *CONFIDENCE LIMITS ON THE CORRELATION COEFFICIENT.*

SAMPLE SIZE = 27

α →	.005 ↓+	.01 ↓+	.025 ↓+	.05 ↓+	.10 ↓+	.25 ↓+	
R ↓							
-0.95	-0.9815	-0.9795	-0.9761	-0.9729	-0.9686	-0.9601	0.95
-0.90	-0.9625	-0.9584	-0.9517	-0.9452	-0.9368	-0.9200	0.90
-0.85	-0.9428	-0.9367	-0.9267	-0.9170	-0.9044	-0.8795	0.85
-0.80	-0.9226	-0.9144	-0.9010	-0.8881	-0.8715	-0.8388	0.80
-0.75	-0.9017	-0.8914	-0.8747	-0.8586	-0.8380	-0.7978	0.75
-0.70	-0.8801	-0.8677	-0.8477	-0.8285	-0.8039	-0.7564	0.70
-0.65	-0.8578	-0.8433	-0.8199	-0.7976	-0.7692	-0.7146	0.65
-0.60	-0.8348	-0.8182	-0.7914	-0.7661	-0.7339	-0.6726	0.60
-0.55	-0.8109	-0.7922	-0.7622	-0.7338	-0.6979	-0.6301	0.55
-0.50	-0.7862	-0.7654	-0.7320	-0.7007	-0.6613	-0.5872	0.50
-0.45	-0.7606	-0.7376	-0.7010	-0.6668	-0.6239	-0.5440	0.45
-0.40	-0.7340	-0.7089	-0.6691	-0.6320	-0.5858	-0.5003	0.40
-0.35	-0.7065	-0.6793	-0.6362	-0.5963	-0.5469	-0.4561	0.35
-0.30	-0.6778	-0.6485	-0.6023	-0.5597	-0.5072	-0.4115	0.30
-0.25	-0.6480	-0.6166	-0.5673	-0.5221	-0.4666	-0.3664	0.25
-0.20	-0.6169	-0.5834	-0.5312	-0.4834	-0.4252	-0.3208	0.20
-0.15	-0.5845	-0.5490	-0.4939	-0.4437	-0.3828	-0.2747	0.15
-0.10	-0.5507	-0.5133	-0.4553	-0.4028	-0.3395	-0.2280	0.10
-0.05	-0.5154	-0.4760	-0.4153	-0.3606	-0.2951	-0.1808	0.05
0.00	-0.4785	-0.4372	-0.3739	-0.3172	-0.2497	-0.1330	0.00
0.05	-0.4398	-0.3967	-0.3309	-0.2725	-0.2032	-0.0845	-0.05
0.10	-0.3992	-0.3544	-0.2864	-0.2262	-0.1555	-0.0354	-0.10
0.15	-0.3566	-0.3101	-0.2401	-0.1785	-0.1066	0.0144	-0.15
0.20	-0.3117	-0.2638	-0.1919	-0.1292	-0.0564	0.0649	-0.20
0.25	-0.2645	-0.2152	-0.1418	-0.0781	-0.0048	0.1161	-0.25
0.30	-0.2146	-0.1641	-0.0895	-0.0253	0.0482	0.1681	-0.30
0.35	-0.1618	-0.1104	-0.0349	0.0295	0.1027	0.2209	-0.35
0.40	-0.1058	-0.0537	0.0221	0.0864	0.1588	0.2745	-0.40
0.45	-0.0464	0.0060	0.0818	0.1455	0.2166	0.3290	-0.45
0.50	0.0169	0.0693	0.1444	0.2069	0.2762	0.3845	-0.50
0.55	0.0845	0.1363	0.2101	0.2709	0.3377	0.4408	-0.55
0.60	0.1568	0.2075	0.2791	0.3376	0.4012	0.4982	-0.60
0.65	0.2343	0.2834	0.3519	0.4072	0.4669	0.5567	-0.65
0.70	0.3179	0.3644	0.4286	0.4800	0.5348	0.6162	-0.70
0.75	0.4082	0.4511	0.5098	0.5562	0.6052	0.6769	-0.75
0.80	0.5061	0.5443	0.5959	0.6361	0.6781	0.7389	-0.80
0.85	0.6127	0.6447	0.6873	0.7201	0.7539	0.8020	-0.85
0.90	0.7295	0.7533	0.7846	0.8084	0.8326	0.8666	-0.90
0.95	0.8579	0.8713	0.8886	0.9016	0.9146	0.9326	-0.95

↑ R

α →	.995 ↑-	.99 ↑-	.975 ↑-	.95 ↑-	.90 ↑-	.75 ↑-	

TABLE 10.26. *CONFIDENCE LIMITS ON THE CORRELATION COEFFICIENT.*

SAMPLE SIZE = 28

α →	.005 ↓+	.01 ↓+	.025 ↓+	.05 ↓+	.10 ↓+	.25 ↓+	
R ↓							
-0.95	-0.9811	-0.9791	-0.9758	-0.9726	-0.9684	-0.9600	0.95
-0.90	-0.9617	-0.9577	-0.9510	-0.9446	-0.9362	-0.9197	0.90
-0.85	-0.9417	-0.9356	-0.9257	-0.9160	-0.9036	-0.8791	0.85
-0.80	-0.9211	-0.9129	-0.8997	-0.8869	-0.8704	-0.8382	0.80
-0.75	-0.8998	-0.8896	-0.8730	-0.8571	-0.8367	-0.7970	0.75
-0.70	-0.8779	-0.8656	-0.8456	-0.8266	-0.8023	-0.7555	0.70
-0.65	-0.8552	-0.8408	-0.8175	-0.7955	-0.7674	-0.7136	0.65
-0.60	-0.8317	-0.8152	-0.7887	-0.7636	-0.7318	-0.6714	0.60
-0.55	-0.8075	-0.7889	-0.7591	-0.7310	-0.6956	-0.6288	0.55
-0.50	-0.7824	-0.7617	-0.7286	-0.6976	-0.6587	-0.5858	0.50
-0.45	-0.7563	-0.7335	-0.6973	-0.6634	-0.6211	-0.5424	0.45
-0.40	-0.7294	-0.7045	-0.6650	-0.6283	-0.5827	-0.4986	0.40
-0.35	-0.7014	-0.6744	-0.6318	-0.5924	-0.5436	-0.4543	0.35
-0.30	-0.6723	-0.6432	-0.5976	-0.5555	-0.5037	-0.4096	0.30
-0.25	-0.6420	-0.6109	-0.5623	-0.5176	-0.4629	-0.3644	0.25
-0.20	-0.6106	-0.5774	-0.5258	-0.4786	-0.4213	-0.3187	0.20
-0.15	-0.5778	-0.5426	-0.4882	-0.4386	-0.3787	-0.2725	0.15
-0.10	-0.5436	-0.5065	-0.4492	-0.3975	-0.3352	-0.2257	0.10
-0.05	-0.5078	-0.4689	-0.4090	-0.3551	-0.2907	-0.1784	0.05
0.00	-0.4705	-0.4297	-0.3673	-0.3115	-0.2451	-0.1305	0.00
0.05	-0.4314	-0.3888	-0.3241	-0.2665	-0.1985	-0.0819	-0.05
0.10	-0.3905	-0.3462	-0.2793	-0.2201	-0.1506	-0.0328	-0.10
0.15	-0.3475	-0.3016	-0.2327	-0.1722	-0.1016	0.0171	-0.15
0.20	-0.3022	-0.2550	-0.1844	-0.1228	-0.0513	0.0676	-0.20
0.25	-0.2546	-0.2061	-0.1340	-0.0716	0.0003	0.1188	-0.25
0.30	-0.2044	-0.1549	-0.0816	-0.0187	0.0533	0.1708	-0.30
0.35	-0.1514	-0.1010	-0.0270	0.0362	0.1078	0.2236	-0.35
0.40	-0.0952	-0.0442	0.0301	0.0930	0.1639	0.2772	-0.40
0.45	-0.0356	0.0157	0.0898	0.1521	0.2217	0.3316	-0.45
0.50	0.0277	0.0789	0.1523	0.2134	0.2811	0.3870	-0.50
0.55	0.0952	0.1458	0.2178	0.2772	0.3425	0.4433	-0.55
0.60	0.1674	0.2169	0.2867	0.3437	0.4058	0.5005	-0.60
0.65	0.2447	0.2924	0.3591	0.4130	0.4711	0.5588	-0.65
0.70	0.3278	0.3730	0.4354	0.4854	0.5387	0.6182	-0.70
0.75	0.4174	0.4590	0.5160	0.5611	0.6087	0.6786	-0.75
0.80	0.5144	0.5513	0.6013	0.6403	0.6812	0.7403	-0.80
0.85	0.6197	0.6506	0.6917	0.7235	0.7563	0.8032	-0.85
0.90	0.7347	0.7577	0.7879	0.8109	0.8344	0.8674	-0.90
0.95	0.8609	0.8737	0.8904	0.9029	0.9155	0.9330	-0.95

↑ R

α →	.995 ↑-	.99 ↑-	.975 ↑-	.95 ↑-	.90 ↑-	.75 ↑-

TABLE 10.27. CONFIDENCE LIMITS ON THE CORRELATION COEFFICIENT.

SAMPLE SIZE = 29

α →	.005 ↓+	.01 ↓+	.025 ↓+	.05 ↓+	.10 ↓+	.25 ↓+	
R ↓							
-0.95	-0.9808	-0.9788	-0.9755	-0.9723	-0.9681	-0.9598	0.95
-0.90	-0.9610	-0.9570	-0.9504	-0.9440	-0.9357	-0.9194	0.90
-0.85	-0.9407	-0.9346	-0.9247	-0.9151	-0.9028	-0.8787	0.85
-0.80	-0.9197	-0.9116	-0.8984	-0.8857	-0.8694	-0.8377	0.80
-0.75	-0.8980	-0.8879	-0.8714	-0.8556	-0.8354	-0.7963	0.75
-0.70	-0.8757	-0.8635	-0.8437	-0.8248	-0.8008	-0.7547	0.70
-0.65	-0.8526	-0.8383	-0.8153	-0.7934	-0.7656	-0.7127	0.65
-0.60	-0.8288	-0.8124	-0.7861	-0.7612	-0.7298	-0.6703	0.60
-0.55	-0.8042	-0.7857	-0.7562	-0.7283	-0.6934	-0.6276	0.55
-0.50	-0.7787	-0.7581	-0.7254	-0.6947	-0.6562	-0.5844	0.50
-0.45	-0.7523	-0.7296	-0.6937	-0.6602	-0.6184	-0.5409	0.45
-0.40	-0.7249	-0.7002	-0.6611	-0.6248	-0.5798	-0.4970	0.40
-0.35	-0.6965	-0.6697	-0.6276	-0.5886	-0.5405	-0.4526	0.35
-0.30	-0.6670	-0.6382	-0.5930	-0.5514	-0.5003	-0.4077	0.30
-0.25	-0.6364	-0.6056	-0.5574	-0.5133	-0.4594	-0.3624	0.25
-0.20	-0.6045	-0.5717	-0.5206	-0.4741	-0.4175	-0.3166	0.20
-0.15	-0.5713	-0.5365	-0.4827	-0.4339	-0.3748	-0.2703	0.15
-0.10	-0.5367	-0.5000	-0.4435	-0.3925	-0.3311	-0.2235	0.10
-0.05	-0.5006	-0.4621	-0.4030	-0.3499	-0.2865	-0.1761	0.05
0.00	-0.4629	-0.4226	-0.3610	-0.3061	-0.2407	-0.1281	0.00
0.05	-0.4234	-0.3814	-0.3176	-0.2609	-0.1940	-0.0795	-0.05
0.10	-0.3821	-0.3385	-0.2725	-0.2143	-0.1460	-0.0303	-0.10
0.15	-0.3388	-0.2936	-0.2258	-0.1663	-0.0969	0.0196	-0.15
0.20	-0.2932	-0.2467	-0.1772	-0.1167	-0.0466	0.0701	-0.20
0.25	-0.2453	-0.1976	-0.1267	-0.0655	0.0051	0.1214	-0.25
0.30	-0.1949	-0.1461	-0.0742	-0.0124	0.0582	0.1733	-0.30
0.35	-0.1416	-0.0921	-0.0195	0.0424	0.1127	0.2261	-0.35
0.40	-0.0852	-0.0352	0.0377	0.0993	0.1687	0.2797	-0.40
0.45	-0.0255	0.0247	0.0973	0.1583	0.2264	0.3341	-0.45
0.50	0.0379	0.0879	0.1598	0.2195	0.2858	0.3893	-0.50
0.55	0.1053	0.1548	0.2251	0.2832	0.3469	0.4455	-0.55
0.60	0.1773	0.2256	0.2937	0.3494	0.4100	0.5027	-0.60
0.65	0.2544	0.3009	0.3658	0.4184	0.4751	0.5608	-0.65
0.70	0.3370	0.3810	0.4417	0.4904	0.5424	0.6200	-0.70
0.75	0.4260	0.4664	0.5218	0.5656	0.6120	0.6803	-0.75
0.80	0.5221	0.5579	0.6063	0.6443	0.6840	0.7417	-0.80
0.85	0.6262	0.6560	0.6959	0.7267	0.7586	0.8043	-0.85
0.90	0.7396	0.7618	0.7910	0.8132	0.8360	0.8682	-0.90
0.95	0.8636	0.8760	0.8921	0.9042	0.9164	0.9334	-0.95

↑ *R*

α →	.995 ↑-	.99 ↑-	.975 ↑-	.95 ↑-	.90 ↑-	.75 ↑-

TABLE 10.28. CONFIDENCE LIMITS ON THE CORRELATION COEFFICIENT.

SAMPLE SIZE = 30

α →	.005 ↓+	.01 ↓+	.025 ↓+	.05 ↓+	.10 ↓+	.25 ↓+	
R ↓							
-0.95	-0.9805	-0.9785	-0.9752	-0.9720	-0.9679	-0.9597	0.95
-0.90	-0.9604	-0.9563	-0.9498	-0.9434	-0.9353	-0.9191	0.90
-0.85	-0.9397	-0.9336	-0.9238	-0.9143	-0.9021	-0.8783	0.85
-0.80	-0.9183	-0.9102	-0.8971	-0.8846	-0.8684	-0.8371	0.80
-0.75	-0.8963	-0.8862	-0.8698	-0.8542	-0.8342	-0.7957	0.75
-0.70	-0.8736	-0.8614	-0.8418	-0.8231	-0.7994	-0.7539	0.70
-0.65	-0.8502	-0.8360	-0.8131	-0.7914	-0.7640	-0.7117	0.65
-0.60	-0.8260	-0.8097	-0.7836	-0.7590	-0.7279	-0.6692	0.60
-0.55	-0.8010	-0.7826	-0.7533	-0.7258	-0.6913	-0.6264	0.55
-0.50	-0.7751	-0.7547	-0.7222	-0.6919	-0.6539	-0.5831	0.50
-0.45	-0.7484	-0.7259	-0.6903	-0.6571	-0.6158	-0.5395	0.45
-0.40	-0.7206	-0.6961	-0.6574	-0.6215	-0.5771	-0.4954	0.40
-0.35	-0.6918	-0.6653	-0.6236	-0.5850	-0.5375	-0.4509	0.35
-0.30	-0.6620	-0.6334	-0.5887	-0.5476	-0.4972	-0.4060	0.30
-0.25	-0.6310	-0.6004	-0.5528	-0.5092	-0.4560	-0.3606	0.25
-0.20	-0.5987	-0.5662	-0.5157	-0.4698	-0.4140	-0.3147	0.20
-0.15	-0.5651	-0.5307	-0.4775	-0.4293	-0.3711	-0.2683	0.15
-0.10	-0.5302	-0.4939	-0.4380	-0.3877	-0.3273	-0.2214	0.10
-0.05	-0.4937	-0.4556	-0.3972	-0.3449	-0.2825	-0.1739	0.05
0.00	-0.4556	-0.4158	-0.3550	-0.3009	-0.2366	-0.1258	0.00
0.05	-0.4158	-0.3743	-0.3114	-0.2556	-0.1897	-0.0772	-0.05
0.10	-0.3742	-0.3311	-0.2661	-0.2088	-0.1417	-0.0279	-0.10
0.15	-0.3305	-0.2860	-0.2192	-0.1607	-0.0925	0.0220	-0.15
0.20	-0.2847	-0.2388	-0.1705	-0.1110	-0.0421	0.0725	-0.20
0.25	-0.2365	-0.1895	-0.1198	-0.0596	0.0097	0.1238	-0.25
0.30	-0.1858	-0.1379	-0.0672	-0.0066	0.0627	0.1758	-0.30
0.35	-0.1323	-0.0836	-0.0124	0.0484	0.1173	0.2285	-0.35
0.40	-0.0758	-0.0267	0.0448	0.1052	0.1733	0.2820	-0.40
0.45	-0.0160	0.0333	0.1044	0.1641	0.2308	0.3364	-0.45
0.50	0.0474	0.0965	0.1668	0.2253	0.2901	0.3916	-0.50
0.55	0.1149	0.1632	0.2320	0.2888	0.3511	0.4476	-0.55
0.60	0.1867	0.2339	0.3004	0.3548	0.4140	0.5047	-0.60
0.65	0.2635	0.3088	0.3722	0.4235	0.4789	0.5627	-0.65
0.70	0.3457	0.3885	0.4477	0.4951	0.5459	0.6217	-0.70
0.75	0.4340	0.4733	0.5272	0.5699	0.6151	0.6818	-0.75
0.80	0.5292	0.5640	0.6111	0.6480	0.6867	0.7430	-0.80
0.85	0.6323	0.6611	0.6998	0.7297	0.7608	0.8053	-0.85
0.90	0.7441	0.7655	0.7938	0.8154	0.8376	0.8689	-0.90
0.95	0.8662	0.8781	0.8937	0.9054	0.9172	0.9338	-0.95

↑ R

α →	.995 ↑-	.99 ↑-	.975 ↑-	.95 ↑-	.90 ↑-	.75 ↑-

TABLE 10.29. *CONFIDENCE LIMITS ON THE CORRELATION COEFFICIENT.*

SAMPLE SIZE = 40

α →	.005 ↓+	.01 ↓+	.025 ↓+	.05 ↓+	.10 ↓+	.25 ↓+	
R ↓							
-0.95	-0.9777	-0.9758	-0.9727	-0.9697	-0.9659	-0.9586	0.95
-0.90	-0.9548	-0.9509	-0.9447	-0.9388	-0.9314	-0.9170	0.90
-0.85	-0.9312	-0.9254	-0.9162	-0.9074	-0.8963	-0.8751	0.85
-0.80	-0.9070	-0.8993	-0.8870	-0.8754	-0.8607	-0.8328	0.80
-0.75	-0.8822	-0.8725	-0.8572	-0.8427	-0.8246	-0.7903	0.75
-0.70	-0.8565	-0.8450	-0.8266	-0.8095	-0.7879	-0.7475	0.70
-0.65	-0.8302	-0.8167	-0.7954	-0.7755	-0.7507	-0.7043	0.65
-0.60	-0.8030	-0.7877	-0.7634	-0.7409	-0.7129	-0.6609	0.60
-0.55	-0.7751	-0.7578	-0.7307	-0.7055	-0.6744	-0.6170	0.55
-0.50	-0.7462	-0.7271	-0.6971	-0.6695	-0.6353	-0.5728	0.50
-0.45	-0.7165	-0.6955	-0.6627	-0.6326	-0.5956	-0.5283	0.45
-0.40	-0.6857	-0.6629	-0.6274	-0.5949	-0.5552	-0.4833	0.40
-0.35	-0.6540	-0.6294	-0.5912	-0.5565	-0.5141	-0.4380	0.35
-0.30	-0.6211	-0.5948	-0.5541	-0.5171	-0.4723	-0.3923	0.30
-0.25	-0.5872	-0.5591	-0.5159	-0.4768	-0.4297	-0.3461	0.25
-0.20	-0.5520	-0.5223	-0.4767	-0.4356	-0.3863	-0.2995	0.20
-0.15	-0.5156	-0.4842	-0.4363	-0.3934	-0.3421	-0.2524	0.15
-0.10	-0.4778	-0.4449	-0.3948	-0.3502	-0.2971	-0.2049	0.10
-0.05	-0.4386	-0.4042	-0.3521	-0.3059	-0.2512	-0.1569	0.05
0.00	-0.3978	-0.3621	-0.3081	-0.2605	-0.2043	-0.1084	0.00
0.05	-0.3555	-0.3184	-0.2628	-0.2139	-0.1566	-0.0593	-0.05
0.10	-0.3114	-0.2732	-0.2160	-0.1661	-0.1078	-0.0098	-0.10
0.15	-0.2654	-0.2262	-0.1677	-0.1169	-0.0580	0.0403	-0.15
0.20	-0.2175	-0.1774	-0.1179	-0.0665	-0.0072	0.0910	-0.20
0.25	-0.1675	-0.1266	-0.0664	-0.0146	0.0448	0.1423	-0.25
0.30	-0.1152	-0.0738	-0.0131	0.0388	0.0979	0.1942	-0.30
0.35	-0.0605	-0.0187	0.0421	0.0937	0.1522	0.2468	-0.35
0.40	-0.0031	0.0387	0.0993	0.1503	0.2079	0.3000	-0.40
0.45	0.0571	0.0987	0.1585	0.2087	0.2648	0.3539	-0.45
0.50	0.1204	0.1614	0.2201	0.2689	0.3231	0.4085	-0.50
0.55	0.1870	0.2270	0.2840	0.3310	0.3829	0.4638	-0.55
0.60	0.2572	0.2959	0.3505	0.3952	0.4442	0.5199	-0.60
0.65	0.3314	0.3682	0.4197	0.4616	0.5072	0.5768	-0.65
0.70	0.4099	0.4442	0.4918	0.5303	0.5717	0.6345	-0.70
0.75	0.4931	0.5242	0.5671	0.6014	0.6381	0.6931	-0.75
0.80	0.5816	0.6087	0.6458	0.6752	0.7063	0.7525	-0.80
0.85	0.6759	0.6981	0.7281	0.7517	0.7765	0.8129	-0.85
0.90	0.7765	0.7927	0.8144	0.8312	0.8488	0.8743	-0.90
0.95	0.8843	0.8931	0.9049	0.9139	0.9232	0.9366	-0.95

↑ *R*

α →	.995 ↑-	.99 ↑-	.975 ↑-	.95 ↑-	.90 ↑-	.75 ↑-

TABLE 10.30. CONFIDENCE LIMITS ON THE CORRELATION COEFFICIENT.

SAMPLE SIZE = 60

α →	.005 ↓+	.01 ↓+	.025 ↓+	.05 ↓+	.10 ↓+	.25 ↓+	
R ↓							
-0.95	-0.9740	-0.9722	-0.9694	-0.9668	-0.9635	-0.9573	0.95
-0.90	-0.9474	-0.9438	-0.9382	-0.9330	-0.9265	-0.9143	0.90
-0.85	-0.9201	-0.9148	-0.9064	-0.8987	-0.8890	-0.8710	0.85
-0.80	-0.8922	-0.8851	-0.8740	-0.8638	-0.8511	-0.8275	0.80
-0.75	-0.8635	-0.8548	-0.8410	-0.8283	-0.8127	-0.7837	0.75
-0.70	-0.8342	-0.8237	-0.8074	-0.7923	-0.7737	-0.7396	0.70
-0.65	-0.8041	-0.7920	-0.7730	-0.7556	-0.7343	-0.6953	0.65
-0.60	-0.7733	-0.7594	-0.7380	-0.7183	-0.6943	-0.6506	0.60
-0.55	-0.7416	-0.7261	-0.7022	-0.6804	-0.6537	-0.6056	0.55
-0.50	-0.7090	-0.6920	-0.6657	-0.6417	-0.6126	-0.5603	0.50
-0.45	-0.6756	-0.6570	-0.6283	-0.6024	-0.5709	-0.5147	0.45
-0.40	-0.6413	-0.6212	-0.5902	-0.5623	-0.5286	-0.4688	0.40
-0.35	-0.6060	-0.5844	-0.5513	-0.5215	-0.4857	-0.4225	0.35
-0.30	-0.5697	-0.5466	-0.5114	-0.4799	-0.4422	-0.3758	0.30
-0.25	-0.5323	-0.5079	-0.4707	-0.4376	-0.3980	-0.3288	0.25
-0.20	-0.4938	-0.4680	-0.4290	-0.3943	-0.3531	-0.2815	0.20
-0.15	-0.4541	-0.4271	-0.3864	-0.3503	-0.3075	-0.2337	0.15
-0.10	-0.4132	-0.3851	-0.3427	-0.3053	-0.2612	-0.1855	0.10
-0.05	-0.3710	-0.3418	-0.2979	-0.2594	-0.2142	-0.1370	0.05
0.00	-0.3274	-0.2972	-0.2521	-0.2126	-0.1664	-0.0880	0.00
0.05	-0.2824	-0.2514	-0.2051	-0.1648	-0.1178	-0.0386	-0.05
0.10	-0.2360	-0.2041	-0.1569	-0.1159	-0.0684	0.0112	-0.10
0.15	-0.1879	-0.1554	-0.1074	-0.0659	-0.0181	0.0615	-0.15
0.20	-0.1381	-0.1051	-0.0566	-0.0149	0.0330	0.1123	-0.20
0.25	-0.0866	-0.0533	-0.0045	0.0373	0.0851	0.1635	-0.25
0.30	-0.0331	0.0003	0.0492	0.0907	0.1380	0.2152	-0.30
0.35	0.0223	0.0557	0.1043	0.1454	0.1920	0.2675	-0.35
0.40	0.0799	0.1130	0.1610	0.2014	0.2469	0.3202	-0.40
0.45	0.1397	0.1723	0.2193	0.2587	0.3029	0.3735	-0.45
0.50	0.2019	0.2338	0.2794	0.3175	0.3600	0.4274	-0.50
0.55	0.2666	0.2975	0.3414	0.3778	0.4182	0.4818	-0.55
0.60	0.3341	0.3636	0.4052	0.4396	0.4775	0.5368	-0.60
0.65	0.4045	0.4322	0.4711	0.5031	0.5381	0.5924	-0.65
0.70	0.4781	0.5036	0.5392	0.5682	0.5999	0.6486	-0.70
0.75	0.5550	0.5778	0.6095	0.6352	0.6630	0.7054	-0.75
0.80	0.6355	0.6551	0.6822	0.7040	0.7274	0.7629	-0.80
0.85	0.7200	0.7358	0.7575	0.7748	0.7933	0.8211	-0.85
0.90	0.8086	0.8200	0.8354	0.8476	0.8607	0.8800	-0.90
0.95	0.9018	0.9079	0.9162	0.9227	0.9295	0.9396	-0.95

↑ R

α →	.995 ↑-	.99 ↑-	.975 ↑-	.95 ↑-	.90 ↑-	.75 ↑-	

TABLE 10.31. CONFIDENCE LIMITS ON THE CORRELATION COEFFICIENT.

SAMPLE SIZE = 100

α →	.005 ↓+	.01 ↓+	.025 ↓+	.05 ↓+	.10 ↓+	.25 ↓+	
R ↓							
-0.95	-0.9698	-0.9682	-0.9658	-0.9636	-0.9608	-0.9558	0.95
-0.90	-0.9389	-0.9358	-0.9310	-0.9266	-0.9212	-0.9114	0.90
-0.85	-0.9075	-0.9029	-0.8957	-0.8892	-0.8813	-0.8668	0.85
-0.80	-0.8754	-0.8693	-0.8599	-0.8513	-0.8408	-0.8219	0.80
-0.75	-0.8427	-0.8351	-0.8235	-0.8129	-0.8000	-0.7768	0.75
-0.70	-0.8093	-0.8003	-0.7865	-0.7739	-0.7587	-0.7314	0.70
-0.65	-0.7752	-0.7648	-0.7488	-0.7344	-0.7170	-0.6858	0.65
-0.60	-0.7404	-0.7286	-0.7106	-0.6944	-0.6748	-0.6400	0.60
-0.55	-0.7048	-0.6917	-0.6717	-0.6537	-0.6321	-0.5939	0.55
-0.50	-0.6684	-0.6541	-0.6322	-0.6125	-0.5890	-0.5475	0.50
-0.45	-0.6313	-0.6157	-0.5919	-0.5707	-0.5454	-0.5008	0.45
-0.40	-0.5933	-0.5765	-0.5510	-0.5283	-0.5012	-0.4539	0.40
-0.35	-0.5544	-0.5365	-0.5093	-0.4852	-0.4566	-0.4067	0.35
-0.30	-0.5147	-0.4956	-0.4669	-0.4415	-0.4114	-0.3592	0.30
-0.25	-0.4740	-0.4539	-0.4238	-0.3972	-0.3657	-0.3114	0.25
-0.20	-0.4323	-0.4113	-0.3798	-0.3521	-0.3194	-0.2634	0.20
-0.15	-0.3896	-0.3678	-0.3351	-0.3063	-0.2726	-0.2150	0.15
-0.10	-0.3459	-0.3233	-0.2894	-0.2598	-0.2252	-0.1663	0.10
-0.05	-0.3011	-0.2778	-0.2430	-0.2126	-0.1772	-0.1172	0.05
0.00	-0.2552	-0.2312	-0.1956	-0.1646	-0.1286	-0.0679	0.00
0.05	-0.2081	-0.1836	-0.1473	-0.1158	-0.0794	-0.0182	-0.05
0.10	-0.1598	-0.1348	-0.0980	-0.0662	-0.0295	0.0318	-0.10
0.15	-0.1102	-0.0849	-0.0477	-0.0157	0.0211	0.0822	-0.15
0.20	-0.0592	-0.0338	0.0036	0.0356	0.0723	0.1329	-0.20
0.25	-0.0069	0.0186	0.0559	0.0878	0.1242	0.1840	-0.25
0.30	0.0469	0.0723	0.1094	0.1409	0.1768	0.2355	-0.30
0.35	0.1022	0.1274	0.1640	0.1950	0.2301	0.2873	-0.35
0.40	0.1591	0.1839	0.2198	0.2500	0.2842	0.3396	-0.40
0.45	0.2177	0.2419	0.2768	0.3061	0.3391	0.3922	-0.45
0.50	0.2781	0.3015	0.3350	0.3632	0.3947	0.4452	-0.50
0.55	0.3402	0.3626	0.3946	0.4214	0.4512	0.4987	-0.55
0.60	0.4044	0.4255	0.4556	0.4806	0.5085	0.5525	-0.60
0.65	0.4705	0.4901	0.5180	0.5411	0.5666	0.6069	-0.65
0.70	0.5387	0.5566	0.5819	0.6027	0.6257	0.6616	-0.70
0.75	0.6092	0.6251	0.6473	0.6656	0.6856	0.7168	-0.75
0.80	0.6821	0.6956	0.7144	0.7297	0.7465	0.7725	-0.80
0.85	0.7575	0.7682	0.7831	0.7952	0.8084	0.8286	-0.85
0.90	0.8355	0.8430	0.8535	0.8620	0.8712	0.8852	-0.90
0.95	0.9163	0.9203	0.9258	0.9303	0.9351	0.9424	-0.95

↑ R

α →	.995 ↑-	.99 ↑-	.975 ↑-	.95 ↑-	.90 ↑-	.75 ↑-

TABLE 10.32. *CONFIDENCE LIMITS ON THE CORRELATION COEFFICIENT.*

<u>*SAMPLE SIZE* = 120</u>

α →	.005 ↓+	.01 ↓+	.025 ↓+	.05 ↓+	.10 ↓+	.25 ↓+	
R ↓							
-0.95	-0.9684	-0.9669	-0.9646	-0.9625	-0.9600	-0.9554	0.95
-0.90	-0.9362	-0.9333	-0.9287	-0.9246	-0.9196	-0.9105	0.90
-0.85	-0.9034	-0.8991	-0.8923	-0.8862	-0.8788	-0.8654	0.85
-0.80	-0.8700	-0.8643	-0.8554	-0.8474	-0.8376	-0.8202	0.80
-0.75	-0.8360	-0.8289	-0.8179	-0.8080	-0.7961	-0.7746	0.75
-0.70	-0.8013	-0.7928	-0.7799	-0.7682	-0.7541	-0.7289	0.70
-0.65	-0.7660	-0.7562	-0.7412	-0.7278	-0.7116	-0.6829	0.65
-0.60	-0.7299	-0.7189	-0.7020	-0.6869	-0.6688	-0.6367	0.60
-0.55	-0.6931	-0.6808	-0.6622	-0.6455	-0.6255	-0.5903	0.55
-0.50	-0.6556	-0.6421	-0.6217	-0.6035	-0.5818	-0.5436	0.50
-0.45	-0.6173	-0.6027	-0.5806	-0.5610	-0.5376	-0.4966	0.45
-0.40	-0.5782	-0.5625	-0.5389	-0.5178	-0.4929	-0.4494	0.40
-0.35	-0.5383	-0.5216	-0.4964	-0.4741	-0.4477	-0.4019	0.35
-0.30	-0.4975	-0.4799	-0.4533	-0.4298	-0.4021	-0.3542	0.30
-0.25	-0.4559	-0.4373	-0.4094	-0.3849	-0.3559	-0.3062	0.25
-0.20	-0.4133	-0.3939	-0.3648	-0.3393	-0.3093	-0.2579	0.20
-0.15	-0.3698	-0.3496	-0.3195	-0.2931	-0.2621	-0.2093	0.15
-0.10	-0.3254	-0.3045	-0.2733	-0.2461	-0.2144	-0.1605	0.10
-0.05	-0.2799	-0.2584	-0.2264	-0.1985	-0.1661	-0.1113	0.05
0.00	-0.2333	-0.2113	-0.1786	-0.1502	-0.1173	-0.0619	0.00
0.05	-0.1857	-0.1632	-0.1300	-0.1012	-0.0679	-0.0121	-0.05
0.10	-0.1370	-0.1141	-0.0805	-0.0514	-0.0180	0.0379	-0.10
0.15	-0.0870	-0.0639	-0.0300	-0.0009	0.0326	0.0883	-0.15
0.20	-0.0358	-0.0126	0.0213	0.0504	0.0838	0.1390	-0.20
0.25	0.0166	0.0398	0.0737	0.1026	0.1356	0.1900	-0.25
0.30	0.0704	0.0934	0.1270	0.1556	0.1881	0.2414	-0.30
0.35	0.1255	0.1483	0.1814	0.2094	0.2412	0.2931	-0.35
0.40	0.1821	0.2044	0.2368	0.2641	0.2950	0.3452	-0.40
0.45	0.2402	0.2619	0.2933	0.3197	0.3495	0.3976	-0.45
0.50	0.2998	0.3208	0.3510	0.3763	0.4047	0.4504	-0.50
0.55	0.3611	0.3811	0.4098	0.4338	0.4606	0.5035	-0.55
0.60	0.4241	0.4430	0.4699	0.4923	0.5173	0.5571	-0.60
0.65	0.4889	0.5064	0.5312	0.5518	0.5747	0.6110	-0.65
0.70	0.5555	0.5714	0.5939	0.6124	0.6330	0.6653	-0.70
0.75	0.6241	0.6381	0.6578	0.6741	0.6920	0.7200	-0.75
0.80	0.6948	0.7066	0.7232	0.7369	0.7518	0.7752	-0.80
0.85	0.7676	0.7770	0.7901	0.8008	0.8126	0.8307	-0.85
0.90	0.8426	0.8493	0.8585	0.8660	0.8741	0.8867	-0.90
0.95	0.9201	0.9236	0.9284	0.9323	0.9366	0.9431	-0.95

↑ *R*

α →	.995 ↑-	.99 ↑-	.975 ↑-	.95 ↑-	.90 ↑-	.75 ↑-

TABLE 10.33. *CONFIDENCE LIMITS ON THE CORRELATION COEFFICIENT.*

SAMPLE SIZE = 150

α →	.005 ↓+	.01 ↓+	.025 ↓+	.05 ↓+	.10 ↓+	.25 ↓+	
R ↓							
-0.95	-0.9668	-0.9654	-0.9633	-0.9614	-0.9591	-0.9548	0.95
-0.90	-0.9331	-0.9303	-0.9261	-0.9223	-0.9178	-0.9095	0.90
-0.85	-0.8987	-0.8947	-0.8885	-0.8829	-0.8761	-0.8639	0.85
-0.80	-0.8638	-0.8585	-0.8503	-0.8429	-0.8340	-0.8182	0.80
-0.75	-0.8283	-0.8217	-0.8116	-0.8025	-0.7916	-0.7722	0.75
-0.70	-0.7922	-0.7843	-0.7724	-0.7617	-0.7488	-0.7261	0.70
-0.65	-0.7554	-0.7464	-0.7326	-0.7204	-0.7056	-0.6797	0.65
-0.60	-0.7179	-0.7078	-0.6923	-0.6785	-0.6620	-0.6331	0.60
-0.55	-0.6798	-0.6685	-0.6514	-0.6362	-0.6180	-0.5862	0.55
-0.50	-0.6410	-0.6286	-0.6100	-0.5934	-0.5736	-0.5392	0.50
-0.45	-0.6014	-0.5881	-0.5679	-0.5500	-0.5288	-0.4919	0.45
-0.40	-0.5612	-0.5468	-0.5252	-0.5061	-0.4835	-0.4444	0.40
-0.35	-0.5201	-0.5048	-0.4819	-0.4617	-0.4378	-0.3966	0.35
-0.30	-0.4782	-0.4621	-0.4380	-0.4167	-0.3917	-0.3486	0.30
-0.25	-0.4356	-0.4187	-0.3934	-0.3712	-0.3451	-0.3004	0.25
-0.20	-0.3921	-0.3744	-0.3481	-0.3250	-0.2980	-0.2519	0.20
-0.15	-0.3477	-0.3294	-0.3021	-0.2783	-0.2505	-0.2031	0.15
-0.10	-0.3024	-0.2835	-0.2554	-0.2309	-0.2024	-0.1541	0.10
-0.05	-0.2562	-0.2368	-0.2080	-0.1830	-0.1539	-0.1048	0.05
0.00	-0.2090	-0.1892	-0.1598	-0.1344	-0.1049	-0.0553	0.00
0.05	-0.1609	-0.1407	-0.1109	-0.0851	-0.0553	-0.0055	-0.05
0.10	-0.1117	-0.0912	-0.0611	-0.0352	-0.0053	0.0446	-0.10
0.15	-0.0615	-0.0408	-0.0106	0.0154	0.0453	0.0950	-0.15
0.20	-0.0101	0.0106	0.0408	0.0668	0.0965	0.1456	-0.20
0.25	0.0424	0.0630	0.0931	0.1188	0.1482	0.1966	-0.25
0.30	0.0960	0.1165	0.1462	0.1716	0.2005	0.2479	-0.30
0.35	0.1509	0.1710	0.2003	0.2251	0.2533	0.2994	-0.35
0.40	0.2070	0.2268	0.2553	0.2794	0.3068	0.3513	-0.40
0.45	0.2645	0.2836	0.3113	0.3346	0.3609	0.4035	-0.45
0.50	0.3233	0.3417	0.3682	0.3905	0.4156	0.4560	-0.50
0.55	0.3836	0.4011	0.4262	0.4473	0.4709	0.5088	-0.55
0.60	0.4453	0.4617	0.4853	0.5049	0.5269	0.5620	-0.60
0.65	0.5086	0.5237	0.5454	0.5634	0.5835	0.6155	-0.65
0.70	0.5734	0.5871	0.6067	0.6228	0.6408	0.6693	-0.70
0.75	0.6399	0.6520	0.6691	0.6832	0.6989	0.7235	-0.75
0.80	0.7082	0.7183	0.7327	0.7445	0.7576	0.7781	-0.80
0.85	0.7782	0.7863	0.7976	0.8068	0.8171	0.8330	-0.85
0.90	0.8501	0.8558	0.8637	0.8702	0.8773	0.8883	-0.90
0.95	0.9240	0.9270	0.9312	0.9345	0.9382	0.9440	-0.95

↑ R

α →	.995 ↑-	.99 ↑-	.975 ↑-	.95 ↑-	.90 ↑-	.75 ↑-	

TABLE 10.34. CONFIDENCE LIMITS ON THE CORRELATION COEFFICIENT.

SAMPLE SIZE = 200

α →	.005 ↓+	.01 ↓+	.025 ↓+	.05 ↓+	.10 ↓+	.25 ↓+	
R ↓							
-0.95	-0.9649	-0.9637	-0.9618	-0.9600	-0.9580	-0.9543	0.95
-0.90	-0.9293	-0.9269	-0.9231	-0.9197	-0.9156	-0.9083	0.90
-0.85	-0.8932	-0.8895	-0.8839	-0.8789	-0.8729	-0.8622	0.85
-0.80	-0.8565	-0.8517	-0.8443	-0.8377	-0.8299	-0.8159	0.80
-0.75	-0.8192	-0.8133	-0.8042	-0.7962	-0.7865	-0.7694	0.75
-0.70	-0.7814	-0.7744	-0.7637	-0.7541	-0.7428	-0.7228	0.70
-0.65	-0.7430	-0.7349	-0.7226	-0.7117	-0.6987	-0.6759	0.65
-0.60	-0.7039	-0.6948	-0.6810	-0.6688	-0.6543	-0.6289	0.60
-0.55	-0.6643	-0.6542	-0.6389	-0.6255	-0.6095	-0.5816	0.55
-0.50	-0.6239	-0.6129	-0.5963	-0.5817	-0.5643	-0.5342	0.50
-0.45	-0.5830	-0.5711	-0.5532	-0.5374	-0.5187	-0.4865	0.45
-0.40	-0.5413	-0.5286	-0.5095	-0.4926	-0.4728	-0.4386	0.40
-0.35	-0.4990	-0.4854	-0.4652	-0.4474	-0.4265	-0.3906	0.35
-0.30	-0.4559	-0.4416	-0.4203	-0.4017	-0.3798	-0.3423	0.30
-0.25	-0.4121	-0.3972	-0.3749	-0.3555	-0.3327	-0.2938	0.25
-0.20	-0.3675	-0.3520	-0.3289	-0.3087	-0.2851	-0.2450	0.20
-0.15	-0.3222	-0.3061	-0.2822	-0.2614	-0.2372	-0.1961	0.15
-0.10	-0.2761	-0.2595	-0.2349	-0.2136	-0.1888	-0.1469	0.10
-0.05	-0.2291	-0.2121	-0.1870	-0.1653	-0.1400	-0.0975	0.05
0.00	-0.1813	-0.1640	-0.1384	-0.1164	-0.0908	-0.0478	0.00
0.05	-0.1326	-0.1151	-0.0892	-0.0669	-0.0411	0.0020	-0.05
0.10	-0.0831	-0.0653	-0.0392	-0.0168	0.0091	0.0522	-0.10
0.15	-0.0326	-0.0147	0.0114	0.0339	0.0597	0.1025	-0.15
0.20	0.0189	0.0367	0.0628	0.0851	0.1107	0.1531	-0.20
0.25	0.0713	0.0890	0.1149	0.1370	0.1623	0.2040	-0.25
0.30	0.1248	0.1423	0.1678	0.1895	0.2144	0.2551	-0.30
0.35	0.1792	0.1965	0.2215	0.2427	0.2669	0.3065	-0.35
0.40	0.2348	0.2516	0.2759	0.2966	0.3200	0.3581	-0.40
0.45	0.2915	0.3077	0.3312	0.3511	0.3735	0.4100	-0.45
0.50	0.3493	0.3649	0.3873	0.4063	0.4276	0.4622	-0.50
0.55	0.4083	0.4231	0.4443	0.4622	0.4823	0.5147	-0.55
0.60	0.4685	0.4824	0.5022	0.5188	0.5375	0.5674	-0.60
0.65	0.5301	0.5428	0.5610	0.5762	0.5932	0.6205	-0.65
0.70	0.5929	0.6043	0.6207	0.6343	0.6495	0.6738	-0.70
0.75	0.6570	0.6671	0.6813	0.6932	0.7064	0.7274	-0.75
0.80	0.7226	0.7310	0.7430	0.7529	0.7639	0.7813	-0.80
0.85	0.7896	0.7963	0.8057	0.8134	0.8220	0.8355	-0.85
0.90	0.8582	0.8628	0.8694	0.8747	0.8807	0.8900	-0.90
0.95	0.9283	0.9307	0.9341	0.9369	0.9400	0.9448	-0.95

↑ R

α →	.995 ↑-	.99 ↑-	.975 ↑-	.95 ↑-	.90 ↑-	.75 ↑-	

TABLE 10.35. *CONFIDENCE LIMITS ON THE CORRELATION COEFFICIENT.*

SAMPLE SIZE = 400

α →	.005 ↓+	.01 ↓+	.025 ↓+	.05 ↓+	.10 ↓+	.25 ↓+	
R ↓							
-0.95	-0.9611	-0.9601	-0.9586	-0.9573	-0.9558	-0.9531	0.95
-0.90	-0.9217	-0.9198	-0.9169	-0.9144	-0.9113	-0.9060	0.90
-0.85	-0.8819	-0.8791	-0.8749	-0.8711	-0.8666	-0.8588	0.85
-0.80	-0.8417	-0.8380	-0.8324	-0.8275	-0.8216	-0.8115	0.80
-0.75	-0.8010	-0.7964	-0.7896	-0.7835	-0.7764	-0.7640	0.75
-0.70	-0.7598	-0.7544	-0.7464	-0.7393	-0.7309	-0.7164	0.70
-0.65	-0.7181	-0.7120	-0.7028	-0.6947	-0.6851	-0.6686	0.65
-0.60	-0.6760	-0.6691	-0.6588	-0.6497	-0.6391	-0.6207	0.60
-0.55	-0.6333	-0.6257	-0.6144	-0.6045	-0.5928	-0.5727	0.55
-0.50	-0.5902	-0.5819	-0.5696	-0.5588	-0.5462	-0.5245	0.50
-0.45	-0.5465	-0.5376	-0.5244	-0.5129	-0.4993	-0.4761	0.45
-0.40	-0.5023	-0.4929	-0.4788	-0.4665	-0.4521	-0.4276	0.40
-0.35	-0.4576	-0.4476	-0.4327	-0.4198	-0.4047	-0.3789	0.35
-0.30	-0.4123	-0.4018	-0.3863	-0.3727	-0.3569	-0.3301	0.30
-0.25	-0.3665	-0.3555	-0.3394	-0.3253	-0.3089	-0.2811	0.25
-0.20	-0.3200	-0.3087	-0.2920	-0.2775	-0.2606	-0.2320	0.20
-0.15	-0.2731	-0.2614	-0.2442	-0.2292	-0.2119	-0.1827	0.15
-0.10	-0.2255	-0.2135	-0.1959	-0.1806	-0.1630	-0.1332	0.10
-0.05	-0.1773	-0.1651	-0.1472	-0.1317	-0.1137	-0.0836	0.05
0.00	-0.1285	-0.1161	-0.0979	-0.0823	-0.0641	-0.0338	0.00
0.05	-0.0791	-0.0666	-0.0482	-0.0325	-0.0142	0.0162	-0.05
0.10	-0.0290	-0.0165	0.0019	0.0178	0.0360	0.0663	-0.10
0.15	0.0217	0.0343	0.0526	0.0684	0.0865	0.1166	-0.15
0.20	0.0731	0.0856	0.1038	0.1195	0.1374	0.1671	-0.20
0.25	0.1252	0.1375	0.1556	0.1710	0.1886	0.2178	-0.25
0.30	0.1780	0.1901	0.2078	0.2229	0.2401	0.2686	-0.30
0.35	0.2315	0.2433	0.2606	0.2753	0.2920	0.3196	-0.35
0.40	0.2857	0.2972	0.3139	0.3281	0.3443	0.3708	-0.40
0.45	0.3407	0.3518	0.3678	0.3814	0.3969	0.4222	-0.45
0.50	0.3965	0.4070	0.4222	0.4351	0.4498	0.4737	-0.50
0.55	0.4530	0.4629	0.4772	0.4894	0.5031	0.5255	-0.55
0.60	0.5103	0.5195	0.5328	0.5441	0.5568	0.5774	-0.60
0.65	0.5685	0.5769	0.5890	0.5993	0.6108	0.6296	-0.65
0.70	0.6274	0.6350	0.6458	0.6550	0.6653	0.6819	-0.70
0.75	0.6873	0.6938	0.7033	0.7112	0.7201	0.7344	-0.75
0.80	0.7480	0.7535	0.7613	0.7679	0.7753	0.7871	-0.80
0.85	0.8096	0.8139	0.8200	0.8251	0.8308	0.8400	-0.85
0.90	0.8721	0.8751	0.8793	0.8829	0.8868	0.8931	-0.90
0.95	0.9356	0.9371	0.9393	0.9412	0.9432	0.9465	-0.95

↑ R

α →	.995 ↑-	.99 ↑-	.975 ↑-	.95 ↑-	.90 ↑-	.75 ↑-

PART C

MATHEMATICAL DERIVATIONS

11. THE NONCENTRAL t-DISTRIBUTION

11.1 THE CUMULATIVE OF THE NONCENTRAL t-DISTRIBUTION

Let X have a normal distribution with mean zero and variance one and let Y be independent of X and have a chi-squared distribution with f degrees of freedom and let δ be any constant. Then $T_f(\delta) = (X+\delta)/\sqrt{Y/f}$ has a noncentral t-distribution with noncentrality parameter δ and f degrees of freedom. Note that if $\delta = 0$ then $T_f(\delta)$ has a (central) Student t-distribution with f degrees of freedom.

The joint density of X and Y is then given by

$$\frac{1}{\sqrt{2\pi}} e^{-X^2/2} \frac{1}{\Gamma(f/2) 2^{f/2}} Y^{(f-2)/2} e^{-Y/2} \qquad \begin{array}{l} Y \geq 0 \\[1mm] -\infty < X < +\infty \end{array}$$

If the change in variables given by $X = (ZU/\sqrt{f}) - \delta$ and $Y = U^2$ is made in this last expression and U is integrated out, then the density of $Z = T_f$ is obtained. That result is then integrated with respect to Z from minus infinity to t to give the cumulative distribution function,

$$\Pr\{T_f \leq t\} = \frac{\sqrt{2\pi}}{\Gamma\left(\dfrac{f}{2}\right) 2^{(f-2)/2}} \int_0^\infty G\left(\frac{tU}{\sqrt{f}} - \delta\right) U^{f-1} G'(U) dU$$

where

$$G'(X) = \frac{1}{\sqrt{2\pi}} e^{-X^2/2}$$

and

$$G(X) = \int_{-\infty}^{X} G'(t) dt$$

If this cumulative distribution function is integrated by parts repeatedly, one obtains for odd values of f

$$\Pr\{T_f \leq t\} = G(-\delta\sqrt{B}) + 2T(\delta\sqrt{B},A) + 2[M_1 + M_3 + \cdots + M_{f-2}]$$

where

$$A = \frac{t}{\sqrt{f}} \quad \text{and} \quad B = \frac{f}{f + t^2}$$

and

$$T(h,a) = \frac{1}{2\pi} \cdot \int_0^a \frac{\exp\left[-\frac{h^2}{2}(1+X^2)\right]}{1+X^2} dX$$

is a function discussed and tabulated in Owen (1956) and Owen (1963). The M's are defined below. For even values of f,

$$Pr\{T_f \leq t\} = G(-\delta) + \sqrt{2\pi} \, [M_0 + M_2 + \cdots + M_{f-2}]$$

where

$$M_{-1} = 0$$

$$M_0 = A\sqrt{B} \; G'(\delta\sqrt{B}) G(\delta A \sqrt{B})$$

$$M_1 = B\left[\delta A M_0 + \frac{A}{\sqrt{2\pi}} G'(\delta)\right]$$

$$M_2 = \frac{1}{2} B[\delta A M_1 + M_0]$$

$$M_3 = \frac{2}{3} B[\delta A M_2 + M_1]$$

$$M_4 = \frac{3}{4} B\left[\frac{1}{2} \delta A M_3 + M_2\right]$$

$$\cdot \qquad \cdot \qquad \cdot$$
$$\cdot \qquad \cdot \qquad \cdot$$
$$\cdot \qquad \cdot \qquad \cdot$$

$$M_k = \frac{k-1}{k} B[a_k \delta A M_{k-1} + M_{k-2}]$$

and where

$$a_k = \frac{1}{(k-2)a_{k-1}} \quad \text{for} \quad k \geq 3 \quad \text{and} \quad a_2 = 1.$$

Some special properties of the noncentral t-distribution are now immediately obvious from the above formulas:

$$Pr\{T_f \leq t \mid \delta\} = 1 - Pr\{T_f \leq -t \mid -\delta\}$$

$$Pr\{T_f \leq 0 \mid \delta\} = G(-\delta)$$

If $f = 1$, $t = 1$ and $G\left(\dfrac{\delta}{\sqrt{2}}\right) = P$, then $\Pr\{T_1 \leq 1\} = 1 - P^2$.

Also if $\delta = 0$, the above noncentral t-distribution reduces to the Student t-distribution. Note that $T(0,A) = (\arctan A)/(2\pi)$. Note also that in this case the M's with odd subscripts can be computed independently of the M's with even subscripts and vice versa.

For odd values of f, therefore, the Student t-cumulative distribution function reduces to

$$\Pr\{\text{Student-}t \leq t\} =$$

$$\frac{1}{2} + (\arctan A)/(\pi) + (AB)/(\pi)\left[b_0 + b_1 B + b_2 B^2 + \cdots + b_{(f-3)/2} B^{(f-3)/2}\right]$$

where $b_0 = 1$ and $b_r = \dfrac{2r}{2r+1} b_{r-1}$; and for even values of f

$$\Pr\{\text{Student-}t \leq t\} = \frac{1}{2} + \frac{A\sqrt{B}}{2}\left[c_0 + c_1 B + c_2 B^2 + \cdots + c_{(f-2)/2} B^{(f-2)/2}\right]$$

where $c_0 = 1$ and $c_r = \dfrac{2r-1}{2r} c_{r-1}$ and, as before,

$$A = t/(\sqrt{f}) \quad \text{and} \quad B = f/(f+t^2)$$

11.2 ONE-SIDED TOLERANCE LIMITS

A method for finding a one-sided tolerance limit will now be given such that at least a proportion P of a normal population will be above (or below) the tolerance limit with probability γ (i.e., if the experimental procedure as described below were repeated an infinite number of times and all of the hypotheses were met exactly, then $100\gamma\%$ of the tolerance limits would cover at least a proportion P of the population).

A sample of n observations is taken at random from a normal distribution with unknown mean μ and unknown standard deviation σ. The sample values are denoted by $x_1, x_2, \cdots, x_n$ and

$$\bar{x} = \frac{1}{n}\sum_{i=1}^{n} x_i \quad \text{and} \quad s^2 = \frac{1}{n-1}\sum_{i=1}^{n} (x_i - \bar{x})^2$$

are computed from these sample values.

In more complicated problems the estimate of σ may be obtained from an analysis of variance computation and be based on f degrees of freedom. The quantity s as defined above has $f = n - 1$ degrees of freedom. In what follows, s will be used for either the analysis of variance estimate or the estimate above. How s or $\bar{x}$ are computed is not important to the problem here so long as fs^2/σ^2 has a chi-squared distribution with f degrees of freedom and $\bar{x}$ has a normal distribution with mean μ and standard deviation equal to $\sigma/\sqrt{n}$ (μ and σ are unknown). The problem then becomes a problem in finding k so that either $\bar{x} + ks$ or $\bar{x} - ks$ is the required tolerance limit.

Mathematically, the problem is to find k such that

$$P_{\bar{x},s}\left\{P_X(X \leq \bar{x} + ks) \geq P\right\} = \gamma$$

where X has a normal distribution with mean μ and standard deviation σ, and P and γ are specified probabilities. Define K_p by:

$$\frac{1}{\sqrt{2\pi}} \int_{-\infty}^{K_P} e^{-t^2/2} \, dt = P$$

and then,

$$P_{\bar{x},s}\left\{P_X(X \leq \bar{x} + ks) \geq P\right\} = \Pr\left\{\frac{\bar{x} + ks - \mu}{\sigma} \geq K_P\right\}$$

Rewriting once more this becomes

$$\Pr\left\{\frac{\dfrac{\bar{x} - \mu}{\sigma}\sqrt{n} - K_P\sqrt{n}}{\dfrac{s}{\sigma}} \geq -k\sqrt{n}\right\} = \gamma$$

This is now in the form of the noncentral t-distribution with f degrees of freedom and with $\delta = -K_P\sqrt{n}$ and $t = -k\sqrt{n}$. Or equivalently, this may be written

$$\Pr\left\{T_f \leq k\sqrt{n} \mid \delta = K_P\sqrt{n}\right\} = \gamma$$

where T_f has a noncentral t-distribution. Hence the quantity k which is desired may be computed from the percentage points of the noncentral t-distribution.

11.3 THE SAMPLING PLAN PROCEDURE

The problem is to find a value of k such that the $\Pr\{\bar{x} + ks \le U\} = 1 - \gamma$, i.e., the probability of accepting a lot is $1 - \gamma$. (The quantity γ is usually taken equal to 0.90 for a consumer's risk of 0.10, and γ is taken equal to 0.95 for a producer's risk of 0.05.) We want to find k so that

$$\Pr\left\{\left(\frac{\bar{x} - \mu}{\sigma}\right)\sqrt{n} - \left(\frac{U - \mu}{\sigma}\right)\sqrt{n} \le -\frac{ks}{\sigma}\sqrt{n}\right\} = 1 - \gamma$$

Next we divide both sides of the inequality by s/σ and the quantity on the left is of the form of the noncentral t. Hence, we have

$$\Pr\left\{T_f \le -k\sqrt{n} \mid \delta = -\left(\frac{U - \mu}{\sigma}\right)\sqrt{n}\right\} = 1 - \gamma$$

where T_f is a noncentral t variable with f degrees of freedom (the number of degrees of freedom of s^2). In most applications $f = n - 1$. From a consumer's point of view we want to accept a lot with no more than $100(1-P)\%$ of the population above U only $100(1-\gamma)\%$ of the time and hence $K_p = \dfrac{U - \mu}{\sigma}$. Then we have

$$\Pr\{T_f \le k\sqrt{n} \mid \delta = K_p\sqrt{n}\} = \gamma$$

This equation is identical with the one given for the tolerance limit problem.

11.4 THE DENSITY OF NONCENTRAL t IN TERMS OF THE CUMULATIVE

The value of k, as defined in Section 11.2, may now be obtained by using Newton's method for interpolating for a root of an equation. To implement this method, the derivative of the distribution function is needed. Let

$$H_f(t) = \Pr\{T_f \le t\}$$

$$= \frac{\sqrt{2\pi}}{\Gamma\left(\frac{f}{2}\right)2^{(f-2)/2}} \int_0^\infty G\left(\frac{tx}{\sqrt{f}} - \delta\right) x^{f-1} G'(x)\,dx$$

Then the derivative with respect to t is given by:

$$H'_f(t) = \frac{f}{t}\left[H_{f+2}\left(\frac{\sqrt{f+2}\ t}{\sqrt{f}}\right) - H_f(t)\right] = \frac{f}{t}M_f(t)\sqrt{2\pi}\quad \text{if}\quad f\quad \text{is even}$$

$$= \frac{2f}{t}M_f(t)\quad \text{if}\quad f\quad \text{is odd}$$

where $M_f(t) = M_k$ is defined in Section 11.1 by a recursion formula and for this derivative means that the recursion formula must be applied two steps more than would be necessary to get the cumulative distribution function. Newton's method as applied to the problem of finding k is to start with an approximate value of k, say k_0, and then obtain $k_1, k_2, k_3, \cdots$ from

$$k_{r+1} = k_r - \frac{H_f(k_r\sqrt{n}) - \gamma}{H'_f(k_r\sqrt{n})},\quad r = 0,1,2,\cdots$$

where

$$H'_f(k\sqrt{n}) = \frac{f}{k}\left[H_{f+2}\left(\frac{\sqrt{f+2}}{\sqrt{f}}\ k\sqrt{n}\right) - H_f(k\sqrt{n})\right]$$

This process was repeated until two consecutive values of k were equal to 8 decimal places or until $H_f(k_r\sqrt{n}) - \gamma$ was zero to 8 decimal places. This is the value of k given in Table 1.

11.5   THE DERIVATIVE OF THE CUMULATIVE WITH RESPECT TO δ

It is convenient to have the derivative of the cumulative distribution function of noncentral t with respect to δ when computing Table 7. Again let

$$H(\delta) = H_f(t\,|\,\delta) = Pr\{T_f \le t\,|\,\delta\} = \frac{\sqrt{2\pi}}{\Gamma\left(\frac{f}{2}\right)2^{(f-2)/2}}\int_0^\infty G\left(\frac{tx}{\sqrt{f}} - \delta\right)x^{f-1}G'(x)dx$$

Putting $f = 1$, the derivative with respect to δ is given by

$$H'(\delta) = H'_1(t\,|\,\delta) = -2G'\left(\frac{\delta}{\sqrt{1+t^2}}\right)G\left(\frac{\delta t}{\sqrt{1+t^2}}\right)\frac{1}{\sqrt{1+t^2}} = \frac{-2}{t}M_0(t)$$

where M_0 is defined in Section 11.1.

The derivative with respect to δ for $f \geq 2$ is given by

$$\frac{dH_f(t|\delta)}{d\delta} = \frac{\sqrt{2f}}{t} \frac{\Gamma\left(\frac{f+1}{2}\right)}{\Gamma\left(\frac{f}{2}\right)} \left[H_{f-1}\left(t\sqrt{\frac{f-1}{f}}\right) - H_{f+1}\left(t\sqrt{\frac{f+1}{f}}\right)\right]$$

$$= \frac{-\sqrt{2f}}{t} \frac{\Gamma\left(\frac{f+1}{2}\right)}{\Gamma\left(\frac{f}{2}\right)} M_{f-1}(t)\sqrt{2\pi} \quad \text{if} \quad f \quad \text{is odd}$$

$$= \frac{-2\sqrt{2f}}{t} \frac{\Gamma\left(\frac{f+1}{2}\right)}{\Gamma\left(\frac{f}{2}\right)} M_{f-1}(t) \quad \text{if} \quad f \quad \text{is even}$$

where $M_{f-1}(t) = M_k$ is defined in Section 11.1 by a recursion formula and for the derivative means that the recursion formula must be applied one step more than would be necessary to get the cumulative distribution function.

Note that $H_f(t|\delta) = H_f(t)$ and hence $H_{f-1}\left(t\sqrt{\frac{f-1}{f}}\right)$ indicates a change in the argument t, but no change in δ. Also note that this derivative is for $f \geq 2$ and that it is always negative.

The above results were used to find confidence limits on $\gamma = \Pr\{Y \geq L\}$ given in Table 7. Owen and Hua (1977) show that for given n, K, and η, the lower confidence limit γ^* can be obtained by solving for δ in the expression

$$H_f(t|\delta) = \Pr\left\{T_f \leq K\sqrt{n} \mid \delta = \sqrt{n} \, K_{\gamma^*}\right\} = \eta$$

where $t = K\sqrt{n}$ and $f = n-1$. Then $G(\delta/\sqrt{n})$ is the value given in Table 7.

The desired value of δ was found by using Newton's method. Given an approximate value of δ, say δ_0, we obtain $\delta_1, \delta_2, \cdots$ from

$$\delta_{r+1} = \delta_r - \frac{H(t|\delta_r) - \eta}{H'(t|\delta_r)} \quad , \quad r = 0,1,2,\cdots$$

This process was repeated until two consecutive values of δ were equal to 8 decimal places.

For given n and η, values of δ_0 are found by first noting that for $K = 0$, the desired value of δ satisfies $H_f(0|\delta) = G(-\delta) = \eta$. This value can be used for δ_0 when $K = 0.2$. The process can be continued using as the value of δ_0 when $K = 0.4$, the value obtained from the iteration above when $K = 0.2$.

11.6 THE DERIVATIVE OF t WITH RESPECT TO δ

Let $H_f(t|\delta) = \varepsilon = \dfrac{\sqrt{2\pi}}{f\left(\dfrac{f}{2}\right) 2^{(f-2)/2}} \int_0^\infty G\left(\dfrac{tx}{\sqrt{f}} - \delta\right) x^{f-1} G'(x)\,dx$

i.e., hold the value of the integral constant and now differentiate t with respect to δ. The result is:

$$\frac{dt}{d\delta} = \frac{\sqrt{f} \displaystyle\int_0^\infty G'\left(\frac{tx}{\sqrt{f}} - \delta\right) x^{f-1} G'(x)\,dx}{\displaystyle\int_0^\infty G'\left(\frac{tx}{\sqrt{f}} - \delta\right) x^{f} G'(x)\,dx}$$

Integration by parts yields

$$\text{for} \quad f = 1, \quad \frac{dt}{d\delta} = \frac{+G'\left(\dfrac{\delta}{\sqrt{1+t^2}}\right) G\left(\dfrac{\delta t}{\sqrt{1+t^2}}\right)\sqrt{1+t^2}}{\dfrac{\delta t}{\sqrt{1+t^2}} G'\left(\dfrac{\delta}{\sqrt{1+t^2}}\right) G\left(\dfrac{\delta t}{\sqrt{1+t^2}}\right) + \dfrac{1}{\sqrt{2\pi}} G'(\delta)} = \frac{M_0(t)}{M_1(t)}$$

$$\text{for} \quad f \geq 2, \quad \frac{dt}{d\delta} = \frac{\sqrt{f}\,\Gamma\left(\dfrac{f+1}{2}\right) H_{f+1}\left(t\sqrt{\dfrac{f+1}{f}}\right) - H_{f-1}\left(t\sqrt{\dfrac{f-1}{f}}\right)}{\sqrt{2}\,\Gamma\left(\dfrac{f+2}{2}\right) H_{f+2}\left(t\sqrt{\dfrac{f+2}{f}}\right) - H_f(t)}$$

$$= \frac{\sqrt{\pi f}\,\Gamma\left(\dfrac{f+1}{2}\right)}{\Gamma\left(\dfrac{f}{2}\right)} \frac{M_{f-1}(t)}{M_f(t)} \qquad \text{if } f \text{ is odd}$$

$$= \frac{2}{\sqrt{\pi f}} \frac{\Gamma\left(\dfrac{f+1}{2}\right)}{\Gamma\left(\dfrac{f}{2}\right)} \frac{M_{f-1}(t)}{M_f(t)} \qquad \text{if } f \text{ is even}$$

where $M_f(t) = M_k$ is defined in Section 11.1 by a recursion formula. Note that $\dfrac{dt}{d\delta}$ is positive. This is most easily seen in the integral from above.

11.7 A CHECK ON k FOR n = 2

As a further check on the case f = 1, the problem of finding k was reformulated in terms of the bivariate normal probability distribution. Let

$$B(\widetilde{h},\widetilde{k};\rho) = \frac{1}{2\pi\sqrt{1-\rho^2}} \int_{-\infty}^{\widetilde{h}} \int_{-\infty}^{\widetilde{k}} \exp\left[-\frac{1}{2}\left(\frac{x^2 - 2\rho xy + y^2}{1-\rho^2}\right)\right] dx\ dy$$

Then our problem reduces first to finding ρ such that

$$B\left(\frac{\delta\sqrt{1+\rho}}{\sqrt{2}}\ ,\ \frac{\delta\sqrt{1+\rho}}{\sqrt{2}}\ ;\ \rho\right) = 1 - \gamma$$

and then obtaining k from

$$k = \frac{1}{\sqrt{n}}\sqrt{\frac{1-\rho}{1+\rho}}$$

In order to get a value of ρ close to the value needed, an approximate value of ρ was computed from

$$\rho \approx \frac{K_{1-\gamma/2}^2 - n\ K_p^2}{K_{1-\gamma/2}^2 + n\ K_p^2}$$

and then the computer iterated on ρ until the value of

$$B\left(\frac{\delta\sqrt{1+\rho}}{\sqrt{2}}\ ,\ \frac{\delta\sqrt{1+\rho}}{\sqrt{2}}\ ;\ \rho\right)$$

was within 10^{-8} of being equal to $1-\gamma$. The resulting k values agreed with those computed by the formulas of Section 11.1. To evaluate $B(\widetilde{h},\widetilde{k};\rho)$ the FORTRAN subroutine given by Donnelly (1973) was used.

11.8 LIMITING VALUES OF k

When the value of f becomes infinite the value of σ is then known and the tolerance limit becomes $\bar{x} + k\sigma$ and k may be obtained from

$$k = K_p + \frac{K_\gamma}{\sqrt{n}}$$

When the value of n becomes infinite the value of μ is then known and the tolerance limit becomes $\mu + ks$ and k may be obtained from

$$k = K_p \sqrt{\frac{f}{\chi^2_{1-\gamma}}}$$

where $\chi^2_{1-\gamma}$ is a percentage point of the chi-squared distribution with f degrees of freedom, i.e.,

$$Pr\left\{\chi^2 \leq \chi^2_{1-\gamma}\right\} = 1 - \gamma$$

When both n and f are infinite the tolerance limit becomes $\mu + k\sigma$ where

$$k = K_p$$

and the result holds with probability $\gamma = 1$.

For $n = \infty$ and $f = 1$, (with σ unknown and μ known) the formula specializes to

$$k = \frac{K_p}{K_{1-\gamma/2}} \, , \quad \text{i.e.,} \quad k \text{ is the ratio of two normal deviates;}$$

and for $n = \infty$ and $f = 2$,

$$k = \frac{K_p}{\sqrt{-\log_e \gamma}}$$

11.9   THE MOMENTS OF THE NONCENTRAL t-DISTRIBUTION

The moments of the noncentral t-distribution have been given by Hogben, Pinkham, and Wilk (1961) and by others previously, e.g., Merrington and Pearson (1958) and are polynomial functions of δ whose coefficients are functions of f. The mean and second, third, and fourth moments about the mean are as follows:

$$\mu = c_{11} \delta \qquad\qquad \mu_2 = c_{22} \delta^2 + c_{20}$$

$$\mu_3 = c_{33} \delta^3 + c_{31} \delta \qquad\qquad \mu_4 = c_{44} \delta^4 + c_{42} \delta^2 + c_{40}$$

where

$$c_{11} = \frac{\sqrt{\frac{f}{2}} \; \Gamma\left(\frac{f-1}{2}\right)}{\Gamma\left(\frac{f}{2}\right)}$$

$$c_{22} = \frac{f}{f-2} - c_{11}^2$$

$$c_{20} = \frac{f}{f-2}$$

$$c_{33} = c_{11}\left[\frac{f(7-2f)}{(f-2)(f-3)} + 2c_{11}^2\right]$$

$$c_{31} = \frac{3f}{(f-2)(f-3)} \; c_{11}$$

$$c_{44} = \frac{f^2}{(f-2)(f-4)} - \frac{2f(5-f)c_{11}^2}{(f-2)(f-3)} - 3c_{11}^4$$

$$c_{42} = \frac{6f}{f-2}\left[\frac{f}{f-4} - \frac{(f-1)c_{11}^2}{f-3}\right]$$

$$c_{40} = \frac{3f^2}{(f-2)(f-4)}$$

Hogben, Pinkham, and Wilk table the exact values of the c's.

Merrington and Pearson (1958) give the following formula for finding the r-th moment about the origin for the noncentral t-distribution:

$$\mu_r' = E(t^r) = \frac{f^{r/2}}{2} \; \frac{\Gamma[(f-r)/2]}{\Gamma(f/2)} \; e^{-\frac{1}{2}\delta^2} \; D^r \; e^{\frac{1}{2}\delta^2}$$

where D^r indicates the r-th derivative of the function following it. Hence,

$$\mu_5' = c_{11}[\delta^5 + 10\delta^3 + 15\delta][f^2/\{(f-3)(f-5)\}]$$

and

$$\mu_6' = [\delta^6 + 15\delta^4 + 45\delta^2 + 15][f^3/\{(f-2)(f-4)(f-6)\}]$$

277

11.10 AN ALTERNATIVE EXPRESSION FOR THE NONCENTRAL t-DISTRIBUTION

On page 387 of the article by Johnson and Welch (1940) a formula is given for even values of f for the cumulative distribution function of noncentral t in terms of Hh functions. This formula has f/2 terms, i.e., is a finite sum and it is easily convertible into a finite sum of confluent hypergeometric functions. Owen and Amos (1963) have another expression involving the hypergeometric function which holds for odd and even values of f.

As before, let

$$H_f(t) = Pr\{T_f \le t\} = \frac{\sqrt{2\pi}}{\Gamma\left(\frac{f}{2}\right) 2^{(f-2)/2}} \int_0^\infty G\left(\frac{tx}{\sqrt{f}} - \delta\right) x^{f-1} G'(x) dx$$

and let $A = \dfrac{t}{\sqrt{f}}$ and $B = \dfrac{f}{f + t^2}$. Then Owen and Amos (1963) give

$$H_f(t) = 1 - G(\delta\sqrt{B}) + \sqrt{2}\ G'(\delta\sqrt{B}) \left[\frac{\Gamma\left(\frac{f-1}{2}\right)}{\Gamma\left(\frac{f-2}{2}\right)} A\sqrt{B}\ S_1 - \frac{\delta}{\sqrt{2}}\ \sqrt{B}\ S_2 \right]$$

where

$$S_1 = \sum_{k=0}^\infty a_k\ \phi\left(-k, \frac{1}{2}; \frac{\delta^2 B}{2}\right)$$

and

$$S_2 = \sum_{k=0}^\infty b_{k+1}\ \phi\left(-k, \frac{3}{2}; \frac{\delta^2 B}{2}\right)$$

and where

$$a_0 = 1, \quad a_k = a_{k-1}\ A^2\ B\ \frac{(2k-f)(2k-1)}{(2k)(2k+1)}$$

$$b_0 = 1, \quad b_k = b_{k-1}\ A^2\ B\ \frac{(2k-1-f)}{(2k)}$$

One series will always terminate when f is an integer; S_1 when f is even and S_2 when f is odd.

The quantity ϕ is a confluent hypergeometric function defined by the series

$$\phi(a,c;x) = \sum_{k=0}^{\infty} \frac{\Gamma(a+k)\ \Gamma(c)}{\Gamma(a)\ \Gamma(c+k)}\ \frac{x^k}{k!}$$

See Erdélyi, Magnus, Oberhettinger, and Tricomi (1953), p. 254, and (1954) for the properties of the confluent hypergeometric function. The values of ϕ may be generated by the following recurrence relations.

$$\phi(-k-1,c;x) = \left(1 + \frac{k-x}{k+c}\right)\phi(-k,c;x) - \frac{k}{k+c}\ \phi(-k+1,c;x)$$

$$\phi(0,c;x) = 1, \quad \phi(-1,c;x) = 1 - \frac{x}{c} \quad \text{with} \quad c = 1/2 \quad \text{or} \quad 3/2$$

Craig (1941) expresses the noncentral t-distribution in terms of Incomplete Beta functions as follows:

$$H_f(t) = e^{-\frac{1}{2}\delta^2} \sum_{k=0}^{\infty} \left(\frac{\delta^2}{2}\right)^k \frac{1}{k!}\ I_{A^2B}\left(k + \frac{1}{2},\frac{1}{2}\right)$$

where

$$I_x(p,q) = \frac{\Gamma(p+q)}{\Gamma(p)\ \Gamma(q)} \int_0^x y^{p-1}(1-y)^{q-1}\ dy$$

12. TWO-SIDED TOLERANCE LIMITS FOR A NORMAL DISTRIBUTION (CONTROL CENTER)

A random sample $x_1,x_2,\cdots,x_n$ from a normal (μ,σ^2) distribution is available. We want to compute k so that with probability γ at least a proportion P of the sampled distribution lies between $\bar{x} - ks$ and $\bar{x} + ks$. In other words we want to find k such that

$$P_{\bar{x},s}\{P_X\{\bar{x} - ks < X < \bar{x} + ks\} \geq P\} = \gamma \tag{12.1}$$

We first note that the desired value of k is independent of μ and σ^2 so we can solve the above problem for $\mu = 0$, $\sigma^2 = 1$.

In their fundamental paper on two-sided tolerance limits Wald and Wolfowitz (1946) showed that k can be computed by the following formula:

Define

$$A(\bar{x},s,k) = G(\bar{x}+ks) - G(\bar{x}-ks)$$

where $G(x)$ is the cumulative normal up to x. Then $A(\bar{x},s,k)$ is the proportion of the sampled normal distribution contained in the interval $\bar{x} - ks$ to $\bar{x} + ks$.

For a given value of $k > 0$ define

$$Q(P,k) = \Pr\{A(\bar{x},s,k) \geq P\}$$

and define

$$Q(P,k|\bar{x}) = \Pr\{A(\bar{x},s,k) \geq P|\bar{x}\}$$

to be the conditional probability that $A(\bar{x},s,k) \geq P$ given the value of the sample mean. Then $Q(P,k)$ is the expected value of $Q(P,k|\bar{x})$, i.e.,

$$Q(P,k) = \frac{\sqrt{n}}{\sqrt{2\pi}} \int_{-\infty}^{\infty} Q(P,k|\bar{x}) e^{-\frac{1}{2}n\bar{x}^2} d\bar{x} \qquad (12.2)$$

Wald and Wolfowitz show that $Q(P,k|\bar{x})$ can be computed from

$$\begin{aligned} Q(P,k|\bar{x}) &= \Pr\{s \geq r(\bar{x},P)/k\} \\ &= \Pr\{\chi_f^2 \geq fr^2/k^2\} \end{aligned} \qquad (12.3)$$

where $r = r(\bar{x},P)$ is the unique root of the equation

$$G(\bar{x}+r) - G(\bar{x}-r) = P \qquad (12.4)$$

and where fs^2 has a chi-squared distribution with f degrees of freedom. (In our application $f = n - 1$, but Wallis (1951) and Weissberg and Beatty (1960) consider applications where f is not necessarily equal to $n - 1$.) Then the desired value of k satisfies the equation

$$Q(P,k) = \frac{2\sqrt{n}}{\sqrt{2\pi}} \int_{0}^{\infty} \Pr\{\chi_f^2 \geq fr^2/k^2\} e^{-\frac{1}{2}n\bar{x}^2} d\bar{x} = \gamma \qquad (12.5)$$

For a given value of n, P, and k, $Q(P,k)$ was evaluated numerically. Simpson's rule integration formula was used with intervals of 0.05 for $\sqrt{n}\,\bar{x}$. Accuracy considerations show that the method yields at least 8 correct decimal places for the evaluation of $Q(P,k)$.

$Q(P,k)$ can be solved iteratively for the value of k which makes $Q(P,k)$ equal to γ to at least 8 decimal places. To start the iteration an approximate value for k is obtained from $k^* = r^*u$, where r^* is determined from $G(1/\sqrt{n} + r^*) - G(1/\sqrt{n} - r^*) = P$, and $u = \sqrt{f/\chi^2_{f,\gamma}}$, where $\chi^2_{f,\gamma}$ is the lower $100(1-\gamma)$ percentage point of a chi-squared distribution with f degrees of freedom, i.e. $Pr\{\chi^2_f \leq \chi^2_{f,\gamma}\} = 1 - \gamma$. Separate tables of r^* and u were computed by Weissberg and Beatty (1960). The adequacy of the approximation is considered by Ellison (1964b). Approximations to r^* and u are given by Gardner and Hull (1966) and a different approximation to k is given by Howe (1969).

13. THE BIVARIATE NONCENTRAL t-DISTRIBUTION

13.1 TWO-SIDED TOLERANCE LIMITS FOR A NORMAL DISTRIBUTION
   (CONTROL BOTH TAILS)

A random sample $x_1, x_2, \cdots, x_n$ from a normal distribution is available and we want to compute $\bar{x} - k_1 s$ and $\bar{x} + k_2 s$ so that with probability γ no more than a proportion p_1 is below $\bar{x} - k_1 s$ and no more than a proportion p_2 is above $\bar{x} + k_2 s$. In other words, we want to find k_1 and k_2 such that

$$P_{\bar{x},s}\left\{P_X\{X \leq \bar{x} - k_1 s\} \leq p_1 \quad \text{and} \quad P_X\{X \geq \bar{x} + k_2 s\} \leq p_2\right\} = \gamma$$

where γ usually takes values in the range $0.75 \leq \gamma < 1$, but may take values anywhere in the range $0 < \gamma < 1$, and p_1 and p_2 are assumed to be less than one-half.

This reduces to

$$Pr\left\{{}_1T_f \leq t_1 \quad \text{and} \quad {}_2T_f \geq t_2\right\} = \gamma$$

where $t_1 = k_1/\sqrt{n}$, $t_2 = -k_2\sqrt{n}$, $\delta_1 = -K_{p_1}\sqrt{n}$, $\delta_2 = K_{p_2}\sqrt{n}$ and $({}_1T_f, {}_2T_f)$ have a joint bivariate noncentral t-distribution with noncentrality parameters δ_1 and δ_2 respectively. See Owen (1965b) for more details. Note that δ_1 is positive and δ_2 is negative since the K_p's are negative. Note also that $f = n - 1$ for the procedure described above.

Let $R = (\delta_1 - \delta_2)/(A_1 - A_2)$. Then for odd degrees of freedom we have

$$\gamma = 1 + G(-\delta_1\sqrt{B_1}) - G(-\delta_2\sqrt{B_2}) + 2T\left(\delta_1\sqrt{B_1}, (\delta_1 A_1 B_1 - R)/(B_1\delta_1)\right)$$

$$- 2T\left(\delta_2\sqrt{B_2}, (\delta_2 A_2 B_2 - R)/(B_2\delta_2)\right) + 2[_1M_1^+ + _1M_3^+ + \cdots + _1M_{f-2}^+]$$

$$- 2[_2M_1^+ + _2M_3^+ + \cdots + _2M_{f-2}^+]$$

and for even values of f

$$\gamma = \sqrt{2\pi}\,[_1M_0^+ + _1M_2^+ + \cdots + _1M_{f-2}^+] - \sqrt{2\pi}\,[_2M_0^+ + _2M_2^+ + \cdots + _2M_{f-2}^+]$$

where the prefix one indicates that

$$\delta_1 = -K_{p1}\sqrt{n}, \quad t_1 = k_1\sqrt{n}, \quad A_1 = t_1/\sqrt{f}, \quad B_1 = 1/(1+A_1^2)$$

would be used in the M's and the prefix two indicates that

$$\delta_2 = K_{p2}\sqrt{n}, \quad t_2 = k_2\sqrt{n}, \quad A_2 = t_2/\sqrt{f}, \quad B_2 = 1/(1+A_2^2)$$

should be used. The M's are defined by

$$M_0^+ = A\sqrt{B}\ G'(\delta\sqrt{B})\ G\left(\frac{\delta AB - R}{\sqrt{B}}\right)$$

$$M_1^+ = B\left[\delta A M_0^+ + AG'(\delta\sqrt{B})G'\left(\frac{\delta AB - R}{\sqrt{B}}\right)\right]$$

$$M_2^+ = \frac{1}{2}\,B[\delta A M_1^+ + M_0^+] + L_1$$

$$\begin{matrix} \cdot & & \cdot \\ \cdot & & \cdot \\ \cdot & & \cdot \end{matrix}$$

$$M_k^+ = \frac{k-1}{k}\,B[a_k \delta A M_{k-1}^+ + M_{k-2}^+] + L_{k-1}$$

where

$$L_1 = \frac{1}{2}\,ABRG'(R)G'(AR-\delta)$$

and

$$L_{k-1} = a_{k+2}\,RL_{k-2} \quad \text{for} \quad k \geq 3,$$

and

$$a_1 = 1, \quad a_2 = 1$$

$$a_k = \frac{1}{(k-2)a_{k-1}} \quad \text{for} \quad k \geq 3.$$

Since k_1 and k_2 are not determined uniquely by this single condition, we first decide that if $p_1 = p_2$ then k_1 should equal k_2. This then gives a unique solution and these values of k are given in Table 4.

When $p_1 \neq p_2$, Owen (1965b) showed that for sample sizes ≥ 5, using the k's as given in Table 4 is satisfactory, i.e., differences are small between the exact values of k and the k's computed for an equal tail split. This is true even with very unequal values of p_1 and p_2 (p_1 up to 1,000 times bigger than p_2). Hence, to keep the tables of manageable size only exact values for the k's when $p_1 = p_2$ are tabled.

13.2 TWO-SIDED SAMPLING PLANS FOR A NORMAL DISTRIBUTION (CONTROL CENTER)

A batch of items manufactured under the same or essentially the same conditions (commonly called a lot) is presented for acceptance inspection. A sample of size n is drawn from the lot and a single characteristic (assumed to follow a normal distribution) is measured on each of the items in the sample and recorded as $x_1, x_2, \cdots, x_n$. Based on these measurements the lot is accepted if $\bar{x} - k_1 s \geq L$ and $\bar{x} + k_2 s \leq U$ where k_1 and k_2 are the quantities tabulated in Table 5, and L is a lower specification limit and U is an upper specification limit for the measured variate. The lot is rejected if either $\bar{x} - k_1 s < L$ or if $\bar{x} + k_2 s > U$.

The constants k_1 and k_2 may be given in advance or may be found so that the probability of acceptance of a lot is some specified value. The probability of accepting the lot is

$$\Pr\{L \leq \bar{x} - k_1 s \quad \text{and} \quad \bar{x} + k_2 s \leq U\} = 1 - \gamma$$

If p_1 is the proportion of the submitted lot which is below L and p_2 is the proportion above U, then

$$\frac{L - \mu}{\sigma} = K_{p_1} \quad \text{and} \quad \frac{U - \mu}{\sigma} = -K_{p_2}$$

This probability may be written in terms of the bivariate noncentral t-distribution as

$$\text{Pr}\{_1T_f \geq t_1 \quad \text{and} \quad _2T_f \leq t_2\} = 1 - \gamma$$

where

$$f = n - 1, \quad t_1 = k_1\sqrt{n}, \quad t_2 = -k_2\sqrt{n}$$

$\delta_1 = -K_{p1}\sqrt{n}$, $\delta_2 = K_{p2}\sqrt{n}$ where $(_1T_f, _2T_f)$ have the joint bivariate noncentral t-distribution introduced by Owen (1965b). Note that δ_1 is usually positive and δ_2 is negative since the K_p's themselves are usually negative.

For odd values of f

$$1 - \gamma = 1 - 2T(\delta_1\sqrt{B_1}, A_1) + 2T(\delta_2\sqrt{B_2}, A_2)$$

$$+ 2T\left(\delta_1\sqrt{B_1}, \; (\delta_1 A_1 B_1 - R)/(B_1\delta_1)\right)$$

$$- 2T\left(\delta_2\sqrt{B_2}, \; (\delta_2 A_2 B_2 - R)/(B_2\delta_2)\right)$$

$$- 2[_1M_1^* + _1M_3^* + \cdots + _1M_{f-2}^*]$$

$$+ 2[_2M_1^* + _2M_3^* + \cdots + _2M_{f-2}^*]$$

where the prefix one on the M's indicates that

$$\delta_1 = -K_{p1}\sqrt{n}, \quad t_1 = k_1\sqrt{n}, \quad A_1 = t_1/\sqrt{f}, \quad B_1 = 1/(1+A_1^2)$$

should be used and the prefix two indicates that

$$\delta_2 = K_{p2}\sqrt{n}, \quad t_2 = k_2\sqrt{n}, \quad A_2 = t_2/\sqrt{f}, \quad B_2 = 1/(1+A_2^2)$$

should be used and $R = (\delta_1 - \delta_2)/(A_1 - A_2)$, and the M^*'s are defined by

$$M_0^* = A\sqrt{B} \; G'(\delta\sqrt{B}) \left[G(\delta A\sqrt{B}) - G\left(\frac{\delta AB - R}{\sqrt{B}}\right) \right]$$

$$M_1^* = B\left[\delta A M_0^* + AG'(\delta A\sqrt{B}) \left\{ G'(\delta A\sqrt{B}) - G'\left(\frac{\delta AB - R}{\sqrt{B}}\right) \right\} \right]$$

$$M_2^* = \frac{1}{2} B[\delta A M_1^* + M_0^*] - L_1$$

$$\begin{matrix} \cdot & \cdot & \cdot \\ \cdot & \cdot & \cdot \\ \cdot & \cdot & \cdot \end{matrix}$$

$$M_k^* = \frac{k-1}{k} B[a_k \delta A M_{k-1}^* + M_{k-2}^*] - L_{k-1}$$

where the L's and the a's were defined in Section 13.1. Note also that if $p_1 = p_2$ and $k_1 = k_2$ there is some adding of terms, e.g.,

$$-2T(\delta_1 \sqrt{B_1}, A_1) + 2T(\delta_2 \sqrt{B_2}, A_2) = -4T(\delta_1 \sqrt{B_1}, A_1)$$

For even values of f

$$1 - \gamma = G(\delta_1) - G(\delta_2) - \sqrt{2\pi}\ [_1M_0^* + {_1}M_2^* + \cdots + {_1}M_{f-2}^*]$$

$$+ \sqrt{2\pi}\ [_2M_0^* + {_2}M_2^* + \cdots + {_2}M_{f-2}^*]$$

We can now compute the probability of acceptance for the true values of p_1 and p_2, but if one wants to pre-assign values of p_1 and p_2 (proportions defective which he might want to guard against) and determine k_1 and k_2 to give a fixed probability of acceptance, then one has to look at the possible ways in which the lots can actually be submitted. Bowker and Goode (1952) pointed out that if the sum of proportions is fixed and equal to p, say, then one can plot the probability of acceptance against p. For different splits of p between the two tails, one generally gets different probabilities of acceptance which means the operating characteristic is a band instead of a single curve. Bowker and Goode considered using $k_1 = k_2 = k'$ determined from a one-sided test for control of a proportion p in one tail and found that in certain cases the pre-assigned probability of acceptance was missed by a considerable amount when in fact $p_1 = \frac{1}{2}p$ and $p_2 = \frac{1}{2}p$ (see p. 135 of Bowker and Goode).

They also stated (p. 134) that the maximum of the operating characteristic band corresponded to the curve computed from the plot $p_1 = p_2 = \frac{1}{2}p$ while the minimum corresponded to the curve computed from the split $p_1 = p$, $p_2 = 0$. Direct computation shows that while these curves corresponded to critical values, they are not always in this relationship. For the following values of n and larger (depending on $p_1 = p_2$) it was found that the situation was as Bowker and Goode described it, while for smaller values of n, the curve corresponding to $p_1 = p_2 = \frac{1}{2}p$ is on the minimum and the curve corresponding to $p_1 = p$, $p_2 = 0$ is on the maximum (near $\gamma = 0.90$).

$p_1 = p_2$	0.10	0.05	0.025	0.01	0.001	0.0001	0.00001
n	6	9	11	14	20	27	34

Actually right at the change-over the maximum may occur at some other combination of p_1, p_2, as there is a continuous function involved. Hence, right at the values of n given care should be used, but computations at other splits of the proportion defective indicate that by taking the sample size 20% larger or 20% smaller than the value tabled above, there is no doubt about the situation being as described.

Owen (1964) considered finding $k_1 = k_2$ for pre-assigned values of $p_1 = p_2$ using the above formulas and tabulated some values of $k_1 = k_2$. The operating characteristic must then be plotted in three dimensions, with axes p_1, p_2, and probability of acceptance. However, the interpretation given by Owen (1964) is in error in that in order to control the proportion in each tail below $\frac{1}{2}p$, the pre-assigned values of the proportions should be $p_1 = 0$ and $p_2 = \frac{1}{2}p$, since the quantity on the right of $1 - \gamma$ in the above formulas is strictly a decreasing function of p_1, for p_2 fixed. Hence the above formulas are not useful for finding k_1 and k_2 with pre-assigned values of p_1 and p_2. The univariate noncentral t-distribution function of Section 11.1 is used. However, once k_1 and k_2 are fixed, the above formulas are useful in finding the operating characteristic of the procedure.

On the other hand, if just the sum of the proportions defective in either tail is to be controlled (control center), then the larger value of k must be used, obtained either from a one-sided test with $p = p_1 + p_2$; or by the use of the above formulas with all possible splits between the tails considered. We have tabulated these values of k in Table 5.

13.3 TWO-SIDED SAMPLING PLANS FOR A NORMAL DISTRIBUTION (CONTROL BOTH TAILS)

We emphasize the result given in Section 13.2 above by giving Table 6. It is, of course, just another version of Table 1 and is based entirely on the univariate noncentral t-distribution.

If we wish to control both tails of a normal distribution with unknown proportions in the tails, and control both tails with a given probability γ it is necessary to choose the worst possible split for each tail, which is that we have to assume all of the defectiveness is in one tail only. This then leads to the univariate noncentral t-distribution.

Even though the problem as given calls for no more than p_1 in the lower tail and no more than p_2 in the upper tail, we have to solve for k_1 assuming at most p_1 in the lower tail and zero in the upper tail; and we solve for k_2 assuming at most p_2 in the upper tail and zero in the lower tail. All of this is predicated on the assumption that we do not want to exceed p_1 in the lower tail nor p_2 in the upper tail with probability γ.

Note that the situation with tolerance limits is entirely different than the situation with sampling plans. This is because tolerance limits formulas allow for the worst case while the sampling plans formulas obscure the worst case.

14. THE CORRELATION COEFFICIENT

14.1 THE DENSITY OF THE SAMPLE CORRELATION COEFFICIENT

Assume X and Y have a joint bivariate normal distribution with means μ_x, μ_y and standard deviations σ_x, σ_y respectively, and correlation ρ. If $(x_1, y_1), \cdots, (x_n, y_n)$ denotes a sample from (X, Y), the sample correlation coefficient r is defined by

$$r = \sum_{i=1}^{n} (x_i - \bar{x})(y_i - \bar{y}) / [(n-1)s_x s_y]$$

For given n, the distribution of r is independent of μ_x, μ_y, σ_x, and σ_y, but depends upon the value of ρ.

The density function for r was derived by Fisher (1915) and can be written as

$$f_n(r, \rho) = \frac{n-2}{\pi} (1-\rho^2)^{\frac{1}{2}(n-1)} (1-r^2)^{\frac{1}{2}(n-4)} I_{n-1}(p) \qquad (14.1.1)$$

where

$$-1 < r < 1, \quad -1 < \rho < 1, \quad p = \rho r$$

and where

$$I_m(p) = \int_0^\infty (\cosh w - p)^{-m} \, dw \quad \text{for} \quad |p| < 1 \qquad (14.1.2)$$

If we let $\cosh w = \dfrac{1 - pz}{1 - z}$ then the integral in (14.1.2) is transformed to

$$I_m(p) = 2^{-\frac{1}{2}}(1-p)^{-m+\frac{1}{2}} \int_0^1 z^{-\frac{1}{2}}(1-z)^{m-1}[1 - \tfrac{1}{2}(1+p)z]^{-\frac{1}{2}} \, dz \qquad (14.1.3)$$

By expanding $[1 - \tfrac{1}{2}(1+p)z]^{-\frac{1}{2}}$ in a uniformly convergent series and integrating term by term we obtain the expression

$$I_m(p) = \sqrt{\frac{\pi}{2}} \; \frac{\Gamma(m)}{\Gamma(m+\frac{1}{2})} \; (1-p)^{-m+\frac{1}{2}} \; F\!\left(\tfrac{1}{2}, \; \tfrac{1}{2}, \; m+\tfrac{1}{2}; \; \frac{1+p}{2}\right) \qquad (14.1.4)$$

where

$$F(a,b,c;z) = 1 + \frac{ab}{c} z + \frac{a(a+1)b(b+1)}{2! \; c(c+1)} z^2 + \cdots$$

is the hypergeometric series.

In particular

$$F\!\left(\tfrac{1}{2}, \; \tfrac{1}{2}, \; m+\tfrac{1}{2}; \; \frac{1+p}{2}\right) = 1 + \frac{(1+p)}{4(2m+1)} + \frac{9}{16} \frac{(1+p)^2}{(2m+1)(2m+3)} + \mathcal{O}\!\left(\frac{1}{m^3}\right)$$

Some useful special properties of $I_m(p)$ are given by the following:

$$I_1(p) = (1-p^2)^{-\frac{1}{2}} \cos^{-1}(-p) \qquad (14.1.5)$$

By repeated differentiation, with respect to p, under the integral sign in (14.1.2) it follows that

$$I_m(p) = \frac{1}{(m-1)!} \frac{\partial^{m-1}}{\partial p^{m-1}} I_1(p) \qquad (14.1.6)$$

which gives for $m = 2$,

$$I_2(p) = \{1 + pI_1(p)\}/(1-p^2) \qquad (14.1.7)$$

For $m \geq 1$ the following recurrence formula is satisfied:

$$I_{m+2}(p) = \{(2m+1)pI_{m+1}(p) + mI_m(p)\}/\{(m+1)(1-p^2)\} \qquad (14.1.8)$$

For $0 \le p < 1$, (14.1.8) can be used to find a sequence of values for $I_3(p), I_4(p), \cdots, I_{n-1}(p)$.

If $-1 < p < 0$ the recurrence formula is numerically unstable. (However, backward recursion can be used. Gautschi (1967) considers three term recurrence formulas in general and gives algorithms for backward recursion.) Since in this case $(1+p) < 1$, the series given by (14.1.4) will converge extremely fast, so that $I_m(p)$ can be evaluated by using (14.1.4).

14.2 TABLES AND EXACT FORMULAS FOR THE CUMULATIVE DISTRIBUTION

Define the cumulative distribution for r by

$$F_n(r,\rho) = \int_{-1}^{r} f_n(u,\rho)\,du$$

Extensive tables of $f_n(r,\rho)$ and $F_n(r,\rho)$ are given to 5 decimal places by David (1954). Tables are given for $r = -1.0(0.05)1.0$, $\rho = 0.0(0.1)0.8$; and for $r = -1.0(0.05)0.60(0.025)0.80(0.01)0.95(0.005)1.00$ for $\rho = 0.9$; for $n = 3(1)25, 50, 100, 200, 400$.

For $n = 3(1)8$ Garwood (1933) gave exact expressions for $F_n(r,\rho)$ in terms of the density functions for smaller n. In particular for $n = 3,4,5,6$ he showed that

$$F_3(r,\rho) = \frac{\cos^{-1}(-r)}{\pi} - \frac{\sqrt{1-r^2}(\rho)}{\pi}\frac{\cos^{-1}(-p)}{\sqrt{1-p^2}} \tag{14.2.1}$$

$$F_4(r,\rho) = \frac{\sqrt{1-\rho^2}\,\sqrt{1-r^2}}{\rho}\,f_3(r,\rho) - \frac{\sqrt{1-\rho^2}}{\pi\rho} + \frac{\cos^{-1}(\rho)}{\pi} \tag{14.2.2}$$

$$F_5(r,\rho) = \frac{\sqrt{1-\rho^2}\,\sqrt{1-r^2}}{2\rho}\,f_4(r,\rho) - \frac{r}{2}(1-r^2)f_3(r,\rho)$$

$$\quad - \frac{\sqrt{1-r^2}(1+\rho^2)}{2\pi\rho}\frac{\cos(-p)}{\sqrt{1-p^2}}\frac{\cos^{-1}(-r)}{\pi} \tag{14.2.3}$$

$$F_6(r,\rho) = \frac{\sqrt{1-\rho^2}\,\sqrt{1-r^2}}{3\rho}\,f_5(r,\rho) + \frac{(1-\rho^2)r}{3\rho^2}\,f_4(r,\rho)$$

$$\quad - \frac{(1-\rho^2)^{3/2}\sqrt{1-r^2}}{3\rho^3}\,f_3(r,\rho) + \frac{\sqrt{1-\rho^2}(1-4\rho^2)}{3\pi\rho^3} + \frac{\cos^{-1}(\rho)}{\pi}$$

$$\tag{14.2.4}$$

Since ρ appears in the denominator of the expressions given by (14.2.2), (14.2.3), and (14.2.4), some caution is necessary when evaluating the expressions when ρ is close to zero. The expressions can be rewritten however. As an example, if in (14.2.2) we first note that

$$f_3(r,\rho) = \frac{1}{\pi} \frac{(1-\rho^2)}{\sqrt{1-r^2}} I_2(\rho)$$

$$= \frac{1}{\pi} \frac{(1-\rho^2)}{\sqrt{1-r^2}} \left(\frac{1 + p\ \cos^{-1}(-p)/\sqrt{1-p^2}}{1-p^2} \right) \tag{14.2.5}$$

then substituting (14.2.5) into (14.2.2) and collecting terms yields

$$F_4(r,\rho) = -\frac{\sqrt{1-\rho^2}}{\pi\rho} \frac{(\rho^2-p^2)}{(1-p^2)} + r\left(\frac{1-\rho^2}{1-p^2}\right)^{3/2} \cos^{-1}(-p)/\pi + \cos^{-1}(\rho)/\pi$$

$$= \frac{-\rho(1-r^2)\sqrt{1-\rho^2}}{(1-p^2)} + r\left(\frac{1-\rho^2}{1-p^2}\right)^{3/2} \cos^{-1}(-p)/\pi + \cos^{-1}(\rho)/\pi$$

$$\tag{14.2.6}$$

14.3 RECURRENCE FORMULAS FOR THE CUMULATIVE DISTRIBUTION

$F_n(r,\rho)$ satisfies the following recurrence formula (see pages 207-208 of Hotelling (1953) for details):

$$(m+2)\rho^2 F_{m+5}(r,\rho) = -[m-2(m+1)\rho^2] F_{m+3}(r,\rho) + m(1-\rho^2) F_{m+1}(r,\rho)$$

$$+ \frac{(1-\rho^2)^{\frac{m}{2}+1}(1-r^2)^{\frac{1}{2}(m-1)}}{\pi(m+1)} \left\{ (m+2)\rho\,[(m+1)(1-r^2)(1-\rho^2) \right.$$

$$- (2m+1)(1-p^2)]I_{m+3}(p) + r[(m+1)^2 + (3m^2+6m+2)\rho^2]I_{m+2}(p) \Big\}$$

$$\tag{14.3.1}$$

For odd n the above formula can be used repeatedly for m = 2,4,$\cdots$,(n-5) to find $F_7,F_9,\cdots,F_n$ from F_3 and F_5 for which exact expressions are given by (14.2.1) and (14.2.3) respectively. For even n, the above formula can be used repeatedly for m = 3,5,$\cdots$,n-5 to find $F_8,F_{10},\cdots,F_n$ from F_4 and F_6 for which exact expressions are given by (14.2.2) and (14.2.4) respectively.

However, the formula can only be applied if $\rho^2 \geq \frac{m}{2(m+1)}$, since if ρ^2 is less than this bound, the first two terms on the right hand side are of opposite sign and the formula is numerically unstable.

14.4 DEVELOPMENT OF A SERIES FOR THE CUMULATIVE DISTRIBUTION

Hotelling (1953, pp. 203-205) defines

$$Q_n(r,\rho) = \int_\rho^r f_n(x,\rho)\,dx \quad \text{and develops} \quad Q_n(r,\rho)$$

in a uniformly convergent series for $-1 < \rho < r < 1$. From the relationships

$$P_n(r,\rho) = \int_r^1 f_n(x,\rho)\,dx = Q_n(1,\rho) - Q_n(r,\rho)$$

$$P_n(-1,\rho) = 1$$

$$F_n(r,\rho) = 1 - P_n(r,\rho)$$

$$F_n(-r,-\rho) = 1 - F_n(r,\rho)$$

we can compute $F_n(r,\rho)$ for any values of r and ρ with $-1 \leq r \leq 1$, $-1 < \rho < 1$.

By combining (14.1.1) and (14.1.4) we may write

$$f_n(x,\rho) = \frac{(m-1)}{\sqrt{2\pi}} \frac{\Gamma(m)}{\Gamma(m+\frac{1}{2})} (1-\rho^2)^{\frac{m}{2}} (1-x^2)^{\frac{1}{2}(m-3)} (1-\rho x)^{-m+\frac{1}{2}} F\left(\tfrac{1}{2}, \tfrac{1}{2}, m+\tfrac{1}{2}; \frac{1+\rho x}{2}\right)$$

where $m = n - 1$.

By using the relationship $(1+\rho x) = 2 - (1-\rho x)$ the term in $(1+\rho x)^j$ in $F\left(\tfrac{1}{2}, \tfrac{1}{2}, m+\tfrac{1}{2}; \frac{1+\rho x}{2}\right)$ can be expanded in terms of $1 - \rho x$ by using the binomial theorem to expand $[2 - (1-\rho x)]^j$. If we define

$$N_k = \int_\rho^r (1-\rho^2)^{\frac{m}{2}} (1-x^2)^{\frac{1}{2}(m-3)} (1-\rho x)^{-m+k+\frac{1}{2}}\,dx, \quad k = 0,1,\cdots$$

then $Q_n(r,\rho)$ can be expressed linearly in terms of $N_0, N_1, \cdots$.
The result is

$$Q_n(r,\rho) = \frac{(m-1)\Gamma(m)}{\sqrt{2\pi}\,\Gamma(m+\frac{1}{2})} \left\{ N_0 + \frac{2N_0 - N_1}{4(2m+1)} + \frac{9(4N_0 - 4N_1 + N_2)}{32(2m+1)(2m+3)} + \cdots \right\}$$

Hotelling shows that the error committed by truncating the series at any point is less than $2/(1-|\rho|)$ times the last term used.

Hotelling also develops N_k in a series involving Incomplete Beta Functions. Define the incomplete Beta function by

$$B_b(p,q) = \int_0^b y^{p-1}(1-y)^{q-1} \, dy$$

A FORTRAN subroutine for evaluating $B_b(p,q)/B_1(p,q)$ is given by Majumder and Bhattacharjee (1973). If the input parameter BETA is set equal to 1.0 then the routine returns $B_b(p,q)$.

In the definition of N_k we take as the variable of integration

$$y = (x-\rho)^2/(1-\rho x)^2$$

Then after algebraic simplification we obtain

$$1 - y = (1-x^2)(1-\rho^2)/(1-\rho x)^2$$

$$1 + \rho y^{\frac{1}{2}} = (1-\rho^2)/(1-\rho x)$$

$$y^{\frac{1}{2}} = (x-\rho)/(1-\rho x)$$

$$(1-x^2) = (1-y)(1-\rho^2)/(1+\rho y^{\frac{1}{2}})^2$$

$$(1-\rho x) = (1-\rho^2)/(1+\rho y^{\frac{1}{2}})$$

$$\frac{dx}{dy} = \frac{1}{2} y^{-\frac{1}{2}}(1-\rho^2)/(1+\rho y^{\frac{1}{2}})^2$$

Substitution of these last three quantities into the definition of N_k yields

$$N_k = \frac{1}{2}(1-\rho^2)^k \int_0^b y^{-\frac{1}{2}}(1+\rho y^{\frac{1}{2}})^{\frac{1}{2}-k}(1-y)^{\frac{1}{2}(m-3)} dy$$

where $b = (r-\rho)^2/(1-\rho r)^2$ and where $0 \le b \le 1$.

By expanding $(1+\rho y^{\frac{1}{2}})^{\frac{1}{2}-k}$ in a uniformly convergent series and integrating term by term N_k may be expressed as

$$N_k = \frac{1}{2}(1-\rho^2)^k \sum_{s=0}^{\infty} \frac{\Gamma(\tfrac{3}{2}-k)}{\Gamma(\tfrac{3}{2}-k-s)s!} \rho^s B_b\{(s+1)/2, \ (m-1)/2\}$$

Hotelling shows that for s large enough the absolute value of the ratio of the term of order $(s+1)$ to the term of order s is bounded by $\left|\rho b^{\frac{1}{2}}\right| < 1$ so that the series is rapidly convergent.

14.5 CONFIDENCE INTERVALS FOR THE CORRELATION COEFFICIENT

The entry in Table 10 for a given sample size n, α, and r, is the value of ρ (given to 4 decimal places) which satisfies $P_n(r,\rho) = \alpha$ or equivalently $F_n(r,\rho) = 1 - \alpha$.

The values of ρ were found by using a modified form of successive approximations which uses an estimate of the derivative of $P_n(r,\rho)$ with respect to ρ. The method is equivalent to using linear interpolation. If ρ_0 and ρ_1 are two initial estimates of ρ then we obtain a sequence of estimates $\rho_2, \rho_3, \cdots$ from

$$\rho_{j+1} = \rho_j + a_j\left[\alpha - P_n(r,\rho_j)\right]$$

where

$$a_j = (\rho_j - \rho_{j-1})/[P_n(r,\rho_j) - P_n(r,\rho_{j-1})], \quad j = 1,2,\cdots$$

This process was repeated until two successive values of ρ agreed to 6 decimal places.

Originally the series developed in Section 14.4 was used to evaluate $P_n(r,\rho)$. The values of ρ obtained were rounded to 4 decimal places and stored. These values were then checked by using (a) the exact formulas given by Garwood for $3 \le n \le 8$ and described in Section 14.2 for $n = 3(1)6$; (b) the recursion formula given by (14.3.1) for values of $n \ge 9$ and $\rho^2 > \frac{1}{2}$; (c) numerical integration using Simpson's rule for values of $n \ge 9$ and $\rho^2 < \frac{1}{2}$.

There is excellent agreement between the above methods. The values of $P_n(r,\rho)$ computed by these various formulas agreed to at least 8 decimal places.

The initial estimates ρ_0 and ρ_1 were obtained as follows: Fisher (1921) showed that the quantity $z' = \tanh^{-1} r = \frac{1}{2} \ell n \frac{1 + r}{1 - r}$ can be treated as a normally distributed random variable with mean $\mu_{z'}$, and variance $\sigma_{z'}^2$

293

given by

$$\mu_{z'} = \tfrac{1}{2}\ln\frac{1+\rho}{1-\rho} + \frac{\rho}{2(n-1)}\left\{1+\frac{5+\rho^2}{4(n-1)}+\cdots\right\}$$

$$\approx \tfrac{1}{2}\ln\frac{1+\rho}{1-\rho}$$

$$\sigma_{z'}^2 = \frac{1}{n-1}\left\{1+\frac{4-\rho^2}{2(n-1)}+\frac{22-6\rho^2-3\rho^4}{6(n-1)^2}+\cdots\right\}$$

$$\approx \frac{1}{n-3}$$

For $n \geq 4$ an initial estimate ρ_0 is the value of ρ which satisfies

$$\tfrac{1}{2}\ln\frac{1+\rho}{1-\rho} = (\tanh^{-1}r - z_\alpha/\sqrt{n-3})$$

That is

$$\rho_0 = \tanh(\tanh^{-1}r - z_\alpha/\sqrt{n-3})$$

where z_α is the upper 100α percentage point of the unit normal distribution. Then z_α satisfies

$$\int_{z_\alpha}^{\infty} \frac{1}{\sqrt{2\pi}} \exp(-t^2/2)\,dt = \alpha$$

To obtain the estimate ρ_1 we first note that $P_n(r,\rho)$ is a monotone increasing function in ρ. Then if

$$P_n(r,\rho_0) < \alpha \quad \text{we set} \quad \rho_1 = \rho_0 + 0.001$$

and if

$$P_n(r,\rho_0) > \alpha \quad \text{we set} \quad \rho_1 = \rho_0 - 0.001$$

15. SCREENING PROCEDURES

15.1 SCREENING BASED ON NORMAL VARIABLES (ALL PARAMETERS KNOWN)

We first assume that we have a bivariate normal population of random variables (X, Y) where Y is called the performance variable and X is called the screening variable. The correlation between X and Y is ρ and the means of X and Y are μ_x and μ_y, respectively. Similarly, the standard deviations are σ_x and σ_y for X and Y, respectively. We assume that we start with a proportion γ of acceptable product, i.e.,

$$\Pr\{Y \geq \mu_y - K_\gamma \sigma_y\} = \gamma$$

where $L = \mu_y - K_\gamma \sigma_y$ is some lower specification limit, and K_γ is defined as before.

We want to select items from our population by choosing those items for which $X \geq \mu_x - K_\beta \sigma_x$ i.e.,

$$\Pr\{X \geq \mu_x - K_\beta \sigma_x\} = \beta$$

provided $\rho > 0$.

Hence the proportion of the population in the selected population which now has $Y \geq L$ is

$$\Pr\{Y \geq \mu_y - K_\gamma \sigma_y \mid X \geq \mu_x - K_\beta \sigma_x\} = \delta$$

Table 8 gives values of β for fixed values of (ρ, γ, δ). Hence, we can raise our proportion of acceptable product from γ to δ by this screening process.

15.2 SCREENING BASED ON NORMAL VARIABLES (UNKNOWN PARAMETERS)

In case no parameters are known, it is necessary to estimate them from a preliminary sample of size n $(x_1, y_1), (x_2, y_2), (x_3, y_3), \cdots, (x_n, y_n)$ from the population, as follows:

$$\bar{x} = \sum_{i=1}^{n} x_i/n, \quad \bar{y} = \sum_{i=1}^{n} y_i/n,$$

$$s_x^2 = (n-1)^{-1} \sum_{i=1}^{n} (x_i - \bar{x})^2, \quad s_y^2 = (n-1)^{-1} \sum_{i=1}^{n} (y_i - \bar{y})^2$$

and

$$r = \sum_{i=1}^{n} (x_i - \bar{x})(y_i - \bar{y}) / [(n-1)s_x s_y]$$

A 100η% lower confidence limit on ρ is obtained from Table 10; call it ρ^*. Also a 100η% lower confidence limit on $\gamma = \Pr\{Y \geq L\}$ is obtained from Table 7; call it γ^*. Then Table 9 is entered with the parameters $(f = n-1, \gamma^*, \rho^*[n/(n+1)]^{\frac{1}{2}}, \delta)$ in place of $(f, \gamma, \rho, \delta)$. Values of t_β are read from Table 9 and we accept (select) all additional items for which $X \geq \bar{x} - t_\beta s_x[(n+1)/n]^{\frac{1}{2}}$. We can then be at least $100(2\eta-1)\%$ sure that at least $100\delta\%$ of the Y's are above L in the selected population.

First, let us evaluate

$$\Pr\{Y \geq L \mid X \geq \mu_x - t_\beta s_x'\} = \delta$$

where $s_x'^2 = \sum_{i=1}^{n} (x_i - \mu_x)^2 / n$.

This may be rewritten

$$\Pr\{Z \geq -K_\gamma \mid T_f \geq -t_\beta\} = \delta$$

where Z is $N(0,1)$ and where $T_f = (X - \mu_x)/s_x'$ has the Student t-distribution with $f = n$ degrees of freedom.

Equivalently, this expression can take the form

$$\Pr\{Z \leq K_\gamma \mid T_f \leq t_\beta\} = \delta$$

where

$$\Pr\{T_f \leq t_\beta\} = \beta$$

and hence

$$\Pr\{Z \leq K_\gamma, T_f \leq t_\beta\} = \delta\beta$$

The value of t_β which satisfies this equation is given in Table 9. In Section 15.3 below we give expressions for evaluating this probability. Now we note that in the situation where all of the parameters are unknown, μ_x would have to be estimated by $\bar{x}$. Then we would screen by accepting product for which

$$X \geq \bar{x} - t_\beta s_x[(n+1)/n]^{\frac{1}{2}}$$

and $(X-\bar{x})[n/(n+1)]^{\frac{1}{2}}/s_x$ has the Student t-distribution with $f = n - 1$ degrees of freedom. The correlation between $X - \bar{x}$ and Y is $\rho \sqrt{\dfrac{n}{n+1}}$ if the correlation between Y and X is ρ since X and Y are independent of $\bar{x}$. These two facts establish the multipliers $[(n+1)/n]^{\frac{1}{2}}$ and $[n/(n+1)]^{\frac{1}{2}}$.

In elementary probability it is established that $\Pr\{A \cup B\} = \Pr\{A\} + \Pr\{B\} - \Pr\{A \cap B\}$, where A and B are any two events. If we let A be the event $\gamma \geq \gamma^*$ and B be the event $\rho \geq \rho^*$, then since $\Pr\{A\} = \Pr\{B\} = \eta$, we have $\Pr\{A \cap B\} = 2\eta - \Pr\{A \cup B\}$. But $\Pr\{A \cup B\} \leq 1$. Combining these results we obtain $\Pr\{\gamma \geq \gamma^* \text{ and } \rho \geq \rho^*\} \geq 2\eta - 1$.

Now we wish to establish that δ is a monotone increasing function of γ and is also a monotone increasing function of ρ. We do this by differentiating

$$\beta\delta = \Pr\{Z \leq K_\gamma, \; T_f \leq t_\beta\}$$

In Section 15.3 we will show that

$$\beta\delta = \frac{\sqrt{2\pi}}{\Gamma\left(\dfrac{f}{2}\right) 2^{(f-2)/2}} \int_0^\infty x^{f-1} \, G'(x) \, B\left(K_\gamma, \, \frac{t_\beta x}{\sqrt{f}} \, ; \rho\right) dx$$

where $B(\widetilde{h}, \widetilde{k}; \rho)$ is the standardized bivariate normal cumulative to $(\widetilde{h}, \widetilde{k})$ with correlation ρ.

The derivative of $B\left(K_\gamma, \, \dfrac{t_\beta x}{\sqrt{f}} \, ; \rho\right)$ with respect to γ is

$$G'(K_\gamma) G\left(\frac{t_\beta x - \rho K_\gamma \sqrt{f}}{\sqrt{f} \; \sqrt{1 - \rho^2}}\right)$$

which is always positive and since the rest of the $\beta\delta$ integral is also always positive and over a positive range the derivative of δ with respect to γ is positive.

Similarly, the derivative of $B\left(K_\gamma, \, \dfrac{t_\beta x}{\sqrt{f}} \, ; \rho\right)$ with respect to ρ is

$$\frac{1}{\sqrt{1-\rho^2}} G'(K_\gamma) G'\left(\frac{t_\beta x - \rho K_\gamma \sqrt{f}}{\sqrt{f} \; \sqrt{1-\rho^2}}\right)$$

which is always positive and by the same reasoning the derivative of δ with respect to ρ is positive.

Hence, when we use $100\eta\%$ lower confidence limits on ρ and on γ we can be at least $100(2\eta-1)\%$ sure that both ρ and γ will be larger than the limits used to obtain β. This then implies that we can be at least $100(2\eta-1)\%$ sure that δ is greater than the delta we selected in entering Table 9.

15.3 MATHEMATICAL DERIVATION OF THE NORMAL CONDITIONED ON t-DISTRIBUTION

We wish to derive representations of

$$\Pr\{Z \le K_\gamma, \ T_f \le t_\beta\} = P_f(K_\gamma, t_\beta)$$

We first start with the joint bivariate (V,Z) normal distribution with zero means and unit variances and correlation ρ. We also need the U variate which has a square root of a chi-squared distribution with f degrees of freedom divided by its degrees of freedom, i.e., V/U has Student's t-distribution. Since the densities of (V,Z) and U are independent we can write the joint density of (U,V,Z) as

$$\frac{1}{\sqrt{1-\rho^2}} \ G'\left(\frac{Z-\rho V}{\sqrt{1-\rho^2}}\right) G'(V) \ \frac{f^{f/2}\sqrt{2\pi}}{\Gamma\left(\frac{f}{2}\right) 2^{(f-2)/2}} \ U^{f-1} \ G'(U\sqrt{f})$$

We now make the change of variables $t = V/U$, $Z = Z$ and $V = V$ and integrate V from $-\infty$ to $+\infty$, Z from $-\infty$ to K_γ and t from $-\infty$ to t_β. If this is done in different orders the following representations are found:

$$P_f(K_\gamma, t_\beta) = \int_{-\infty}^{K_\gamma} \Pr\left\{T_f \le \frac{t_\beta}{\sqrt{(1-\rho^2)}} \ \Big| \ \lambda\right\} G'(X) \ dX \qquad (15.3.1)$$

where T_f has a noncentral t-distribution with f degrees of freedom and noncentrality parameter $\lambda = \rho X/\sqrt{1-\rho^2}$.

$$P_f(K_\gamma, t_\beta) = \frac{\Gamma\left(\frac{f+1}{2}\right)}{\Gamma\left(\frac{f}{2}\right)\sqrt{\pi f}} \int_{-\infty}^{t_\beta} \left(1 + \frac{X^2}{f}\right)^{-(f+1)/2} \Pr\left\{T_{f+1} \le \frac{-\rho X\sqrt{f+1}}{\sqrt{1-\rho^2}\sqrt{f+X^2}} \ \Big| \ \lambda\right\} dX$$

$$(15.3.2)$$

where T_{f+1} has a noncentral t-distribution with $f+1$ degrees of freedom and noncentrality parameter $\lambda = -K_\gamma / \sqrt{1-\rho^2}$.

$$P_f(K_\gamma, t_\beta) = \int_{-\infty}^{+\infty} G'(X)\, G\left(\frac{K_\gamma - c_1 X}{\sqrt{1-c_1^2}}\right) \Pr\left\{T_f \leq \frac{t_\beta}{\sqrt{1-c_2^2}}\middle|\ \lambda\right\} dX \qquad (15.3.3)$$

where $G(X) = \dfrac{1}{\sqrt{2\pi}} \displaystyle\int_{-\infty}^{X} G'(t)\,dt$ and T_f has a noncentral t-distribution with f degrees of freedom and noncentrality parameter $\lambda = c_2 X / \sqrt{1-c_2^2}$, and $c_1 c_2 = \rho$ with $0 \leq c_i^2 \leq 1$.

$$P_f(K_\gamma, t_\beta) = \frac{\sqrt{2\pi}}{\Gamma\!\left(\dfrac{f}{2}\right) 2^{(f-2)/2}} \int_0^\infty X^{f-1}\, G'(X)\, B\left(K_\gamma,\ \frac{t_\beta X}{\sqrt{f}};\rho\right) dX \qquad (15.3.4)$$

where $B(\tilde{h}, \tilde{k}; \rho)$ is the standardized bivariate normal cumulative to $(\tilde{h}, \tilde{k})$ with correlation ρ.

$$P_f(K_\gamma, t_\beta) = \int_{-\infty}^{+\infty} G'(X)\, G\left(\frac{K_\gamma - \rho X}{\sqrt{1-\rho^2}}\right) \Pr\left\{\chi_f^2 \geq \frac{fX^2}{t_\beta^2}\right\} dX \qquad (15.3.5)$$

where χ_f^2 has the chi-squared distribution with f degrees of freedom.

For one degree of freedom, we have

$$P_1(K_\gamma, t_\beta) = 2C\left(0,\ K_\gamma,\ 0;\ \rho_{12} = \frac{\rho}{\sqrt{1+t_\beta^2}},\ \rho_{13} = \frac{t_\beta}{\sqrt{1+t_\beta^2}},\ \rho_{23} = 0\right)$$

$$(15.3.6)$$

where $C(\tilde{h}, \tilde{k}, \tilde{m};\ \rho_{12}, \rho_{13}, \rho_{23})$ is the cumulative trivariate normal distribution to $(\tilde{h}, \tilde{k}, \tilde{m})$ with correlations $(\rho_{12}, \rho_{13}, \rho_{23})$.

For two degrees of freedom, we have

$$P_2(K_\gamma, t_\beta) = B(0, K_\gamma; \rho) + \frac{t_\beta}{\sqrt{t_\beta^2 + 2}}\, B\left(0,\ \frac{K_\gamma \sqrt{t_\beta^2 + 2}}{\sqrt{2(1-\rho^2) + t_\beta^2}};\ -\frac{\rho t_\beta}{\sqrt{2(1-\rho^2) + t_\beta^2}}\right)$$

$$(15.3.7)$$

As checks for special cases we have for $\rho = 0$

$$P_f(K_\gamma, t_\beta \mid \rho = 0) = G(K_\gamma) \; Pr\{T_f \le t_\beta \mid \lambda = 0\}$$

and for $t_\beta = 0$

$$P_f(K_\gamma, 0) = B(K_\gamma, 0; \rho)$$

In order to compute this function for any number of degrees of freedom we derived the following recursion formula by integrating (15.3.1) above by parts and making use of the fact that for the noncentral t-distribution

$$Pr\{T_f \le t \mid \lambda\} = \frac{\sqrt{2\pi}}{\Gamma\left(\frac{f}{2}\right) 2^{(f-2)/2}} \int_0^\infty G\left(\frac{tX}{\sqrt{f}} - \lambda\right) X^{f-1} \; G'(X) \; dX$$

The result can be expressed as

$$P_f(K_\gamma, t_\beta) = P_{f-2}\left(K_\gamma, t_\beta\sqrt{\frac{f-2}{f}}\right) + \frac{t_\beta \Gamma\left(\frac{f-1}{2}\right)}{2\Gamma\left(\frac{f}{2}\right)\sqrt{\pi f}} \left(\frac{f}{t_\beta^2 + f}\right)^{(f-1)/2}$$

$$\cdot Pr\left\{T_{f-1} \le \frac{-\rho t_\beta \sqrt{f-1}}{\sqrt{1-\rho^2}\sqrt{t_\beta^2 + f}} \;\middle|\; \lambda\right\}$$

where $\lambda = -K_\gamma/\sqrt{1-\rho^2}$ for $f > 2$.

Owen and Ju (1977) give representations for $Pr\{Z \le K_\gamma, \; T_f \le t_\beta\}$ when $\rho = 1$. In case some of the parameters are known, the reader is referred to articles by Owen and Boddie (1976) and by Owen and Su (1977). In case more than one screening variable is available the reader is referred to Thomas, Owen and Gunst (1977). In case there are two-sided specification limits on the performance variable, the reader is referred to Li and Owen (1979).

SOURCES OF THE TABLES

A version of Table 1 appeared in Owen (1963). There is one more column (corresponding to $P = 0.99999$) in Owen (1963) and some additional sample sizes. Otherwise the Table 1 given here is the same as given by Owen (1963).

Table 2 was computed especially for this volume and has not appeared elsewhere.

Table 3 was computed especially for this volume. The tables were published in Odeh (1978) for $N = 2(1)98,100$. The exact values given here differ from the approximate values which appeared originally in Eisenhart, Hastay, and Wallis (1947) and which were widely copied. Additional values for $P = 0.975$, 0.995, and tables for $\gamma = 0.50$, 0.975, 0.995 have been added here.

Table 4 is an expansion of Owen and Frawley (1971). Additional values for $P = 0.010$, 0.100, and tables for $\gamma = 0.5$, 0.975, 0.995 have been added here.

Table 5 is a recomputation of Table II in Owen (1967) with the addition of a column for $P = 0.005$ and additional sample sizes.

Table 6 is a recomputation of Table III in Owen (1967) with the addition of a column for $P = 0.005$ and additional sample sizes.

Table 7 is an expansion of Tables I, II and III in Owen and Hua (1977). Additional tables for $\eta = 0.5$, 0.75, 0.975, 0.995 have been added here.

Table 8 is a recomputation of Tables 1, 2, and 3 in Owen, McIntire and Seymour (1975) with a slight change in the range of the parameters.

Table 9 is an expansion of Table 1 in Owen and Haas (1978) and of Owen and Ju (1977) when $\rho = 1.0$. Additional tables for $\delta = 0.90$, 0.995 have been added here.

Table 10 was computed especially for this volume and has not appeared elsewhere.

BIBLIOGRAPHY

Airey, J.R. (1931). Tables of Hh Functions, Table XV. *British Assoc. Math. Tables, Vol. I.* British Association, London. pp. 60-72. (Second edition 1946, Third edition 1951).

Aitchison, J. (1964). Bayesian tolerance regions, *J. R. Statist. Soc. B,* 26:161-175.

Aitchison, J. (1966). Expected-cover and linear-utility tolerance intervals, *J. R. Statist. Soc. B,* 28:57-62.

Albert, G.E. and Johnson, R.B. (1951). On the estimation of certain intervals which contain assigned proportions of a normal univariate population, *Ann. Math. Statist.,* 22:596-599.

Altman, Irving B. (1957). The new MIL-STD-414 sampling inspection by variables, *Industr. Qual. Contr.,* 14:23-26.

American Statistical Association, Washington, B.C. (1950). *Acceptance Sampling, a Symposium.* 105th Annual Meeting, ASA, held at Cleveland, Ohio, Jan. 27, 1946.

Amos, D.E. (1964). Representations of the central and non-central t-distributions, *Biometrika,* 51:451-458.

Amos, D.E. (1976). Computation of the central and noncentral F distribution, *Commun. Statist. Theor. Meth. A,* 5:261-281.

Amos, D.E. (1978). Evaluation of some cumulative distribution functions by numerical quadrature, *SIAM Rev.,* 20:778-800.

Anderson, Robert H. (1971). Applications of MIL-STD-414 for quality assurance of nuclear materials, *Trans. 8th Ann. Western Quality Conf.,* Anaheim, Calif.

Barton, D.E. (1961). Unbiased estimation of a set of probabilities, *Biometrika,* 48:227-229.

Borth, D.M. (1973). A modification of Owen's method for computing the bivariate normal integral, *Appl. Statist.,* 22:82-85.

Bowden, David C. (1968). Tolerance interval in regression, *Technometrics,* 10:207-209.

Bowker, A.H. and Goode, H.P. (1952). *Sampling Inspection by Variables.* McGraw-Hill, New York.

Bowker, A.H. and Lieberman, G.J. (1955). *Handbook of Industrial Statistics.* Prentice-Hall, New Jersey.

Bowker, A.H. and Lieberman, G.J. (1959). *Engineering Statistics.* Prentice-Hall, New Jersey.

Cacoullos, T. (1965). A relation between t and F-distributions, *J. Amer. Statist. Assoc.,* 60:528-531. Correction 60:1249.

Chernoff, H. and Lieberman, G.J. (1957). Sampling inspection by variables with no calculations, *Industr. Qual. Contr.,* 13(7):5-7.

Chew, Victor (1966). Confidence, prediction, and tolerance regions for the multivariate normal distribution, *J. Amer. Statist. Assoc.,* 61:605-617.

Cooper, B.E. (1968a). Algorithm AS 4: An auxiliary function for distribution integrals, *Appl. Statist.,* 17:190-192.

Cooper, B.E. (1968b). Algorithm AS 5: The integral of the non-central t-distribution, *Appl. Statist.,* 17:193-194.

Craig, Cecil C. (1941). Note on the distribution of noncentral t with an application, *Ann. Math. Statist.,* 12:224-228.

Croarkin, Mary C. (1962). Graphs for determining the power of Student's t-test, *J. Res. Nat. Bur. Stand.,* 66B:59-70. Table Errata *Math. Comp.,* 17:334.

Cucconi, O. (1962). On a simple relation between the number of degrees of freedom and the critical values of Student's t, *Mem. Accad. Pat.,* 74:179-187.

Daley, D.J. (1974). Computation of bi- and tri-variate normal integrals, *Appl. Statist.,* 23:435-438.

Das, N.G. and Mitra, S.K. (1964). Effect of non-normality on plans for sampling inspection by variables, *Sankhyā A,* 26:169-176.

David, F.N. (1954). *Tables of the Ordinates and Probability Integral of the Distribution of the Correlation Coefficient in Small Samples.* Cambridge at the University Press.

Diviney, T.E. and David, N.A. (1963). A graphical application of Military Standard 414, *Industr. Qual. Contr.,* 19:13-14.

Donnelly, T.G. (1973). Algorithm 462: Bivariate normal distribution, *Commun. ACM,* 16:638.

Duncan, Acheson J. (1958). Design and operation of a double-limit variable sampling plan, *J. Amer. Statist. Assoc.,* 53:543-550.

Eisenhart, C., Hastay, M.W. and Wallis, W.A. (1947). *Techniques of Statistical Analysis.* McGraw-Hill, New York. (A.H. Bowker wrote Chapter 2).

Ellison, Bob E. (1964a). Two theorems for inferences about the normal distribution with applications in acceptance sampling, *J. Amer. Statist. Assoc.,* 59:89-95.

Ellison, Bob E. (1964b). On two-sided tolerance intervals for a normal distribution, *Ann. Math. Statist.*, 35:762-772.

Erdélyi, A., Magnus, W., Oberhettinger, F. and Tricomi, F.G. (1953). *Higher Transcendental Functions*. Bateman Manuscript Project 1 and 2. McGraw-Hill, New York. Table Errata *Math. Comp.*, 16:308, 17:338, 18:360, 19:361,527, 20:641.

Erdélyi, A., Magnus, W., Oberhettinger, F. and Tricomi, F.G. (1954). *Tables of Integral Transforms*. Bateman Manuscript Project 1 and 2. McGraw-Hill, New York. Table Errata *Math. Comp.*, 18:353, 19:361, 20:641.

Faulkenberry, G. David and Weeks, David L. (1968). Sample size determination for tolerance limits, *Technometrics*, 10:343-348.

Faulkenberry, G.D. and Daly, J.C. (1970). Sample size for tolerance limits on a normal distribution, *Technometrics*, 12:813-821.

Fisher, R.A. (1915). Frequency distribution of the values of the correlation coefficient in samples from an indefinitely large population, *Biometrika*, 10:507-521.

Fisher, R.A. (1921). On the 'probable error' of a coefficient of correlation deduced from a small sample, *Metron*, 1:1-32.

Folks, J.L., Pierce, D.A. and Stewart, C. (1965). Estimating the fraction of acceptable product, *Technometrics*, 7:43-50.

Fraser, D.A.S. and Guttman, I. (1956). Tolerance regions, *Ann. Math. Statist.*, 27:162-179.

Gardiner, D.A. and Bombay, Barbara F. (1965). An approximation to Student's t, *Technometrics*, 7:71-72.

Gardiner, D.A. and Hull, Norma C. (1966). An approximation to two-sided tolerance limits for normal populations, *Technometrics*, 8:115-122.

Garner, Norman R. (1958). Curtailed sampling for variables, *J. Amer. Statist. Assoc.*, 53:862-867.

Garwood, F. (1933). The probability integral of the correlation coefficient in samples from a normal bi-variate population, *Biometrika*, 25:71-78.

Gautschi, W. (1967). Computational aspects of three-term recurrence relations, *SIAM Review*, 9:24-82.

Goldberg, H. and Levine, H. (1946). Approximate formulas for the percentage points and normalization of t and χ^2, *Ann. Math. Statist.*, 17:216-225.

Guenther, William C. (1975). A sample size formula for a non-central t test, *The American Statistician*, 29:120-121.

Guttman, Irwin (1957). On the power of optimum tolerance regions when sampling from normal distributions, *Ann. Math. Statist.*, 28:773-778.

Guttman, Irwin (1970). *Statistical Tolerance Regions: Classical and Bayesian*. Charles Griffin & Co., Ltd., London.

Hader, R.J. (1959). Some basic aspects of sampling, *Textile Quality Control Papers*, Am. Soc. Quality Control, 6:67-77.

Hald, A. (1952). *Statistical Theory with Engineering Applications*. John Wiley, New York.

Halperin, Max (1963). Approximations to the non-central t, with applications, *Technometrics*, 5:295-305. Errata 6:482.

Hamaker, H.C. (1961). A sampling table with adjustable scales, *EOQC Bull.*, No. 14, 440-443.

Hamaker, Hugo C. (1979). Acceptance sampling for percent defective by variables and by attributes, *J. Quality Tech.*, 11:139-148.

Hanson, D.L. and Koopmans, L.H. (1964). Tolerance limits for the class of distributions with increasing hazard rates, *Ann. Math. Statist.*, 35:1561-1570.

Harley, Betty I. (1957). Relation between the distributions of noncentral t and of a transformed correlation coefficient, *Biometrika*, 44:219-224.

Harris, M., Horvitz, D.G. and Mood, A.M. (1948). On the determination of sample sizes in designing experiments, *J. Amer. Statist. Assoc.*, 43:391-402.

Hawkins, Douglas M. (1975). From the noncentral t to the normal integral. *The American Statistician*, 29:42-43.

Hendricks, Walter A. (1936). An approximation to "Student's" distribution, *Ann. Math. Statist.*, 7:210-221.

Hogben, D., Pinkham, R.S. and Wilk, M.B. (1961). The moments of the non-central t-distribution, *Biometrika*, 48:465-468.

Hogben, D., Pinkham, R.S. and Wilk, M.B. (1964a). The moments of a variate related to non-central t, *Ann. Math. Statist.*, 35:298-314.

Hogben, D., Pinkham, R.S. and Wilk, M.B. (1964b). An approximation to the distribution of Q (a variate related to the non-central t), *Ann. Math. Statist.*, 35:315-318.

Hotelling, Harold (1953). New light on the correlation coefficient and its transforms, *J. R. Stat. Soc. B*, 2:193-225. (Discussion 225-232).

Howe, W.G. (1969). Two-sided tolerance limits for normal populations — some improvements, *J. Amer. Statist. Assoc.*, 64:610-620.

Ireson, W., Resnikoff, G.J. and Smith, B.E. (1961). Statistical tolerance limits for determining process capability, *J. Indust. Eng.*, 12:126-131.

Jennett, W.J. and Welch, B.L. (1939). The control of proportion defective as judged by a single quality characteristic varying on a continuous scale, *J. R. Statist. Soc. Suppl.*, 6:80-88.

Jílek, M. and Likar, O. (1959). Coefficients for the determination of one-sided tolerance limits of normal distributions, *Ann. Inst. Statist. Math.*, 11:45-48.

Jílek, M. and Líkar, O. (1960a). Tolerance regions of the normal distribution with known variance and unknown mean, *Aust. J. Statist.*, 2:78-83.

Jílek, M. and Líkar, O. (1960b). Tolerance limits of the normal distribution with known μ and unknown σ, *Biom. Zeit.*, 2:204-209.

John, S. (1963). A tolerance region for multivariate normal distributions, *Sankhyā A*, 25:363-368.

Johnson, N.L. and Welch, B.L. (1940). Applications of the noncentral *t*-distribution, *Biometrika*, 31:362-389.

Johnson, N.L. (Ed.)(1968). Query: Tolerance interval in regression, *Technometrics*, 10:207-209.

Kabe, D.G. (1976). On confidence bands for quantiles of a normal population, *J. Amer. Statist. Assoc.*, 71:417-419.

Kirkpatrick, R.L. (1970a). Confidence limits on a percent defective characterized by two specification limits, *J. Quality Tech.*, 2:150-155.

Kirkpatrick, R.L. (1970b). Solutions for the WAGR sequential *t*-test; sequential variables sampling plans to control the percent defective, *Trans. 24th Annual Technical Conference, ASQC.* pp. 375-386.

Kirkpatrick, R.L. (1972). Quantitative adjustments for percent defectives in nonnormal distributions, *Trans. 26th Annual Technical Conference, ASQC.* pp. 439-450.

Kirkpatrick, R.L. (1977). Sample sizes to set tolerance limits, *J. Quality Tech.*, 9:6-12.

Koth, H. and Schmidt, P.W. (1962). Graphical applications of Military Standard 414, Form B, *Qualitätskontrolle*, 7:71-72.

Kraemer, Helena Chmura and Paik, Minja (1979). A central *t* approximation to the noncentral *t* distribution, *Technometrics*, 21:357-360.

Kramer, C.Y. (1966). Approximation to the cumulative *t*-distribution, *Technometrics*, 8:358-359.

Laubscher, Nico F. (1960). Normalizing the noncentral *t* and *F* distributions, *Ann. Math. Statist.*, 31:1105-1112.

Levert, C. (1959). A nomogram for confidence intervals and exceedance probabilities, *Statist. Neerlandica*, 13:3-14.

Li, Loretta and Owen, D.B. (1979). Two-sided screening procedures in the bivariate case, *Technometrics*, 21:79-85.

Lieberman, Alfred (1957). Tables for the determination of two-sided tolerance limits for the normal distribution, Report No. 373-17(55), Bureau of Ships, Navy Department, Washington.

Lieberman, Gerald J. and Resnikoff, George J. (1955). Sampling plans for inspection by variables, *J. Amer. Statist. Assoc.*, 50:457-516.

Lieberman, Gerald J. (1958). Tables for one-sided statistical tolerance limits, *Industr. Qual. Contr.*, 14(10):7-9.

Lieberman, Gerald J. and Miller, Rupert G., Jr. (1963). Simultaneous confidence intervals in regression, *Biometrika*, 50:155-168.

Locks, M.O., Alexander, M.J. and Byars, B.J. (1963). New tables of the non-central *t*-distribution, ARL Technical Report No. 63-19, Wright-Patterson Air Force Base.

Majumaer, K.L. and Bhattacharjee, G.P. (1973). Algorithm AS 63: The incomplete Beta integral, *Appl. Statist.*, 22:409-411.

Merrington, M. and Pearson, E.S. (1958). An approximation to the distribution of non-central *t*, *Biometrika*, 45:484-491.

Mitra, S.K. (1957). Table for tolerance limits for a normal population based on sample mean and range or mean range, *J. Amer. Statist. Assoc.*, 52:88-94.

Nelson, Lloyd S. (1977). Tolerance factors for normal distributions, *J. Quality Tech.*, 9:198-199.

Neyman, J., Iwazkiewicz, K. and Kolodziejczyk, S. (1935). Statistical problems in agricultural experimentation, *J. R. Stat. Soc. Suppl.*, 2:107-180.

Neyman, J. and Tokarska, B. (1936). Errors of the second kind in testing Student's hypothesis, *J. Amer. Statist. Assoc.*, 31:318-326.

Odeh, Robert E. (1978). Tables of two-sided tolerance factors for a normal distribution, *Commun. Statist. Simul. Comp. B*, 7:183-201.

Office of the Assistant Secretary of Defense (Supply and Logistics) (1957). Sampling procedures and tables for inspection by variables for percent defective, MIL-STD-414.

Office of the Assistant Secretary of Defense (Supply and Logistics) (1958). Mathematical and Statistical principles underlying MIL-STD-414.

Owen, Donald B. (1956). Tables for computing bivariate normal probabilities, *Ann. Math. Statist.*, 27:1075-1090.

Owen, Donald B. (1957). The bivariate normal probability distribution, Report No. SC-3831 (TR), Sandia Corporation, Albuquerque.

Owen, D.B. and Wiesen, J.M. (1959). A method of computing bivariate normal probabilities with an application to handling errors in testing and measuring, *Bell System Tech. J.*, 38:553-572.

Owen, D.B. (1962). *Handbook of Statistical Tables*. Addison-Wesley, Massachusetts.

Owen, Donald B. (1963). Factors for one-sided tolerance limits and for variables sampling plans, Monograph No. SCR-607, Sandia Corporation, Albuquerque.

Owen, D.B. and Amos, D.E. (1963). Programs for computing percentage points of the noncentral t-distribution, Monograph No. SCR-551, Sandia Corporation, Albuquerque.

Owen, D.B. (1964). Control of percentages in both tails of the normal distribution, *Technometrics*, 6:377-387. Errata 8:570.

Owen, D.B. (1965a). The power of Student's t-test, *J. Amer. Statist. Assoc.*, 60:320-333.

Owen, D.B. (1965b). A special case of a bivariate non-central t-distribution, *Biometrika*, 52:437-446.

Owen, D.B. (1966). One-sided variables sampling plans, *Industr. Qual. Contr.*, 22:450-456.

Owen, D.B. (1967). Variables sampling plans based on the normal distribution, *Technometrics*, 9:417-423.

Owen, D.B. (1968). A survey of properties and applications of the noncentral t-distribution, *Technometrics*, 10:445-478. Republished in *Cuadernos de Estadistica Aplicada e Investigacion Operativa*, VI (1969).

Owen, D.B. (1969). Summary of recent work on variables acceptance sampling with emphasis on non-normality, *Technometrics*, 11:631-637.

Owen, D.B. and Frawley, W.H. (1971). Factors for tolerance limits which control both tails of the normal distribution, *J. Quality Tech.*, 3:69-79.

Owen, D.B., McIntire, D. and Seymour, E. (1975). Tables using one or two screening variables to increase acceptable product under one-sided specifications, *J. Quality Tech.*, 7:127-138.

Owen, D.B. (1976). Discussion on The draft standard BS 6002: sampling procedures and charts for inspection by variables, by J.C. Gascoigne and I.D. Hill, *J. R. Stat. Soc. A*, 139:315.

Owen, D.B. and Boddie, J.W. (1976). A screening method for increasing acceptable product with some parameters unknown, *Technometrics*, 18:195-199.

Owen, D.B. and Haas, R.W. (1977). Tables of the normal conditioned on t-distribution. *Contributions to Survey Sampling and Applied Statistics*, H.A. David (Ed.). Academic Press, New York. pp. 295-318.

Owen, D.B. and Hua, Tsushung A. (1977). Tables of confidence limits on the tail area of the normal distribution, *Commun. Statist. Simul. Comp. B*, 6:285-311.

Owen, D.B. and Ju, Faming (1977). The normal conditioned on t-distribution when the correlation is one, *Commun. Statist. Simul. Comp. B*, 6:167-179.

Owen, D.B. and Su, Yueh-ling Hsiao (1977). Screening based on normal variables, *Technometrics*, 19:65-68.

Patnaik, P.B. (1955). Hypotheses concerning the means of observations in normal samples, *Sankhyā*, 15:343-372.

Paulson, Edward (1943). A note on tolerance limits, *Ann. Math. Statist.*, 14:90-93.

Pearson, E.S. and Hartley, H.O. (1951). Charts of the power function for analysis of variance tests, derived from the non-central *F*-distribution, *Biometrika*, 38:112-130.

Peiser, A.M. (1943). Asymptotic formulas for significance levels of certain distributions, *Ann. Math. Statist.*, 14:56-62, 20:128-129.

Proschan, Frank (1953). Confidence and tolerance intervals for the normal distribution, *J. Amer. Statist. Assoc.*, 48:550-564.

Rao, J.N.K., Subrahmaniam, K. and Owen, D.B. (1972). Effect of non-normality on tolerances limits which control percentages in both tails of the normal distribution, *Technometrics*, 14:571-575.

Resnikoff, George J. and Lieberman, G.J. (1957). *Tables of the Non-central t-Distribution*. Stanford University Press, Stanford.

Resnikoff, George J. (1962). Tables to facilitate the computation of percentage points of the non-central *t*-distribution, *Ann. Math. Statist.*, 33:580-586.

Robbins, Herbert (1948). The distribution of Student's *t* when the population means are unequal, *Ann. Math. Statist.*, 19:406-410.

Satterthwaite, Franklin E. (1946). An approximate distribution of estimates of variance components, *Biometrics*, 2:110-114.

Scheuer, Ernest M. and Spurgeon, Robert A. (1963). Some percentage points of the noncentral *t*-distribution, *J. Amer. Statist. Assoc.*, 58:176-182.

Seshadri, V. and Odell, P.L. (1959). A method for determining a confidence bound on unreliability when time-to-failure is normally distributed. US Naval Nuclear Ordnance Evaluation Unit, Albuquerque.

Siotani, M. (1964). Tolerance regions for a multivariate normal population, *Ann. Inst. Statist. Math.*, 16:135-153.

Sowden, R.R. and Ashford, J.R. (1969). Computation of the bivariate normal integral, *Appl. Statist.*, 18:169-180.

Srivastava, A.B.L. (1958). Effect of non-normality on the power function of *t*-test, *Biometrika*, 45:421-430.

Srivastava, A.B.L. (1961). Variables sampling inspection for non-normal samples, *J. Sci. Engrg. Res.*, 5:145-152.

Stange, K. (1958,1960,1961). Quantitative Inspection, *Metrika*, 1:111-129, 3:151-165, 4:1-29.

Stange, K. (1960). Variable inspection design, *Qualitätskontrolle und Operational Research*, 5:1-7,29-34,75-80,155-161.

Steck, G.P. and Owen, D.B. (1959). Percentage points for the distribution of outgoing quality, *J. Amer. Statist. Assoc.*, 54:689-694.

Storer, R.L. and Davison, W.R. (1955). Simplified procedures for sampling inspection by variables, *Industr. Qual. Contr.*, 12(1):15-18.

Summers, Robert D. (1965). An inequality for the sample coefficient of variation and an application of variables sampling, *Technometrics*, 7:67-68.

Thomas, Jeannie Gouras, Owen, D.B. and Gunst, R.F. (1977). Improving the use of educational tests as evaluation tools, *J. Educational Stat.*, 2:55-77.

van Eeden, Constance (1961). Some approximations to the percentage points of the noncentral *t*-distribution, *Rev. Int. Statist. Inst.*, 29:4-31.

Wald, Abraham (1942). Setting of tolerance limits when the sample is large, *Ann. Math. Statist.*, 13:389-399.

Wald, A. and Wolfowitz, J. (1946). Tolerance limits for a normal distribution, *Ann. Math. Statist.*, 17:208-215.

Wallis, W. Allen (1951). Tolerance intervals for linear regression. *Proc. Second Berkeley Symp. on Math. Statist. Prob.*, J. Neyman (Ed.), University of California Press, Los Angeles and Berkeley. pp. 43-51.

Walsh, John E. (1948). On the use of the non-central *t*-distribution for comparing percentage points of normal populations, *Ann. Math. Statist.*, 19:93-94.

Walsh, John E. (1949). On the power function of the 'best' *t*-test solution of the Behrens-Fisher problem, *Ann. Math. Statist.*, 20:616-618.

Walsh, John E. (1956). Validity of approximate normality values for $\mu \pm k\sigma$ areas of practical type continuous populations, *Ann. Inst. Statist. Math.*, 8:79-86.

Walsh, John E. (1957). Further consideration of normality values for $\mu \pm k\sigma$ areas of continuous populations, *Ann. Inst. Statist. Math.*, 9:127-129.

Walsh, John E., David, H.T. and Fay, E.A. (1958). Acceptance inspection by variables when the measurements are subject to error, *Ann. Inst. Statist. Math.*, 10:107-129.

Wampler, Roy H. (1976). One-sided tolerance limits for the normal distribution, $P = 0.80$, $\gamma = 0.80$, *J. Res. Nat. Bur. Stand.*, 80B:343-346.

Weissberg, A. and Beatty, G.H. (1960). Tables of tolerance-limit factors for normal distributions, *Technometrics*, 2:483-500. Errata 3:576-577.

Wilks, Samuel S. (1941). Determination of sample sizes for setting tolerance limits, *Ann. Math. Statist.*, 12:91-96.

Wilson, A.L. (1967). An approach to simultaneous tolerance intervals in regression, *Ann. Math. Statist.*, 38:1536-1540.

Wolfowitz, Jacob (1946). Confidence limits for the fraction of a normal population which lies between two given limits, *Ann. Math. Statist.*, 17:483-488.

Woods, Walter M. (1960). Variables sampling inspection procedures which guarantee acceptance of perfectly screened lots, Report No. 47, Applied Math. and Statist. Lab., Stanford University.

Young, J.C. and Minder, Ch. E. (1974). Algorithm AS 76: An integral useful in calculating non-central *t* and bivariate normal probabilities, *Appl. Statist.*, 23:455-457.

Zăludová, Agnes H. (1958). Statistical quality control of production processes using individual sample values, *Bull. Int. Statist. Inst.*, 36:573-578.

INDEX OF SYMBOLS

INDEX